T0234614

Lecture Notes in Computer Science 9843

Commenced Publication in 1973
Founding and Former Series Editors:
Gerhard Goos, Juris Hartmanis, and Jan van Leeuwen

Editorial Board

David Hutchison
  Lancaster University, Lancaster, UK
Takeo Kanade
  Carnegie Mellon University, Pittsburgh, PA, USA
Josef Kittler
  University of Surrey, Guildford, UK
Jon M. Kleinberg
  Cornell University, Ithaca, NY, USA
Friedemann Mattern
  ETH Zurich, Zürich, Switzerland
John C. Mitchell
  Stanford University, Stanford, CA, USA
Moni Naor
  Weizmann Institute of Science, Rehovot, Israel
C. Pandu Rangan
  Indian Institute of Technology, Madras, India
Bernhard Steffen
  TU Dortmund University, Dortmund, Germany
Demetri Terzopoulos
  University of California, Los Angeles, CA, USA
Doug Tygar
  University of California, Berkeley, CA, USA
Gerhard Weikum
  Max Planck Institute for Informatics, Saarbrücken, Germany

More information about this series at http://www.springer.com/series/7407

Veli Mäkinen · Simon J. Puglisi
Leena Salmela (Eds.)

Combinatorial Algorithms

27th International Workshop, IWOCA 2016
Helsinki, Finland, August 17–19, 2016
Proceedings

Springer

Editors
Veli Mäkinen
University of Helsinki
Helsinki
Finland

Leena Salmela
University of Helsinki
Helsinki
Finland

Simon J. Puglisi
University of Helsinki
Helsinki
Finland

ISSN 0302-9743 ISSN 1611-3349 (electronic)
Lecture Notes in Computer Science
ISBN 978-3-319-44542-7 ISBN 978-3-319-44543-4 (eBook)
DOI 10.1007/978-3-319-44543-4

Library of Congress Control Number: 2016931327

LNCS Sublibrary: SL1 – Theoretical Computer Science and General Issues

© Springer International Publishing Switzerland 2016
This work is subject to copyright. All rights are reserved by the Publisher, whether the whole or part of the material is concerned, specifically the rights of translation, reprinting, reuse of illustrations, recitation, broadcasting, reproduction on microfilms or in any other physical way, and transmission or information storage and retrieval, electronic adaptation, computer software, or by similar or dissimilar methodology now known or hereafter developed.
The use of general descriptive names, registered names, trademarks, service marks, etc. in this publication does not imply, even in the absence of a specific statement, that such names are exempt from the relevant protective laws and regulations and therefore free for general use.
The publisher, the authors and the editors are safe to assume that the advice and information in this book are believed to be true and accurate at the date of publication. Neither the publisher nor the authors or the editors give a warranty, express or implied, with respect to the material contained herein or for any errors or omissions that may have been made.

Printed on acid-free paper

This Springer imprint is published by Springer Nature
The registered company is Springer International Publishing AG Switzerland

Preface

This volume contains revised versions of papers presented at the 27th International Workshop on Combinatorial Algorithms (IWOCA 2016), held August 17–19, 2016, in Helsinki, Finland.

IWOCA 2016 continued a long and well-established tradition of encouraging high-quality research in theoretical computer science, providing an opportunity to bring together specialists and young researchers working in the area. The IWOCA conference series grew out of a 17-year history of the Australasian Workshop on Combinatorial Algorithms (AWOCA). Previous AWOCA and IWOCA meetings have been held in Australia, Indonesia, South Korea, Japan, Czech Republic, Canada, UK, India, France, the USA, and Italy.

We solicited papers in the broad area of combinatorial algorithms. The Program Committee decided to accept 35 papers, out of a total of 87 submissions. Each submission received at least three reviews. Papers were submitted and reviewed using the EasyChair online system. Authors of accepted papers come from 21 countries, across three continents (Asia, Europe, North America).

The scientific program included three invited lectures, given by:

- Leslie Anne Goldberg on "Approximately Counting list H-Colourings"
- Giuseppe F. Italiano on "2-Connectivity Problems in Directed Graphs"
- Petteri Kaski on "Polynomial Representations in Algorithm Design"

We thank the invited speakers for accepting our invitation and for their excellent presentations at the conference. The program also included an open problem session, chaired by Gabriele Fici. The open problems presented can be found at the open problem collection of IWOCA at http://iwoca.org. This year for the second year running, IWOCA had a Best Student Paper Award, sponsored by the European Association for Theoretical Computer Science (EATCS). It was decided to assign this award to the paper "Online Chromatic Number Is PSPACE-Complete" by Martin Böhm and Pavel Veselý.

We thank all authors who submitted their work for consideration to IWOCA 2016. We wish to thank the Program Committee and the external reviewers, whose many thorough reviews helped us select the papers presented. The success of the scientific program is due to their hard work. We also thank the EATCS (European Association for Theoretical Computer Science), Federation of Finnish Learned Societies, and the Helsinki Institute for Information Technology for their support of the conference.

IWOCA 2016 was organized by the Department of Computer Science of the University of Helsinki, whose administrative and financial support we gratefully acknowledge.

August 2016

Simon Puglisi
Veli Mäkinen
Leena Salmela

Organization

Steering Committee

Costas S. Iliopoulos   King's College London
William F. Smyth   McMaster University

Program Committee

Golnaz Badkobeh	University of Warwick, UK
Hideo Bannai	Kyushu University, Japan
Petra Berenbrink	Simon Fraser University, Canada
Christina Boucher	Colorado State University, USA
Charles Colbourn	Arizona State University, USA
Vida Dujmovic	University of Ottawa, Canada
Gabriele Fici	University of Palermo, Italy
Travis Gagie	University of Helsinki, Finland
Roberto Grossi	University of Pisa, Italy
Pinar Heggernes	University of Bergen, Norway
Seokhee Hong	University of Sydney, Australia
Costas Iliopoulos	King's College London, UK
Jesper Jansson	Kyoto University, Japan
Telikepalli Kavitha	Tata Institute of Fundamental Research, Mumbai, India
Ralf Klasing	CNRS, France
Christian Komusiewicz	Friedrich-Schiller-Universität Jena, Germany
Jan Kratochvil	Charles University, Czech Republic
Daniela Kühn	Birmingham University, UK
Zsuzsanna Lipták	University of Verona, Italy
Martin Milanič	University of Primorska, Slovenia
Petra Mutzel	University of Dortmund, Germany
Veli Mäkinen (Co-chair)	University of Helsinki, Finland
Christophe Paul	CNRS, France
Solon Pissis	King's College London, UK
Simon Puglisi (Co-chair)	University of Helsinki, Finland
Oliver Schaudt	Universität zu Köln, Germany
Michiel Smid	Carleton University, Canada
Tatiana Starikovskaya	University of Bristol, UK
Jukka Suomela	Aalto University, Finland
Alexandru Tomescu	University of Helsinki, Finland
Przemysław Uznański	ETH Zurich, Switzerland
Stéphane Vialette	CNRS, France

Dorothea Wagner	Karlsruhe Institute of Technology, Germany
Oren Weimann	University of Haifa, Israel
Sue Whitesides	University of Victoria, Canada
Christos Zaroliagis	University of Patras, Greece

Organizing Committee

Leena Salmela, chair	University of Helsinki, Finland
Veli Mäkinen	University of Helsinki, Finland
Simon J. Puglisi	University of Helsinki, Finland
Alexandru Tomescu	University of Helsinki, Finland
Daniel Valenzuela	University of Helsinki, Finland

Student Volunteers

Jarno Alanko	University of Helsinki, Finland
Riku Walve	University of Helsinki, Finland
Bella Zhukova	University of Helsinki, Finland

Additional Reviewers

Ásgeirsson, Eyjólfur Ingi
Ahn, Hee-Kap
Amani, Mahdi
Angelini, Patrizio
Axtmann, Michael
Bampas, Evangelos
Barbero, Florian
Baum, Moritz
Bergamini, Elisabetta
Bevern, René Van
Bhattacharya, Pritam
Blanchet-Sadri, Francine
Boeckenhauer,
    Hans-Joachim
Bonomo, Flavia
Bosman, Thomas
Bougeret, Marin
Bousquet, Nicolas
Brandstädt, Andreas
Brankovic, Ljiljana
Brettell, Nick
Bryant, Randal
Burcsi, Péter
Bärtschi, Andreas

Bökler, Fritz
Cicalese, Ferdinando
Courcelle, Bruno
Cucu-Grosjean, Liliana
de Carvalho,
    Marcelo Henriques
Dittmann, Christoph
Droschinsky, Andre
Ekim, Tinaz
Faro, Simone
Fertin, Guillaume
Fountoulakis, Nikolaos
Frid, Anna
Friedetzky, Tom
Gawrychowski, Pawel
Graf, Daniel
Guay-Paquet, Mathieu
Gudmundsson, Joachim
Hamann, Michael
Harutyunyan, Hovhannes
Hermelin, Danny
Huang, Chien-Chung
Inenaga, Shunsuke
Islam, A.S.M. Sohidull

Iwata, Yoichi
Kavitha, Telikepalli
Kling, Peter
Koivisto, Mikko
Kudahl, Christian
Kumar, Mithilesh
Kundu, Ritu
Lampis, Michael
Larsen, Kasper Green
Lecroq, Thierry
Liedloff, Mathieu
Lo, Allan
Luosto, Kerkko
Mallmann-Trenn, Frederik
Manea, Florin
Marino, Andrea
Matijevic, Domagoj
Megow, Nicole
Mercas, Robert
Mertzios, George
Michail, Othon
Morin, Pat
Moseley, Benjamin
Mouawad, Amer

Mozes, Shay
Mustafa, Nabil
Mycroft, Richard
Mydlarz, Marcelo
Mömke, Tobias
Nakashima, Yuto
Nichterlein, André
Niedermann, Benjamin
Noellenburg, Martin
Pajak, Dominik
Patel, Viresh
Paterson, Mike
Perarnau, Guillem
Perkins, Will
Pirola, Yuri

Popa, Alexandru
Previtali, Marco
Prutkin, Roman
Rajendraprasad, Deepak
Rampersad, Narad
Rapaport, Ivan
Raymond, Jean-Florent
Rizzi, Romeo
Rusu, Irena
Sanita, Laura
Sau, Ignasi
Sawada, Joe
Seki, Shinnosuke
Shur, Arseny
Sikora, Florian

Sorge, Manuel
Stewart, Lorna
Storandt, Sabine
Strasser, Ben
Stuckey, Peter J.
Takes, Frank
Thomas, Robin
Tiskin, Alexander
Treglown, Andrew
Uchizawa, Kei
Valicov, Petru
Versari, Luca
Wang, Haitao
Zhang, Jingru
Zhang, Peng

Abstracts of Invited Talks

Approximately Counting List H-Colourings

Leslie Ann Goldberg

Department of Computer Science, University of Oxford, Oxford, UK
leslie.goldberg@cs.ox.ac.uk

An *H*-colouring of a graph *G* is a homomorphism from *G* to *H* (a map from the vertices of *G* to the vertices of *H* that maps edges of *G* to edges of *H*). The "classification programme" in computational complexity aims to classify graphs *H* according to the difficulty of algorithmic problems, for example, the problem of constructing a homomorphism from an input graph *G* to *H*, or the problem of counting homomorphisms from *G* to *H* or (more recently) the problem of approximately counting these homomorphisms. I will explain the classifications that are known, focussing especially on "list H-colouring," which generalises *H*-colouring in the same way that "list colouring" generalises ordinary (proper) vertex colouring. We still don't know a complete classification for approximately counting *H*-colourings, but for approximately counting list *H*-colourings, there is more progress. Here it turns out that there is a trichotomy in the approximation complexity, based on hereditary graph classes. The talk will describe joint work with Andreas Galanis and Mark Jerrum.

2-Connectivity Problems in Directed Graphs

Giuseppe F. Italiano

Università di Roma "Tor Vergata," Rome, Italy
giuseppe.italiano@uniroma2.it

We survey some recent results on 2-edge and 2-vertex connectivity problems in directed graphs. Despite being complete analogs of the corresponding notions on undirected graphs, in digraphs 2-vertex and 2-edge connectivity have a much richer and more complicated structure. It is thus not surprising that 2-connectivity problems on directed graphs appear to be more difficult than on undirected graphs. For undirected graphs it has been known for over 40 years how to compute all bridges, articulation points, 2-edge- and 2-vertex-connected components in linear time, by simply using depth first search. In the case of digraphs, however, the very same problems have been much more challenging and have been tackled only recently.

Polynomial Representations in Algorithm Design

Petteri Kaski

Department of Computer Science,
Helsinki Institute for Information Techonlogy HIIT,
Aalto University, Helsinki, Finland
petteri.kaski@aalto.fi

Currently the asymptotically fastest known algorithm designs for a number of *a priori* purely combinatorial problems are based on algebraic techniques. This talk gives a brief survey on the use of polynomials and implicit polynomial representations in such designs. We start by recalling some of the classics and proceed towards recent multivariate polynomial sieving and batch evaluation frameworks that yield the state of the art for a range of problems including k-clique counting, graph coloring, Hamiltonian path, motif search, and so forth. Designs based on polynomials not only can give the fastest known and often embarrassingly parallel algorithms, the polynomial representation in itself may serve as a *proof* that the computation was correctly executed.

Contents

Combinatorics

Probabilistics

Computational Complexity

On the Complexity of Computing Treebreadth

Guillaume Ducoffe[1,2(✉)], Sylvain Legay[3], and Nicolas Nisse[1,2]

[1] Univ. Nice Sophia Antipolis, CNRS, I3S, UMR 7271,
06900 Sophia Antipolis, France
guillaume.ducoffe@inria.fr
[2] Inria, Sophia Antipolis, France
[3] LRI, Univ. Paris Sud, Université Paris-Saclay, 91405 Orsay, France

Abstract. During the last decade, metric properties of the *bags* of tree-decompositions of graphs have been studied. Roughly, the *length* and the *breadth* of a tree-decomposition are the maximum diameter and radius of its bags respectively. The *treelength* and the *treebreadth* of a graph are the minimum length and breadth of its tree-decompositions respectively. *Pathlength* and *pathbreadth* are defined similarly for path-decompositions. In this paper, we answer open questions of [Dragan and Köhler, Algorithmica 2014] and [Dragan, Köhler and Leitert, SWAT 2014] about the computational complexity of treebreadth, pathbreadth and pathlength. Namely, we prove that computing these graph invariants is NP-hard. We further investigate graphs with treebreadth one, i.e., graphs that admit a tree-decomposition where each bag has a dominating vertex. We show that it is NP-complete to decide whether a graph belongs to this class. We then prove some structural properties of such graphs which allows us to design polynomial-time algorithms to decide whether a bipartite graph, resp., a planar graph, has treebreadth one.

1 Introduction

Tree-decompositions [20] aim at decomposing graphs into pieces, called *bags*, organized in a tree-like manner (formal definitions are postponed to Sect. 1.3). Roughly, the *width* of a tree-decomposition is the maximum size of its bags. A lot of work has been dedicated to compute tree-decompositions with small width since such decompositions can be efficiently exploited for algorithmic purposes [4]. Computing the corresponding graph invariant, the *treewidth* of a graph G (i.e., the minimum width among all tree-decompositions of G), is NP-hard [2] and no constant-approximation algorithm is likely to exist [22]. Moreover, real-life networks generally have a large treewidth [11]. These drawbacks motivated the study of other optimization criteria for tree-decompositions.

In particular, the metric properties of the bags have been studied. Roughly, the *length* and the *breadth* of a tree-decomposition are the maximum diameter

---

This work is partially supported by ANR project Stint under reference ANR-13-BS02-0007 and ANR program "Investments for the Future" under reference ANR-11-LABX-0031-01.

© Springer International Publishing Switzerland 2016
V. Mäkinen et al. (Eds.): IWOCA 2016, LNCS 9843, pp. 3–15, 2016.
DOI: 10.1007/978-3-319-44543-4_1

and radius of its bags respectively. The corresponding graph parameters are the *treelength* [13] and the *treebreadth* [14] respectively. Recent studies suggest that some classes of real-life networks – including biological networks and social networks – have bounded treebreadth [1]. This metric tree-likeness can be exploited in algorithms. For instance, bounded treebreadth graphs admit a PTAS for the TRAVELING SALESMAN problem [18]. They also admit compact distance labeling schemes [12]. Furthermore, the diameter and the radius of bounded treebreadth graphs can be approximated up to an additive constant in linear time [9]. In contrast to the above result, we emphasize that under classical complexity assumptions the diameter of general graphs *cannot* be approximated up to an additive constant in subquadratic time, that is prohibitive for large graphs [8].

On the computational side, it is known that computing the treelength is NP-hard [19]. However, contrary to the treewidth, there exists a 3-approximation algorithm for computing the treelength [13]. In [14], a 3-approximation algorithm for computing the treebreadth is presented but the computational complexity of this problem is left open. Note that, because treelength and treebreadth differ by at most a factor 2 [14], any polynomial-time algorithm for computing the treebreadth, or an α-approximation algorithm for some $\alpha < 3/2$, would improve the 3-approximation algorithm for treelength [13].

A *path-decomposition* of a graph is a tree-decomposition where the bags are organized according to a path structure. Treelength and treebreadth have their "path counterpart", namely the *pathlength* and the *pathbreadth*. In [15], they have been shown to be useful in the design of approximation algorithms for bandwidth and line-distortion. A 2-approximation (resp., a 3-approximation) algorithm is given for computing the pathlength (resp., the pathbreadth) but the computational complexity of both problems is left open.

The main contributions of this paper are to answer the open problems of [14] and [15]. Namely, we prove that computing the treebreadth, pathlength and pathbreadth of graphs are all NP-hard problems.

1.1   Related Work

In contrast with treewidth [5], deciding whether a graph has treelength at most k is NP-complete for every fixed $k \geq 2$ [19]. However, the reduction used for treelength goes through weighted graphs and then goes back to unweighted graphs using rather elegant gadgets. It does not seem to us these gadgets can be easily generalized in order to apply to the treebreadth.

Relationship between treewidth and treelength (and so, treebreadth) has been investigated in [10]. The two parameters are uncomparable in general graphs. For instance, cycles have treewidth at most two but treelength $\lceil n/3 \rceil$, while cliques have treewidth $n - 1$ but treelength equal to one [13]. However, they differ by at most a constant ratio in the graphs with bounded genus and bounded isometric cycles [10]. Hence we are also motivated in this work to better understand the structure of tree-decompositions with small width for bounded genus graphs, and to improve their computation.

Recently, the MINIMUM ECCENTRICITY SHORTEST-PATH problem – close to the problem of computing the pathlength and pathbreadth – has been proved NP-hard [16]. Let us point out that for every fixed k, it can be decided in polynomial time whether a graph admits a shortest-path with eccentricity at most k [16]. Our results will show the situation is different for pathlength and pathbreadth.

1.2   Our Contributions

On the negative side, we prove in Sect. 2 that computing the treebreadth is NP-hard. More precisely, we first prove that recognizing graphs with treebreadth one is NP-complete. The latter may be a bit surprising since in comparison, graphs with treelength one are exactly the chordal graphs [19], and so, they can be recognized in linear time. Our reduction has distant similarities with the one for treelength. However, it does not need any detour through weighted graphs. Then, we show that the problem of deciding whether a graph has treebreadth one is polynomially equivalent to the problem of deciding whether a graph has treebreadth at most k, for every fixed $k \geq 1$.

Next, we show that deciding if a graph has pathlength at most 2 is NP-hard even in the class of graphs with pathlength at most 3. We also show that deciding if a graph has pathbreadth at most 1 is NP-hard even in the class of graphs with pathbreadth at most 2. Hence, for any $\epsilon > 0$, the pathlength and the pathbreadth cannot be approximated within a factor $\frac{3}{2} - \epsilon$ and $2 - \epsilon$ respectively unless $P = NP$.

On the positive side, we present polynomial-time algorithms for deciding whether a graph has treebreadth at most one, in the class of bipartite graphs and in the class of planar graphs. Precisely, we prove that a bipartite graph has treebreadth one if and only if it can be clique-decomposed in *tree-convex* bipartite graphs [21]. Furthermore, while the planar graphs of treebreadth one are quite specific (in particular, we prove that they have treewidth at most 4), the algorithm is intricate and relies on structural properties of graphs with treebreadth one.

Due to lack of space, several proofs are only sketched or even omitted. They can be found in our technical report [17].

1.3   Definitions and Notations

Graphs in this study are finite, simple, connected and unweighted. Given a graph $G = (V, E)$, the set $N_G(v)$ denotes the set of neighbors of $v \in V$ in G. Furthermore, let $N_G[v] = N_G(v) \cup \{v\}$. The distance $dist_G(u, v)$ between two vertices $u, v \in V$ in G is the minimum length (number of edges) of a path between u and v in G. We will omit the subscript when no ambiguity occurs.

A *tree-decomposition* $(T, \mathcal{X})$ of G is a pair consisting of a tree T and of a family $\mathcal{X} = (X_t)_{t \in V(T)}$ of subsets of V indexed by the nodes of T and satisfying:

- $\bigcup_{t \in V(T)} X_t = V$;
- for any edge $e = \{u, v\} \in E$, there exists $t \in V(T)$ such that $u, v \in X_t$;
- for any $v \in V$, $\{t \in V(T) \mid v \in X_t\}$ induces a subtree, denoted by T_v, of T.

The sets X_t are called *the bags* of the decomposition. For any $t \in V(T)$, the *diameter* of the bag X_t equals $\max_{v,w \in X_t} dist_G(v, w)$. We emphasize that the distance is the one in G (not in $G[X_t]$). The *radius* of X_t equals $\min_{v \in V} \max_{w \in X_t} dist_G(v, w)$. We point out that the vertex v in previous definition does not necessarily belong to X_t. The *length* of $(T, \mathcal{X})$ is the maximum diameter of its bags, while the *breadth* of $(T, \mathcal{X})$ is the maximum radius of its bags.

The *treelength* and the *treebreadth* of G, respectively denoted by $tl(G)$ and $tb(G)$, are the minimum length and breadth of its tree-decompositions, respectively. Pathlength and pathbreadth are defined similarly in the case of path decompositions, that is, when T is a path. It has been observed in [14,15] that the four above parameters are contraction-closed invariants.

A tree-decomposition is called *reduced* if no bag is included in another one. Starting from any tree-decomposition, a reduced tree-decomposition can be obtained in polynomial time by contracting any two adjacent bags with one contained in the other until it is no more possible to do that. Note that such a process does not modify the width, the length nor the breadth of the decomposition.

In the following we will make use of the well-known *Helly property* in our proofs: any family of pairwise intersecting subtrees in a tree has a nonempty intersection.

2   Hardness of Treebreadth, Pathlength and Pathbreadth

The main result of this section is the NP-completeness of deciding whether $tb(G) \leq k$, for any fixed $k \geq 1$. We first prove that the problem is NP-complete for $k = 1$. Then, we show that the problem of deciding the treebreadth of a graph is polynomially equivalent to the problem of recognizing graphs with treebreadth one. Using similar techniques, we prove that computing pathlength, resp., pathbreadth, is NP-hard.

We start by a structural result on graphs with treebreadth one which will be a key lemma used throughout the paper. A tree-decomposition $(T, \mathcal{X})$ of a graph is a *star-decomposition* if for each $t \in V(T)$, $X_t \subseteq N[v]$ for some $v \in X_t$. That is, star-decompositions are similar to decompositions of breadth one, but the dominator of each bag has to belong to the bag itself. Lemma 1 shows that both definitions are actually equivalent.

Lemma 1. *For any graph G with $tb(G) \leq 1$, every reduced tree-decomposition of G of breadth one is a star-decomposition.*

Proof. Let $(T, \mathcal{X})$ be any reduced tree-decomposition of G of breadth one. We will prove it is a star-decomposition. To prove it, let $X_t \in \mathcal{X}$ be arbitrary and let $v \in V$ be such that $\max_{w \in X_t} dist_G(v, w) = 1$, which exists because X_t has radius one. We now show that $v \in X_t$. Indeed, since the subtree T_v and the subtrees $T_w, w \in X_t$, pairwise intersect, then it comes by the Helly Property that $T_v \cap \left(\bigcap_{w \in X_t} T_w \right) \neq \emptyset$ *i.e.*, there is some bag containing $\{v\} \cup X_t$. As a result, we have that $v \in X_t$ because $(T, \mathcal{X})$ is a reduced tree-decomposition. The latter implies that $(T, \mathcal{X})$ is a star-decomposition because X_t is arbitrary.   □

We then show the main result of this section.

Theorem 1. *Deciding whether a graph has treebreadth one is NP-complete.*

In order to prove Theorem 1, we reduce the following particular instance of CHORDAL SANDWICH (proved to be NP-hard in [6]) to our problem. In [19], the author also proposed a reduction from CHORDAL SANDWICH in order to prove that computing treelength is NP-hard. However, we will need different gadgets than in [19], and we will need different arguments to prove correctness of the reduction.

---

*Problem 1 (*CHORDAL SANDWICH WITH $\overline{nK_2}$*).*

Input: graphs $G_1 = (V, E_1)$ and $G_2 = (V, E_2)$ such that $E_1 \subseteq E_2$, $|V|$ is even and the complementary $\bar{G}_2$ of G_2 induces a perfect matching.
Question: Is there a chordal graph $H = (V, E)$ such that $E_1 \subseteq E \subseteq E_2$?

Perhaps surprisingly, the restriction on the structure of $\bar{G}_2$ is a key element in our reduction. Indeed, we will need the following technical lemma whose proof can be found in [17].

Lemma 2. *Let $G_1 = (V, E_1)$, $G_2 = (V, E_2)$ such that $E_1 \subseteq E_2$ and $\bar{G}_2$ is a perfect matching. Suppose that $\langle G_1, G_2 \rangle$ is a yes-instance of CHORDAL SANDWICH WITH $\overline{nK_2}$.*

There exists a tree-decomposition $(T, \mathcal{X})$ of G_1 with $|\mathcal{X}| = |V|/2 + 1$ bags such that for every $\{u, v\} \notin E_2$, $T_u \cap T_v = \emptyset$ and there are two adjacent bags $B_u \in T_u$ and $B_v \in T_v$ such that $B_u \backslash u = B_v \backslash v$.

Proof of Theorem 1. The problem is in NP. To prove the NP-hardness, let $\langle G_1, G_2 \rangle$ be any instance of CHORDAL SANDWICH WITH $\overline{nK_2}$. Let G' be the graph constructed from G_1 as follows. First, a clique V' of $2n = |V|$ vertices is added to G_1. Vertices $v \in V$ are in one-to-one correspondance with vertices $v' \in V'$. Then, for every $\{u, v\} \notin E_2$, u and v are respectively made adjacent to all vertices in $V' \backslash v'$ and $V' \backslash u'$. Finally, we add a copy of the gadget F_{uv}, depicted in Fig. 1(a), and the vertices s_{uv} and t_{uv} are made adjacent to the four vertices u, v, u', v'.

We will prove $tb(G') = 1$ if and only if $\langle G_1, G_2 \rangle$ is a yes-instance of CHORDAL SANDWICH WITH $\overline{nK_2}$.

In one direction, assume $tb(G') = 1$, let $(T, \mathcal{X})$ be a star-decomposition of G' (which exists by Lemma 1). We prove that the triangulation of G_1 obtained from this star-decomposition is the desired chordal sandwich. Let $H = (V, \{\{u, v\} \mid T_u \cap T_v \neq \emptyset\})$. H is a chordal graph such that $E_1 \subseteq E(H)$. To prove that

$\langle G_1, G_2 \rangle$ is a yes-instance of CHORDAL SANDWICH WITH $\overline{nK_2}$, it suffices to prove that $T_u \cap T_v = \emptyset$ for every $\{u, v\} \notin E_2$. We claim that it is implied by $T_{s_{uv}} \cap T_{t_{uv}} \neq \emptyset$. Indeed, assume $T_{s_{uv}} \cap T_{t_{uv}} \neq \emptyset$ and $T_u \cap T_v \neq \emptyset$. Since $s_{uv}, t_{uv} \in N(u) \cap N(v)$, $T_u, T_v, T_{s_{uv}}, T_{t_{uv}}$ pairwise intersect, there is a bag with u, v, s_{uv}, t_{uv} by the Helly property. The latter contradicts that $(T, \mathcal{X})$ is a star-decomposition because no vertex dominates the four vertices. Hence the claim is proved. So, let us prove that $T_{s_{uv}} \cap T_{t_{uv}} \neq \emptyset$. By contradiction, if $T_{s_{uv}} \cap T_{t_{uv}} = \emptyset$ then every bag B onto the path between $T_{s_{uv}}$ and $T_{t_{uv}}$ must contain c_{uv}, x_{uv}. Since $N[c_{uv}] \cap N[x_{uv}] = \{s_{uv}, t_{uv}\}$ and $(T, \mathcal{X})$ is a star-decomposition, it implies either $s_{uv} \in B$ and $B \subseteq N[s_{uv}]$ or $t_{uv} \in B$ and $B \subseteq N[t_{uv}]$. So, there are two adjacent bags $B_s \in T_{s_{uv}}, B_t \in T_{t_{uv}}$ such that $B_s \subseteq N[s_{uv}]$ and $B_t \subseteq N[t_{uv}]$. In particular, $B_s \cap B_t$ must intersect the path (y_{uv}, w_{uv}, z_{uv}) because $y_{uv} \in N(s_{uv})$ and $z_{uv} \in N(t_{uv})$. However, $N[s_{uv}] \cap N[t_{uv}] \cap \{y_{uv}, w_{uv}, z_{uv}\} = \emptyset$, that is a contradiction. As a result, $T_{s_{uv}} \cap T_{t_{uv}} \neq \emptyset$ and so, $T_u \cap T_v = \emptyset$ for any $\{u, v\} \notin E_2$.

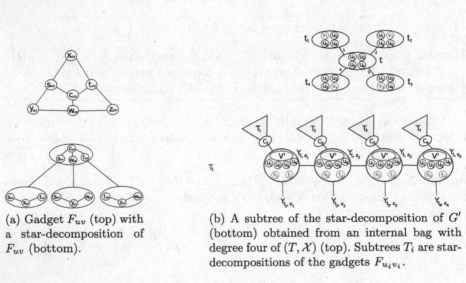

(a) Gadget F_{uv} (top) with a star-decomposition of F_{uv} (bottom).

(b) A subtree of the star-decomposition of G' (bottom) obtained from an internal bag with degree four of $(T, \mathcal{X})$ (top). Subtrees T_i are star-decompositions of the gadgets $F_{u_i v_i}$.

Fig. 1. Construction of the star-decomposition in the proof of Theorem 1

Conversely, assume that $\langle G_1, G_2 \rangle$ is a yes-instance of CHORDAL SANDWICH WITH $\overline{nK_2}$. Since $\bar{G}_2$ is a perfect matching by the hypothesis, let $(T, \mathcal{X})$ be as stated in Lemma 2. We will modify $(T, \mathcal{X})$ in order to obtain a star-decomposition of G'. To do so, we will use the fact that there are $|V|/2 = n$ edges in $E(T)$ and the properties stated by Lemma 2. Indeed, this implies that there is a one-to-one mapping $\alpha : E(T) \to E(\bar{G}_2)$ between the edges of T and the non-edges of G_2. Precisely, for any edge $e = \{t, s\} \in E(T)$, let $\alpha(e) = \{u, v\} \in E(\bar{G}_2)$ be the non-edge of G_2 such that $u \in X_t, v \in X_s$ and $X_t \backslash u = X_s \backslash v$.

Intuitively, the star-decomposition $(T', \mathcal{X}')$ of G' is obtained as follows. For any $t \in V(T)$ with incident edges $e_1, \cdots, e_d$, we first replace X_t by a path-decomposition $(Y_{t,e_1}, \cdots, Y_{t,e_d})$. Then, for any edge $e = \{t, s\} \in E(T)$, an edge is added between $Y_{t,e}$ and $Y_{s,e}$. Finally, the center-bag of some star-decomposition of the gadget $F_{\alpha(e)}$ is made adjacent to $Y_{t,e}$ (see Fig. 1(b) for an illustration).

More formally, let $t \in V(T)$ and $e \in E(T)$ incident to t, and let $\{u,v\} = \alpha(e)$. Let $Y_{t,e} = V' \cup X_t \cup \{s_{uv}, t_{uv}\}$ (note that $Y_{t,e}$ is dominated by $u' \in V'$). Let $e_1, \cdots, e_d$ be the edges incident to t in T, in any order. For $1 \le i < d$, add an edge between Y_{t,e_i} and $Y_{t,e_{i+1}}$. For any edge $e = \{t,s\} \in E(T)$, add an edge between $Y_{t,e}$ and $Y_{s,e}$. Finally, add the star-decomposition $(T^e, \mathcal{X}^e)$ for the gadget $F_{\alpha(e)}$ as depicted in Fig. 1(a) and add an edge between its center and $Y_{t,e}$.

The resulting $(T', \mathcal{X}')$ is a star-decomposition of G', hence $tb(G') = 1$. $\square$

We next show that computing the treebreadth is polynomially equivalent to the recognition of graphs with treebreadth one.

Lemma 3. *For every graph G, for every positive integer r, there exists a graph G'_r computable in polynomial time such that $tb(G) \le r$ if and only if $tb(G'_r) \le 1$.*

Proof. Let G have vertices $v_1, v_2, \ldots, v_n$, and let $r > 0$. The graph G'_r is obtained from G by adding a clique $U = \{u_1, u_2, \ldots, u_n\}$ so that for every $1 \le i \le n$, u_i is adjacent to all vertices in $B_G(v_i, r) = \{x \in V(G) \mid dist_G(v_i, x) \le r\}$.

If $tb(G) \le r$ then we claim that given a tree-decomposition $(T, \mathcal{X})$ of G with breadth at most r, one obtains a star-decomposition of G'_r by adding the clique U in every bag in $\mathcal{X}$. Indeed, for every bag $X_t \in \mathcal{X}$, by the hypothesis there is $v_i \in V(G)$ such that $\max_{x \in X_t} dist_G(v_i, x) \le r$, hence $X_t \cup U \subseteq N_{G'_r}[u_i]$. Conversely, if $tb(G'_r) \le 1$ then we claim that given a star-decomposition $(T', \mathcal{X}')$ of G'_r, one obtains a tree-decomposition of G with breadth at most r by removing every vertex of the clique U from every bag in $\mathcal{X}'$. Indeed, for every bag $X'_t \in \mathcal{X}'$, by the hypothesis there is $y \in X'_t$ such that $X'_t \subseteq N_{G'_r}[y]$. Furthermore, $y \in \{u_i, v_i\}$ for some $1 \le i \le n$, and so, since $N_{G'_r}[v_i] \subseteq N_{G'_r}[u_i]$ by construction, $X'_t \backslash U \subseteq N_{G'_r}(u_i) \backslash U = \{x \in V(G) \mid dist_G(v_i, x) \le r\}$. $\square$

Lemma 4. *For every graph G, for every positive integer r, there exists a graph G' computable in polynomial time such that $tb(G) \le 1$ if and only if $tb(G') \le r$.*

Proof. For every $\{u,v\} \in E(G)$, let F^r_{uv} be obtained from F_{uv} in Fig. 1(a) by adding an edge $\{s_{uv}, t_{uv}\}$ then subdividing each edge $r-1$ times. The graph G' is obtained from G by substituting each edge $\{u,v\} \in E(G)$ with a distinct copy of F^r_{uv} then identifying u, v with s_{uv}, t_{uv}.

If $tb(G) \le 1$ then let us modify a star-decomposition $(T, \mathcal{X})$ of G in a tree-decomposition $(T', \mathcal{X}')$ of G' of breadth at most r. Clearly, every bag in $\mathcal{X}$ has radius at most r in G'. Furthermore, let $(T^{uv}, \mathcal{X}^{uv})$ be the star-decomposition of F_{uv} in Fig. 1(a), with three leaf-bags and one central bag. It can be modified in a tree-decomposition of F^r_{uv} by (i) adding in each bag containing both end-vertices of an edge in F_{uv} the $r-1$ vertices in F^r_{uv} that result from its subdivision, and (ii) adding a new leaf-bag with $\{u,v\}$ and the $r-1$ vertices that result from its subdivision. Finally, let $(T', \mathcal{X}')$ be obtained from $(T, \mathcal{X})$ by adding an edge between some bag in $T_u \cap T_v$ and the central bag of T^{uv} for every $\{u,v\} \in E(G)$. Since $(T', \mathcal{X}')$ has breadth r, $tb(G') \le r$.

Conversely, if $tb(G') \leq r$ then we claim that given a tree-decomposition $(T', \mathcal{X}')$ of G' of breadth at most r, one obtains a tree-decomposition of G of breadth one by removing every vertex of $V(G')\backslash V(G)$ from the bags in $\mathcal{X}'$. Before proving the claim, observe that no vertex in $V(G')\backslash V(G)$ can be at distance at most r from three vertices in $V(G)$, and in case it is at distance at most r from two vertices $u, v \in V(G)$ then $\{u, v\} \in E(G)$. Therefore, in order to prove the claim it suffices to prove that $u = s_{uv}$ and $v = t_{uv}$ are in a common bag of $\mathcal{X}'$ for every $\{u, v\} \in E(G)$. The latter can be proved by elaborating on the same arguments as for Theorem 1.   □

From Lemmas 3, 4 and Theorem 1, it follows that:

Theorem 2. *For any fixed $k \geq 1$, deciding whether a graph G has treebreadth at most k is NP-complete.*

To conclude this section, we consider pathlength and pathbreadth. Due to lack of space, the proofs are postponed in [17].

Theorem 3. *For any $\epsilon > 0$, the pathlength (resp., the pathbreadth) cannot be approximated within a factor $\frac{3}{2} - \epsilon$ (resp., $2 - \epsilon$) unless $P = NP$.*

3   Graphs with Treebreadth One: Some Polynomial Cases

In this section, we investigate further the class of graphs with treebreadth one. It strictly contains chordal graphs and dually chordal graphs, well-studied graph classes in algorithmic graph theory [7]. We first show some useful lemmas that somehow state that we can restrict our study on graphs without clique-separator. Then, we show that the problem of recognizing graphs with treebreadth one can be solved in polynomial time in the class of bipartite graphs and in the class of planar graphs.

Let $G = (V, E)$ be a connected graph. Recall that a set $S \subset V$ is a *separator* if $G \backslash S$ is disconnected. It is called a *clique-separator* if S induces a complete graph. A *full component* for S is any connected component C of $G \backslash S$ such that $N(C) = S$. If C is a full component for S then we call the induced subgraph $G[C \cup S]$ a *block*. Finally, S is a *minimal separator* if there exist at least two full components for S.

Our objective is to prove that if a graph G has treebreadth one then so do all its blocks. In fact, we will prove a slightly more general result:

Lemma 5. *Let $G = (V, E)$, S be a separator and W be the union of some connected components of $G \backslash S$. If $tb(G) = 1$ and W contains a full component for S, then $tb(G[W \cup S]) = 1$.*

Proof. Let $(T, \mathcal{X})$ be a star-decomposition of G. We remove vertices in $V \backslash (W \cup S)$ from bags in $\mathcal{X}$, that yields a tree-decomposition $(T, \mathcal{X}')$ of $G[W \cup S]$. We will prove that $(T, \mathcal{X}')$ has breadth one (but is not necessarily a star-decomposition). Indeed, let $X'_t \in \mathcal{X}'$. By construction, $X'_t \subseteq X_t$ with $X_t \in \mathcal{X}$. Let $v \in X_t$ satisfy

$X_t \subseteq N_G[v]$. If $v \in X'_t$, then we are done. Else, since for all $x \notin S \cup W, N(x) \cap (S \cup W) \subseteq S$ (because S is a separator by the hypothesis), we must have that $X_t \subseteq S$. Let $A \subseteq W$ be a full component for S, that exists by the hypothesis, let T_A be induced by the bags intersecting A. Since T_A and the subtrees $T_x, x \in X_t$ pairwise intersect — because for all $x \in X_t$, $x \in S$ and so, x has a neighbour in A —, then by the Helly property there is a bag in $\mathcal{X}$ containing X_t and intersecting A. Furthermore, any $u \in V$ dominating this bag must be either in S or in A, so, in particular there is $u \in A \cup S$ such that $X_t \subseteq N[u]$. $\square$

The converse of Lemma 5 does not hold in general (see Fig. 2), yet there are interesting cases when it does.

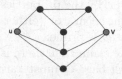

Fig. 2. $S = \{u, v\}$ separates G in two subgraphs of treebreadth 1. However, $tb(G) = 2$.

Lemma 6. *Let $G = (V, E)$ with a minimal clique-separator S and A be a full component. Then, $tb(G) = 1$ if and only if $tb(G[A \cup S]) = 1$ and $tb(G[V \setminus A]) = 1$.*

The proof of Lemma 6 is deferred to [17]. Recall that computing the clique-minimal-decomposition of a graph G takes $\mathcal{O}(nm)$-time, where m denotes the number of edges [3]. By doing so, one replaces a graph G with the maximal subgraphs of G that have no clique-separator, *a.k.a. atoms*. So, in the following we will only consider graphs without a clique-separator, *a.k.a., prime graphs*.

3.1   Bipartite Graphs

Bipartite graphs with treebreadth one are an interesting subclass of their own since they contain the convex bipartite graphs and the chordal bipartite graphs (*i.e.*, bipartite graphs with no induced cycle of length at least six). In this section, we present a linear-time algorithm that decides whether a prime bipartite graph has treebreadth one, and computes a corresponding decomposition if any. Since the clique-decomposition of a given bipartite graph can be computed in linear time, this proves combined with Lemma 6 that it can be decided in linear time whether a bipartite graph has treebreadth one.

More precisely, we show that prime bipartite graphs with treebreadth one coincide with *tree-convex* bipartite graphs, a generalization of convex bipartite graphs [21]. A bipartite graph is called tree-convex if it admits a tree-decomposition where the bags are the close neighbourhoods of any one side of its bipartition. By definition, tree-convex graphs have treebreadth one. The following lemma is a converse of this result.

Lemma 7. *Let $G = (V_0 \cup V_1, E)$ be a prime bipartite graph with treebreadth one. There is $(T, \mathcal{X})$ a star-decomposition of G such that either $\mathcal{X} = \{N[v_0] \mid v_0 \in V_0\}$, or $\mathcal{X} = \{N[v_1] \mid v_1 \in V_1\}$.*

Proof. Let $(T, \mathcal{X})$ be a star-decomposition of G minimizing $|\mathcal{X}|$. Suppose there is some $v_0 \in V_0$, there is $t \in V(T)$ such that $X_t \subseteq N_G[v_0]$ (the case when there is some $v_1 \in V_1$, there is $t \in V(T)$ such that $X_t \subseteq N_G[v_1]$ is symmetrical to this one). We claim that for every $t' \in V(T)$, there is $v_0' \in V_0$ such that $X_{t'} \subseteq N_G[v_0']$. By contradiction, let $v_0 \in V_0, v_1 \in V_1$, let $t, t' \in V(T)$ be such that $X_t \subseteq N_G[v_0], X_{t'} \subseteq N_G[v_1]$. By connectivity of the tree T we may assume w.l.o.g. that $\{t, t'\} \in E(T)$. Moreover, $N_G(v_0) \cap N_G(v_1) = \emptyset$ because G is bipartite. Therefore, $X_t \cap X_{t'} \subseteq \{v_0, v_1\}$, and in particular if $X_t \cap X_{t'} = \{v_0, v_1\}$ then v_0, v_1 are adjacent in G. However, by the properties of a tree-decomposition this implies that $X_t \cap X_{t'}$ is a clique-separator (either an edge or a single vertex), thus contradicting the fact that G is prime.

Let $v_0 \in V_0$ be arbitrary. We claim that there is a unique bag X_t, $t \in V(T)$, containing v_0. Indeed, any such bag X_t must satisfy $X_t \subseteq N_G[v_0]$, hence the subtree T_{v_0} can be contracted into a single bag $\bigcup_{t \in T_{v_0}} X_t$ without violating the property for the tree-decomposition to be a star-decomposition. As a result, the uniqueness of the bag X_t follows from the minimality of $|\mathcal{X}|$. Since X_t is unique and $X_t \subseteq N_G[v_0]$, therefore $X_t = N_G[v_0]$ and so, $\mathcal{X} = \{N[v_0] \mid v_0 \in V_0\}$. $\quad\square$

As shown in [21], tree-convex graph recognition can be reduced to hyper-tree recognition, that can be done in linear time [7]. Altogether, we obtain the following characterization of bipartite graphs with treebreadth one.

Corollary 1. *A bipartite graph has treebreadth one if and only if every of its atoms is tree-convex, which can be decided in linear time.*

3.2   Planar Graphs

In this section, we sketch a quadratic algorithm to recognize prime planar graphs of treebreadth one. Combined with Lemma 6, this shows that planar graphs of treebreadth one can be recognized in quadratic time. Our algorithm also allows to compute a corresponding decomposition in cubic time. Since the full analysis is lengthy, all proofs in this section are deferred to [17].

Our work in this section brings more insights on tree-decompositions with small width for planar graphs. Indeed, we prove the following.

Lemma 8. *For every planar graph G, $tb(G) \le 1$ implies $tw(G) \le 4$.*

The algorithm is recursive. Given $G = (V, E)$, we search for a specific vertex, called a *leaf-vertex*, whose closed neighborhood must be a leaf-bag of a star-decomposition if $tb(G) = 1$. Basing on Lemma 5 and a delicate case-by-case analysis of the structure of star-decompositions, we define three types of leaf-vertices (*e.g.*, see Fig. 3). A vertex v is a *leaf-vertex* if one of the following conditions hold.

Type 1. $N(v)$ induces an $a_v b_v$-path for some $a_v, b_v \in V \backslash \{v\}$, denoted by Π_v, of length at least 3 and there is $d_v \in V \backslash \{v\}$ such that $N(v) \subseteq N(d_v)$.

Type 2. $N(v)$ induces a path, denoted by $\Pi_v = (a_v, b_v, c_v)$, of length 2.

Type 3. $N(v)$ consists of two non adjacent vertices a_v and c_v, and there is $b_v \in (N(a_v) \cap N(c_v)) \backslash \{v\}$.

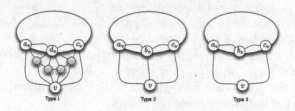

Fig. 3. The three kinds of leaf-vertices.

Ideally, we would like to remove v from G and apply recursively our algorithm on $G \backslash v$. However, in some case $tb(G \backslash v) = 1$ while $tb(G) > 1$ (see Fig. 2). So, we must also add edges between vertices that must be in a common bag of a star-decomposition of G if $tb(G) = 1$[1]. The choice of the edges to add is made more difficult by the need for the resulting graph G' to stay prime and planar in order to apply our algorithm recursively on G'. To show that $tb(G) = 1$ if and only if the resulting graph has treebreadth one also requires tedious lemmas.

Theorem 4. *Recognizing planar graphs of treebreadth one can be done in quadratic time. Moreover, a star-decomposition (if any) can be computed in cubic time.*

Sketch proof. Let $G = (V, E)$ be a prime planar graph. We can assume $|V| \geq 8$ and G has no star-decomposition with two bags (both cases are treated separately by exhaustive search). In such case, $tb(G) = 1$ implies there exists a leaf-vertex v, that can be found in linear time.

If $G \backslash v$ is prime then we prove $tb(G) = 1$ if and only if $tb(G \backslash v) = 1$, except in the special case when v is of Type 2 or 3 and $|(N(a_v) \cap N(c_v)) \backslash v| \leq 2$. Furthermore, we prove for the latter case that a_v, c_v must have two common neighbours u_v, b_v in $G \backslash v$ (else, $tb(G) > 1$) and G', obtained from G by adding the edges $\{v, u_v\}, \{v, b_v\}$, is planar and prime, and it satisfies $tb(G) = 1$ if and only if $tb(G') = 1$. So, we call the algorithm either on G' or on $G \backslash v$[2].

The most difficult situation is when $G \backslash v$ contains a clique-separator. This case is reduced to the one when v is of Type 2, there is an edge-separator (b_v, u_v) of $G \backslash v$, and $\{a_v, u_v\} \notin E$. Then, we aim at applying the algorithm recursively

---

[1] We aim at turning the separator $N(v)$ into a clique. However, we cannot do that directly since it would break the distances in G, and the graph needs to stay planar.

[2] When v is of Type 1 we call the algorithm on G', obtained from $G \backslash v$ by contracting the internal nodes of Π_v to an edge, in order to obtain a quadratic complexity.

on G', obtained from $G\backslash v$ by adding the edge $\{a_v, c_v\}$. However, $tb(G') = 1$ does not imply $tb(G) = 1$ in general. We prove it is the case if u_v, c_v are nonadjacent or $N(u_v) \cap N(a_v)$ does not disconnect a_v from u_v in $G\backslash(c_v, v)$.

Else, we compute a plane embedding of G, and a vertex $x \in N(a_v) \cap N(u_v)$ such that: v, c_v and all other common neighbours of a_v, u_v are in a same region $\mathcal{R}$, bounded by (a_v, x, u_v, b_v). We wish to create an $a_v u_v$-path in $V\backslash\mathcal{R}$ by adding edges in $N(b_v) \cap N(x)$. In doing so, we go back to the previous subcase as now $N(a_v) \cap N(u_v)$ is no more a $a_v u_v$-separator of $G\backslash(c_v, v)$. However, we have to ensure that it is possible to add such a path in $V\backslash\mathcal{R}$, and that its addition does not affect the value of treebreadth for the graph. We prove it is the case unless $V \subseteq \mathcal{R}$ (in which case we apply the algorithm recursively on G', obtained from G by identifying b_v with x), or if there is a leaf-vertex $l \in N(b_v) \cap N(x)$. Furthermore, in the latter case we replace v with l in the above analysis, *i.e.*, l becomes the actual leaf-vertex to be considered.

Additional properties are needed in order to prove the algorithm terminates, and that it does so in a linear number of steps.   □

Conclusion. We conclude this paper by some questions that remain open. First, it would be interesting to know the complexity of deciding the treebreadth of planar graphs. Second, all the reductions presented in this paper rely on constructions containing large clique or clique-minor. We left open the problem of recognizing graphs with tree-breadth one in the class of graphs with bounded treewidth or bounded clique-number. More generally, is the problem of computing the treebreadth Fixed-Parameter Tractable when it is parameterized by the treewidth or by the size of a largest clique-minor?

References

1. Abu-Ata, M., Dragan, F.: Metric tree-like structures in real-world networks: an empirical study. Networks **67**, 49–68 (2016)
2. Arnborg, S., Corneil, D., Proskurowski, A.: Complexity of finding embeddings in a k-tree. SIAM J. Algebraic Discrete Methods **8**(2), 277–284 (1987)
3. Berry, A., Pogorelcnik, R., Simonet, G.: An introduction to clique minimal separator decomposition. Algorithms **3**(2), 197–215 (2010)
4. Bodlaender, H.: Treewidth: Characterizations, applications, and computations (2006)
5. Bodlaender, H.: A linear-time algorithm for finding tree-decompositions of small treewidth. SIAM J. Comput. **25**(6), 1305–1317 (1996)
6. Bodlaender, H., Fellows, M., Warnow, T.: Two strikes against perfect phylogeny. In: ICALP 1992, Vienna, Austria, pp. 273–283 (1992)
7. Brandstädt, A., Dragan, F., Chepoi, V., Voloshin, V.: Dually chordal graphs. SIAM J. Discrete Math. **11**(3), 437–455 (1998)
8. Chechik, S., Larkin, D., Roditty, L., Schoenebeck, G., Tarjan, R., Williams, V.V.: Better approximation algorithms for the graph diameter. In: ACM SODA 2014, pp. 1041–1052. SIAM (2014)
9. Chepoi, V., Dragan, F., Estellon, B., Habib, M., Vaxès, Y.: Diameters, centers, and approximating trees of δ-hyperbolic geodesic spaces and graphs. In: SCG 2008, New York, NY, USA, pp. 59–68. ACM (2008)

10. Coudert, D., Ducoffe, G., Nisse, N.: To approximate treewidth, use treelength! SIAM J. Discrete Math. (to appear, 2016)
11. de Montgolfier, F., Soto, M., Viennot, L.: Treewidth and hyperbolicity of the internet. In: 2011 10th IEEE International Symposium on Network Computing and Applications (NCA), pp. 25–32, August 2011
12. Dourisboure, Y., Dragan, F., Gavoille, C., Chenyu, Y.: Spanners for bounded tree-length graphs. Theor. Comput. Sci. **383**(1), 34–44 (2007)
13. Dourisboure, Y., Gavoille, C.: Tree-decompositions with bags of small diameter. Discrete Math. **307**(16), 2008–2029 (2007)
14. Dragan, F., Köhler, E.: An approximation algorithm for the tree t-spanner problem on unweighted graphs via generalized chordal graphs. Algorithmica **69**(4), 884–905 (2014)
15. Dragan, F.F., Köhler, E., Leitert, A.: Line-distortion, bandwidth and path-length of a graph. In: Ravi, R., Gørtz, I.L. (eds.) SWAT 2014. LNCS, vol. 8503, pp. 158–169. Springer, Heidelberg (2014)
16. Dragan, F.F., Leitert, A.: On the minimum eccentricity shortest path problem. In: Dehne, F., Sack, J.-R., Stege, U. (eds.) WADS 2015. LNCS, vol. 9214, pp. 276–288. Springer, Heidelberg (2015)
17. Ducoffe, G., Legay, S., Nisse, N.: On computing tree and path decompositions with metric constraints on the bags. Technical Report RR-8842 (2016)
18. Krauthgamer, R., Lee, J.: Algorithms on negatively curved spaces. In: FOCS 2006, pp. 119–132. IEEE (2006)
19. Lokshtanov, D.: On the complexity of computing treelength. Discrete Appl. Math. **158**(7), 820–827 (2010)
20. Robertson, N., Seymour, P.: Graph minors. II: algorithmic aspects of tree-width. J. Algorithms **7**(3), 309–322 (1986)
21. Wang, C., Liu, T., Jiang, W., Xu, K.: Feedback vertex sets on tree convex bipartite graphs. In: Lin, G. (ed.) COCOA 2012. LNCS, vol. 7402, pp. 95–102. Springer, Heidelberg (2012)
22. Wu, Y., Austrin, P., Pitassi, T., Liu, D.: Inapproximability of treewidth and related problems. J. Artif. Intell. Res. (JAIR) **49**, 569–600 (2014)

Online Chromatic Number is PSPACE-Complete

Martin Böhm[(✉)] and Pavel Veselý[(✉)]

Computer Science Institute of Charles University, Prague, Czech Republic
{bohm,vesely}@iuuk.mff.cuni.cz

Abstract. In the online graph coloring problem, vertices from a graph G, known in advance, arrive in an online fashion and an algorithm must immediately assign a color to each incoming vertex v so that the revealed graph is properly colored. The exact location of v in the graph G is not known to the algorithm, since it sees only previously colored neighbors of v. The *online chromatic number* of G is the smallest number of colors such that some online algorithm is able to properly color G for any incoming order. We prove that computing the online chromatic number of a graph is PSPACE-complete.

1   Introduction

In the classical graph coloring problem we assign a color to each vertex of a given graph such that the graph is properly colored, i.e., no two adjacent vertices have the same color. The chromatic number χ of a graph G is the smallest k such that G can be colored with k distinct colors. Deciding whether the chromatic number of a graph is at most k is well known to be NP-complete, even in the case with three colors.

The online variant of graph coloring can be defined as follows: The vertices of G arrive one by one, and an online algorithm must color vertices as they arrive so that the revealed graph is properly colored at all times. When a vertex arrives, the algorithm sees edges to previously colored vertices. The online algorithm may use additional knowledge of the whole graph G; more precisely, a copy of G is sent to the algorithm at the start of the input. However, the exact correspondence between the incoming vertices and the vertices of the copy of G is not known to the algorithm. This problem is called ONLINE GRAPH COLORING.

In this paper we focus on a graph parameter called the *online chromatic number* $\chi^O(G)$ of a graph G. This parameter is analogous to the standard chromatic number of a graph: It denotes the smallest number k such that there exists a deterministic online algorithm which is able to color the specified graph G using k colors.

The online chromatic number has been studied since 1990 [3]. One of the main open problems in the area is the computational complexity of deciding

M. Böhm–Supported by CE-ITI under grant P202/12/G061 of GA ČR and by the GAUK project 548214.
P. Veselý–Supported by the GAUK project 548214.

© Springer International Publishing Switzerland 2016
V. Mäkinen et al. (Eds.): IWOCA 2016, LNCS 9843, pp. 16–28, 2016.
DOI: 10.1007/978-3-319-44543-4_2

whether $\chi^O(G) \leq k$ for a specified simple graph G, given G and k on input; see e.g. Kudahl [10]. We denote this decision problem as ONLINE CHROMATIC NUMBER. In this paper, we fully resolve this problem:

Theorem 1. *The decision problem* ONLINE CHROMATIC NUMBER *is PSPACE-complete.*

As is usual in the online computation model, we can view ONLINE GRAPH COLORING as a game between two players, which we call PAINTER (representing the online algorithm) and DRAWER (often called ADVERSARY in the online algorithm literature). In each round DRAWER chooses an uncolored vertex v from G and sends it to PAINTER without telling him to which vertex of G it corresponds, only revealing the edges to the previously sent vertices. Then PAINTER must properly color ("paint") v, i.e., PAINTER cannot use a color of a neighbor of v. We stress that in this paper PAINTER is restricted to be deterministic. The game continues with the next round until all vertices of G are colored.

Deciding the outcome of many two-player games is PSPACE-complete; among those are Amazons, Checkers and Hex, to name a few. However, in most of these games both players have roughly the same power. This does not hold for ONLINE GRAPH COLORING which is highly asymmetric, since DRAWER has perfect information (knows which vertices are sent and how they are colored), but PAINTER does not. PAINTER may only guess to which part of the graph does the colored subgraph really belong. This is the main difficulty in proving PSPACE-hardness.

Examples. Consider a path P_4 on four vertices. Initially, DRAWER sends two nonadjacent vertices. If PAINTER assigns different colors to them, then these are the first and the third vertex of P_4, thus the second vertex must get a third color; otherwise they obtained the same color a and they are the endpoints of P_4, therefore the second and the third vertex get different colors which are not equal to a. In both cases, there are three colors on P_4 and thus $\chi^O(P_4) = 3$, while $\chi(P_4) = 2$.

Note also that we may think of DRAWER deciding where an incoming vertex belongs at some time *after* it is colored provided that the choice still allows for at least one isomorphism to the original G. This is possible only for a deterministic PAINTER.

A particularly interesting class of graphs in terms of χ^O is the class of binomial trees. A binomial tree of order k is defined inductively: The binomial tree of order 0 is a single vertex (the root) and the binomial tree of order k is created by taking two disjoint copies of binomial trees of order $k-1$, adding an edge between their roots and choosing one of their roots as the root for the resulting tree. Thus P_4 is a binomial tree of order 2 with root on the second vertex of P_4.

It is not hard to generalize the example of P_4 and show that the online chromatic number of the binomial tree of order k is $k+1$ [3]. This shows that the ratio between χ^O and χ may be arbitrarily large even for the class of trees.

History and Related Work. The online problem ONLINE GRAPH COLORING has been known since 1976 [1], originally studied in the variant where the algorithm has no extra information at the start of the input. Bean [1] showed that no online algorithm that is compared to an offline algorithm can perform well under this metric. The notion of online chromatic number was first defined in 1990 by [3].

For the online problem, Lovász, Saks and Trotter [11] show an algorithm with a *competitive ratio* $O(n/\log^* n)$, where the competitive ratio is a ratio of the number of colors used by the online algorithm to the (standard) chromatic number. This was later improved to $O(n \log \log \log n / \log \log n)$ by Kierstad [8] using a deterministic algorithm. There is a better $O(n/\log n)$-competitive randomized algorithm against an oblivious adversary by Halldórsson [5]. A lower bound on the competitive ratio of $\Omega(n/\log^2 n)$ was shown by Halldórsson and Szegedy [7].

Our variant of ONLINE GRAPH COLORING, where the algorithm receives a copy of the graph at the start, was suggested by Halldórsson [6], where it is shown that the lower bound $\Omega(n/\log^2 n)$ also holds in this model. (Note that the previously mentioned algorithmic results are valid for this model also.)

Kudahl [9] recently studied ONLINE CHROMATIC NUMBER as a complexity problem. The paper shows that the problem is coNP-hard and lies in PSPACE. Later [10] he proved that if some part of the graph is precolored, i.e., some vertices are assigned some colors prior to the coloring game between DRAWER and PAINTER and DRAWER also reveals edges to the precolored vertices for each incoming vertex, then deciding whether $\chi^O(G) \leq k$ is PSPACE-complete. We call this decision problem ONLINE CHROMATIC NUMBER WITH PRECOLORING. The paper [10] conjectures that ONLINE CHROMATIC NUMBER (with no precolored part) is PSPACE-complete too. Interestingly, it is possible to decide $\chi^O(G) \leq 3$ in polynomial time [4].

Keep in mind that while ONLINE GRAPH COLORING is an online problem, ONLINE CHROMATIC NUMBER is an (offline) decision problem of checking whether $\chi^O(G) \leq k$.

Proof Outline. It is not hard to see that ONLINE CHROMATIC NUMBER belongs to PSPACE: The online coloring is represented by a game tree which is evaluated using the Minimax algorithm. This can be done in polynomial space, since the number of rounds in the game is bounded by n, i.e., the number of vertices, and possible moves of each player can be enumerated in polynomial space: PAINTER has at most n possible moves, because it either uses a color already used for a vertex, or it chooses a new color, and DRAWER has at most 2^s moves where s is the number of colored vertices, since it chooses which colored vertices shall be adjacent to the incoming vertex. DRAWER must ensure that sent vertices form an induced subgraph of G, but this can be checked in polynomial space.

Inspired by [10], we prove the PSPACE-hardness of ONLINE CHROMATIC NUMBER by a reduction to Q3DNF-SAT, i.e., the satisfiability of a fully quantified formula in the 3-disjunctive normal form (3-DNF). An example of such a formula is

$$\forall x_1 \exists x_2 \forall x_3 \exists x_4 ... : (x_1 \wedge x_2 \wedge \neg x_3) \vee (\neg x_1 \wedge x_2 \wedge \neg x_4) \vee ...$$

The similar problem of satisfiability of a fully quantified formula in the 3-conjunctive normal form is well known to be PSPACE-complete. Since PSPACE is closed under complement, Q3DNF-SAT is PSPACE-complete as well. Note that by an easy polynomial reduction, we can assume that each 3-DNF clause contains exactly three literals.

We show the hardness in several iterative steps. First, in Sect. 2, we present a new, simplified proof of the PSPACE-hardness of ONLINE CHROMATIC NUMBER WITH PRECOLORING in which the sizes of both precolored and non-precolored parts of our construction are linear in the size of the formula.

Then, in Sect. 3, we strengthen the result by reducing the size of the precolored part to be logarithmic in the size of the formula. This is achieved by adding linearly many vertices to our construction.

Finally, in Sect. 4, we show how to remove one precolored vertex and replace it by a non-precolored part, while keeping the PSPACE-hardness proof valid. The cost for removing one vertex is that the size of the graph is multiplied by a constant, but since we apply it only logarithmically many times, we obtain a graph of polynomial size and with no precolored vertex. This will complete the proof of Theorem 1.

We remark that removing the last precolored vertex is the most difficult part of proving PSPACE-hardness of ONLINE CHROMATIC NUMBER. Still, our technique for removing a precolored vertex can be used for any graph satisfying a few assumptions.

We omit some proofs and some technical aspects of our construction due to space restrictions. A preprint version [2] with full details can be found at https://arxiv.org/abs/1604.05940.

In our analysis, PAINTER often uses the natural greedy algorithm FIRSTFIT, which is ubiquitous in the literature (see [6,11]):

Definition 1. *The online algorithm* FIRSTFIT *colors an incoming vertex u using the smallest color not present among colored vertices adjacent to u.*

2   Construction with a Large Precolored Part

Our first construction will reduce the PSPACE-complete problem Q3DNF-SAT to ONLINE COLORING WITH PRECOLORING with a large precolored part. Given a fully quantified formula Q in the 3-disjunctive normal form, we will create a graph G_1 that will simulate this formula. We assume that the formula contains n variables $x_i, (1 \leq i \leq n)$ and m clauses $C_a, (1 \leq a \leq m)$, and that variables are indexed in the same order as they are quantified.

Our main resource will be a large precolored clique K_{col} on k vertices and naturally using k colors; the number k will be specified later. Using such a precolored clique, we can restrict the allowed colors on a given uncolored vertex v by connecting it with the appropriate vertices in K_{col}, i.e., we connect v to all vertices in K_{col} which do not have a color allowed for v.

For simplicity we use the precoloring in the strong sense, i.e., PAINTER is able to recognize which vertex in K_{col} is which. We use this to easily recognize colors.

However, it is straightforward to avoid the strong precoloring by modifying the precolored part; for example by creating i independent and identical copies of the i-th vertex in K_{col}, each having the same color and the same edges to other vertices in K_{col} and the rest of the graph. With such a modification, PAINTER would able to recognize each color by the number of its vertices in K_{col}.

Each vertex in K_{col} thus corresponds to a color. Colors used by PAINTER are naturally denoted by numbers $1, 2, 3, \ldots, k$, but we shall also assign meaningful names to them.

We want to construct a graph G_1 that has the online chromatic number k if and only if the quantified 3-DNF formula can be satisfied. See Fig. 1 for an example of G_1 and an overview of our construction. We use the following gadgets for variables and clauses:

1. For a variable x_i which is quantified universally, we will create a gadget consisting of *universal* vertices $x_{i,t}$ and $x_{i,f}$, connected by an edge. The vertex $x_{i,t}$ represents the positive literal x_i, while $x_{i,f}$ represents the negative literal $\neg x_i$. Both vertices have exactly two allowed colors: set_i and $unset_i$. If $x_{i,t}$ is assigned the color set_i, it corresponds to setting the variable x_i to 1, and vice versa.

   Note that if DRAWER presents a vertex $x_{j,t}$ to PAINTER, PAINTER is able to recognize that it is a vertex corresponding to the variable x_j, but it is not able to recognize whether it is the vertex $x_{j,t}$ or $x_{j,f}$.

2. For a variable x_j which is quantified existentially, we will create a gadget consisting of three *existential* vertices $x_{j,t}$ (for the positive literal x_j), $x_{j,f}$ (for the literal $\neg x_j$) and $x_{j,h}$ (the helper vertex), connected as a triangle.

   Coloring of the first two variables also corresponds to setting the variable x_j to true or false, but in a different way: $x_{j,t}$ has allowed colors $set_{j,t}$ and $unset_j$, while $x_{j,f}$ has allowed colors $set_{j,f}$ and $unset_j$. We want to avoid both $x_{j,t}$ and $x_{j,f}$ to have the color of type set, and so the "helper" vertex $x_{j,h}$ can be colored only by $set_{j,t}$ or $set_{j,f}$.

   Note that the color choice for the vertices of x_j means that if PAINTER is presented any vertex of this variable, PAINTER can recognize it and decide whether to set x_j to 1 (and color accordingly) or to 0.

   We call existential and universal vertices together *variable vertices*.

3. For each clause C_a, we will add four new vertices. First, we create a vertex $l_{a,i}$ for each literal in the clause, which is connected to one of the vertices $x_{i,t}$ and $x_{i,f}$ corresponding to the sign of the literal. For example if $C_a = (x_i \wedge \neg x_j \wedge x_k)$, then $l_{a,i}$ is connected to $x_{i,t}$, $l_{a,j}$ is connected to $x_{j,f}$ and $l_{a,k}$ to $x_{k,t}$. The allowed colors on a vertex $l_{a,i}$ are $\{f_a, unset_i\}$.

   Finally, we add a fourth vertex d_a connected to the three vertices $l_{a,i}, l_{a,j}, l_{a,k}$. This vertex can be colored only using the color f_a or the color $false_a$. The color $false_a$ is used to signal that this particular clause is evaluated to be 0. If the color f_a is used for the vertex d_a, this means that the clause is evaluated to 1, because f_a is not present on any of $l_{a,i}, l_{a,j}, l_{a,k}$, thus they have colors of type $unset_i$ and their neighbors corresponding to literals have colors of type set.

4. The last vertex we add to the construction will be F, a final vertex. The vertex F is connected to all the vertices d_a corresponding to the clauses. The allowed colors of the vertex F are $\text{false}_1, \text{false}_2, \text{false}_3, ..., \text{false}_m$. This final vertex corresponds to the final evaluation of the formula. If all clauses are evaluated to 0, the vertex F has no available color left and must use a new color.

We have listed all the vertices and colors in our graph G_1 and the functioning of our gadgets, but we will need slightly more edges. The reasoning for the edges is as follows: If DRAWER presents any vertex of the type $l_{a,i}, d_a$ or F before presenting the variable vertices, or in the case when the variable vertices are presented out of the quantifier order, we want to give an advantage to PAINTER so it can finalize the coloring.

This will be achieved by allowing PAINTER to treat all remaining universal vertices as existential vertices, i.e., PAINTER can recognize which of the two universal vertices $x_{j,t}, x_{j,f}$ corresponds to setting x_j to 1.

To be precise, we add the following edges to G_1:

- Every existential vertex $x_{j,t}, x_{j,f}, x_{j,h}$ is connected to all previous universal vertices $x_{i,t}$, that is to all such $x_{i,t}$ for which $i < j$.
- Every universal vertex $x_{j,t}, x_{j,f}$ is connected to all previous universal vertices $x_{i,t}$ such that $i < j$.
- Every vertex of type $l_{a,i}$ is connected to all the universal vertices $x_{i't}$ for $i' \neq i$. Note that $l_{a,i}$ is connected either to $x_{i,t}$, or to $x_{i,f}$; we do not add an edge to such vertices.
- Every vertex of type d_a is connected to all universal vertices $x_{i,t}$ for all i.
- The vertex F is connected to all the universal vertices $x_{i,t}$ for all i.

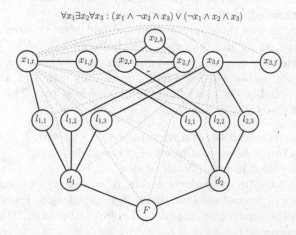

$$\forall x_1 \exists x_2 \forall x_3 : (x_1 \wedge \neg x_2 \wedge x_3) \vee (\neg x_1 \wedge x_2 \wedge x_3)$$

Fig. 1. The construction for a sample formula. The thick black edges are the normal edges of the construction, and the dashed orange edges are the additional edges that guarantee precedence of vertices. The lists of allowed colors of each vertex are not listed in the figure. (Color figure online)

We call all non-precolored vertices the *gadgets* for variables and clauses.

The number of colors allowed for PAINTER (the same as the size of K_{col}) is $k = 2m + 2n_\forall + 3n_\exists$ where m is the number of clauses, $n_\forall$ the number of universally quantified variables and $n_\exists$ the number of existentially quantified variables.

The analysis of our construction is fairly straightforward (see [2] for details).

3   Construction with a Precolored Part of Logarithmic Size

We now make a step to the general case without precoloring by reducing the size of the precolored part so that it has only logarithmic size. Our construction is based on the one with a large precolored part; namely, all the vertices $x_{i,t}, x_{i,f}, x_{j,t}, x_{j,f}, x_{j,h}, l_{a,i}, d_a, F$ (the gadgets for variables and clauses) and the whole color clique K_{col} will be connected the same way. Let G_1 denote the gadgets for variables and clauses and K_{col}.

Since K_{col} is now not precolored and DRAWER may send it after the gadgets, we help PAINTER by a structure for recognizing vertices in G_1 or for saving colors.

We remark that there is also a simpler construction with a logarithmic number of precolored vertices. If we just add precolored vertices to recognize vertices in G_1, the following proof would work and be easier. However, when we replace a precolored vertex v by some non-precolored graph in Sect. 4, we will use some conditions on the graph G_2 that this construction would not satisfy.

3.1   Nodes

Our structure will consist of many small *nodes*, all of them have the same internal structure, only their adjacencies with other vertices vary.

Each node consists of three vertices and a single edge; vertices are denoted by p_1, p_2, p_3 and the edge leads between p_2 and p_3. We call the vertices p_1 and p_2 the *lower partite set* of the node, p_3 form the *upper partite set*. See Fig. 2 for an illustration of a node. Clearly, the online chromatic number of a node is two. The intuition behind the nodes is as follows:

Fig. 2. Node

- If DRAWER presents vertices of a node in the correct way, PAINTER needs to use two colors in the lower partite set of every node.
- No color can be used in two different nodes.
- Each vertex $v \in G_1$ (in the gadgets and in K_{col}) has its own associated node A. If the vertex p_3 from A does not arrive before v is sent, PAINTER can color p_3 and v with the same color, thus save a color. Otherwise, PAINTER can use the node to recognize v.
- Universal vertices $x_{i,t}, x_{i,f}$ for each universally quantified variable x_i should be distinguishable only by the same vertices as in the previous section. Therefore they are both associated with the same two nodes.

Let N be the number of vertices in G_1. We create N nodes, denoted by $A_1, \ldots, A_N$, one for each vertex in G_1. For any two distinct nodes A_i and A_j ($i \neq j$), there is an edge between each vertex in A_i and each vertex in A_j. Therefore, no color can be used in two nodes.

We have noted above that each node is associated with a vertex; we now make the connection precise. Let $v_1, \ldots, v_N$ be the vertices in G_1 (in an arbitrary order). Then we say that A_i *identifies* the vertex v_i. Moreover, if v_i is a vertex $x_{k,t}$ or $x_{k,f}$ for a universally quantified variable x_k and v_j is the other vertex, then A_j also identifies v_i and A_i also identifies v_j. Thus each node identifies one or two vertices and each vertex is identified by one or two nodes.

Edges between a vertex v in the original construction G_1 and a node depend on whether the node identifies v, or not. For a vertex $v \in G_1$ and for a node A, if A identifies v, we connect only the whole lower partite set of A to v, i.e., we add two edges from v to both p_1 and p_2 of A. Otherwise, we add three edges – one between v and every vertex in A.

3.2   Precolored Vertices

The only precolored part P of the graph is intended for distinguishing nodes. Since there are N nodes in total, we have $p = \lceil \log_2 N \rceil$ precolored vertices $z_1, z_2, \ldots z_p$ with no edges among them. Precolored vertices have a color that may be used later for coloring G_1 (the gadgets and K_{col}). For simplicity, we again use the precoloring in the strong sense, i.e., PAINTER is able to recognize which precolored vertex is which.

We connect all vertices in the node A_i to z_j if the j-th bit in the binary notation of i is 1; otherwise z_j is not adjacent to any vertex in A_i.

Clearly, the node to which an incoming vertex belongs can be recognized by its adjacency to the precolored vertices. Note that a vertex from nodes is connected to at least one precolored vertex and there is no edge between G_1 and precolored vertices.

So far, we have introduced all vertices and edges in our construction of the graph G_2. We omit the rest of the analysis due to space restrictions; see [2] for details.

4   Removing Precoloring

In this section we show how to replace one precolored vertex by a large nonprecolored graph whose size is a constant factor of the size of the original graph, while keeping PAINTER's winning strategy in the case of a satisfiable formula. DRAWER's winning strategy in the other case is of course preserved also and easier to see. We prove the following lemma which holds for all graphs with precolored vertices satisfying a few assumptions.

Lemma 1. *Let G be a graph with precolored subgraph G_p created from a fully quantified formula ϕ, and let $v_p \in G_p$ be a precolored vertex of G.*

Let D be the induced subgraph with all nonprecolored vertices that are not connected to v_p and let E be the induced subgraph with all nonprecolored vertices that are connected to v_p.

Let k be an integer. Assume that the following holds:

1. *$\chi^O(G) \leq k$ if and only if ϕ is satisfiable,*
2. *in the winning strategy of PAINTER in the case if ϕ is satisfiable, PAINTER can color E using FIRSTFIT before two nonadjacent vertices from D arrive. Moreover, in this case if FIRSTFIT assigns the same color to a vertex in D and to a vertex in E before two nonadjacent vertices from D arrive, PAINTER can still color G using k colors.*

Then there exists an integer k' and a graph G' with the following properties:

- *G' has only $|V(G_p)| - 1$ precolored vertices, and $|V(G')| \leq 25|V(G)|$,*
- *G' can be constructed from G in polynomial time,*
- *it holds that $\chi^O(G') \leq k'$ if and only if ϕ is satisfiable.*

Theorem 1 follows by an iterative application of Lemma 1; the details of this application can be found in [2].

Construction of G'. Let N be the total number of vertices in D and E and let $S = 8N$. Our graph G' consists of precolored part $G'_p := G_p \backslash \{v_p\}$, graphs D and E and three huge cliques A, B and C of size S; cliques A, B and C together form a *supernode*. We keep the edges inside and between D and E and the edges between G'_p and $D \cup E$ as they are in G.

We add a complete bipartite graph between cliques B and C, i.e., $B \cup C$ forms a clique of size $2S$. No vertex in A is connected to B or C. In other words, the supernode is created from a node by replacing each vertex by a clique of size S and the only edge in the node by a complete bipartite graph.

There are no edges between the supernode (cliques A and $B \cup C$) and a precolored vertex in G'_p. It remains to add edges between the supernode and $D \cup E$. There is an edge between each vertex in E and each vertex in the supernode, while every vertex in D is connected only to the whole A and B, but not to any vertex in C. The fact that D and C are not adjacent at all is essential in our analysis. Our construction is depicted in Fig. 3.

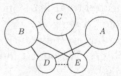

Fig. 3. Our construction G'. (The remaining precolored vertices are not shown.)

Proof (Proof of Lemma 1). Let G' be the graph defined as above. Note that the number of vertices in G' is at most $25|V(G)|$, G' can be constructed from G in polynomial time and G' has only $|V(G_p)| - 1$ precolored vertices. Therefore, it remains to prove $\chi^O(G') \leq k'$ if and only if ϕ is satisfiable for some k'. We set k' to $k + 2S$, since there will be at most $2S$ colors used in the supernode.

Assuming that ϕ is not satisfiable, it is straightforward to design a winning strategy for DRAWER on G'; we only need to adapt the approach of Sects. 2 and 3. See [2] for a full description of the strategy.

In the rest of this section we focus on the opposite direction: assuming that ϕ is satisfiable, we show that PAINTER can color G' with k' colors regardless of the strategy of DRAWER. In the following, when we refer to the colored part of G', we do not take precolored vertices into account. PAINTER actually does not look at precolored vertices unless it uses its winning strategy for coloring G, which exists by the assumptions.

Intuition. At the beginning PAINTER has too little data to infer anything about the vertices. Therefore, PAINTER shall wait for two nonadjacent vertices from D and for two large cliques (larger than $S/2$) with a small intersection. Before such vertices arrive, it will color greedily.

Note that the greedy coloring algorithm eventually stops before everything is colored. Having two large cliques, one mostly from A and the other mostly from $B \cup C$, and two nonadjacent vertices from D, PAINTER is able to recognize where an incoming vertex belongs. Therefore, PAINTER can use the supernode like a precolored vertex and colors the remaining vertices from D and E by its original winning strategy on G.

This approach may fail if a part of D is already colored by PAINTER's application of the greedy strategy. To remedy this, we prove that colors used on D so far are also used in C or E, or will be used on C later.

The other obstacle is that PAINTER might not be able to distinguish between one clique from D and vertices in A if nothing from B arrives. Nevertheless, each vertex u in such a "hidden" clique is connected to all other colored vertices in D and to the whole colored part of E, otherwise it would be distinguishable from vertices in A. Hence, it does not matter on the color of u.

In summation, the sheer size of the supernode should allow PAINTER to be able to use it as if it would be precolored. Still, we need to allow for some small margin of error. This leads us to the following definition:

Definition 2. *Let N be the number of vertices of $D \cup E$ as in the construction of G'. For subgraphs $X, Y \subseteq G'$, we say that X is practically a subgraph of Y if $|V(X) \backslash V(Y)| \leq N$, and X is practically disjoint with Y if $|V(X) \cap V(Y)| \leq N$.*

We also say that a vertex v is practically universal to a subgraph $X \subseteq G'$ if it is adjacent to all vertices in X except at most N of them. Similarly, we say a vertex v is practically independent of a subgraph $X \subseteq G'$ if v has at most N neighbors in X.

At first, the player PAINTER uses the following algorithm for coloring incoming vertices, which may stop when it detects two useful vertices d_1 and d_2:

Algorithm WAITFORD: For an incoming vertex u sent by DRAWER:

1. Let G'_A be the revealed part of G' (i.e., colored vertices and u, but not precolored vertices)
2. Find a maximum clique in G'_A and denote it as K_1.
3. Find a maximum clique in G'_A practically disjoint with K_1 and denote it as K_2.
4. If $|K_2| \geq S/2$ and there are two nonadjacent vertices d_1 and d_2 in G'_A which are both *not* practically universal to K_1 or both *not* practically universal to K_2:
5.   Stop the algorithm.
6. Otherwise, color u using FIRSTFIT.

While the algorithm may seem to use a huge amount of computation for one step, we should realize that we are not concerned with time complexity when designing the strategy for PAINTER. In fact, even a non-constructive proof of existence of a winning strategy would be enough to imply existence of a PSPACE algorithm for finding it – we have observed already in Sect. 1 that ONLINE CHROMATIC NUMBER lies in PSPACE.

Let v be the incoming vertex u when WAITFORD stops; note that v is not colored by the algorithm and v can be from any part of G'.

One of the cliques K_1 and K_2 is practically a subgraph of $B \cup C$ and we denote this clique by K_{BC}. The other clique must be practically a subgraph of A and we denote it by K_A. (Keep in mind that both cliques may contain up to N vertices from $D \cup E$.) We remark that some vertices from C must have arrived, as A and B alone are indistinguishable by Step 4 of WAITFORD. By the same argument, the player PAINTER knows whether $K_1 = K_A$ or $K_1 = K_{BC}$.

Let d_1 and d_2 be the nonadjacent vertices that caused the algorithm to stop. We observe that $d_1, d_2 \in D$ by eliminating all other possibilities:

- Neither of d_1 and d_2 can be from E, since any vertex of E is practically universal to both cliques.
- Both d_1 and d_2 cannot be from $B \cup C$ or both from A, as they would be adjacent.
- If d_1 is in $B \cup C$ and d_2 in A, then we have a contradiction with the fact that d_1 and d_2 are not practically universal to the same clique.
- If $d_1 \in D$ and d_2 would be from A or B, then d_1 and d_2 are adjacent.
- Finally, if $d_1 \in D$ and $d_2 \in C$, then the clique to which they are not practically universal cannot be the same for both, since d_1 is universal to the whole A and d_2 to the whole $B \cup C$.

Having cliques K_A and K_{BC} and vertices $d_1, d_2 \in D$, PAINTER uses the following rules to recognize where an incoming or a colored vertex u belongs:

- If u is practically universal to both K_{BC} and K_A, then $u \in E$.
- If u is practically universal to K_{BC} and practically independent of K_A and u is adjacent to d_1, then $u \in B$.
- If u is practically universal to K_{BC} and practically independent of K_A, but there is no edge between d_1 and u, then $u \in C$.
- If u is not practically universal to K_{BC}, but it is practically universal to K_A, then $u \in A$ or $u \in D$.
  - Among such vertices, if there is a vertex not adjacent to u or u is not adjacent to a vertex in E or u is adjacent to a vertex in B, then $u \in D$; we say that such u is *surely in* D.
  - Otherwise, PAINTER cannot yet recognize whether $u \in A$ or $u \in D$.

The reader should take a moment to verify that indeed, the set of rules covers all possible cases for u.

Let $\tilde{A}, \tilde{B}, \tilde{C}, \tilde{D}, \tilde{E}$ be the colored parts of G' when WAITFORD stops. We observe that in the last case of the recognition the vertices from $\tilde{D}$ which are indistinguishable from A form a clique; we denote it by K_D. Note that all vertices in K_D are connected to all vertices surely from D that arrived and K_D contains all vertices in $\tilde{D}$ that are not surely in D. We stress that PAINTER does not know K_D or even its size.

Intuition for the Next Step. Since PAINTER can now recognize the parts of the construction (with an exception of K_D), we may basically use the winning strategy for PAINTER on G and FIRSTFIT on the rest. More precisely, PAINTER creates a virtual copy of G, adds vertices into it and simulates the winning strategy on this virtual graph.

Our main problem is that some part of D (namely $\tilde{D}$) is already colored. We shall prove that if $\tilde{D}$ is not a clique, PAINTER can ignore colors used in $\tilde{D}$ (but not the colors that it will use on D), as they are already present in C or E or they may be used later in C. If $\tilde{D}$ is a clique, it may be the case that C and A arrived fully and have the same colors, thus PAINTER cannot ignore colors on $\tilde{D}$.

Another obstacle in the simulation is K_D, the hidden part of D. To overcome this, PAINTER tries to detect vertices in K_D and reclassify them as surely in D. PAINTER shall keep that all vertices in K_D are connected to all currently colored vertices in D and E, therefore it does not matter much on colors in K_D.

When PAINTER discovers a vertex from K_D, it adds the vertex immediately to its simulation of G. On the other hand, the size of K_D increases when DRAWER sends a vertex from D which is indistinguishable from A.

The details of the algorithm used by PAINTER to finish the coloring of G' using k' colors are omitted due to space restrictions and can be found in [2]. □

Acknowledgments. The authors thank Christian Kudahl and their supervisor Jiří Sgall for useful discussions on the problem.

References

1. Bean, D.R.: Effective coloration. J. Symbolic Logic **41**(2), 469–480 (1976)
2. Böhm, M., Veselý, P.: Online chromatic number is PSPACE-complete, arXiv preprint (2016). https://arxiv.org/abs/1604.05940
3. Gyárfás, A., Lehel, J.: First fit and on-line chromatic number of families of graphs. Ars Combinatoria **29C**, 168–176 (1990)
4. Gyárfás, A., Kiraly, Z., Lehel, J.: On-line graph coloring and finite basis problems. In: Combinatorics: Paul Erdos is Eighty, vol. 1, pp. 207–214 (1993)
5. Halldórsson, M.M.: Parallel and on-line graph coloring. J. Algorithms **23**, 265–280 (1997)
6. Halldórsson, M.M.: Online coloring known graphs. Electron. J. Combinatorics **7**(1), R7 (2000)
7. Halldórsson, M.M., Szegedy, M.: Lower bounds for on-line graph coloring. Theor. Comput. Sci. **130**(1), 163–174 (1994)
8. Kierstad, H.: On-line coloring k-colorable graphs. Israel J. Math. **105**, 93–104 (1998)
9. Kudahl, C.: On-line graph coloring. Master's thesis, University of Southern Denmark (2013)

10. Kudahl, C.: Deciding the on-line chromatic number of a graph with pre-coloring is PSPACE-complete. In: Paschos, V.T., Widmayer, P. (eds.) CIAC 2015. LNCS, vol. 9079, pp. 313–324. Springer, Heidelberg (2015)
11. Lovász, L., Saks, M., Trotter, W.T.: An on-line graph coloring algorithm with sublinear performance ratio. Ann. Discrete Math. **43**, 319–325 (1989)

Computational Geometry

Bounded Embeddings of Graphs in the Plane

Radoslav Fulek[(✉)]

IST Austria, Am Campus 1, 3400 Klosterneuburg, Austria
radoslav.fulek@gmail.com

Abstract. A drawing in the plane ($\mathbb{R}^2$) of a graph $G = (V, E)$ equipped with a function $\gamma : V \to \mathbb{N}$ is *x-bounded* if (i) $x(u) < x(v)$ whenever $\gamma(u) < \gamma(v)$ and (ii) $\gamma(u) \leq \gamma(w) \leq \gamma(v)$, where $uv \in E$ and $\gamma(u) \leq \gamma(v)$, whenever $x(w) \in x(uv)$, where $x(.)$ denotes the projection to the x-axis. We prove a characterization of isotopy classes of embeddings of connected graphs equipped with γ in the plane containing an x-bounded embedding. Then we present an efficient algorithm, which relies on our result, for testing the existence of an x-bounded embedding if the given graph is a forest. This partially answers a question raised recently by Angelini et al. and Chang et al., and proves that c-planarity testing of flat clustered graphs with three clusters is tractable when the underlying abstract graph is a forest.

Keywords: Graph planarity testing · Weakly simple embedding · c-planarity · PQ-tree · Algebraic crossing number

1 Introduction

Testing planarity of graphs with additional constraints is a popular theme in the area of graph visualizations abundant with open problems mainly of algorithmic nature. One of the most exciting open problems in the area is to determine the complexity status, i.e., P, NP-hard, or IP, of deciding for a pair of (planar) graphs G_1 and G_2, whose edge sets possibly intersect, if there exists a drawing of $G_1 \cup G_2$ in the plane, whose restriction to both graphs, G_1 and G_2, is an embedding (edge-crossing free drawing). The problem, also known as SEFE-2, was introduced in 2003 by Brass et al. in [8] and its prominence was realized by Schaefer in [34], where polynomial time reductions of many problems in the area to SEFE-2 is given, see Fig. 2 therein.

Among the problems reducible to SEFE-2 in a polynomial time is a notoriously difficult open problem raised under the name of *c-planarity* in 1995 by Feng, Cohen and Eades [16,17]. The problem asks for a given planar graph G equipped with a hierarchical structure on its vertex set, i.e., clusters, to decide if a planar embedding G with the following property exists: the vertices in each

The research leading to these results has received funding from the People Programme (Marie Curie Actions) of the European Union's Seventh Framework Programme (FP7/2007-2013) under REA grant agreement no [291734].

© Springer International Publishing Switzerland 2016
V. Mäkinen et al. (Eds.): IWOCA 2016, LNCS 9843, pp. 31–42, 2016.
DOI: 10.1007/978-3-319-44543-4_3

cluster are drawn inside a disc so that the discs form a laminar set family corresponding to the given hierarchical structure and the embedding has the least possible number of edge-crossings with the boundaries of the discs. Again we are interested in the complexity status of the problem.

On the other hand, quite well understood from the algorithmic perspective are upward embeddings of directed acyclic planar graphs [4,22] and closely related various layered drawings of leveled graphs [3,29]. In the setting of layered drawings we place the vertices on, e.g., parallel lines or concentric circles, corresponding to the levels of G. Furthermore, we require that edges lie between the levels of their endpoints and that edges are monotone in the sense that they intersect any line (circle) parallel to (concentric with) the chosen lines (circles) at most once. Also these easier planarity variants are reducible in a polynomial time to SEFE-2 [34]. The layered drawings with parallel lines representing levels are called *level drawings*. The x-bounded embeddings treated in this work generalize level planarity and constitute a special case of c-planarity as we will see later.

The algorithmic problems in the area can be classified according to whether the isotopy class of a desired embedding of the given graph is a part of the input, where often prescribing an isotopy class makes the problem much easier. If the isotopy class is a part of the input the problem often reduces to a flow/ matching problem, see e.g., [1,2,4]. If not, the main building block of polynomial-time algorithms is often a data structure that can efficiently store all isotopy classes of admissible embeddings throughout the execution of the algorithm. The two widely used data structures (and their variants) that serve this purpose are PQ-tree [7,26] (or its un-rooted variant PC-tree) and SPQR-tree [14]. While the two mentioned techniques, the one using a flow/matching and the one using a tree-based data structure, are very popular in the field of graph drawing we are not aware of any previous work combining them.

Let (G, γ) denote a pair of a planar graph $G = (V, E)$ and a function $\gamma :$ $V \to \mathbb{N}$. Similarly as in [21], a drawing in the plane ($\mathbb{R}^2$) of G is x-bounded if (i) $x(u) < x(v)$ whenever $\gamma(u) < \gamma(v)$ and (ii) $\gamma(u) \leq \gamma(w) \leq \gamma(v)$, where $uv \in E$ and $\gamma(u) \leq \gamma(v)$, whenever $x(w) \in x(uv)$, where $x(.)$ denotes the projection to the x-axis, see Fig. 1a for an illustration. By [19, Lemma 2] and the corresponding variant of the weak Hanani–Tutte theorem [19, Theorem 1] we have the following.

Lemma 1. *There exists an x-bounded embedding of (G, γ) in which the projection $x(e)$ of every edge $e \in E$ is injective, i.e., x-monotone, if there exists an arbitrary x-bounded embedding of (G, γ).*

Hence, the question of deciding whether an x-bounded embedding of (G, γ) exists is equivalent to deciding the existence of a *strip clustered embedding* of Angelini et al. [1]. For that reason we call an x-bounded drawing an *x-bounded embedding* if it is edge-crossing free and $x(e)$ is injective for every edge $e \in E$, see Fig. 1b for an illustration. Moreover, by [33, Theorem 2] edges in such embedding can be turned into straight-line segments. We prove a characterization of isotopy classes of embeddings of G in the plane containing an x-bounded embedding,

Theorem 1. We remark that the characterization easily implies [19, Theorem 1], see the proof in [18, Theorem 1.3]. In fact, the characterization was extracted from its proof in [19].

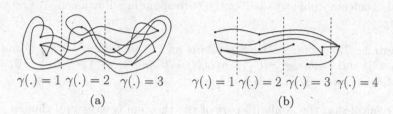

$$\gamma(.) = 1 \; \gamma(.) = 2 \quad \gamma(.) = 3 \qquad \gamma(.) = 1 \; \gamma(.) = 2 \; \gamma(.) = 3 \; \gamma(.) = 4$$

(a)   (b)

Fig. 1. (a) An x-bounded drawing of a pair (G, γ), each vertical strip contains vertices whose γ value is the same; (b) An x-bounded embedding of a pair (G, γ), x-monotone as required by our definition.

We use the characterization to prove the correctness of a PQ-tree based algorithm to test if an x-bounded embedding of (G, γ), where G is a forest, exists. Since the proof of our characterization proceeds by a reduction to the matching problem in the spirit of a recent work of Angelini et al. [1], the proof of our result combines the two techniques from the above.

Só far, the most general partial results on SEFE-2 [6] and c-planarity [11,24] are based on PQ/SPQR-tree style data structures, or Hanani–Tutte variants [34]. However, we suspect that a resolution of the tractability status of SEFE-2 or c-planarity must use the topology of the plane in an "essential way", e.g., by using its topological invariants such as Euler characteristic that are usually not exploited in approaches based on PQ-tree style data structures or in the proofs of Hanani–Tutte variants. The proof of our characterization relies crucially besides, Hall's theorem [15], on Euler's formula for planar graphs.

The characterization turns the problem of the existence of an x-bounded embedding into a problem that can be solved efficiently by employing a PQ-tree at least in the case of trees. [1] We suspect that with additional twists the problem can be solved efficiently for any graph. Moreover, our recent work [20] hints at the possibility that Theorem 1 could be generalized to the setting of cyclic c-planarity or even c-planarity with pipes [13].

1.1   Results

Refer to Sect. 2 for the definitions. Suppose that we have (G, γ) as above, where G is connected, and let $\mathcal{E}$ denote the isotopy class of an embedding of G in the plane. Let us treat $\mathcal{E}$ as an embedded two-dimensional polytopal complex [35], and let $\mathcal{C} = (\mathcal{E}, \mathbb{Z}_2)$ be the corresponding chain complex, i.e., in $\mathcal{C}$ two-dimensional chains are generated by the inner faces of $\mathcal{E}$, one-dimensional chains by the

---

[1] By employing the technique of Bläsius and Rutter [6] also in the case of a union of internally disjoint paths between a pair of vertices [18].

edges, etc. The boundary operator $\partial(.)$ is defined as usual, and we put $\partial(v) = \emptyset$, for any $v \in V$, and hence, $\gamma(\partial(v)) = \emptyset$. Let $i_{\mathcal{E}}(C_1, C_2)$ denote the algebraic intersection number [31] of the supports of chains C_1 and C_2 in $\mathcal{E}$ such that $dim(C_1) + dim(C_2) = 2$, where $dim(.)$ is dimension, and the support of both C_1 and C_2 is homeomorphic to a ball of the corresponding dimension. We prove the following.

Theorem 1. *The isotopy class $\mathcal{E}$ contains an x-bounded embedding if and only if $i_{\mathcal{E}}(C_1, C_2) = 0$ whenever $\gamma(C_1) \cap \gamma(\partial C_2) = \emptyset$ and $\gamma(\partial C_1) \cap \gamma(C_2) = \emptyset$, where $\gamma(.)$ is extended over $\mathbb{R}$ linearly to edges.*

We remark that the "only if" part of the theorem is easy, and thus, it is the "if" part that is interesting. Instead of proving Theorem 1 we prove its equivalent reformulation, Theorem 3 (stated only for trees due to space limitations), that is less conceptual, but more convenient to work with. Theorem 3 allows us to employ the PQ-tree data structure to show the following result.

Theorem 2. *We can test in cubic time if (G, γ) admits an x-bounded embedding when the underlying abstract graph G is a forest.*

Theorem 2 extends a recent result of Angelini et al. [1] and partially answers an open problem asked by Chang et al. in the arxiv version of [10, Appendix D.2]. In particular, our results imply that we can test in a polynomial time if a straight-line drawing of a tree into a line is weakly simple.

Flat Clustered Graph. A *flat clustered graph*, shortly *c-graph*, is a pair (G, T), where $G = (V, E)$ is a graph and $T = \{V_0, \ldots, V_{c-1}\}$, $\biguplus_i V_i = V$, is a partition of the vertex set into *clusters*. A c-graph (G, T) is *clustered planar* (or briefly *c-planar*) if G has an embedding in the plane such that (i) for every $V_i \in T$ there is a topological disc $D(V_i)$, where interior$(D(V_i)) \cap$ interior$(D(V_j)) = \emptyset$, if $i \neq j$, containing all the vertices of V_i in its interior, and (ii) every edge of G intersects the boundary of $D(V_i)$ at most once for every $D(V_i)$. A c-graph (G, T) with the given isotopy class of an embedding of G is *c-planar* if additionally the embedding belongs to the given class. A *clustered drawing and embedding* of a flat clustered graph (G, T) is a drawing and embedding, respectively, of G satisfying (i) and (ii). In 1995 Feng, Cohen and Eades [16,17] introduced the notion of clustered planarity for clustered graphs, shortly c-planarity, (using a more general hierarchical clustering) as a natural generalization of graph planarity. (Under a different name Lengauer [30] studied a similar concept in 1989.)

By [18, Lemma 1.2] we obtain the following corollary of Theorem 2.

Corollary 1. *Let G be a forest. We can test in cubic time if a c-graph (G, T) with three clusters is c-planar.*

To illustrate the difficulty of c-planarity we mention that already in the case of three clusters [12], if G is a cycle, the polynomial time algorithm for c-planarity is not trivial, while if G can be any graph, its existence is still open. Biedl [5] gave a polynomial time algorithm for c-planarity with two clusters. A different approach for two clusters was considered by Hong and Nagamochi [25] and the result also follows from a work by Gutwenger et al. [24]. Beyond two clusters a polynomial time algorithm for c-planarity was obtained only in special cases, e.g., [11,23,24,27,28].

Organization. Due to space limitation, we only state Theorem 3 for trees (Sect. 2) and establish Theorem 2 only in the case when the underlying abstract graph is a subdivided star (Sect. 3). Solving this case already illustrates all the main ideas. The arxiv submission [18] contains the full version of the paper together with various extensions of our results.

2   Preliminaries

Notation. Let $G = (V, E)$ denote a connected planar graph possibly with multi-edges. A *drawing* of G is a representation of G in the plane where every vertex in V is represented by a unique point and every edge $e = uv$ in E is represented by a Jordan arc joining the two points that represent u and v. We assume that in a drawing no edge passes through a vertex, no two edges touch and every pair of edges cross in finitely many points. An *embedding* of G is an edge-crossing free drawing. If it leads to no confusion, we do not distinguish between a vertex or an edge and its representation in the drawing and we use the words "vertex" and "edge" in both contexts. Since in the problem we study connected components of G can be treated separately, we can afford to assume that G is connected throughout the paper. The *rotation* at a vertex is the counter-clockwise cyclic order of the end pieces of its incident edges in a drawing of G. The *rotation system* of a graph is the set of rotations at all its vertices. The *isotopy class* of an embedding of G is described by the rotations at its vertices and choice of the (unbounded) outer face. When talking about sub-graphs of G in an isotopy class we mean it w.r.t. an embedding in the class.

Given a pair (G, γ) we naturally associate with it a partition of the vertex set into the *clusters* V_i's such that v belongs to $V_{\gamma(v)}$. We refer to the cluster whose vertices get label i as to the ith cluster.

Characterization. We present a necessary and sufficient condition for the isotopy class $\mathcal{E}$ of an embedding of (G, γ), where G is a tree, to contain an x-bounded embedding. For the remainder of this section we assume that G is given by the isotopy class of its embedding $\mathcal{E}$.

We use the following definition of $i_{\mathcal{E}}(P_1, P_2)$ of a pair of oriented paths P_1 and P_2 in $\mathcal{E}$ [9]. We orient P_1 and P_2 arbitrarily. Let P denote the sub-graph of G that is the union of P_1 and P_2. We define $cr_{P_1,P_2}(v) = +1$ $(cr_{P_1,P_2}(v) = -1)$ if v is a vertex of degree four in P such that the paths P_1 and P_2 alternate in the rotation at v and at v the path P_2 crosses P_1 from left to right (right to left) with respect to the chosen orientations of P_1 and P_2. We define $cr_{P_1,P_2}(v) = +1/2$ $(cr_{P_1,P_2}(v) = -1/2)$ if v is a vertex of degree three in P such that at v the path P_2 is oriented towards P_1 from left, or from P_1 to right (towards P_1 from right,

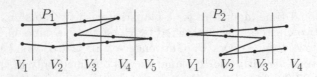

Fig. 2. A path P_1 that is a 1-cap (left); and a path P_2 that is a 4-cup (right).

or from P_1 to left) in the direction of P_1. The algebraic intersection number of P_1 and P_2, $i_\mathcal{E}(P_1, P_2)$, is then the sum of $cr_{P_1,P_2}(v)$ over all vertices of degree three and four in P.

Let $G' \subseteq G$. Let $\max(G')$ and $\min(G')$, respectively, denote the maximal and minimal value of $\gamma(v)$, $v \in V(G')$.

Definition of an i-cap and i-cup. A path P in G is an *i-cap* and a *j-cup* if for the end vertices u, v of P and all $w \neq u, v$ of P we have $\min(P) = \gamma(u) = \gamma(v) = i \neq \gamma(w)$ and $\max(P) = \gamma(u) = \gamma(v) = j \neq \gamma(w)$, respectively, (see Fig. 2). A pair of an *i-cap* P_1 and *j-cup* P_2 is *interleaving* if (i) $\min(P_1) < \min(P_2) \leq \max(P_1) < \max(P_2)$; and (ii) P_1 and P_2 intersect in a path (or a single vertex). An interleaving pair of an oriented *i-cap* P_1 and a *j-cup* P_2 is *infeasible*, if $i_\mathcal{E}(P_1, P_2) \neq 0$, and *feasible*, otherwise. Thus, feasibility does not depend on the orientation. Note that $i_\mathcal{E}(P_1, P_2)$ can be either $0, 1$ or -1. Throughout the paper by an infeasible and feasible pair of paths we mean an infeasible and feasible, respectively, interleaving pair of an *i-cap* and *j-cup*.

Theorem 3. *The isotopy class $\mathcal{E}$ of a tree G contains an x-bounded embedding of (G, γ) if and only if $\mathcal{E}$ does not contain an infeasible interleaving pair of paths.*

3   Subdivided Stars

In this section we give an algorithm proving Theorem 2, when G is a subdivided star. Throughout the present section we assume $|\gamma(u) - \gamma(v)| \leq 1$ for every edge $uv \in E(G)$. This can be assumed without loss of generality as the edges uv such that $|\gamma(u) - \gamma(v)| > 1$ can be subdivided by $|\gamma(u) - \gamma(v)| - 1 = k$ vertices $u_1, \ldots, u_k$ extending $\gamma(.)$ so that $|\gamma(u_i) - \gamma(u_{i+1})| = 1$, for $i = 0, \ldots, k$, where $u_0 = u$ and $u_{k+1} = v$.

In the sequel $G = (V, E)$ is a subdivided star. Thus, G is a tree that contains a special vertex v, *the center of the star*, of an arbitrary degree and all the other vertices in G are either of degree one or two.

In what follows we show how to use Theorem 3 for a polynomial-time x-bounded embeddability testing if the underlying abstract graph is a subdivided star. The algorithm is based on testing in polynomial time whether the columns of a 0–1 matrix can be ordered so that, in every row, either the ones or the zeros are consecutive. We, in fact, consider matrices containing $0, 1$ and also an ambiguous symbol $*$. A matrix containing $0,1$ and $*$ as its elements has the *circular-ones property* if there exists a permutation of its columns such that in every row, either the ones or the zeros are consecutive among undeleted symbols after we delete all $*$. Then each row in the matrix corresponds to a constraint imposed on the rotation at v by Theorem 3 simultaneously for many pairs of paths.

By Theorem 3 it is enough to decide if there exists a rotation at v so that every interleaving pair of an s-cap P_1 and b-cup P_2 meeting at v is feasible. Note that if either P_1 or P_2 does not contain v in its interior the corresponding pair is feasible. Since G is a subdivided star, an interleaving pair P_1 and P_2 passing through v restricts the set of all rotations at v in an x-bounded embedding of (G, γ). Namely, if e_i and f_i are edges incident to P_i at v then in an x-bounded embedding of (G, γ) in the rotation at v the edges e_1, f_1 do not alternate with the edges e_2, f_2, i.e., e_1 and f_1 are consecutive when we restrict the rotation to e_1, f_1, e_2, f_2. We denote such a restriction by $\{e_1 f_1\}\{e_2 f_2\}$. Given a cyclic order $\mathcal{O}$ of edges incident to v, we can interpret $\{e_1 f_1\}\{e_2 f_2\}$ as a Boolean predicate which is "true" if and only if e_1, f_1 do not alternate with the edges e_2, f_2 in $\mathcal{O}$. Of course, for a given cyclic order we have $\{ab\}\{cd\}$ if and only if $\{cd\}\{ab\}$, and $\{ab\}\{cd\}$ if and only if $\{ba\}\{cd\}$. Then our task is to decide in polynomial time if the rotation at v can be chosen so that the predicates $\{e_1 f_1\}\{e_2 f_2\}$ of all the interleaving pairs P_1 and P_2 are "true". The problem of finding a cyclic ordering satisfying a given set of Boolean predicates of the form $\{e_1 f_1\}\{e_2 f_2\}$ is NP-complete in general, since the problem of total ordering [32] can be easily reduced to it in polynomial time. However, in our case the instances satisfy the following structural properties making the problem tractable (as we see later).

Observation 1. *If $\{ab\}\{cd\}$ is false and $\{ab\}\{de\}$ is true then $\{ab\}\{ce\}$ is false.*

The restriction on rotations at v by the pair of an s-cap P_1 and b-cup P_2 is *witnessed* by an ordered pair (s, b), where $s < b$. We treat such pair as an interval in $\mathbb{N}$. Let $I = \{(s, b)| \ (s, b)$ witnesses *a restriction on rotations at v by a pair of paths*$\}$.

Observation 2. *If an s-cap P contains v then P contains an s'-cap P' containing v as a sub-path for every s' such that $s < s' < \gamma(v)$. Similarly, if a b-cup P contains v then P contains a b'-cup P' containing v as a sub-path for every b' such that $\gamma(v) < b' < b$.*

Observation 3. *Let $s < s' < b < b'$, $s, s', b, b' \in \mathbb{N}$. If the set I contains both (s, b) and (s', b'), it also contains (s, b') and (s', b).*

We would like to reduce the question of determining if we can choose a rotation at v making all the interleaving pairs feasible to the following problem. Let $S = \{e_1, \ldots, e_n\}$ be the set of edges incident to v. Let $\mathcal{S}' = \{L_i', R_i'| \ i = 1, \ldots\}$ be a set of polynomial size in n such that $R_i', L_i' \subseteq S$ and $|L_i'|, |R_i'| \geq 2$, $L_{i+1}' \cup R_{i+1}' \subseteq L_i' \cup R_i'$. Can we cyclically order S so that both R_i' and L_i', for every $R_i', L_i' \in \mathcal{S}'$, appear consecutively, when restricting the order to $R_i' \cup L_i'$? Once we accomplish the reduction, we end up with the problem of testing the circular-ones property on matrices containing $0, 1$ and $*$ as elements, where each $*$ has only $*$ symbol underneath. This problem is solvable in polynomial time as we will see later. We construct an instance for this problem which is a matrix $M = (m_{ij})$ as follows. The ith row of M corresponds to the pair L'_i and R'_i and each column corresponds to an element of S. For each pair L'_i, R'_i we have $m_{ij} = 0$ if $j \in L'_i$, $m_{ij} = 1$ if $j \in R'_i$, and $m_{ij} = *$ otherwise. Note that our desired condition on S' implies that in M each $*$ has only $*$ symbols underneath. The equivalence of both problems is obvious.

In order to reduce our problem of deciding if a "good" rotation at v exists, we first linearly order intervals in I. Let $(s_0, b_0) \in I$ be the inclusion-wise minimal interval in I maximizing s_0, and similarly let $(s'_0, b'_0) \in I$ be the inclusion-wise minimal interval in I minimizing b'_0. By Observation 3 we have $s_0 = s'_0$ and $b_0 = b'_0$. Thus, let $(s_0, b_0) \in I$ be such that s_0 is the biggest and b_0 is the smallest one. Inductively we relabel elements in I as follows. Let $(s_{i+1}, b_{i+1}) \in I$ be such that $s_{i+1} < s_i < b_i < b_{i+1}$ and subject to that condition s_{i+1} is the biggest and b_{i+1} is the smallest one. By Observation 3 all the elements in I can be ordered as follows

$$(\mathbf{s_0}, \mathbf{b_0}), (s_{0,1}, b_0), \ldots, (s_{0,i_0}, b_0), (s_0, b_{0,1}), \ldots (s_0, b_{0,j_0}), (\mathbf{s_1}, \mathbf{b_1}), \ldots, \quad (1)$$

where $s_{k,i+1} < s_{k,i} < s_k$ and $b_{k,i+1} > b_{k,i} > b_k$. For example, the ordering corresponding to the graph in Fig. 3 is $(4,6), (3,6), (2,6), (4,7), (3,7), (2,7)$. Let $E(s, b)$ and $E'(s, b)$ denote the set of all the edges incident to v contained in an s-cap and b-cup, respectively, induced by $V_s \cup V_{s+1} \ldots \cup V_b$, where $(s, b) \in I$, Thus, $E(s, b) \cup E'(s, b)$ contain edges incident to v contained in an interleaving pair that yields a restriction on rotations at v witnessed by (s, b). Note that $E(s, b) \cap E'(s, b) = \emptyset$. By Observation 2, $E(s_{k,j+1}, b_k) \subseteq E(s_{k,j}, b_k)$ and $E'(s_k, b_{k,j+1}) \subseteq E'(s_k, b_{k,j})$. The restrictions witnessed by (s, b) correspond to the following condition. In the rotation at v the edges in $E(s, b)$ follow the edges in $E'(s, b)$. Indeed, otherwise we have a four-tuple of edges e_1, e_2, f_1 and f_2 incident to v, such that $e_1, f_1 \in P_1$ and $e_2, f_2 \in P_2$, where P_1 and P_2 form an interleaving pair of an s_i-cap and b_i-cup, violating the restriction $\{e_1 f_1\}\{e_2 f_2\}$ on the rotation at v. However, such a four-tuple is not possible in an embedding by Theorem 3.

Let $L_i = E(s, b)$ and $R_i = E'(s, b)$, respectively, for $(s, b) \in I$, where i is the index of the position of (s, b) in (1). Note that $E(s_{i+1}, b_{i+1}) \cup E'(s_{i+1}, b_{i+1}) \subseteq E(s_i, b_i) \cup E'(s_i, b_i)$. Our intermediate goal of reducing our problem to the circular-ones property testing would be easy to accomplish if I consisted only of intervals of the form (s_i, b_i) defined above. However, in I there might be intervals of the form (s_i, b), $b \neq b_i$, or (s, b_i), $s \neq s_i$. Hence, we cannot just put $L'_i := L_i$ and $R'_i := R_i$ for all i, since we do not necessarily have the condition $L_{i+1} \cup R_{i+1} \subseteq L_i \cup R_i$ satisfied for all i.

Definition of S'. Let $S = \{L_i, R_i | i = 1, \ldots\}$. We obtain S' from S by deleting the least number of elements from L_i's and R_i's so that $L'_{i+1} \cup R'_{i+1} \subseteq L'_i \cup R'_i$ for every i. More formally, S' is defined recursively as S'_m, where $S'_1 = \{L'_1, R'_1 | L'_1 = L_1, R'_1 = R_1\}$ and $S'_j = S_{j-1} \cup \{L'_j, R'_j | L'_j = L_j \cap (L'_{j-1} \cup R'_{j-1}), R'_j = R_j \cap (L'_{j-1} \cup R'_{j-1})\}$. Luckily, the following lemma lying at the heart of the proof of our result shows that the information contained in S' is all we need.

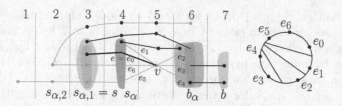

Fig. 3. A subdivided star (on the left) with the center v, and some restrictions on the set of rotation at v (on the right) corresponding to the intervals (s_α, b_α), $(s_{\alpha,1}, b_\alpha)$ and (s, b). We have $\{e_0, e_5, e_6\} = E(s_{\alpha,1}, b_\alpha) \subseteq E(s, b) = \{e_0, e_2, e_5, e_6\}$ and $\{e_3, e_4\} = E'(s, b) \subseteq E'(s_{\alpha,1}, b_\alpha) = \{e_3, e_4, e_1, e_2\}$. Thus, by removing e_0 from $E(s, b)$ we obtain the same restrictions on the rotation at v.

Lemma 2. *We can cyclically order the elements in S so that every pair L_i', R_i' in S' gives rise to two disjoint cyclic intervals if and only if (G, γ) admits an x-bounded embedding.*

Proof. The proof of the lemma is by a double-induction. In the "outer–loop" we induct over $|S'|/2$ while respecting the order of pairs L_i, R_i given by (1). In the "inner–loop" we induct over the size of S, where in the base case of the jth step of the "outer–loop" we have $S_{j,0} = L_j' \cup R_j'$. In each kth step of the "inner–loop" we assume by induction hypothesis that a cyclic ordering $\mathcal{O}$ of S satisfies all the restrictions imposed by $\{L_i, R_i | i = 1, \ldots, j-1\}$ and $L_j \cap S_{j,k-1}, R_j \cap S_{j,k-1}$. Clearly, once we show that $\mathcal{O}$ satisfies restrictions imposed by $L_j \cap S_{j,k}, R_j \cap S_{j,k}$, where $S_{j,k} = S_{j,k} \cup \{e\}$ and $e \in (L_j \cup R_j) \setminus S_{j,k-1}$ we are done.

Refer to Fig. 3. Roughly speaking, by (1) a "problematic" edge e is an initial edge on a path starting at v that never visits a cluster b_α after passing through the cluster s_α such that $e \in E(s_\alpha, b_\alpha)$ (or vice-versa with $E'(s_\alpha, b_\alpha)$). The edge e is an (α, β)-*lower trim* (or (α, β)-⊶) if the lowest index i for which $e \notin L_i' \cup R_i'$ corresponds to $E(s_{\alpha,\beta}, b_\alpha) \cup E'(s_{\alpha,\beta}, b_\alpha)$, where $\beta > 0$. Analogously, the edge e is an (α, β)-*upper trim* (or (α, β)-⊸) if the lowest index i for which $e \notin L_i' \cup R_i'$ corresponds to $E(s_\alpha, b_{\alpha,\beta}) \cup E'(s_\alpha, b_{\alpha,\beta})$, where $\beta > 0$. By (1) and symmetry (reversing the order of clusters) we can assume that e is an (α, β)-⊸, and $e \in E(s_{\alpha,\beta-\beta'}, b_\alpha)$, for some $\beta' > 0$, where $s_{\alpha,0} = s_\alpha$, and $e \in E(s, b) = L_j$, where $s = s_{\alpha,\beta-\beta'}$ and $b > b_\alpha$, following $E(s_{\alpha,\beta}, b_\alpha)$ in our order. Moreover, we pick e so that e maximizes i for which $e \in L_i' \cup R_i'$. We say that e was "trimmed" at the $(i+1)^{\text{th}}$ step.

Thus, e is contained in $E(s, b)$ for some s, b such that $E(s, b), E'(s, b)$ follows $E(s_{\alpha,\beta}, b_\alpha), E'(s_{\alpha,\beta}, b_\alpha)$ in our order. However, it must be that

$$E(s_{\alpha,\beta-\beta'} = s, b_\alpha) \subseteq E(s, b) \text{ and } E'(s, b) \subseteq E'(s_{\alpha,\beta-\beta'} = s, b_\alpha), \quad (2)$$

where the first relation follows directly from the fact $b > b_\alpha$ and the second relation is a direct consequence of Observation 2. In what follows we show that (2) implies that $\mathcal{O}$ satisfies all the required restrictions involving e. We consider an arbitrary four-tuple of edges $e_1', e_2', e_3' \in S_{j,k-1}$ that together with e gives rise to a restriction $\{e'_1e'_2\}\{e'_3e\}$ on $\mathcal{O}$ witnessed by (s,b). The incriminating four-tuple must also contain an element from $E(s,b) \setminus E(s,b_\alpha)$, let us denote it by $f = e'_3$. Indeed, otherwise by (2) the restriction is witnessed by (s,b_α) and we are done by induction hypothesis. Then $e'_1, e'_2 \in E'(s,b)$. For the sake of contradiction we suppose that the order $\mathcal{O}$ violates the restriction $\{e'_1e'_2\}\{ef\}$. Let $g \in L'_{i'} \subseteq E(s_{\alpha,\beta}, b_\alpha)$, for some i'. Note that g exists (see Fig. 4), for if an edge $g' \in E(s_{\alpha,\beta}, b_\alpha)$ is not in $L'_{i'}$, it means that g' was "trimmed" before e and we can put g to be an arbitrary element from $E(s'', b'')$ minimizing s'' appearing before $E(s_\alpha, b_\alpha)$ in our order.

Here, the reasoning goes as follows. Let $P_{g'}$ denote the path from v passing through g' and ending in a leaf. Recall that s_i's are decreasing and b_i's are increasing as i increases. Thus, if we "trimmed" g' before e, it had to be a $\leftarrow$ by $s_{\alpha,\beta} < s_\alpha$, but then there exists a path starting at v that reaches a cluster with a smaller index than a cluster reached by $P_{g'}$ before it reaches even the cluster $b_{\alpha-1} < b_\alpha$. Note that the edge g can be also chosen as an edge in $E(s_{\alpha,\beta}, b_\alpha)$ minimizing i such that the path starting at v passing through g has a vertex in the ith cluster. This choice of g plays a crucial role in our proof of the extension of the lemma for trees.

Thus, $g \in S_{j,k-1}$ by the choice of e, since $e \notin L'_{i'}$. Note that $g \in E(s,b_\alpha)$, and hence, $g \in E(s,b)$ by (2). By Observation 1 a restriction $\{e'_1e'_2\}\{fg\}$ is violated as well due to the restriction $\{e'_1e'_2\}\{eg\}$, that $\mathcal{O}$ satisfies by induction hypothesis, witnessed by (s,b_α). However, by (2) $\{e'_1e'_2\}\{fg\}$ is witnessed by (s,b) and we reach a contradiction with induction hypothesis. ∎

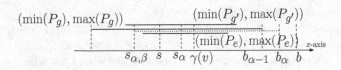

Fig. 4. Three intervals of clusters corresponding to three paths that start at v: P_e that passes through e and ends in the first vertex in the cluster $s_{\alpha,\beta-1}$, $P_{g'}$ that passes through g' and ends in a leaf, and P_g that ends in the first vertex of the cluster s''. (An alternative interval for P_g is dotted.) Here, g' was "trimmed" before e.

By Lemma 2 we successfully reduced our question to the problem stated above. The problem slightly generalizes the algorithmic question considered by Hsu and McConnell [26] about testing 0–1 matrices for circular ones property. An almost identical problem of testing 0–1 matrices for consecutive ones property was already considered by Booth and Lueker [7] in the context of interval and planar graphs' recognition. A matrix has the *consecutive ones* property if it admits a permutation of columns resulting in a matrix in which ones are consecutive in every row. Our generalization is a special case of the related problem of simultaneous PQ-ordering considered recently by Bläsius and Rutter [6], and thus, tractable.

References

1. Angelini, P., Da Lozzo, G., Di Battista, G., Frati, F.: Strip planarity testing. In: Wismath, S., Wolff, A. (eds.) GD 2013. LNCS, vol. 8242, pp. 37–48. Springer, Heidelberg (2013)
2. Angelini, P., Da Lozzo, G., Di Battista, G., Di Donato, V., Kindermann, P., Rote, G., Rutter, I., Planarity, W.: Embedding graphs with direction-constrained edges, Chap. 70, pp. 985–996 (2016)
3. Bachmaier, C., Brandenburg, F.J., Forster, M.: Radial level planarity testing and embedding in linear time. In: Liotta, G. (ed.) GD 2003. LNCS, vol. 2912, pp. 393–405. Springer, Heidelberg (2004)
4. Bertolazzi, P., Di Battista, G., Liotta, G., Mannino, C.: Upward drawings of tri-connected digraphs. Algorithmica **12**(6), 476–497 (1994)
5. Biedl, T.C.: Drawing planar partitions III: two constrained embedding problems. Rutcor Res. Rep. **13–98**, 13–98 (1998)
6. Bläsius, T., Rutter, I.: Simultaneous PQ-ordering with applications to constrained embedding problems. In: Proceedings of the Twenty-Fourth Annual ACM-SIAM Symposium on Discrete Algorithms, SODA 2013, New Orleans, 6–8 January 2013, pp. 1030–1043 (2013). http://arxiv.org/abs/1112.0245
7. Booth, K.S., Lueker, G.S.: Testing for the consecutive ones property, interval graphs, and graph planarity using PQ-tree algorithms. J. Comput. Syst. Sci. **13**(3), 335–379 (1976)
8. Brass, P., Cenek, E., Duncan, C.A., Efrat, A., Erten, C., Ismailescu, D.P., Kobourov, S.G., Lubiw, A., Mitchell, J.S.B.: On simultaneous planar graph embeddings. Comput. Geometry **36**(2), 117–130 (2007)
9. Cairns, G., Nikolayevsky, Y.: Bounds for generalized thrackles. Discrete Comput. Geom. **23**(2), 191–206 (2000)
10. Chang, H.C., Erickson, J., Xu, C.: Detecting weakly simple polygons. In: Proceedings of the Twenty-Sixth Annual ACM-SIAM Symposium on Discrete Algorithms, pp. 1655–1670 (2015). arXiv:1407.3340
11. Cortese, P.F., Di Battista, G., Frati, F., Patrignani, M., Pizzonia, M.: C-planarity of c-connected clustered graphs. J. Graph Algorithms Appl. **12**(2), 225–262 (2008)
12. Cortese, P.F., Di Battista, G., Patrignani, M., Pizzonia, M.: Clustering cycles into cycles of clusters. J. Graph Algorithms Appl. **9**(3), 391–413 (2005)
13. Cortese, P.F., Di Battista, G., Patrignani, M., Pizzonia, M.: On embedding a cycle in a plane graph. Discrete Math. **309**(7), 1856–1869 (2009)
14. Di Battista, G., Tamassia, R.: Incremental planarity testing. In: 30th Annual Symposium on Foundations of Computer Science, pp. 436–441, October 1989
15. Diestel, R.: Graph Theory. Graduate Texts in Mathematics, vol. 173, 3rd edn. Springer, Heidelberg (2005)
16. Feng, Q.-W., Cohen, R.F., Eades, P.: How to draw a planar clustered graph. In: Ding-Zhu, D., Li, M. (eds.) COCOON 1995. LNCS, vol. 959, pp. 21–30. Springer, Heidelberg (1995)
17. Feng, Q.-W., Cohen, R.F., Eades, P.: Planarity for clustered graphs. In: Spirakis, P. (ed.) Algorithms–ESA 1995. LNCS, vol. 979, pp. 213–226. Springer, Heidelberg (1995)
18. Radoslav Fulek. Toward the Hanani-Tutte theorem for clustered graphs. 2014. arXiv:1410.3022v2
19. Fulek, R.: Towards the Hanani-Tutte theorem for clustered graphs. In: Kratsch, D., Todinca, I. (eds.) WG 2014. LNCS, vol. 8747, pp. 176–188. Springer, Heidelberg (2014)

20. Fulek, R.: C-planarity of embedded cyclic c-graphs (2016). arXiv:1602.01346v2
21. Fulek, R., Pelsmajer, M.J., Schaefer, M., Štefankovič, D.: Hanani-Tutte, monotone drawings and level-planarity. In: Pach, J. (ed.) Thirty Essays on Geometric Graph Theory, pp. 263–288. Springer, New York (2012)
22. Garg, A., Tamassia, R.: On the computational complexity of upward and rectilinear planarity testing. SIAM J. Comput. **31**(2), 601–625 (2002)
23. Goodrich, M.T., Lueker, G.S., Sun, J.Z.: C-planarity of extrovert clustered graphs. In: Healy, P., Nikolov, N.S. (eds.) GD 2005. LNCS, vol. 3843, pp. 211–222. Springer, Heidelberg (2006)
24. Gutwenger, C., Jünger, M., Leipert, S., Mutzel, P., Percan, M., Weiskircher, R.: Advances in c-planarity testing of clustered graphs. In: Goodrich, M.T., Kobourov, S.G. (eds.) GD 2002. LNCS, vol. 2528. Springer, Heidelberg (2002)
25. Hong, S.-H., Nagamochi, H.: Simpler algorithms for testing two-page book embedding of partitioned graphs. Theoretical Computer Science (2016)
26. Hsu, W.-L., McConnell, R.M.: PC-trees and circular-ones arrangements. Theoret. Comput. Sci. **296**(1), 99–116 (2003)
27. Jelínek, V., Jelínková, E., Kratochvíl, J., Lidický, B.: Clustered planarity: embedded clustered graphs with two-component clusters. In: Tollis, I.G., Patrignani, M. (eds.) GD 2008. LNCS, vol. 5417, pp. 121–132. Springer, Heidelberg (2009)
28. Jelınková, E., Kára, J., Kratochvıl, J., Pergel, M., Suchỳ, O., Vyskocil, T.: Clustered planarity: small clusters in cycles and Eulerian graphs. J. Graph Algorithms Appl. **13**(3), 379–422 (2009)
29. Jünger, Michael, Leipert, Sebastian, Mutzel, Petra: Level planarity testing in linear time. In: Whitesides, Sue H. (ed.) GD 1998. LNCS, vol. 1547, pp. 224–237. Springer, Heidelberg (1999)
30. Lengauer, T.: Hierarchical planarity testing algorithms. J. ACM **36**(3), 474–509 (1989)
31. Mabillard, I., Wagner, U. Eliminating Tverberg points, I. An analogue of the Whitney trick. In: Proceedings of theThirtieth Annual Symposium on Computational Geometry, SOCG 2014, pp. 171:171–171:180 (2014)
32. Opatrny, J.: Total ordering problem. SIAM J. Comput. **8**(1), 111–114 (1979)
33. Pach, J., Tóth, G.: Monotone drawings of planar graphs. J. Graph Theory **46**(1), 39–47 (2004). arXiv:1101.0967
34. Schaefer, M.: Toward a theory of planarity: Hanani-Tutte and planarity variants. J. Graph Algorithms Appl. **17**(4), 367–440 (2013)
35. Ziegler, G.M.: Lectures on Polytopes, vol. 152. Springer Science & Business, New York (1995)

Crushing Disks Efficiently

Stefan Funke[1]([✉]), Filip Krumpe[1]([✉]), and Sabine Storandt[2]([✉])

[1] University of Stuttgart, Stuttgart, Germany
{funke,krumpe}@fmi.uni-stuttgart.de
[2] Universität Würzburg, Würzburg, Germany
storandt@informatik.uni-wuerzburg.de

Abstract. Given a set of prioritized disks with fixed centers in $\mathbb{R}^2$ whose radii grow linearly over time, we are interested in computing an *elimination order* of these disks assuming that when two disks touch, the one with lower priority is 'crushed'. A straightforward algorithm has running time $O(n^2 \log n)$ which we improve to expected $O(n(\log^6 n + \Delta^2 \log^2 n + \Delta^4 \log n))$ where Δ is the ratio between largest and smallest radii amongst the disks. For a very natural application of this problem in the map rendering domain, we have $\Delta = O(1)$.

1 Introduction

Given a set of points/centers $P = \{c_1, \ldots, c_n\}$, $c_i \in \mathbb{R}^2$ with associated radii $r_1, \ldots, r_n$, $r_i \in \mathbb{R}^+$ and priorities $p_1, \ldots p_n$, $p_i \in \mathbb{N}$, $p_i \neq p_j$ for $i \neq j$, we are interested in the following dynamic process: Time $t \in \mathbb{R}_0^+$ progresses continuously starting with $t = 0$. For a given time t, point c_i induces a disk $D(c_i, r_i t)$ centered at c_i and with radius $R_i(t) = r_i t$. As time progresses, at some point two disks touch for the first time; we discard/eliminate the one with lower priority and let time continue further until the next two disks touch. The process finishes once only one disk remains.

To keep the presentation simple we assume non-degenerate position of the input which in our case prohibits more than one pair of disks touching at the same time. Hence for given P, radii, and priorities, the *elimination order* of the disks is uniquely determined. The challenge is to compute this elimination order for given P, radii, and priorities *efficiently*.

1.1 Motivation

A common task when rendering a map is the labelling of points of interest (POIs) like countries, cities, amenities, etc. on the map. Typically one is after a hierarchical labelling scheme, where in a very coarse view of the map only the most important POIs are labelled, but when zooming in more and more labels appear until in a very zoomed-in view all labels are present. If labels are not to change or overlap when the map is rotated, it is very convenient to allocate a disk-like shape for each label. Furthermore, we expect that while zooming in, no previously present labels disappear again. This and several other consistency

© Springer International Publishing Switzerland 2016
V. Mäkinen et al. (Eds.): IWOCA 2016, LNCS 9843, pp. 43–54, 2016.
DOI: 10.1007/978-3-319-44543-4_4

criteria were postulated by Been et al. in [1]. So considering the reverse process, starting from a fully zoomed-in view we have all POIs labeled with their label written in a small disk around the POI, all scaled that none of the disks overlap. The relative radii of the disks depend on the label sizes and the font sizes. Zooming out while keeping the font sizes corresponds to blowing up the disks. As soon as two disks intersect, this might induce an overlap of the respective labels, so one of the labels has to be eliminated for that zoom level and all levels further zoomed out. Naturally one would eliminate the label of the less important POI.

Computing an elimination order as defined at the beginning yields a consistent level-of-detail hierarchy of map labels, in particular avoiding spurious appearance and disappearance of labels as it is still often experienced with common map renderings like Google/Bing/Yahoo maps. See Fig. 1 for an example of two zoom levels of the region around Stuttgart (not computed with the algorithm of this paper, though).

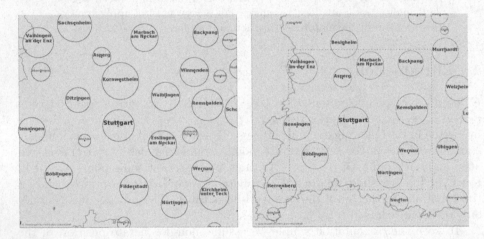

Fig. 1. Labelings of the region Stuttgart in Germany with the disks reserved for each labels to prevent overlaps when rotating. Zoomed-in (left) and further zoomed-out (right). Map data ©OpenStreetMap contributors

1.2 Related Work

Eppstein and Erickson in [6] considered the following problem: Given n motorcycles, each motorcycle i starting at some position $c_i \in \mathbb{R}^2$ and moving in some direction d_i at speed s_i, a motorcycle crashes as soon as it hits the track of another motorcycle (similar to the light cycles in the movie TRON). The goal is to determine which motorcycle crashes into which track and which motorcycles survive. This is expressed in the so-called *motorcycle graph*. Construction of the latter given initial starting positions, directions and speeds of the motorcycles is surprisingly difficult. While a naive algorithm (similar to the naive algorithm for our problem) takes $O(n^2 \log n)$ time, the best known algorithm as presented in [6] has a running time of $O(n^{17/11+\epsilon})$ for any $\epsilon > 0$. In spite of the simplicity of motorcycle graphs, they have so far resisted construction in near-linear time.

If all radii are equal, our problem becomes quite easy. Maintaining the closest pair under deletions using a data structure like [4] (explained in more detail later on) allows for computing the elimination order in near-linear time.

In [2] the authors consider a related family of *active range optimization (ARO)* problems. One variant — the 2-dimensional simple ARO — is quite similar to our problem at hand. Phrased in terms of our problem, their goal is to identify an elimination sequence which *maximizes the sum* of life times of all circles. In fact they do not consider circles but squares, show NP-hardness and a $1/24$ approximation algorithm with near-linear running time. The hardness crucially relies on the freedom to choose which of two collision partners survives. In our concrete application setting, this is not desired, as we typically have strong preferences which label should survive, e.g., the label for the city of Munich should not be eliminated by the label of the small town Unterschleißheim when zooming out. From a complexity theoretical point of view, our problem formulation is simpler as the $O(n^2 \log n)$ naive algorithm already shows. The aim of this work is to improve upon the naive algorithm such that millions or even billions of points (as implicit in the OpenStreetMap data set) can be labelled. Following [2], a more practical direction is taken in [10], formulating the 2-dimensional ARO problem as an integer linear program and proposing greedy heuristics with $O(n^2)$ running time.

Contribution. We devise an efficient algorithm for computing the elimination order for given P, radii, and priorities in expected $O(n(\log^6 n + \Delta^2 \log^2 n + \Delta^4 \log n))$ time improving upon the $O(n^2 \log n)$ running time of the naive algorithm. Here Δ is the ratio between the largest and smallest radius in the problem instance. Our approach has the potential to be generalized to higher (fixed) dimensions by replacing the data structure due to Chan [4] by respective approximate variants, e.g. [3,9].

2   Crushing Disks

In this section we will first briefly sketch a naive algorithm with roughly quadratic running time and outline some problems that discard straightforward improvement strategies. We then introduce our new algorithm, and prove its correctness and improved running time.

2.1   Naive Algorithm

A naive algorithm to solve the problem is simply computing/predicting all pairwise collision times and throwing these potential events into a min-priority queue organized according to the collision time. Then the events are popped one-by-one from the priority queue. If the two centers c_i, c_j of the current event are still alive, the one with smaller priority is discarded (and appended to the elimination sequence). If at least one of them has already been discarded, the event is ignored. See Algorithm 1 for the pseudo-code. It takes $\Theta(n^2)$ time to compute all potential collision events and $O(n^2 \log n)$ to process them through the priority queue (e.g. an ordinary heap), resulting in an $O(n^2 \log n)$ overall running time.

Algorithm 1. Naive algorithm

for all $(c_i, c_j)|i, j = 1, \ldots, n \wedge i < j$ do
  $t_{ij} \leftarrow \frac{|c_i c_j|}{r_i + r_j}$
  insert $CollisionEvent(c_i, c_j, t_{ij})$ into event queue Q
end for
while $Q \neq \emptyset$ do
  $curEvent(c_i, c_j, t_{ij}) \leftarrow Q.popMin()$
  if $alive(c_i) \wedge alive(c_j)$ then
    if $p_i < p_j$ then
      discard and output c_i
    else
      discard and output c_j
    end if
  end if
end while

It is clear that such a running time is prohibitive for an application domain as the one outlined in the introduction where we have millions or billions of POIs labels. Unfortunately, in this very naive algorithm, this running time does not only occur for pathological problem instances, but in fact all the time.

2.2 Algorithmic Tools

To come up with a more efficient solution we need some tools from computational geometry, more concretely we will employ data structures for efficient proximity queries. Let us define the types of queries that are relevant to us. First of all, we are interested in a point from a specified set which is closest to a given query point:

Definition 1. *For a point set $P \subset \mathbb{R}^d$ and a query point q, a nearest neighbor of q is a point $c \in P$ with $|qc| \leq |qc'|$, for all $c' \in P$.*

Furthermore, we are also interested in points within a certain distance of a query point:

Definition 2. *For a point set $P \subset \mathbb{R}^d$, a query point q, and a distance r, a range reporting query returns the set $S := \{c \in P : |qc| \leq r\}$.*

In dimension $d = 2$, Chan in [4] has devised a dynamic data structure that allows for nearest neighbor and range reporting queries in polylogarithmic time.

Theorem 1 (Chan [4]). *For n points in $\mathbb{R}^2$ one can construct a data structure in expected $O(n \log^2 n)$ time which supports deletions in expected amortized $O(\log^6 n)$ time, nearest neighbor queries in $O(\log^2 n)$, and range reporting queries in time $O(\log^2 n + k \log n)$, where $k = |S|$ is the size of the output of the range reporting query.*

Note that simply maintaining a Delaunay triangulation/Voronoi diagram does not suffice since deletion might be very expensive for high-degree vertices.

In dimensions $d > 2$ things become much harder and one typically resort to an approximate notion of proximity. While our algorithms can probably be generalized to that setting as well, we will restrict to the 2-dimensional scenario in this paper. We will sketch some of the necessary modifications in the conclusions. Efficient schemes for approximate proximity queries even in a dynamic setting were devised e.g. in [3,9].

2.3   First Ideas Towards a More Efficient Algorithm

An idea for improvement of the naive algorithm that immediately comes to mind is to avoid for one center c_i the inspection of all other centers, by looking only locally for collision partners. It seems very likely, that the disk around c_i collides only with 'nearby' disks and is discarded before interacting with far away disks. So a natural implementation of this idea lets every center c_i at the very beginning look at a neighborhood with size proportional to the distance to its nearest neighbor and check for collisions with other centers in this neighborhood.

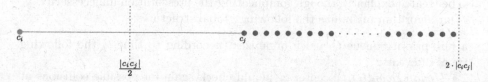

Fig. 2. Example where a neighborhood of a center c_i that is proportional to the distance to its nearest neighbor contains many other centers.

Unfortunately, it is not difficult to come up with examples where for a center c_i the number of other centers in its neighborhood is $\Omega(n)$, see Fig. 2. Furthermore, the actual first collision that happens for a center c_i might in fact be with the furthest other center, see Fig. 3, since all closer centers might have been eliminated before a collision can occur.

The algorithm we propose in the following is based on the idea of inspecting local neighborhoods, but by appropriately delaying the inspections we can make sure that the number of other centers to inspect remains small. Furthermore we do not insist on each center actually knowing its next collision, but make only sure that before every actual collision a neighborhood inspection of at least one of the collision partners takes place predicting the collision.

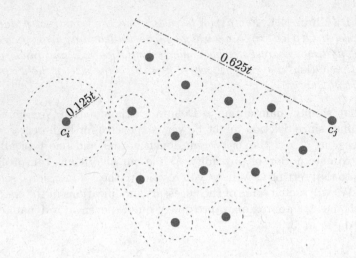

Fig. 3. Example where the first collision for c_i happens with the furthest center if p_j is the maximum priority.

2.4 The Algorithm

We describe our algorithm under a common non-degeneracy assumption, that is, there is no $t > 0$ where more than two disks with respective time-varying radii collide. It is not difficult to enforce this assumption by actual perturbations (e.g. [7,8]) or symbolic perturbation [5]. Alternatively these degenerate situations can also be treated explicitly, though complicating the presentation unnecessarily.

Our algorithm maintains the following global structures:

- a min-priority queue Q which organizes (according to time t) the following types of events:
    - $UpdateEvent(c_i, t)$: center c_i should check again for possible collisions at time t
    - $CollisionEvent(c_i, c_j, t)$: a collision between the disks around c_i and c_j is predicted at time t
- for each center c_i the following sets:
    - $NN(c_i)$: the set of currently alive centers which have c_i as their nearest neighbor
    - $CP(c_i)$: the set of currently alive centers which have c_i as their currently predicted next collision

Within the priority queue Q we resolve ambiguities (with identical times t) by considering UpdateEvents 'smaller' than CollisionEvents and finally resorting to lexicographical ordering.

In Algorithm 2 we have depicted the main loop of our algorithm. For every not yet discarded/'crushed' center c_i there is either an $UpdateEvent(c_i, t)$ or a predicted collision $CollisionEvent(c_i, ., t)$ in the priority queue Q. Our algorithm does not require a collision between disks around centers c_i and c_j to be predicted by both c_i and c_j, but in fact only the center with larger radius, that is, if $r_i \geq r_j$, our algorithm makes sure that c_i predicts the collision between c_i and c_j. There is no harm, though, if both predict the respective collision.

The event queue is initialized with update events as follows: for every c_i its nearest neighbor c_j is determined. Clearly, the disk around c_i cannot collide with another disk with radius less than r_i earlier than $|c_i c_j|/2r_i$, so it suffices for c_i to check for collisions at that time. Then the algorithm processes the events one-by-one, occasionally calling the central collision prediction routine, see Algorithm 3. It is invoked as $PredictCollision(c_i, t)$ for a center c_i at some time t. After determining the nearest neighbor of c_i it checks whether the current radius $(r_i t)$ of the disk around c_i is at least half the distance to its nearest neighbor. If so, a neighborhood of twice the distance to its nearest neighbor is examined for collision. If not, we know that c_i is not responsible for detecting a collision in the near future and send it 'to sleep' via an update event at a time roughly corresponding to its nearest neighbor distance.

2.5   Analysis

We will first argue that our new algorithm indeed computes the same elimination order as the naive algorithm and then show that for a bounded ratio $\Delta = \frac{\max_i r_i}{\min_i r_i}$ our algorithm runs in near-linear time.

Correctness.

Lemma 1. *Algorithm 2 computes the same elimination order as Algorithm 1.*

Proof. We show this by induction, making sure that the i-th collision detected by Algorithm 1 is also detected by Algorithm 2 for $i = 1, \ldots n - 1$. For the base case $i = 1$, w.l.o.g. let (c_1, c_2) be the first collision with $r_1 \geq r_2$. When the update event for c_1 is processed, a neighborhood of twice the distance to c_1's nearest neighbor is examined which also includes c_2 (otherwise c_1 would collide with the disk around its nearest neighbor before hitting the disk around c_2), hence the collision with c_2 is predicted. As no other collision can happen earlier, the base case is settled. Now assume the first i collisions were correctly detected and the respective deletions have taken place. Algorithm 1 produces (c_k, c_l) as the next collision, w.l.o.g. $r_k \geq r_l$. Consider the state of Algorithm 2 right after processing the i-th deletion. At this point c_k must be present in the priority queue. If an $UpdateEvent(c_k, .)$ is in Q, the nearest neighbor (at the time of creation of the update event) of c_k is also still alive. When this update event is processed, the collision (c_k, c_l) will be predicted and processed (as in the base case). On the other hand, if $CollisionEvent(c_k, c_{l'}, .)$ is present in Q, we must have $c_{l'} = c_l$ as when $PredictCollision(c_k, .)$ was called for the last time, c_l was amongst the considered centers in the neighborhood (again same argument as in the base case).   □

Algorithm 2. Main algorithm loop

for all c_i, $i = 1, \ldots n$ do
    determine nearest neighbor c_j of c_i
    $NN(c_j) \leftarrow NN(c_j) \cup \{c_i\}$
    insert UpdateEvent(c_i, $|c_i c_j|/(2r_i)$) into event queue Q
end for
while $Q \neq \emptyset$ do
    $curEvent \leftarrow Q.popMin()$
    if $curEvent = UpdateEvent(c_i, t)$ then
        $PredictCollisions(c_i, t)$
    else if $curEvent = CollisionEvent(c_i, c_j, t)$ and $alive(c_i) \wedge alive(c_j)$ then
        if $p_i < p_j$ then
            discard and output c_i
            for all $c_k \in NN(c_i) \cup CP(c_i)$ do
                remove c_i from $NN(c_k)$ or $CP(c_k)$ (if present)
                $PredictCollisions(c_k, t)$
            end for
        else
            discard and output c_j
            for all $c_k \in NN(c_j) \cup CP(c_j)$ do
                remove c_j from $NN(c_k)$ or $CP(c_k)$ (if present)
                $PredictCollisions(c_k, t)$
             end for
        end if
    end if
end while

Algorithm 3. PredictCollision(c_i, t)

remove existing event UpdateEvent($c_i, .$) or CollisionEvent($c_i, ., .$) from Q
determine nearest neighbor c_j of c_i
$NN(c_j) \leftarrow NN(c_j) \cup \{c_i\}$
$t' \leftarrow |c_i c_j|/(2r_i)$
if $t < t'$ then
    insert UpdateEvent(c_i, t') into event queue Q
else
    $C \leftarrow \{c | c \in D(c_i, 2|c_i c_j|)\}$
    determine element $c_k \in C$ with minimum collision time t_k
    insert CollisionEvent(c_i, c_k, t_k) into event queue Q
    $CP(c_k) \leftarrow CP(c_k) \cup \{c_i\}$
end if

Running Time. Bounding the running time of the algorithm consists of essentially two steps. We need to bound the size of the result of the range reporting query (determination of the set C in the *PredictCollision* subroutine) and the number of times *PredictCollision* is invoked. Let us first turn to the former. In the following Lemmas, t' refers to $|c_i c_j|/(2r_i)$, that is, the adjusted nearest neighbor distance (line 4 in *PredictCollision*).

Lemma 2. *At any time t and for any center c_i we have during an invocation of PredictCollision with $t \geq t'$ that for $C = \{c \in D(c_i, \alpha r_i t)\}$, $|C| = O(\alpha^2 \Delta^2)$.*

Proof. The disks with radius $r_{min} t$ around the centers in C have to be all disjoint and completely contained in a disk of radius $(\alpha r_i + r_{min})t$ around c_i. Hence the size of C is upper bounded by:

$$O\left(\frac{((\alpha r_i + r_{min})t)^2}{(r_{min} t)^2}\right) = O\left(\frac{((\alpha \Delta r_{min} + r_{min})t)^2}{(r_{min} t)^2}\right) = O(\alpha^2 \Delta^2)$$

The Lemma follows. □

So essentially as long as Δ is not too big, the cost for an individual range reporting query is quite moderate.

Lemma 3. *Whenever PredictCollision is called, it takes expected $O(\log^2 n + \Delta^2 \log n)$ time.*

Proof. Determining the nearest neighbor takes expectedly $O(\log^2 n)$ time using Chan's data structure [4]. In case $t < t'$ we only have to remove and insert a single event into Q which can be done in $O(\log n)$ time. Otherwise, according to Lemma 2 with $\alpha = 2$ we have to report and inspect $O(\Delta^2)$ other centers which can be done in expected $O(\log^2 n + \Delta^2 \log n)$ time. Having predicted the earliest collision, insertion into the event queue takes $O(\log n)$ time. □

Now let us make sure that *PredictCollision* is not called too often. We partition the calls to *PredictCollision* according to the calling context, that is:

- $PredictCollision(c_i, t)$ was called because c_i had c_j as nearest neighbor, and c_j was discarded.
- $PredictCollision(c_i, t)$ was called because c_i had c_j as collision partner, and c_j was discarded.
- $PredictCollision(c_i, t)$ was called due to an UpdateEvent (having slept)

We will first argue that the number of invocations of *PredictCollision* triggered by discarded nearest neighbors and collision partners is in $O(n)$. To that end we need a small Lemma bounding the number of other centers which have a specific center as nearest neighbor.

Lemma 4. *The number of other centers that have a certain center c as nearest neighbor is $O(1)$.*

Proof. Consider the set $NN(c)$, that is, the set of centers having c as nearest neighbor. Let r be the minimum distance of any point in $NN(c)$ to c. For every point in $NN(c)$ we construct its projection on the boundary of $D(c, r)$ i.e. the disk centered at c with radius r. Let $NN'(c)$ be the described set of projected points. The projection does not change the nearest neighbor relation (compare Fig. 4). Clearly, for neighbors $q_1, q_2 \in NN'(c)$, we have $\angle q_1 p q_2 \geq 60°$, otherwise $|q_1 q_2| < |q_i p|$. It follows that $|NN'(c)| \leq 6$ and hence $|NN(c)| \leq 6$. □

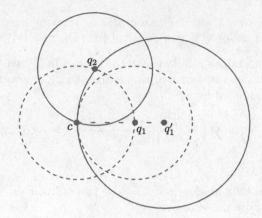

Fig. 4. Setting where 2 points q_1' and q_2 have different distances to the common nearest point c. Shifting q_1' to q_1 on the straight line towards c then $|cq_1| < |cq_1'|$ and disk $D(q_1, |cq_1|)$ is contained in $D(q_1', |cq_1'|)$.

Lemma 5. *The number of times PredictCollision is called from a nearest neighbor context is $O(n)$.*

Proof. According to Lemma 4, whenever a center is discarded, only $O(1)$ calls to *PredictCollision* can be triggered because of that center being a nearest neighbor for the respective other center. Hence in total $O(n)$ such calls happen.   □

Lemma 6. *The number of times PredictCollision is called from a collision partner context is $O(\Delta^2 n)$.*

Proof. Whenever a center is discarded, all other centers that had this center as collision partner must be within distance $2r_{\max}t$ of the discarded center. According to Lemma 2, there are only a constant number of them. Hence in total only $O(n)$ such calls can happen.   □

So the number of calls to *PredictCollision* due to nearest neighbors or collision partners being discarded is fine. Bounding the number of calls due to an update event is not that straightforward. Observe, that for a single center c_i, $\Theta(n)$ such calls might be issued, see Fig. 5. Fortunately, we can prove that in total, this number cannot exceed $O(n)$.

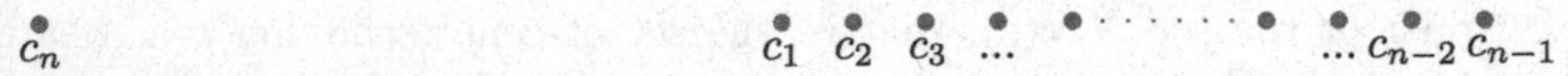

$\overset{\bullet}{c_n}$ $\overset{\bullet}{c_1}\ \overset{\bullet}{c_2}\ \overset{\bullet}{c_3}\ ...\ \bullet\ \bullet\ \cdot\ \cdot\ \cdot\ \cdot\ \cdot\ \cdot\ \cdot\ \bullet\ \ ...\ \overset{\bullet}{c_{n-2}}\ \overset{\bullet}{c_{n-1}}$

Fig. 5. Example where $\Theta(n)$ calls to *PredictCollision* are issued due to update events if $p_1 < p_2 < ... < p_{n-1} < p_n$.

Lemma 7. *The number of times PredictCollision is called from an update event context is $O(n)$.*

Proof. Consider a call to $PredictCollision(c_i, t)$ triggered by an update event. Note that in this case $t = t'$, as the distance to the nearest neighbor of c_i can only change when the nearest neighbor of c_i is discarded which would have triggered a non-update-event related call of $PredictCollision$. Having $t = t'$ makes sure that the exploration of $D(c_i, 2|c_i c_j|)$ takes place and the next call to $PredictCollision$ is not update-related. So, this shows that we can never have two consecutive calls to $PredictCollision(c_i, .)$ being triggered by update events, and hence the total number of calls triggered by update events is $O(n)$ via Lemmas 5 and 6.   □

2.6   Summary

So we have shown that the number of calls to $PredictCollision$ is $O(\Delta^2 n)$ and every call to $PredictCollision$ takes expected $O(\log^2 n + \Delta^2 \log n)$ time, which yields the following theorem about the total running time of our algorithm when employing the dynamic nearest neighbor data structure due to Chan [4].

Theorem 2. *We can compute the elimination order for given P, radii and priorities in expected $O(n(\log^6 n + \Delta^2 \log^2 n + \Delta^4 \log n))$ time.*

Proof. The cost for calling $PredictCollision$ has been bounded by the previous Lemmas. The $n - 1$ deletions can be performed in expected amortized $O(\log^6 n)$ using Chan's data structure. The running time follows.   □

In the real-world application sketched at the beginning, we naturally have $\Delta = O(1)$, since it is essentially determined by the ratio between shortest and longest label string as well as smallest and largest font size used for labelling.

3   Outlook and Future Work

It is natural to consider the problem of computing elimination sequences also for dimensions higher than 2. For our application scenario, a spherical setting might also be of interest, in particular if POIs on a globe are to be labelled.

Unfortunately, no data structure for exact proximity queries like the one by Chan [4] that we have employed is known for dimensions $d > 2$ (and probably also unlikely to exist). In such a case one typically resorts to data structures which answer proximity queries in an *approximate* fashion. At first it seems as we could simply replace Chan's data structure by one of these data structures, e.g. [3,9]. Unfortunately, Lemma 4 does not hold anymore, in particular, a center c_i might be approximate nearest neighbor for many other centers.

Another (more challenging) direction of future work would be trying to get rid of the dependency of the runtime on Δ.

References

1. Been, K., Daiches, E., Yap, C.: Dynamic map labeling. IEEE Trans. Visual Comput. Graphics **12**(5), 773–780 (2006)
2. Been, K., Nöllenburg, M., Poon, S.-H., Wolff, A.: Optimizing active ranges for consistent dynamic map labeling. Comput. Geom. **43**(3), 312–328 (2010)
3. Chan, T.M.: Closest-point problems simplified on the RAM. In: Proceedings of the Thirteenth Annual ACM-SIAM Symposium on Discrete Algorithms, SODA 2002, pp. 472–473. Society for Industrial and Applied Mathematics, Philadelphia (2002)
4. Chan, T.M.: A dynamic data structure for 3-D convex hulls, 2-D nearest neighbor queries. J. ACM **57**(3), 16:1–16:15 (2010)
5. Edelsbrunner, H., Mücke, E.P.: Simulation of simplicity: a technique to cope with degenerate cases in geometric algorithms. ACM Trans. Graph. **9**(1), 66–104 (1990)
6. Eppstein, D., Erickson, J.: Raising roofs, crashing cycles, and playing pool: applications of a data structure for finding pairwise interactions. Discrete Comput. Geometry **22**(4), 569–592 (1999)
7. Funke, S., Klein, C., Mehlhorn, K., Schmitt, S.: Controlled perturbation for delaunay triangulations. In: Proceedings of the Sixteenth Annual ACM-SIAM Symposium on Discrete Algorithms, SODA 2005, pp. 1047–1056. Society for Industrial and Applied Mathematics, Philadelphia (2005)
8. Halperin, D., Shelton, C.R.: A perturbation scheme for spherical arrangements with application to molecular modeling. Comput. Geometry **10**(4), 273–287 (1998). Special Issue on Applied Computational Geometry
9. Mount, D.M., Park, E.: A dynamic data structure for approximate range searching. In: Proceedings of the Twenty-sixth Annual Symposium on Computational Geometry, SoCG 2010, pp. 247–256. ACM, New York (2010)
10. Schwartges, N., Allerkamp, D., Haunert, J., Wolff, A.: Optimizing active ranges for point selection in dynamic maps. In: Proceeding 16th ICA Generalisation Workshop (ICAGW 2013), 10 pages (2013)

Essential Constraints of Edge-Constrained Proximity Graphs

Prosenjit Bose[1], Jean-Lou De Carufel[2], Alina Shaikhet[1(✉)], and Michiel Smid[1]

[1] School of Computer Science, Carleton University, Ottawa, Canada
{jit,michiel}@scs.carleton.ca, alina.shaikhet@carleton.ca
[2] School of Electrical Engineering and Computer Science, University of Ottawa,
Ottawa, Canada
jdecaruf@uottawa.ca

Abstract. Given a plane forest $F = (V, E)$ of $|V| = n$ points, we find the minimum set $S \subseteq E$ of edges such that the edge-constrained minimum spanning tree over the set V of vertices and the set S of constraints contains F. We present an $O(n \log n)$-time algorithm that solves this problem. We generalize this to other proximity graphs in the constraint setting, such as the *relative neighbourhood graph*, *Gabriel graph*, *β-skeleton* and *Delaunay triangulation*.

We present an algorithm that identifies the minimum set $S \subseteq E$ of edges of a given plane graph $I = (V, E)$ such that $I \subseteq CG_\beta(V, S)$ for $1 \leq \beta \leq 2$, where $CG_\beta(V, S)$ is the constraint β-skeleton over the set V of vertices and the set S of constraints. The running time of our algorithm is $O(n)$, provided that the constrained Delaunay triangulation of I is given.

Keywords: Proximity graphs · Constraints · Visibility · MST · Delaunay · β-skeletons

1 Introduction

This paper was inspired by topics in geometric compression. In particular, Devillers et al. [3] investigate how to compute the minimum set $S \subseteq E$ of a given plane triangulation $T = (V, E)$, such that T is a constrained Delaunay triangulation (DT) of the graph (V, S). They show that S and V is the only information that needs to be stored. The graph T can be successfully reconstructed from S and V. Experiments on real data sets (such as terrain models and meshes) show that the size of S is less than 3.4 % of the total number of edges of T, which yields an effective compression of the triangulation.

Our goal is to broaden this research and investigate geometric compression of other neighbourhood graphs. We study minimum spanning trees, relative neighbourhood graphs, Gabriel graphs and β-skeletons for $1 \leq \beta \leq 2$. We give a definition of each of those graphs in the constraint setting (refer to Sect. 2).

This work was partially supported by NSERC.

© Springer International Publishing Switzerland 2016
V. Mäkinen et al. (Eds.): IWOCA 2016, LNCS 9843, pp. 55–67, 2016.
DOI: 10.1007/978-3-319-44543-4_5

Minimum spanning trees (MST) have been studied for over a century and have numerous applications. We study the problem of finding the minimum set S of constraint edges in a given plane forest $F = (V, E)$ such that the edge-constrained MST over the set V of vertices and the set S of constraints contains F. If F is a plane tree then the edge-constrained MST over (V, S) is equal to F. We give an $O(n \log n)$-time algorithm that solves this problem.

Gabriel graphs (GG) were introduced by Gabriel and Sokal in [4]. Toussaint introduced the notion of relative neighbourhood graphs (RNG) in his research on pattern recognition [10]. Both graphs were studied extensively.

Jaromczyk and Kowaluk showed that RNG of a set V of points can be constructed from the Delaunay triangulation of V in time $O(n\alpha(n, n))$, where $\alpha(\cdot)$ is the inverse of the Ackerman function [5]. These two authors, together with Yao, improved the running time of their algorithm to linear [6]. They achieved it by applying a static variant of the *Union-Find* data structure. They also generalized their algorithm to construct the β-skeleton (G_β) for $1 \le \beta \le 2$ in linear time from the Delaunay triangulation of V under the L_p-metric, $1 < p < \infty$. We provide the definition of β-skeleton in Sect. 2. For now, note that the 1-skeleton corresponds to the Gabriel graph and the 2-skeleton corresponds to the relative neighbourhood graph. In this paper, we use two geometric structures: elimination path and elimination forest, introduced by Jaromczyk and Kowaluk [5].

Neighbourhood graphs are known to form a nested hierarchy, one of the first versions of which was established by Toussaint [10]: for any $1 \le \beta \le 2$, $MST \subseteq RNG \subseteq G_\beta \subseteq GG \subseteq DT$. We show that the neighbourhood graphs in the constraint setting form the same hierarchy. Moreover, we show that the minimum set of constraints required to reconstruct a given plane graph (as a part of each of those neighbourhood graphs) form an inverse hierarchy.

In Sect. 2, we present notations and definitions. In Sect. 3, we give some observations concerning constrained MST, show worst-case examples and present an $O(n \log n)$-time algorithm that identifies the minimum set $S \subseteq E$ of constraint edges given a plane forest $F = (V, E)$ such that the edge-constrained MST over the set V of vertices and the set S of constraints contains F. Section 4 presents an algorithm that identifies the minimum set S of edges of a given plane graph $I = (V, E)$ such that $I \subseteq CG_\beta(V, S)$ for $1 \le \beta \le 2$, where $CG_\beta(V, S)$ is a constrained β-skeleton on the set V of vertices and set S of constraints. The hierarchy of the constrained neighbourhood graphs together with the hierarchy of the minimum sets of constraints are given in Sect. 5.

2   Basic Definitions

Let V be a set of n points in the plane and $I = (V, E)$ be a plane graph representing the constraints. Each pair of points $u, v \in V$ is associated with a neighbourhood defined by some property $P(u, v, I)$ depending on the proximity graph under consideration. An edge-constrained neighbourhood graph $G_P(I)$ defined by the property P is a graph with a set V of vertices and a set E_P of edges such that $\overline{uv} \in E_P$ if and only if $\overline{uv} \in E$ or $\overline{uv}$ satisfies $P(u, v, I)$.

For clarity and to distinguish between different types of input graphs, if I is a forest, we will denote I by $F = (V, E)$, to emphasize its properties.

In this paper, we assume that the points in V are in general position (no three points are collinear and no four points are co-circular).

Two vertices u and v are *visible to each other with respect to E* provided that $\overline{uv} \in E$ or the line segment $\overline{uv}$ does not intersect the interior of any edge of E. For the following definitions, let $I = (V, E)$ be a plane graph.

Definition 1 (Visibility Graph of I). *The* visibility graph of I *is the graph* $VG(I) = (V, E')$ *such that* $E' = \{(u, v) : u, v \in V,\ u$ *and* v *are* visible *to each other with respect to* $E\}$. *It is a simple and unweighted graph.*

In Definition 1, we may think of I as of the set of obstacles. The nodes of $VG(I)$ are the vertices of I, and there is an edge between vertices u and v if they can *see* each other, that is, if the line segment $\overline{uv}$ does not intersect the interior of any obstacle in I. We say that the endpoints of an obstacle edge see each other. Hence, the obstacle edges form a subset of the edges of $VG(I)$, and thus $I \subseteq VG(I)$.

Definition 2 (Euclidean Visibility Graph of I). *The* Euclidean visibility graph $EVG(I)$ *is the visibility graph of I, where each edge $\overline{uv}$ $(u, v \in V)$ is assigned weight $w(u, v)$ that is equal to the Euclidean distance between u and v.*

Definition 3 (Constrained Visibility Graph of I). *The* constrained visibility graph $CVG(I)$ *is $EVG(I)$, where each edge of E is assigned weight 0.*

We use the notation $MST(G)$ to refer to a minimum spanning tree of the graph G. We assume that each edge of G has weight equal to its Euclidean length, unless the edge was specifically assigned the weight 0 by our algorithm. If none of the edges of G are assigned the weight 0 then $MST(G)$ is a Euclidean minimum spanning tree of G.

Definition 4 (Constrained Minimum Spanning Tree of F). *Given a plane forest $F = (V, E)$, the* constrained minimum spanning tree $CMST(F)$ *is the minimum spanning tree of $CVG(F)$.*

We assume that all the distances between any two vertices of V are distinct, otherwise, any ties can be broken using lexicographic ordering. This assumption implies that there is a unique MST and a unique $CMST$.

Since each edge of a plane forest F has weight zero in $CVG(F)$, by running Kruskal on $CVG(F)$, we get $F \subseteq CMST(F)$. Notice also that if F is a plane tree, then $F = CMST(F)$.

Definition 5 (Locally Delaunay criterion). *Let G be a triangulation, $\overline{v_1 v_2}$ be an edge in G (but not an edge of the convex hull of G), and $\triangle(v_1, v_2, v_3)$ and $\triangle(v_1, v_2, v_4)$ be the triangles adjacent to $\overline{v_1 v_2}$ in G. We say that $\overline{v_1 v_2}$ is a locally Delaunay edge if the circle through $\{v_1, v_2, v_3\}$ does not contain v_4 or equivalently if the circle through $\{v_1, v_2, v_4\}$ does not contain v_3. Every edge of the convex hull of G is also considered to be locally Delaunay [3].*

Definition 6 (Constrained Delaunay Triangulation of I). *The con-strained Delaunay triangulation $CDT(I)$ is the unique triangulation of V such that each edge is either in E or locally Delaunay. It follows that $I \subseteq CDT(I)$.*

This definition is equivalent to the classical definition used for example by Chew in [1]: $CDT(I)$ is the unique triangulation of V such that each edge e is either in E or there exists a circle C with the following properties:

1. The endpoints of edge e are on the boundary of C.
2. Any vertex of I in the interior of C is not visible to at least one endpoint of e.

The equivalence between the two definitions was shown by Lee and Lin [8].

When considering edge weights of $CDT(I)$, we assume that the weight of each edge is equal to the Euclidean distance between the endpoints of this edge.

The relative neighbourhood graph (RNG) was introduced by Toussaint in 1980 as a way of defining a structure from a set of points that would match human perceptions of the shape of the set [10]. An RNG is an undirected graph defined on a set of points in the Euclidean plane by connecting two points u and v by an edge if there does not exist a third point p that is closer to both u and v than they are to each other. Formally, we can define RNG through the concept of a *lune*. Let $D(x, r)$ denote an open disk centered at x with radius r, i.e., $D(x, r) = \{y : dist(x, y) < r\}$. Let $L_{u,v} = D(u, dist(u, v)) \cap D(v, dist(u, v))$; $L_{u,v}$ is called a *lune*.

Definition 7 (Relative Neighbourhood graph of V). *Given a set V of points, the* Relative Neighbourhood graph of V, $RNG(V)$, *is the graph with vertex set V and the edges of $RNG(V)$ are defined as follows: $\overline{uv}$ is an edge if and only if $L_{u,v} \cap V = \emptyset$.*

Definition 8 (Constrained Relative Neighbourhood graph of I). *The constrained Relative Neighbourhood graph, $CRNG(I)$, is defined as the graph with vertices V and the set E' of edges such that each edge $e = \overline{uv}$ is either in E or, u and v are visible to each other and $L_{u,v}$ does not contain points in V visible from both u and v. It follows that $I \subseteq CRNG(I)$.*

Gabriel graphs were introduced by Gabriel and Sokal in the context of geographic varia-tion analysis [4]. The Gabriel graph of a set V of points in the Euclidean plane expresses the notion of proximity of those points. It is the graph with vertex set V in which any points u and v of V are connected by an edge if $u \neq v$ and the closed disk with $\overline{uv}$ as a diameter contains no other point of V.

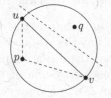

Fig. 1. Removal of p makes $\overline{uv}$ locally Gabriel.

Definition 9 (Locally Gabriel criterion). *The edge $\overline{uv}$ of the plane graph $G = (V, E)$ is said to be locally Gabriel if the vertices u and v are visible to each other and the circle with $\overline{uv}$ as a diameter does not contain any points in V which are visible from both u and v. Refer to Fig. 1.*

Definition 10 (Constrained Gabriel graph of I). *The* constrained Gabriel graph $CGG(I)$ *is defined as the graph with vertices V and the set E' of edges such that each edge is either in E or locally Gabriel. It follows that $I \subseteq CGG(I)$.*

Relative neighbourhood and Gabriel graphs are special cases of a parametrized family of neighbourhood graphs called β-skeletons (defined by Kirkpatrick and Radke in [7]). The neighbourhood $U_{u,v}(\beta)$ is defined for any fixed β $(1 \le \beta < \infty)$ as the intersection of two disks (refer to Fig. 7):

$$U_{u,v}(\beta) = D\left(\left(1 - \frac{\beta}{2}\right)u + \frac{\beta}{2}v, \frac{\beta}{2}dist(u,v)\right) \cap D\left(\left(1 - \frac{\beta}{2}\right)v + \frac{\beta}{2}u, \frac{\beta}{2}dist(u,v)\right)$$

Definition 11 ((lune-based) β-skeleton of V). *Given a set V of points in the plane, the* (lune-based) β-skeleton of V, *denoted $G_\beta(V)$ is the graph with vertex set V and the edges of $G_\beta(V)$ are defined as follows: $\overline{uv}$ is an edge if and only if $U_{u,v}(\beta) \cap V = \emptyset$.*

Notice that $RNG(V)$ is a β-skeleton of V for $\beta = 2$; namely $RNG(V) = G_2(V)$. Similarly, $GG(V) = G_1(V)$.

Definition 12 (Constrained β-skeleton of I). *The* constrained β-skeleton of I, $CG_\beta(I)$ *is the graph with vertex set V and edge set E' defined as follows: $e = \overline{uv} \in E'$ if and only if $e \in E$ or u and v are visible to each other and $U_{u,v}(\beta)$ does not contain points in V which are visible from both u and v.*

3   CMST Algorithm

Problem 1: Let a plane forest $F = (V, E)$ with $|V| = n$ points be given. Find the minimum set $S \subseteq E$ of edges such that $F \subseteq CMST(V, S)$.

Putting it differently, we want to find the smallest subset S of edges of F such that $CMST(F)$ is *equal* to $CMST(V, S)$, although the weights of the two trees may be different. Recall, that $CMST(F) = CMST(V, E)$ is the minimum spanning tree of the weighted graph $CVG(V, E)$ where each edge of E is assigned weight 0, and every other edge is assigned a weight equal to its Euclidean length.

Let us begin by considering an example. We are given a tree $F = (\{v_1, v_2, v_3\}, \{\overline{v_1 v_2}, \overline{v_2 v_3}\})$ (refer to Fig. 2(a)). Figure 2(b) shows $CDT(F)$. Observe that $CDT(F) = DT(\{v_1, v_2, v_3\})$. In other words, $CDT(F) = CDT(\{v_1, v_2, v_3\}, \emptyset)$ and thus no constraints are required to construct $CDT(F)$. However, this is not the case with $CMST(F)$. Obviously $MST(CDT(F)) \ne CMST(F)$ (refer to Fig. 2(c)), because $F \nsubseteq MST(CDT(F))$. We need to identify the minimum set $S \subseteq F$ of edges such that $F = CMST(V, S)$. In this example $S = \{\overline{v_1 v_2}, \overline{v_2 v_3}\}$.

A first idea is to construct an MST of $EVG(V, \emptyset)$. Every edge of F that is not part of $MST(EVG(V, \emptyset))$ should be forced to appear in $CMST(F)$. If we do this by adding each such edge of F to S (recall that every edge in S has weight 0) then, unfortunately, some edges of F, that were part of $MST(EVG(V, \emptyset))$,

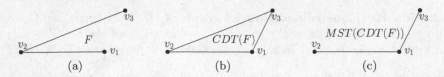

Fig. 2. Example showing relationship between input graph, its CDT and $MST(CDT)$. **(a)** Input graph F. **(b)** No constraints are required to construct $CDT(F)$. **(c)** $F \neq MST(CDT(F))$. We need two constraints $S = \{\overline{v_1 v_2}, \overline{v_2 v_3}\}$.

will no longer be part of the MST of the updated graph. A correct approach is to start with $MST(EVG(V, \emptyset))$ and *eliminate* every edge that is not part of F and does not connect two disconnected components of F. Each such edge $e \in MST(EVG(V, \emptyset))$ creates a cycle c_e in $F \cup \{e\}$ and we have that $c_e \subseteq EVG(V, \emptyset)$. If e becomes the heaviest edge of c_e then it will no longer be part of MST. Thus, we add to S every edge of c_e that is heavier than e. Although this approach gives us a set S such that $F \subseteq CMST(V, S)$, the set S of edges with weight 0 may not be minimal. Consider the example of Fig. 3. We are given a tree $F = (\{v_1, v_2, v_3, v_4\}, \{\overline{v_1 v_2}, \overline{v_2 v_3}, \overline{v_3 v_4}\})$ (refer to Fig. 3(a)). Every edge on the path from v_1 to v_4 in F is heavier than $\overline{v_1 v_4}$ - an edge of $MST(EVG(V, \emptyset)) \backslash F$. In order to eliminate $\overline{v_1 v_4}$ from the MST we assign the weight 0 to all the edges of the path $c_{\overline{v_1 v_4}} \backslash \overline{v_1 v_4}$, i.e. $S = \{\overline{v_1 v_2}, \overline{v_2 v_3}, \overline{v_3 v_4}\}$. However, it is sufficient to assign weight 0 only to the edge $\overline{v_2 v_3}$. In this case, $CMST(V, \{\overline{v_2 v_3}\}) = F$.

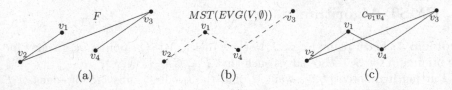

Fig. 3. Counterexample to the optimality of $S = \{\overline{v_1 v_2}, \overline{v_2 v_3}, \overline{v_3 v_4}\}$. The set $S' = \{\overline{v_2 v_3}\}$ is optimal. **(a)** Input graph $F = (V, E)$. **(b)** MST of V. **(c)** Cycle $c_{\overline{v_1 v_4}}$ of the graph $F \cup \{\overline{v_1 v_4}\}$. Every edge on the path from v_1 to v_4 in F is heavier than $\overline{v_1 v_4}$.

Nevertheless, this approach is correct when applied to the MST of a different graph. Instead of considering edges of $MST(EVG(V, \emptyset))$ we apply our idea to $MST(EVG(F))$. Notice that $EVG(F)$ does not have edges that intersect edges of F, and thus we will not encounter cases similar to the example of Fig. 3. Now it may look like we will be missing important information by considering only a subset of $VG(V, \emptyset)$. Can we guarantee that $CMST(V, S)$ will not contain edges that intersect edges of $F \setminus S$? To answer this question, we prove the following statement: $CMST(V, S) \subseteq VG(F)$ (Lemma 1). The basic algorithm for constructing S is given below. We prove its optimality by showing minimality of S (Lemma 3). Later, we present an efficient implementation of this algorithm.

Algorithm 1. S construction for $CMST$

Input: plane forest $F = (V, E)$
Output: minimum set $S \subseteq E$ of constraints such that $F \subseteq CMST(V, S)$

1 Construct $T' = MST(EVG(F))$;   // **we show** $T' = MST(CDT(F))$
2 Initialize $S = \emptyset$;
3 **foreach** $e' \in T'$ **do**
4   **if** $F \cup \{e'\}$ *creates a cycle* $c_{e'}$ **then**
5   **foreach** $e \in c_{e'}$ **do**
6   **if** $w(e) > w(e')$ **then**
7   set $S \leftarrow S \cup \{e\}$

We show the correctness of Algorithm 1 by proving the following lemmas. We start by observing an interesting property of the edges of F that were **not** added to S during the execution of the algorithm. The proof of the property can be found in our full paper.

Property 1. Let S be the output of Algorithm 1 on the input plane forest $F = (V, E)$. Let $T' = MST(EVG(F))$. If $e = \overline{uv} \in F$ and $e \notin S$ then $e \in T'$.

Lemma 1. *Let S be the output of Algorithm 1 on the input plane forest $F = (V, E)$. We have $CMST(V, S) \subseteq VG(F)$.*

Proof. Let $e^* = \overline{ab}$ be an arbitrary edge of $CMST(V, S)$. Assume to the contrary that $e^* \notin VG(F)$ (and hence $e^* \notin F$ and $e^* \notin S$). Thus there exists an edge of F that intersects e^*. Notice, that this edge cannot be in S.

Let k ($1 \le k \le n$) be the number of edges of F that intersect e^*. Let $e_i = \overline{c_i d_i} \in F$ be the edge that intersects e^* at point x_i, where i ($0 \le i < k$) represents an ordering of edges e_i according to the length $|\overline{ax_i}|$. In other words, the intersection point between e^* and e_0 is the closest to a among other edges of F that intersect e^*. Refer to Fig. 4.

We prove this lemma in three steps. First we derive some properties of e_i. Then we show that both endpoints of e^* are outside the disk with e_i as a diameter for every $0 \le i < k$. We finalize the proof by showing that $e^* \notin CGG(V, S)$ and thus by Lemma 9 (establishing that $CMST(V, S) \subseteq CGG(V, S)$, refer to Sect. 5) we have $e^* \notin CMST(V, S)$. This contradicts the definition of e^* which leads to the conclusion that $e^* \in VG(F)$ (meaning that the intersection between e^* and an edge of F is not possible). Refer to our full paper for detailed proof.   □

Lemma 2. *Let S be the output of Algorithm 1 on the input plane forest $F = (V, E)$. We have $F \subseteq CMST(V, S)$.*

Lemma 3. *Let S be the output of Algorithm 1 on the input plane forest $F = (V, E)$. The set S is minimal and minimum.*

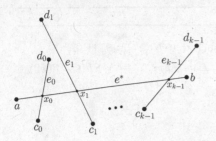

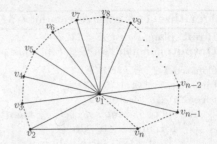

Fig. 4. Intersection between $e^* \in CMST(V, S)$ and k edges of F. Notice that the points $x_0, \ldots, x_{k-1}$ do not belong to V.

Fig. 5. Worst case example showing $n - 1$ constraints. The input tree F is drawn using solid lines. The MST of the set $\{v_1, v_2, \ldots v_n\}$ is dashed.

Lemmas 2 and 3 show the correctness of Algorithm 1 (see proofs in our full paper). However, we said nothing about our strategy of finding cycles in the graph. With a naive approach lines 3–7 of the algorithm could be quadratic in n. Also, the size of the visibility graph of F can be quadratic in the size of V, leading to the complexity of line 1 of the algorithm equal to $O(n^2 \log n)$. Our first step to improve the running time is to reduce the size of the graph we construct MST for. We prove that $MST(EVG(F)) \subseteq CDT(F)$ (refer to our full paper). It is then sufficient to construct $CDT(F)$, whose size is $O(n)$. The running time of line 1 then becomes $O(n \log n)$. Moreover, if F is a plane tree then the construction of $CDT(F)$ can be performed in $O(n)$ time [2].

We use the Link/Cut Tree of Sleator and Tarjan [9] to develop an efficient solution for lines 3–7 of Algorithm 1. Refer to our full paper for a detailed description and implementation. The complexity of the algorithm becomes $O(n \log n)$.

If $F \subseteq MST(V)$ then $F \subseteq CMST(V, \emptyset)$. In other words we do not require constraints at all to obtain Constrained MST that will contain F. It is interesting to consider the opposite problem. How big can the set of constraints be? Fig. 5 shows the worst-case example, where the set S of constraints contains all the edges of F, thus $|S| = n - 1$.

4   Constrained β-skeleton Algorithm

Problem 2: We are given a plane graph $I = (V, E)$ of $|V| = n$ points and $1 \leq \beta \leq 2$. Find the minimum set $S \subseteq E$ of edges such that $I \subseteq CG_\beta(V, S)$. In other words, we are interested in the minimum S such that $CG_\beta(V, S) = CG_\beta(I)$.

For the constrained Gabriel graph, the problem can be solved in a simpler way. We can decide in constant time whether or not the edge $e \in I$ should be a constraint in CGG that contains I by considering at most two triangles adjacent to e in $CDT(I)$. We exploit the fact that S, constructed of all non locally Gabriel edges of I, is necessary and sufficient. Refer to our full paper for details.

Let u, v and p be a triple of vertices of V. Recall the definition of $U_{u,v}(\beta)$; see Sect. 2 and Fig. 6(a). If $p \in U_{u,v}(\beta)$ and p is visible to both u and v, then we say that the vertex p *eliminates* line segment $\overline{uv}$. We prove in Lemma 9 (refer to Sect. 5) that $CRNG(I) = CG_{\beta=2}(I) \subseteq CG_{1 \leq \beta \leq 2}(I) \subseteq CG_{\beta=1}(I) = CGG(I) \subseteq CDT(I)$. The following lemmas further explain a relationship between $CG_\beta(I)$ and $CDT(I)$. Proofs of all the Lemmas of this section are in our full paper.

Lemma 4. *Given a plane graph $I = (V, E)$ and $1 \leq \beta \leq 2$. Let $p \in V$ be the closest vertex to the edge $\overline{uv} \in CDT(I)$ that eliminates $\overline{uv}$. Then $\triangle(u, v, p) \cap V = \emptyset$. Refer to Fig. 6.*

The above lemma implies that if there exists an edge $e^* \in CDT(I)$ that lies between p and $\overline{uv}$, i.e. e^* intersects the interior of $\triangle(u, v, p)$, then e^* intersects both line segments $\overline{up}$ and $\overline{vp}$. Refer to Fig. 6(c). Notice, that e^* cannot intersect $\overline{uv}$, since $e^*, \overline{uv} \in CDT(I)$.

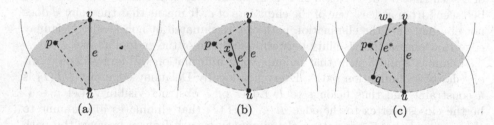

Fig. 6. The neighbourhood $U_{u,v}(\beta = 2)$ is highlighted in gray. (a) The vertex $p \in V$ eliminates $\overline{uv}$. (b) If p is the closest vertex to $\overline{uv} \in CDT(I)$ that eliminates $\overline{uv}$, then $\triangle(u, v, p) \cap V = \emptyset$. (c) If p is the closest vertex to $\overline{uv} \in CDT(I)$ that eliminates $\overline{uv}$, then $p \in U_{e^*}(\beta)$ for every edge of $e^* \in CDT(I)$ that lies between p and $\overline{uv}$.

Lemma 5. *Given a plane graph $I = (V, E)$ and $1 \leq \beta \leq 2$. Let $p \in V$ be the closest vertex to the edge $\overline{uv} \in CDT(I)$ that eliminates $\overline{uv}$. Let $e^* = \overline{qw}$ be an edge of $CDT(I)$ that intersects $\triangle(u, v, p)$. Then $p \in U_{q,w}(\beta)$. Ref. to Fig. 6(c).*

To solve our main problem we will use two geometric structures: *elimination path* and *elimination forest*, introduced by Jaromczyk and Kowaluk in [5]. The elimination path for a vertex p (starting from an adjacent triangle $\triangle(p, u, v) \in CDT(I)$) is an ordered list of edges, such that $p \in U_e(\beta)$ for each edge e of this list. In the work [5] an edge e belongs to the elimination path induced by some point p only if e is eliminated by p. In our problem this is not the case. The point p eliminates e if and only if $p \in U_e(\beta)$ and p is visible to both endpoints of e. We show how to adapt the original elimination forest to our problem later in this section. See our full paper for detailed definition of the elimination path.

Jaromczyk and Kowaluk show that a vertex cannot belong to the β-neighbourhood $U(\beta)$ of more than two edges for a particular triangle. Thus elimination paths do not split. Moreover, they also show that if two elimination paths have a common edge e and they both reached e via the same triangle, then starting from this edge one of the two paths is completely included in another one. This property is very important-it guarantees that the elimination forest can be constructed in linear time. Refer to [5,6] for further details.

Since we are dealing with CDT, the elimination paths defined via the original construction may split at non-locally Delaunay edges. To overcome this problem we terminate the propagation of the elimination path after a non-locally Delaunay edge is encountered and added to the path. Thus, the elimination forest for our problem can also be constructed in linear time. It is shown in Lemma 2.3 in [5] that if two points eliminate a common edge of a triangle in DT (such that both points are external to this triangle) then the two points can eliminate at most one of the remaining edges of this triangle. Similarly, we can show that if two points of V eliminate a common locally Delaunay edge e of an external triangle in CDT then they can eliminate at most one of the remaining edges of the triangle. It is due to the fact that there exists a circle that contains the endpoints of e such that if any vertex v of V is in the interior of the circle then it cannot be "seen" from at least one of the endpoints of e. It means that the point v does not eliminate e,– the elimination path of v terminated at non-locally Delaunay edge that obstructed visibility between v and one of the endpoints of e.

Lemmas 4 and 5 show that no important information will be lost as a result of "shorter" elimination paths. Every non-locally Delaunay edge of $CDT(I)$ is a constraint and thus belongs to I. Edges of I obstruct visibility. Let $p \in V$ be the closest vertex to the edge $\overline{uv} \in CDT(I)$ that eliminates $\overline{uv}$. Assume to the contrary that $\overline{uv}$ is not on the elimination path from p because the path terminated at non-locally Delaunay edge e^* before the path could reach $\overline{uv}$. By Lemma 4 the edge e^* intersects both line segments $\overline{pu}$ and $\overline{pv}$. Refer to Fig. 6(c). Since $e^* \in I$, neither u nor v are visible to the point p. This contradicts the fact that p eliminates $\overline{uv}$. Lemma 6 further shows that if some edge e of I must be a constraint in $CG_\beta(I)$ then it will belong to at least one elimination path, and in particular, to the path induced by the closest point to e that eliminates e.

Lemma 6. *Given a plane graph $I = (V, E)$ and $1 \leq \beta \leq 2$. Let $p \in V$ be the closest vertex to the edge $\overline{uv} \in CDT(I)$ that eliminates $\overline{uv}$. Then $\overline{uv}$ belongs to the elimination path induced by p.*

Each elimination path starts with a special node (we call it a leaf) that carries information about the vertex that induced the current elimination path. A node that corresponds to the last edge of a particular elimination path also carries information about the vertex that started this path. The elimination forest is build from bottom (leaves) to top (roots).

The elimination forest (let us call it ElF) gives us a lot of information, but we still do not know how to deal with visibility. The elimination path induced by point p can contain locally Delaunay edges of I that may obstruct visibility between p and other edges that are further on the path. We want to identify edges that not only belong to the elimination path of some vertex p but whose both endpoints are also visible to p. Observe, that visibility can only be obstructed by edges of I. Let us contract all the nodes of the ElF that correspond to edges not in I. If a particular path is completely contracted, we delete its corresponding leaf as well. Now the ElF contains only nodes of edges that belong to I together with leaves, that identify elimination paths, that originally had at least one edge of I. The correctness of our approach is supported by the following lemma.

Lemma 7. *Given a plane graph $I = (V, E)$, $1 \leq \beta \leq 2$ and a contracted ElF of $CDT(I)$. There exists a vertex of V that eliminates $\overline{uv} \in I$ **if and only if** the node $n_{\overline{uv}}$ of the contracted ElF has a leaf attached to it.*

We are ready to present an algorithm that finds the minimum set $S \subseteq E$ of edges such that $I \subseteq CG_\beta(V, S)$ for constrained β-skeletons ($1 \leq \beta \leq 2$):

Algorithm 2. S construction for constrained β-skeletons

Input: plane graph $I = (V, E)$ and $1 \leq \beta \leq 2$
Output: minimum set $S \subseteq E$ of constraints such that $I \subseteq CG_\beta(V, S)$

1 Construct $CDT(I)$;
2 Initialize $S = \emptyset$;
3 Construct Elimination Forest (ElF) of $CDT(I)$;
4 **foreach** $e \notin E$ **do**
5 $\quad$ contract the node that corresponds to e in ElF;
6 $\quad$ **if** *a particular path is about to be contracted in full* **then**
7 $\quad\quad$ delete its corresponding leaf from ElF

8 **foreach** *node n_e (that corresponds to edge e) of Contracted ElF* **do**
9 $\quad$ **if** n_e *has an immediate leaf attached to it* **then**
10 $\quad\quad$ set $S \leftarrow S \cup \{e\}$;

Let us discuss the correctness of Algorithm 2. Let S be the output of the algorithm on the input plane graph $I = (V, E)$. Notice that the following is true: $I \subseteq CG_\beta(V, S)$. Every edge of E that belongs to S also belongs to $CG_\beta(V, S)$ by definition. According to the algorithm, every edge of $E \setminus S$ does not have a leaf attached to a corresponding node in contracted ElF. By Lemma 7 none of those edges has a vertex that eliminates it.

Lemma 8. *Let S be the output of Algorithm 2 on the input plane graph $I = (V, E)$. The set S is minimum.*

The running time of Algorithm 2 depends on the complexity of the first line. Lines 2–10 can be performed in $O(n)$ time. In the worst case the construction of $CDT(I)$ can take $O(n \log n)$ time. But for some types of input graph I this time can be reduced. If I is a tree or a polygon, then $CDT(I)$ can be constructed in $O(n)$ time. If I is a triangulation, then $I = CDT(I)$ and thus the first line is accomplished in $O(1)$ time.

5 Hierarchy

Proximity graphs form a nested hierarchy, a version of which was established in [10]:

Theorem 1 (Hierarchy). *In any L_p metric, for a fixed set V of points and $1 \leq \beta \leq 2$, the following is true: $MST \subseteq RNG \subseteq G_\beta \subseteq GG \subseteq DT$.*

We show that proximity graphs preserve the above hierarchy in the constraint setting (refer to Lemma 9). We also show that the minimum set of constraints required to reconstruct a given plane graph (as a part of each of those neighbourhood graphs) form an inverse hierarchy (refer to Lemma 10). See the proofs of both lemmas in our full paper.

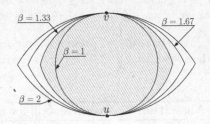

Fig. 7. Lune-based neighbourhoods for $1 \leq \beta \leq 2$, where $U_{u,v}(\beta = 1.33)$ is dashed.

Lemma 9. *Given a plane forest $F = (V, E)$ and $1 \leq \beta \leq 2$, $CMST(F) \subseteq CRNG(F) \subseteq CG_\beta(F) \subseteq CGG(F) \subseteq CDT(F)$. Given a plane graph $I = (V, E)$ and $1 \leq \beta \leq 2$, $CRNG(I) \subseteq CG_\beta(I) \subseteq CGG(I) \subseteq CDT(I)$.*

Lemma 10. *Let S_G denote the the minimum set of constraints of G. Given a plane graph $I = (V, E)$ and $1 \leq \beta \leq 2$, $S_{CRNG(I)} \supseteq S_{CG_\beta(I)} \supseteq S_{CGG(I)} \supseteq S_{CDT(I)}$. Given a plane forest $F = (V, E)$ and $1 \leq \beta \leq 2$, $S_{CMST(F)} \supseteq S_{CRNG(F)} \supseteq S_{CG_\beta(F)} \supseteq S_{CGG(F)} \supseteq S_{CDT(F)}$.*

References

1. Chew, L.P.: Constrained delaunay triangulations. Algorithmica **4**(1), 97–108 (1989)
2. Chin, F., Wang, C.A.: Finding the constrained Delaunay triangulation and constrained Voronoi diagram of a simple polygon in linear time. SIAM J. Comput. **28**(2), 471–486 (1998)
3. Devillers, O., Estkowski, R., Gandoin, P.-M., Hurtado, F., Ramos, P.A., Sacristán, V.: Minimal set of constraints for 2D constrained Delaunay reconstruction. Int. J. Comput. Geometry Appl. **13**(5), 391–398 (2003)
4. Gabriel, K.R., Sokal, R.R.: A new statistical approach to geographic variation analysis. Syst. Zool. **18**(3), 259–278 (1969)
5. Jaromczyk, J.W., Kowaluk, M.: A note on relative neighborhood graphs. In: SoCG, pp. 233–241 (1987)
6. Jaromczyk, J.W., Kowaluk, M., Yao, F.: An optimal algorithm for constructing β-skeletons in l_p metric. Manuscript (1989)
7. Kirkpatrick, D.G., Radke, J.D.: A framework for computational morphology. In: Computational Geometry, vol. 2, pp. 217–248. Machine Intelligence and Pattern Recognition, North-Holland (1985)

8. Lee, D.T., Lin, A.K.: Generalized Dalaunay triangualtion for planar graphs. Discrete Comput. Geometry **1**, 201–217 (1986)
9. Sleator, D.D., Tarjan, R.E.: A data structure for dynamic trees. In: STOC, pp. 114–122. ACM (1981)
10. Toussaint, G.T.: The relative neighbourhood graph of a finite planar set. Pattern Recogn. **12**(4), 261–268 (1980)

Plane Bichromatic Trees of Low Degree

Ahmad Biniaz[⊠], Prosenjit Bose, Anil Maheshwari, and Michiel Smid

Carleton University, Ottawa, Canada
ahmad.biniaz@gmail.com

Abstract. Let R and B be two disjoint sets of points in the plane such that $|B| \leqslant |R|$, and no three points of $R \cup B$ are collinear. We show that the geometric complete bipartite graph $K(R, B)$ contains a non-crossing spanning tree whose maximum degree is at most $\max\left\{3, \left\lceil \frac{|R|-1}{|B|} \right\rceil + 1\right\}$; this is the best possible upper bound on the maximum degree. This solves an open problem posed by Abellanas *et al.* at the Graph Drawing Symposium, 1996.

1 Introduction

Let R and B be two disjoint sets of points in the plane. We assume that the points in R are colored red and the points in B are colored blue. Problems related to bichromatic inputs have been studied extensively in computational geometry, e.g., red-blue intersection [3,15], red-blue separation [4–6,9], and red-blue connection problems [2,6,8,10,11,14]. As for an overview, see the excellent survey of Kaneko and Kano [12]. In this paper we study non-crossing bichromatic spanning trees, which is a red-blue connection problem.

We assume that $R \cup B$ is in *general position*, i.e., no three points of $R \cup B$ are collinear. The *geometric complete bipartite graph* $K(R, B)$ is the graph whose vertex set is $R \cup B$ and whose edge set consists of all the straight-line segments connecting a point in R to a point in B. A *bichromatic tree* on $R \cup B$ is a spanning tree in $K(R, B)$. A *plane bichromatic tree* is a bichromatic tree whose edges do not intersect each other in their interior. A *d-tree* is defined to be a tree whose maximum vertex degree is at most d.

If $R \cup B$ is in general position, then it is possible to find a plane bichromatic tree on $R \cup B$ as follows. Take any red point and connect it to all the blue points. Extend the resulting edges from the blue endpoints to partition the plane into cones. Then, connect the remaining red points in each cone to a suitable blue point on the boundary of that cone without creating crossings. This simple solution produces trees possibly with large vertex degree. In this paper we are interested in computing a plane bichromatic tree on $R \cup B$ whose maximum vertex degree is as small as possible. This problem was first mentioned by Abellanas *et al.* [1] in the Graph Drawing Symposium in 1996:

Problem 1. *Given two disjoint sets R and B of points in the plane, with $|B| \leqslant |R|$, find a plane bichromatic tree on $R \cup B$ having maximum degree $O(|R|/|B|)$.*

Research supported by NSERC.

© Springer International Publishing Switzerland 2016
V. Mäkinen et al. (Eds.): IWOCA 2016, LNCS 9843, pp. 68–80, 2016.
DOI: 10.1007/978-3-319-44543-4_6

Assume $|B| \leqslant |R|$. Any bichromatic tree on $R \cup B$ has $|R| + |B| - 1$ edges. Moreover, each edge is incident on exactly one blue point. Thus, the sum of the degrees of the blue points is $|R| + |B| - 1$. This implies that any bichromatic tree on $R \cup B$ has a blue point of degree at least $\frac{|R|+|B|-1}{|B|} = \frac{|R|-1}{|B|} + 1$. Since the degree is an integer, $\left\lceil \frac{|R|-1}{|B|} \right\rceil + 1$ is the best possible upper bound on the maximum degree.

For cases when $|R| = |B|$ or $|R| = |B| + 1$ it may not always be possible to compute a plane bichromatic tree of degree $\left\lceil \frac{|R|-1}{|B|} \right\rceil + 1 = 2$, i.e., a plane bichromatic path; see the example in the figure on the right which is borrowed from [2]; by adding one red point to the top red chain, an example for the case when $|R| = |B| + 1$ is obtained. In 1998, Kaneko [11] posed the following conjecture.

Conjecture 1 (Kaneko [11]). Let R and B be two disjoint sets of points in the plane such that $|B| \leqslant |R|$ and $R \cup B$ is in general position, and let $k = \left\lceil \frac{|R|}{|B|} \right\rceil$. If $k \geqslant 2$, then there exists a plane bichromatic tree on $R \cup B$ whose maximum vertex degree is at most $k + 1$.

1.1 Previous Work

Assume $|B| \leqslant |R|$ and let $k = \left\lceil \frac{|R|}{|B|} \right\rceil$. Abellanas *et al.* [2] proved that there exists a plane bichromatic tree on $R \cup B$ whose maximum vertex degree is $O(k+\log |B|)$. Kaneko [11] proved the existence of such a tree of degree at most $3k$. For the case where $|R| = |B|$, i.e. $k = 1$, Kaneko [11] showed how to construct a plane bichromatic tree of maximum degree three.

Abellanas *et al.* [2] considered the problem of computing a low degree plane bichromatic tree on some restricted point sets. They proved that if $R \cup B$ is in convex position and $|R| = k|B|$, with $k \geqslant 1$, then $R \cup B$ admits a plane bichromatic tree of maximum degree $k + 2$. If R and B are linearly separable and $|R| = k|B|$, with $k \geqslant 1$, they proved that $R \cup B$ admits a plane bichromatic tree of maximum degree $k + 1$. They also obtained a degree of $k + 1$ for the case when B is equal to the set of points on the convex hull of $R \cup B$.

Kano et al. [13] considered the problem of computing a spanning tree (not necessarily plane) of low degree in a (not necessarily complete) connected bipartite graph G with bipartition (R, B). They showed that if $|B| \leqslant |R| \leqslant k|B| + 1$, with $k \geqslant 1$, and $\sigma(G) \geqslant |R|$, then G has a spanning $(k + 1)$-tree, where $\sigma(G)$ denotes the minimum degree sum of $k + 1$ independent vertices of G.

The problem of computing a plane tree of low degree on multicolored point sets (with more than two colors) is considered in [6,14].

A related problem is to compute a plane bichromatic Euclidean minimum spanning tree on $R \cup B$. This problem is NP-hard [8]. The best polynomial-time algorithm known so far for this problem has approximation factor $O(\sqrt{n})$, where n is the total number of points [8].

1.2 Our Contribution

Our main result is the following theorem that is even sharper than Conjecture 1.

Theorem 1. *Let R and B be two disjoint sets of points in the plane such that $|B| \leqslant |R|$ and $R \cup B$ is in general position, and let $\delta = \left\lceil \frac{|R|-1}{|B|} \right\rceil$. Then, there exists a plane bichromatic tree on $R \cup B$ whose maximum vertex degree is at most $\max\{3, \delta + 1\}$; this is the best possible upper bound on the maximum degree.*

The core of our contribution is given in Sect. 2, where we partially prove Conjecture 1: If $|R| = k|B|$, with $k \geqslant 2$, and $R \cup B$ is in general position, then there exists a plane bichromatic tree on $R \cup B$ whose maximum degree is $k + 1$. We present a constructive proof for obtaining such a tree. Based on the algorithm of Sect. 2, we prove the full Conjecture 1 and Theorem 1 in the full version of this paper (see [7]). As we will see, the proofs are simple for $\delta \geqslant 4$. However, for smaller values of δ, the proofs are much more involved.

2 Plane Bichromatic $(k + 1)$-trees

In this section we prove Conjecture 1 for the case when $|R| = k|B|$ and $k \geqslant 2$:

Theorem 2. *Let R and B be two disjoint sets of points in the plane, such that $|R| = k|B|$, with $k \geqslant 2$, and $R \cup B$ is in general position. Then, there exists a plane bichromatic tree on $R \cup B$ whose maximum vertex degree is at most $k + 1$.*

Note that any bichromatic tree on $R \cup B$ has $|B| + |R| - 1 = (k+1)|B| - 1$ edges. Since each edge is incident to exactly one blue point, the sum of the degrees of the blue points is $(k + 1)|B| - 1$. This implies the following observation:

Observation 1. *Let R and B be disjoint sets of points in the plane such that $|R| = k|B|$, with $k \geqslant 1$ is an integer. Then, in any bichromatic $(k + 1)$-tree on $R \cup B$, one point of B has degree k and each other point of B has degree $k + 1$.*

In order to prove Theorem 2 we provide some notation and definitions. Let P be a set of points in the plane. We denote by $CH(P)$ the convex hull of P. For two points p and q in the plane, we denote by (p, q) the line segment whose endpoints are p and q. Moreover, we denote by $\ell(p, q)$ the line passing through p and q. For a node p in a tree T we denote by $d_T(p)$ the degree of p in T. Let p be a vertex of $CH(P)$. The *radial ordering* of $P - \{p\}$ around p is obtained as follows. Let p_1 and p_2 be the two vertices of $CH(P)$ adjacent to p such that the clockwise angle $\angle p_1 p p_2$ is less than π. For each point q in $P - \{p\}$, define its angle around p—with respect to p_1— to be the clockwise angle $\angle p_1 p q$. Then the desired radial ordering is obtained by ordering the points in $P - \{p\}$ by increasing angle around p.

We start by proving two lemmas that play important roles in the proof of Theorem 2.

Lemma 1. *Let R and B be two sets of red and blue points in the plane, respectively, such that $|B| \geqslant 1$, $k(|B| - 1) < |R| \leqslant k|B|$, with $k \geqslant 2$, and $R \cup B$ is in general position. Define $\alpha = |R| - k(|B| - 1)$, hence $0 < \alpha \leqslant k$. Let b_1, b, b_2 be blue points that are counter clockwise consecutive on $CH(R \cup B)$. Then, in the radial ordering of $R \cup B - \{b\}$ around b, there are α consecutive red points, $r_1, \ldots, r_\alpha$, such that $|R_1| = k|B_1| + 1$ and $|R_2| = k|B_2| + 1$, where R_1 (resp. B_1) is the set of red points (resp. blue points) of $R \cup B - \{b\}$ lying on or to the left of $\ell(b, r_1)$, and R_2 (resp. B_2) is the set of red points (resp. blue points) of $R \cup B - \{b\}$ lying on or to the right of $\ell(b, r_\alpha)$.*

Proof. By a suitable rotation of the plane, we may assume that b is the lowest point of $CH(R \cup B)$, and b_1 (resp. b_2) is to the left (resp. right) of the vertical line passing through b. Note that b_1 is the first point and b_2 is the last point in the clockwise radial ordering of $R \cup B - \{b\}$ around b. See Fig. 1. We define the function f as follows: For every point x in this radial ordering,

$$f(x) = k \cdot (\text{the number of points of } B - \{b\} \text{ lying on or to the left of } \ell(b, x))$$
$$- (\text{the number of points of } R \text{ lying on or to the left of } \ell(b, x)).$$

Based on this definition, we have $f(b_1) = k \geqslant$ 2 and $f(b_2) = k(|B| - 1) - |R| = -\alpha \leqslant -1$. Along this radial ordering, the value of f changes by $+k$ at every blue point and by -1 at every red point. Since $f(b_1) > 0 > f(b_2)$, there exists a point in the radial ordering for which f equals 0. Let v be the last point in the radial ordering where $f(v) = 0$. Since b_2 increases f by $+k$ and $f(b_2) \leqslant -1$, there are at least $k + 2$ points strictly after v in the radial ordering. Let $S = (r_1, \ldots, r_k)$ be the sequence of k points strictly after v in the radial ordering.

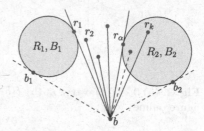

Fig. 1. k consecutive red points in the radial ordering around b. (Color figure online)

Claim: The points of S are red. For the sake of contradiction assume some points of S are blue. Let i be the minimum index in $\{1, \ldots, k\}$ where r_i is blue. Then, the $(i - 1)$ points of S that are before r_i are red. Thus,

$$f(r_i) = f(v) - (i - 1) + k = k - i + 1 > 0.$$

Since $f(r_i) > 0 > f(b_2)$, there exists a point between r_i and b_2 in the radial ordering for which f equals 0. This contradicts the fact that v is the last point in the radial ordering with $f(v) = 0$. This proves the claim.

Thus, each $r_i \in S$ is red. Moreover, $f(r_i) = -i$. We show that the subsequence $S' = (r_1, \ldots, r_\alpha)$ of S satisfies the statement of the lemma; note that, by definition, $\alpha \leqslant k$. Having r_1 and r_α, we define R_1, B_1, R_2 and B_2 as in the statement of the lemma. See Figure 1. By definition of f, we have $f(r_1) = k|B_1| - |R_1| = -1$, and hence $|R_1| = k|B_1| + 1$. Moreover,

$$|R_2| = |R| - |R_1| - |S'| + 2 = |R| - (k|B_1| + 1) - (|R| - k(|B| - 1)) + 2$$
$$= k(|B| - |B_1| - 1) + 1 = k|B_2| + 1,$$

where $|S'|$ is the number of elements in the sequence S'. Note that $R_2 = (R - (R_1 \cup S')) \cup \{r_\alpha\}$. Since r_1 belongs to both R_1 and S', and r_α belongs to R_2, the term "+2" in the first equality is necessary (even for the case when $r_1 = r_\alpha$). The last equality is valid because $B_2 = B - (B_1 \cup \{b\})$. This completes the proof of the lemma. □

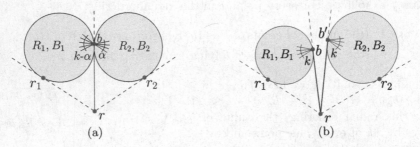

Fig. 2. (a) $0 < f(b) = \alpha < k$, (b) $f(b) = 0$, and b, b' are consecutive blue points. (Color figure online)

Lemma 2. *Let R and B be two sets of red and blue points in the plane, respectively, such that $|B| \geqslant 1$, $|R| = k|B| + 1$, with $k \geqslant 2$, and $R \cup B$ is in general position. Let r_1, r, r_2 be red points that are counter clockwise consecutive on $CH(R \cup B)$. Then, one of the following statements holds:*

1. *There exists a blue point b in the radial ordering of $(R \cup B) - \{r\}$ around r, such that $|R_1| = k(|B_1| - 1) + k - \alpha$ and $|R_2| = k(|B_2| - 1) + \alpha$, where R_1 (resp. B_1) is the set of red points (resp. blue points) of $(R \cup B) - \{r\}$ lying on or to the left of $\ell(r, b)$, R_2 (resp. B_2) is the set of red points (resp. blue points) of $(R \cup B) - \{r\}$ lying on or to the right of $\ell(r, b)$, and $0 < \alpha < k$.*
2. *There exist two consecutive blue points b and b' in the radial ordering of $(R \cup B) - \{r\}$ around r, such that $|R_1| = k|B_1|$ and $|R_2| = k|B_2|$, where R_1 (resp. B_1) is the set of red points (resp. blue points) of $(R \cup B) - \{r\}$ lying on or to the left of $\ell(r, b)$ and R_2 (resp. B_2) is the set of red points (resp. blue points) of $(R \cup B) - \{r\}$ lying on or to the right of $\ell(r, b')$.*

Proof. By a suitable rotation of the plane, we may assume that r is the lowest point of $CH(R \cup B)$, and r_1 (resp. r_2) is to the left (resp. right) of the vertical line passing through r. Note that r_1 is the first point and r_2 is the last point in the clockwise radial ordering of $(R \cup B) - \{r\}$ around r. See Fig. 2. We define the function f as follows: For every point x in this radial ordering,

$$f(x) = k \cdot (\text{the number of points of } B \text{ lying on or to the left of } \ell(r, x))$$
$$- (\text{the number of points of } R - \{r\} \text{ lying on or to the left of } \ell(r, x)).$$

Based on this definition, we have $f(r_1) = -1$ and $f(r_2) = k|B| - (|R| - 1) = 0$. Along this radial ordering, the value of f changes by $+k$ at every blue point and by -1 at every red point. Let v be the point before r_2 in the radial ordering. Since r_2 decreases f by -1 and $f(r_2) = 0$, we have $f(v) = 1$. Since $f(r_1) < 0 < f(v)$, there exists a point b between r_1 and v in the radial ordering such that $f(b) \geqslant 0$ and f is negative at b's predecessor. Let b be the last such point between r_1 and v; it may happen that $b = v$. Observe that b is blue and $f(b) < k$. We consider two cases, depending on whether $0 < f(b) < k$ or $f(b) = 0$.

- $0 < f(b) < k$. Since b is blue and r_2 is red, there is at least one point after b in the radial ordering. Define R_1, B_1, R_2 and B_2 as in the first statement of the lemma. See Fig. 2(a). Let $\alpha = f(b)$. By definition of f, we have $\alpha = f(b) = k|B_1| - |R_1|$, and hence $|R_1| = k|B_1| - \alpha = k(|B_1| - 1) + k - \alpha$. Moreover,

$$|R_2| = |R| - |R_1| - 1 = (k|B| + 1) - (k|B_1| - \alpha) - 1$$
$$= k(|B| - |B_1|) + \alpha = k(|B_2| - 1) + \alpha,$$

where the last equality is valid because b belongs to both B_1 and B_2.
- $f(b) = 0$. In the radial ordering there are at least $k + 1$ points after b since a red point only decreases the value of f and $f(r_2) = 0$. Let b' be the successor of b in the radial ordering. The point b' is blue: If b' is red, we have $f(b') = -1 < 0 < f(v)$ and, thus, there exists a point b'' between b' and v such that $f(b'') \geqslant 0$ and f is negative at the predecessor of b'', contradicting our choice of b. Define R_1, B_1, R_2 and B_2 as in the second statement of the lemma (Fig. 2(b)). Since $f(b) = k|B_1| - |R_1| = 0$, $|R_1| = k|B_1|$. Moreover,

$$|R_2| = |R| - |R_1| - 1 = (k|B| + 1) - k|B_1| - 1 = k(|B| - |B_1|) = k|B_2|,$$

which completes the proof of the lemma.

2.1   Proof of Theorem 2

We use Lemmas 1 and 2 to prove Theorem 2. Let R and B be two disjoint sets of points in the plane, such that $|R| = k|B|$, with $k \geqslant 2$, and $R \cup B$ is in general position. We will present an algorithm, plane-tree, that constructs a plane bichromatic tree of maximum degree $k + 1$ on $R \cup B$ such that each red vertex has degree at most 3. This algorithm uses two procedures, proc1 and proc2:

---

plane-tree(R, B)

Input: A set R of red points and a non-empty set B of blue points, where $|R| = k|B|$, with $k \geqslant 2$, and $R \cup B$ is in general position.

Output: A plane bichromatic $(k + 1)$-tree on $R \cup B$ such that each red vertex has degree at most 3.

---

proc1(R, B, b)

Input: A set R of red points, a non-empty set B of blue points, and a point $b \in B$, where $k(|B| - 1) < |R| \leqslant k|B|$, with $k \geqslant 2$, and b is on $CH(R \cup B)$.

Output: A plane bichromatic $(k+1)$-tree T on $R \cup B$ where $d_T(b) = |R| - k(|B| - 1)$ and each red vertex has degree at most 3.

---

proc2(R, B, r)

Input: A set R of red points, a non-empty set B of blue points, and a point $r \in R$, where $|R| = k|B| + 1$, with $k \geqslant 2$, and r is on $CH(R \cup B)$.

Output: A plane bichromatic $(k + 1)$-tree T on $R \cup B$ where $d_T(r) \in \{1, 2\}$ and each other red vertex has degree at most 3.

First we describe each of the procedures proc1 and proc2. Then we describe algorithm plane-tree. The procedures proc1 and proc2 will call each other. As we will see in the description of these procedures, when proc1 or proc2 is called recursively, the call is always on a smaller point set. We now describe the base cases for proc1 and proc2.

The base case for proc1 happens when $|B| = 1$, i.e., $B = \{b\}$. In this case, we have $1 \leqslant |R| \leqslant k$ and $2 \leqslant |R \cup B| \leqslant k + 1$. We connect all points of R to b, and return the resulting star as a desired tree T where $d_T(b) = |R|$ and each red vertex has degree 1.

The base case for proc2 happens when $|B| = 1$; let b be the only point in B. In this case, we have $|R| = k + 1$ and $|R \cup B| = k + 2$. We connect all points of R to b, and return the resulting star as a desired tree T where $d_T(b) = k + 1$ and each red vertex has degree 1.

In Sect. 2.1.1 we describe proc1(R, B, b), whereas proc2(R, B, r) will be described in Sect. 2.1.2. In these two sections, we assume that both proc1 and proc2 are correct for smaller point sets.

2.1.1   Procedure proc1

The procedure proc1(R, B, b) takes as input a set R of red points, a set B of blue points, and a point $b \in B$, where $|B| \geqslant 2$, $k(|B| - 1) < |R| \leqslant k|B|$, with $k \geqslant 2$, and b is on $CH(R \cup B)$. Let $\alpha = |R| - k(|B| - 1)$, and notice that $1 \leqslant \alpha \leqslant k$. This procedure computes a plane bichromatic $(k + 1)$-tree T on $R \cup B$ such that $d_T(b) = \alpha$ and each red vertex has degree at most 3. Depending on whether or not two points of $CH(R \cup B)$ adjacent to b belong to B, we have two cases.

Case 1: Both vertices of $CH(R \cup B)$ adjacent to b belong to B. We apply Lemma 1 on R, B, and b. Consider the α consecutive red points, $r_1, \ldots, r_\alpha$, and the sets R_1, R_2, B_1, and B_2 in the statement of Lemma 1. Note that r_1 is a red point on $CH(R_1 \cup B_1)$ and r_α is a red point on $CH(R_2 \cup B_2)$. We distinguish between two cases: $1 < \alpha \leqslant k$ and $\alpha = 1$.

Case 1.1: $1 < \alpha \leqslant k$. In this case $r_1 \neq r_\alpha$. Moreover $CH(R_1 \cup B_1)$ and $CH(R_2 \cup B_2)$ are disjoint. Let T_1 be the plane bichromatic $(k+1)$-tree obtained by running proc2 on R_1, B_1, r_1; note that $d_{T_1}(r_1) \in \{1, 2\}$, all other red points in T_1 have degree at most 3, and $|R_1 \cup B_1| < |R \cup B|$. Similarly, let T_2 be the plane bichromatic $(k+1)$-tree obtained by running proc2 on R_2, B_2, r_α; note that $d_{T_2}(r_\alpha) \in \{1, 2\}$, all other red points in T_2 have degree at most 3, and $|R_2 \cup B_2| < |R \cup B|$. Let S be the star obtained by connecting the vertices $r_1, \ldots, r_\alpha$ to b. Then, we obtain a desired tree $T = T_1 \cup T_2 \cup S$. See Fig. 1. T is a plane bichromatic $(k+1)$-tree on $R \cup B$ with $d_T(r_1) \in \{2, 3\}$, $d_T(r_\alpha) \in \{2, 3\}$, $d_T(b) = \alpha$, and $d_T(r_i) = 1$ where $1 < i < \alpha$.

Case 1.2: $\alpha = 1$. In this case $r_1 = r_\alpha$ and $|R| = k(|B| - 1) + 1$. Moreover, $r_1 \in R_1 \cap R_2$. If we handle this case as in the previous case, then it is possible for r_1 to be incident on two edges in each of T_1 and T_2, and incident on one edge in S. This makes $d_T(r_1) = 5$. If $k \geqslant 4$, then T is a desired $(k+1)$-tree. But, if $k = 2, 3$, then T would not be a $(k+1)$-tree. Thus, we handle the case when $\alpha = 1$ differently.

Let x_1 and y_1 be the two blue neighbors of b on $CH(R \cup B)$. By a suitable rotation of the plane, we may assume that b is the lowest point of $CH(R \cup B)$, and x_1 (resp. y_1) is to the left (resp. right) of the vertical line passing through b. Let $C_1 = (x_1, \ldots, x_j = r_1)$ be the sequence of points on the boundary of $CH(R_1 \cup B_1)$ from x_1 to r_1 that are visible from b. Similarly, define $C_2 = (y_1, \ldots, r_1)$ on $CH(R_2 \cup B_2)$. See Fig. 3. Let x_s be the first red point in the sequence C_1, and let y_t be the first red point in the sequence C_2. Note that $s, t \geqslant 2$. It is possible for x_s or y_t or both to be r_1. Consider the subsequences $C_1' = (x_1, \ldots, x_s)$ and $C_1'' = (x_s, \ldots, r_1)$ of C_1 as depicted in Fig. 3(a). Similarly, consider the subsequences $C_2' = (y_1, \ldots, y_t)$ and $C_2'' = (y_t, \ldots, r_1)$ of C_2. Let l_1 and l_2 be the lines passing through (x_{s-1}, x_s) and (y_{t-1}, y_t), respectively. l_1 is tangent to $CH(R_1 \cup B_1)$ and l_2 is tangent to $CH(R_2 \cup B_2)$.

We consider two cases, depending on whether or not l_1 intersects C_2 and l_2 intersects C_1.

Case 1.2.1: l_1 *does not intersect* C_2, *or* l_2 *does not intersect* C_1. Because of symmetry, we assume that l_1 does not intersect C_2. Note that in this case l_1 does not intersect the interior of $CH(R \cup B - \{b\})$. Let $R' = R$ and $B' = B - \{b\}$; note that $|R'| = |R| = k(|B| - 1) + 1 = k|B'| + 1$. In addition, x_s is on $CH(R' \cup B')$. See Fig. 3(a). Let T' be the plane bichromatic $(k+1)$-tree obtained by proc2(R', B', x_s). Note that $d_{T'}(x_s) \in \{1, 2\}$, all other red points in T' have degree at most 3, and $|R' \cup B'| < |R \cup B|$. We obtain a desired tree $T = T' \cup \{(b, x_s)\}$. T is a plane bichromatic $(k+1)$-tree on $R \cup B$ with $d_T(b) = \alpha = 1$ and $d_T(x_s) \in \{2, 3\}$.

Case 1.2.2: l_1 *intersects* C_2, *and* l_2 *intersects* C_1. We consider two cases:

Case 1.2.2.1: l_1 *intersects* C_2', *or* l_2 *intersects* C_1'. Because of symmetry, we assume that l_1 intersects C_2'. Let (y_i, y_{i+1}), with $1 \leqslant i < t$, be the leftmost edge of C_2' that is intersected by l_1 (note that l_1 may intersect two edges of C_2'). Observe that y_i is a blue point. Let $R_1' = R_1 - \{x_s\}$, $B_1' = B_1$, $R_2' = R_2 - \{r_1\}$ and $B_2' = B_2$ as shown in Fig. 3(b). Note that $|R_1'| = k|B_1'|$ and $|R_2'| = k|B_2'|$. In addition, $CH(R_1' \cup B_1')$ and $CH(R_2' \cup B_2')$ are disjoint, x_{s-1} is a blue point on $CH(R_1' \cup B_1')$, and y_i is a blue point on $CH(R_2' \cup B_2')$. Let T_1 be the plane bichromatic $(k+1)$-tree obtained by the recursive call $\mathsf{proc1}(R_1', B_1', x_{s-1})$, and let T_2 be the plane bichromatic $(k+1)$-tree obtained by the recursive call $\mathsf{proc1}(R_2', B_2', y_i)$. Note that $d_{T_1}(x_{s-1}) = k$, $d_{T_2}(y_i) = k$, all red points in T_1 and T_2 have degree at most 3, $|R_1' \cup B_1'| < |R \cup B|$, and $|R_2' \cup B_2'| < |R \cup B|$. We obtain a desired tree $T = T_1 \cup T_2 \cup \{(b, x_s), (x_{s-1}, x_s), (y_i, x_s)\}$; see Fig. 3(b). T is a plane bichromatic $(k+1)$-tree on $R \cup B$ with $d_T(b) = \alpha = 1$, $d_T(x_s) = 3$, $d_T(x_{s-1}) = k+1$ and $d_T(y_i) = k+1$.

Case 1.2.2.2: l_1 intersects C_2'', and l_2 intersects C_1''. In this case $Q = (x_{s-1}, x_s, y_t, y_{t-1})$ is a convex quadrilateral because $l_1 \cap (y_{t-1}, y_t) = \emptyset$ and $l_2 \cap (x_{s-1}, x_s) = \emptyset$. Moreover, Q does not have any point of $R \cup B$ in its interior and it has no intersection with the interiors of $CH(R_1 \cup B_1)$ and $CH(R_2 \cup B_2)$. We handle this case as in Case 1.2.2.1 with the blue point y_{t-1} playing the role of y_i. Observe that this construction gives a valid tree even if $x_s = y_t = r_1$.

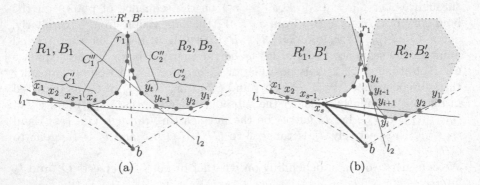

(a)   (b)

Fig. 3. (a) l_1 does not intersect the interior of $CH(R \cup B)$, and (b) l_1 intersects C_2'. (Color figure online)

Case 2: At least one of the vertices on $CH(R \cup B)$ adjacent to b does not belong to B. Let x_1 be such a vertex that belongs to R. Initialize $X = \{x_1\}$. If at least one of the vertices of $CH((R - X) \cup B)$ adjacent to b does not belong to B, let x_2 be such a red point. Add x_2 to the set X. Repeat this process on $CH((R - X) \cup B)$ until $|X| = \alpha$ or both neighbors of b on $CH((R - X) \cup B)$ are blue points. Let $x_1, \ldots, x_\beta$ be the sequence of red points added to X in this process. After this process we have $|X| = \beta$, where $1 \leqslant \beta \leqslant \alpha$. Let S_1 be the star obtained by connecting all points of X to b. See Fig. 4 where S_1 is shown with green bold edges. Observe that $d_{S_1}(b) = \beta$. We distinguish between two cases: $\beta = \alpha$ and $1 \leqslant \beta < \alpha$.

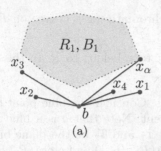 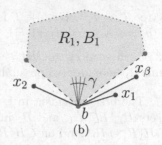

Fig. 4. The edges in S_1 are in bold where (a) $\beta = \alpha$, and (b) $1 \leqslant \beta < \alpha$. (Color figure online)

Case 2.1: $\beta = \alpha$. Let $R_1 = (R - X) \cup \{x_\alpha\}$ and $B_1 = B - \{b\}$. See Fig. 4(a). Note that x_α is a red point on $CH(R_1 \cup B_1)$ and

$$|R_1| = |R| - \alpha + 1 = |R| - (|R| - k(|B| - 1)) + 1 = k|B_1| + 1.$$

Let T_1 be the plane bichromatic $(k + 1)$-tree obtained by $\mathsf{proc2}(R_1, B_1, x_\alpha)$ with $d_{T_1}(x_\alpha) \in \{1, 2\}$ and all other red points in T_1 are of degree at most 3. Note that $|R_1 \cup B_1| < |R \cup B|$. We obtain a desired tree $T = T_1 \cup S_1$. T is a plane bichromatic $(k + 1)$-tree on $R \cup B$ with $d_T(x_\alpha) \in \{2, 3\}$ and $d_T(b) = \alpha = |R| - k(|B| - 1)$ as required.

Case 2.2: $1 \leqslant \beta < \alpha$. In this case both vertices of $CH((R - X) \cup B)$ adjacent to b are blue points. Let $R_1 = R - X$ and $B_1 = B$. See Fig. 4(b). Let $\gamma = \alpha - \beta$ and note that $1 \leqslant \gamma < \alpha \leqslant k$. Then,

$$|R_1| = |R| - \beta = (k(|B| - 1) + \alpha) - \beta = k(|B_1| - 1) + \gamma.$$

Thus, $k(|B_1| - 1) < |R_1| \leqslant k|B_1|$. Let T_1 be the plane bichromatic $(k + 1)$-tree obtained by the recursive call $\mathsf{proc1}(R_1, B_1, b)$ with $d_{T_1}(b) = \gamma$ and all red points of T_1 are of degree at most 3. Note that $|R_1 \cup B_1| < |R \cup B|$. We obtain a desired tree $T = T_1 \cup S_1$. T is a plane bichromatic $(k + 1)$-tree on $R \cup B$ with $d_T(b) = \beta + \gamma = \alpha$.

2.1.2 Procedure proc2

The procedure $\mathsf{proc2}(R, B, r)$ takes as input a set R of red points, a set B of blue points, and a point $r \in R$, where $|B| \geqslant 2$, $|R| = k|B| + 1$, with $k \geqslant 2$, and r is on $CH(R \cup B)$. This procedure computes a plane bichromatic $(k + 1)$-tree T on $R \cup B$ where $d_T(r) \in \{1, 2\}$ and each other red vertex has degree at most 3. We consider two cases, depending on whether or not both points of $CH(R \cup B)$ adjacent to r belong to R.

Case 1: At least one of the vertices on $CH(R \cup B)$ adjacent to r does not belong to R. Let b be such a point belonging to B. Let $R_1 = R - \{r\}$, $B_1 = B$. Note that $|R_1| = k|B_1|$, and b is on $CH(R_1 \cup B_1)$. Let T_1 be the plane bichromatic $(k + 1)$-tree obtained by $\mathsf{proc1}(R_1, B_1, b)$ with $d_{T_1}(b) = k$ and all red points of T_1 are of degree at most 3. Note that $|R_1 \cup B_1| < |R \cup B|$. Then, we obtain a desired tree $T = T_1 \cup \{(r, b)\}$. T is a plane bichromatic $(k + 1)$-tree on $R \cup B$ with $d_T(b) = k + 1$ and $d_T(r) = 1$.

Case 2: Both vertices of $CH(R \cup B)$ adjacent to r belong to R. In this case, by Lemma 2 there are two possibilities:

Case 2.1: The first statement in Lemma 2 holds. Consider the blue point b and the sets R_1, R_2, B_1, and B_2 in this statement. Note that b is a blue point on $CH(R_1 \cup B_1)$ and on $CH(R_2 \cup B_2)$. Let T_1 and T_2 be the plane bichromatic $(k+1)$-trees obtained by running $\mathsf{proc1}(R_1, B_1, b)$ and $\mathsf{proc1}(R_2, B_2, b)$, respectively. Note that $d_{T_1}(b) = k - \alpha$, $d_{T_2}(b) = \alpha$, all red points of T_1 and T_2 have degree at most 3, $|R_1 \cup B_1| < |R \cup B|$, and $|R_2 \cup B_2| < |R \cup B|$. We obtain a desired tree $T = T_1 \cup T_2 \cup \{(r, b)\}$. See Fig. 2(a). T is a plane bichromatic $(k + 1)$-tree on $R \cup B$ with $d_T(r) = 1$ and $d_T(b) = (k - \alpha) + \alpha + 1 = k + 1$.

Case 2.2: The second statement in Lemma 2 holds. Consider the blue points b, b' and the sets R_1, R_2, B_1, and B_2 in this statement. Note that b is a blue point on $CH(R_1 \cup B_1)$ and b' is a blue point on $CH(R_2 \cup B_2)$. Let T_1 and T_2 be the plane bichromatic $(k + 1)$-trees obtained by running $\mathsf{proc1}(R_1, B_1, b)$ and $\mathsf{proc1}(R_2, B_2, b')$. Note that $d_{T_1}(b) = k$, $d_{T_2}(b') = k$, all red points of T_1 and T_2 have degree at most 3, $|R_1 \cup B_1| < |R \cup B|$, and $|R_2 \cup B_2| < |R \cup B|$. We obtain a desired tree $T = T_1 \cup T_2 \cup \{(r, b), (r, b')\}$. See Fig. 2(b). T is a plane bichromatic $(k+1)$-tree on $R \cup B$ with $d_T(b) = k + 1$, $d_T(b') = k + 1$, and $d_T(r) = 2$.

2.1.3   Algorithm plane-tree

Algorithm plane-tree(R, B) takes as input a set R of red points and a non-empty set B of blue points, where $|R| = k|B|$, with $k \geqslant 2$, and $R \cup B$ is in general position. This algorithm constructs a plane bichromatic $(k+1)$-tree T on $R \cup B$ such that each red vertex has degree at most 3. By Observation 1, T has one blue vertex of degree k and the other blue vertices are of degree $k + 1$. We consider two cases, depending on whether or not all vertices of $CH(R \cup B)$ belong to R.

Case 1: At least one of the vertices of $CH(R \cup B)$ belongs to B. Let b be such a vertex. Let T be the tree obtained by running $\mathsf{proc1}(R, B, b)$. T is a plane bichromatic $(k + 1)$-tree on $R \cup B$ with $d_T(b) = k$ and all red vertices of T are of degree at most 3. Notice that b is the only blue vertex of degree k in T.

Case 2: All vertices of $CH(R \cup B)$ belong to R. Let a be an arbitrary red point on $CH(R \cup B)$. By a suitable rotation of the plane, we may assume that a is the lowest point of $CH(R \cup B)$. We add a dummy red point r' at a sufficiently small distance ϵ to the left of a such that the radial ordering of the points in $(R \cup B) - \{a\}$ around r' is the same as their radial ordering around a. See Fig. 5. Now we consider the radial ordering of

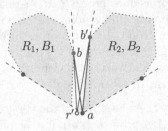

Fig. 5. A dummy red point r' is placed very close to a. (Color figure online)

the points in $R \cup B$ (including a) around r'. We apply Lemma 2 with r' playing the role of r. There are two possibilities:

Case 2.1: The first statement in Lemma 2 holds. Consider the blue point b and the sets R_1, R_2, B_1, and B_2 as in the first statement of Lemma 2. Note that $a \in R_2$, $r' \notin R_1 \cup R_2$, and b is a blue point on $CH(R_1 \cup B_1)$ and on $CH(R_2 \cup B_2)$. Let T_1 and T_2 be the plane bichromatic $(k+1)$-trees obtained by $\mathsf{proc1}(R_1, B_1, b)$ and $\mathsf{proc1}(R_2, B_2, b)$, respectively. Note that $d_{T_1}(b) = k - \alpha$, $d_{T_2}(b) = \alpha$, and all red vertices of T_1 and T_2 have degree at most 3. We obtain a desired tree $T = T_1 \cup T_2$ with $d_T(b) = k - \alpha + \alpha = k$; b is the only blue vertex of degree k in T.

Case 2.2: The second statement in Lemma 2 holds. Consider the blue points b, b' and the sets R_1, R_2, B_1, and B_2 as in the second statement of Lemma 2. Note that $a \in R_2$, $r' \notin R_1 \cup R_2$, b is a blue point on $CH(R_1 \cup B_1)$, and b' is a blue point on $CH(R_2 \cup B_2)$. If we compute trees on $R_1 \cup B_1$ and $R_2 \cup B_2$ and discard r', as we did in the previous case, then the resulting graph is not connected and hence it is not a tree. Thus, we handle this case in a different way. First we remove a from R_2 as shown in Fig. 5; this makes $|R_2| = k|B_2| - 1 = k(|B_2| - 1) + (k - 1)$. Note that $CH(R_1 \cup B_1)$ and $CH(R_2 \cup B_2)$ are disjoint. Let T_1 and T_2 be the plane bichromatic $(k + 1)$-trees obtained by $\mathsf{proc1}(R_1, B_1, b)$ and $\mathsf{proc1}(R_2, B_2, b')$, respectively. Note that $d_{T_1}(b) = k$, $d_{T_2}(b') = k - 1$, and red vertices of T_1 and T_2 have degree at most 3. We obtain a desired tree $T = T_1 \cup T_2 \cup \{(a, b), (a, b')\}$ with $d_T(a) = 2$ and $d_T(b) = k + 1$ and $d_T(b') = k$; b' is the only vertex of degree k in T.

This concludes the description of algorithm $\mathsf{plane\text{-}tree}$. The pseudo code for $\mathsf{proc1}$, $\mathsf{proc2}$, and $\mathsf{plane\text{-}tree}$ are given in the full version of the paper (see [7]).

A simple reduction from the convex hull problem shows that the computation of a plane bichromatic spanning tree has an $\Omega(n \log n)$ lower bound. Using a worst-case deletion-only convex hull data structure, we can compute the tree in Theorems 2, and hence the tree in Theorem 1, in $O(n \cdot \mathrm{polylog}(n))$ time.

References

1. Abellanas, M., Garcia-Lopez, J., Hernández-Peñalver, G., Noy, M., Ramos, P.A.: Bipartite embeddings of trees in the plane. In: Proceedings of the 4th International Symposium on Graph Drawing (GD), pp. 1–10 (1996)
2. Abellanas, M., Garcia-Lopez, J., Hernández-Peñalver, G., Noy, M., Ramos, P.A.: Bipartite embeddings of trees in the plane. Disc. Appl. Math. **93**(2–3), 141–148 (1999)
3. Agarwal, P.K.: Partitioning arrangements of lines I: an efficient deterministic algorithm. Disc. Comp. Geom. **5**, 449–483 (1990)
4. Arora, S., Chang, K.L.: Approximation schemes for degree-restricted MST and red-blue separation problems. Algorithmica **40**(3), 189–210 (2004)
5. Bespamyatnikh, S., Kirkpatrick, D.G., Snoeyink, J.: Generalizing ham sandwich cuts to equitable subdivisions. Disc. Comp. Geom. **24**(4), 605–622 (2000)

6. Biniaz, A., Bose, P., Maheshwari, A., Smid, M.: Plane geodesic spanning trees, Hamiltonian cycles, and perfect matchings in a simple polygon. Comput. Geom. **57**, 27–39 (2016)
7. Biniaz, A., Bose, P., Maheshwari, A., Smid, M.H.M.: Plane bichromatic trees of low degree. CoRR, abs/1512.02730 (2015)
8. Borgelt, M.G., van Kreveld, M.J., Löffler, M., Luo, J., Merrick, D., Silveira, R.I., Vahedi, M.: Planar bichromatic minimum spanning trees. J. Disc. Alg. **7**(4), 469–478 (2009)
9. Demaine, E.D., Erickson, J., Hurtado, F., Iacono, J., Langerman, S., Meijer, H., Overmars, M.H., Whitesides, S.: Separating point sets in polygonal environments. Int. J. Comp. Geom. Appl. **15**(4), 403–420 (2005)
10. Hoffmann, M., Tóth, C.D.: Vertex-colored encompassing graphs. Graphs Comb. **30**(4), 933–947 (2014)
11. Kaneko, A.: On the maximum degree of bipartite embeddings of trees in the plane. In: Akiyama, J., Kano, M., Urabe, M. (eds.) JCDCG 1998. LNCS, vol. 1763, pp. 166–171. Springer, Heidelberg (2000)
12. Kaneko, A., Kano, M.: Discrete geometry on red and blue points in the plane a survey. In: Aronov, B., Basu, S., Pach, J., Sharir, M. (eds.) Discrete and Computational Geometry. Algorithms and Combinatorics, vol. 25, pp. 551–570. Springer, Heidelberg (2003)
13. Kano, M., Ozeki, K., Suzuki, K., Tsugaki, M., Yamashita, T.: Spanning k-trees of bipartite graphs. Electr. J. Comb. **22**(1), P1.13 (2015)
14. Kano, M., Suzuki, K., Uno, M.: Properly colored geometric matchings and 3-trees without crossings on multicolored points in the plane. In: Akiyama, J., Ito, H., Sakai, T. (eds.) JCDCGG 2013. LNCS, vol. 8845, pp. 96–111. Springer, Heidelberg (2014)
15. Mairson, H.G., Stolfi, J.: Reporting and counting intersections between two sets of line segments. In: Earnshaw, R.A. (ed.) Theoretical Foundations of Computer Graphics and CAD, pp. 307–325. Springer, Heidelberg (1988)

Networks

Directing Road Networks
by Listing Strong Orientations

Alessio Conte[1], Roberto Grossi[1(✉)], Andrea Marino[1],
Romeo Rizzi[2], and Luca Versari[3]

[1] Università di Pisa, Pisa, Italy
{conte,grossi,marino}@di.unipi.it
[2] Università di Verona, Verona, Italy
romeo.rizzi@univr.it
[3] Scuola Normale Superiore, Pisa, Italy
luca.versari@sns.it

Abstract. A connected road network with N nodes and L edges has $K \leq L$ edges identified as one-way roads. In a feasible direction, these one-way roads are assigned a direction each, so that every node can reach any other [Robbins '39]. Using $O(L)$ preprocessing time and space usage, it is shown that all feasible directions can be found in $O(K)$ amortized time each. To do so, we give a new algorithm that lists all the strong orientations of an undirected connected graph with m edges in $O(m)$ amortized time each, using $O(m)$ space. The cost can be deamortized to obtain $O(m)$ delay with $O(m^2)$ preprocessing time and space.

1   Introduction

Consider a road network as a connected network with N nodes that correspond to road intersections, and L edges that correspond to road traits. Of the latter, $K \leq L$ are tagged as one-way roads whose direction must be decided, whereas the rest are two-way roads taken in both directions. The network has a feasible direction if there is an assignment of direction to each one-way road, so that from every node it is possible to reach all the other ones in the network. The problem of finding a feasible direction in a road network has been studied since Robbins' theorem, which gives the necessary and sufficient conditions [21]. In particular, the problem is named *one-way street problem* in [22].

Problem Definition. This paper addresses the problem of discovering *all* the feasible directions in the one-way street problem, which might find application in situations where no clear apriori optimality criterion is available for directing the network, and multiple criteria must be tailored for the special situation at hand (e.g. some populous areas of big cities, which contain many narrow one-way roads). We reduce the problem of finding feasible directions in the road

Work partially supported by the Italian Ministry of Education, University, and Research (MIUR) under PRIN 2012C4E3KT national research project AMANDA—Algorithmics for MAssive and Networked DAta.

© Springer International Publishing Switzerland 2016
V. Mäkinen et al. (Eds.): IWOCA 2016, LNCS 9843, pp. 83–95, 2016.
DOI: 10.1007/978-3-319-44543-4_7

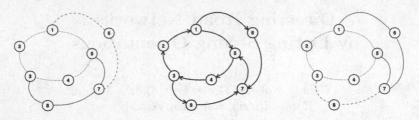

Fig. 1. Two ear decompositions (left and right) and a SO obtained from both (center)

network to the problem of finding *strong orientations* of an undirected graph G with $n \leq 2K$ nodes and $m \leq 2K$ edges, where each strong orientation (SO afterwards) of G produces a distinct directed graph that is strongly connected, that is, every node can reach any other node.

Related Work. Several papers by Roberts and Xu deal with these feasible directions [23–26] in the one-way street problem. The results reduce the latter to the problem of finding a SO of a *mixed multigraph*, which is a multigraph where both directed and undirected edges coexist. Robbins' theorem has been extended by Boesch and Tindel [3] accordingly, and Chung et al. [7] describe a linear time algorithm for finding a strong orientation in a mixed multigraph. In our reduction to listing SOs, however, we have the additional requirement of preserving all feasible directions in the reduction (see Sect. 2).

Some variations of the one-way street problem have been considered with the purpose of minimizing the average [11] or the maximum [12,13,15,19] distance among all pairs of nodes, both of which are NP-hard problems [8,20] (see [18] for a survey). Moreover, the minimum diameter among all the strong orientations of a given graph has been shown to be related with its domination number [12]. Other variations consider, for instance, the distance stretch for each pair of nodes [16], other connectivity constraints [1], cost-based constraints [5], degree-based constraints [2], and forced orientations [6].

The previous works mentioned above do not extend efficiently to our problem. By Robbins' theorem [21] the graphs that admit SOs are exactly the 2-edge connected graphs: in these graphs, for every pair of nodes there are two edge-disjoint paths connecting them; hence, if G is not 2-edge connected, the corresponding road network has no feasible direction. Its proof contains the following remarkable hint to find all the SOs, but it has some issues. Given an ear decomposition of G, it is possible to produce a SO by orienting each ear as a directed path, thus obtaining 2^k SOs from an ear decomposition with k ears. In general, listing ear decompositions and then obtaining SOs seems a natural approach to our problem. However, two different ear decompositions can lead to the same SO. Figure 1 shows two possible ear decompositions of a graph yielding the same SO: first orient the cycle $\{1, 2, 3, 4, 5\}$ clock-wise in both orientations, then in the left one orient the ears as $(1, 6, 7)$ and $(3, 8, 7, 5)$, whereas in the right one as $(3, 8, 7)$ and $(1, 6, 7, 5)$. It is easy to generalize this example, so that the same SO is obtained by many distinct ear decompositions.

A possible way to list *once* all the SOs would be to consider one edge at a time and employ the algorithm in [7] to check which orientations of that edge will lead to a solution. This approach would yield a recursive algorithm taking $O(m^2)$ time per solution because of the $O(m)$ recursion depth. It is natural to ask whether $O(m)$ time is possible, as each solution requires $O(m)$ to be output.

Our Contribution. We present the first algorithm for efficiently listing once all the SOs in a graph G with m edges, with a cost of $O(m)$ time per solution and using $O(m)$ preprocessing time and total space. The cost can be deamortized to obtain $O(m)$ delay with $O(m^2)$ preprocessing time and space, where the delay is the maximum time elapsed between any two consecutive outputs. Using this result, we are able to find all the feasible directions of the road network in $O(K)$ amortized time per solution, using $O(L)$ preprocessing time and total space; also, the cost can be deamortized to obtain $O(K)$ delay using $O(K^2 + L)$ preprocessing time and total space. Furthermore, our approach easily extends to the enumeration of totally cyclic orientations, which are orientations in which every edge is part of a cycle. On a connected graph, these orientations are exactly the SOs [4], otherwise they are combinations of the SOs of each component. Note that SOs are not related to acyclic and cyclic orientations [9,10], orientations with respectively no cycles or at least one, which require different algorithmic techniques.

In the paper we adopt the following notation for an undirected connected graph $G = (V, E)$ with $|V| = n$ nodes and $|E| = m$ edges. An orientation of G is the directed graph $\vec{G} = (V, A)$ where for any pair $\{u, v\} \in E$ either $(u, v) \in A$ or $(v, u) \in A$. The orientation $\vec{G}$ is strong if $\vec{G}$ is strongly connected. For the sake of clarity, we call *edges* the unordered pairs $\{x, y\}$ (undirected graph), while we call *arcs* the two possible orientations (x, y) and (y, x) (directed graphs).

2   From One-Way Streets to Strong Orientations

We show how to list solutions for the one-way street problem by a reduction to the problem of listing strong orientations, as this gives a cleaner proof of our results. As in [22], we use the notion of mixed graph $G = (V, E, A)$, i.e. a graph with vertices (in V) linked by the edges in E and by the arcs in A. Clearly, both directed and undirected graphs are special cases of mixed graphs, in which $E = \emptyset$ or $A = \emptyset$ respectively. Given the mixed graph $G = (V, E, A)$, we say that node x reaches node y if there is a path from x to y that uses directed edges in their correct orientation and/or undirected edges. G is strongly connected if u reaches v for every pairs of nodes $u, v \in V$, and is 2-edge connected if there are two edge-disjoint paths connecting u and v for every pair of distinct nodes $u, v \in V$. We refer to G as a mixed multigraph when E or A are multisets.

Consider a road network R with N intersections, K one-way roads and $L - K$ two-way roads. We thus model R as a mixed multigraph $M = (V_M, E_M, A_M)$ in which every node in V_M represents a road intersection, E_M is the multiset of edges corresponding to the one-way roads, and A_M is the multiset of directed arcs, that contains (x, y) and (y, x) for each two-way road linking the intersections modeled by x and y (hence, $|V_M| = N$, $|E_M| = K$, and $|A_M| = 2(L - K)$). A strong orientation of M is a direction assignment for the edges in E_M such that the resulting directed multigraph is strongly connected. Any edge $\{u, v\} \in E_M$ has two possible orientations (u, v) and (v, u), representing how the corresponding road is directed. We consider this to hold for self-loops as well.

It is straightforward to see how a strong orientation of M corresponds to a feasible way of directing R. We will map strong orientations of the mixed multigraph M to strong orientations of a suitable graph G.

To this aim, we introduce the following operation on mixed multigraphs:

Definition 1 (contraction of a directed cycle). *Given a mixed multigraph $M = (V_M, E_M, A_M)$ and a set of nodes $C \subseteq V_M$ which form a directed cycle, the contraction of C as a node c modifies M as follows: $V_M = (V_M \setminus C) \cup \{c\}$; for each edge $e \in E_M$ and each arc $a \in A_M$, any endpoint of e and a in C is replaced by c; finally, any oriented self-loop on c created this way is removed.*

Note that a contraction can create unoriented selfloops that we preserve along with their endpoints before the contraction.

Lemma 1 shows a useful property of the contraction of a directed 2-cycle, while Lemma 2 shows how to neglect undirected self-loops as well.

Lemma 1. *Let M be a mixed multigraph, and x, y a pair of nodes such that both arcs (x, y) and (y, x) exist in M. Let M' be the mixed multigraph obtained by contracting the directed 2-cycle $C = \{x, y\}$ as a node c. There is a one-to-one correspondence between the strong orientations of M and the ones of M'.*

Proof. Let us show that any SO of M induces a unique SO of M' and vice versa. We remark that all undirected edges of M are preserved in M', although some might have become undirected selfloops, thus we have a mapping from each undirected edge of M to a distinct one of M'. Note that this gives us a bijective mapping of the orientations of M and M', as each orientation is defined by the direction assignment of the undirected edges. Consider now a strong orientation of M: each node can reach/be reached by both x and y, thus in the correspondent orientation of M' each node will reach/be reached by c by construction, making M' strongly connected. Similarly a strong orientation of M' induces a strong orientation M. Thus we have a one-to-one correspondence between strong orientations of M and M'.   □

Lemma 2. *Let M' be the multigraph obtained by removing all the k unoriented self-loops in the mixed multigraph M. Each strong orientation of M' corresponds to 2^k unique strong orientations of M, and all strong orientations of M can be found this way.*

Proof. Strong connectivity is not influenced by the removal of self-loops. Thus, removing all self-loop from a strong orientation of M gives us a strong orientation of M'. Moreover, given a strong orientation of M', we can obtain 2^k unique strong orientations of M by assigning arbitrary orientations to any self-loop (recall that each edge, including self-loops, has two possible orientations). Since two orientations obtained in this way from different orientations of M' are clearly distinct, the statement follows.   □

Using Lemmas 1 and 2 we transform $M = (V_M, E_M, A_M)$ in a graph $G = (V, E)$, exploiting the fact that all the arcs in A_M form a set of directed cycles of size 2 by construction. Our transformation proceeds as described next.

1. We contract every directed cycle in M to obtain an undirected multigraph M' according to Lemma 1. Note that M' contains only unoriented self-loops.
2. We remove all the self loops in M' according to Lemma 2.
3. From the resulting multigraph M'' we obtain $G = (V, E)$ as follows: for each edge $\{x, y\}$ in M'', we have edges $\{x, z\}$ and $\{z, y\}$ in E, where z is a new dummy node, and V is made of the nodes of M'' plus the new dummy nodes.

Note that $|V| = n \leq |V_M| + |E_M| = 2K$ and, similarly, $|E| = m \leq 2K$ by construction. We now show that this transformation is correct.

Lemma 3. *Let G be the graph obtained by applying the above transformation to a mixed multigraph M modelling a road network. Each strong orientation of G corresponds to 2^k unique strong orientations of M, where k is the number of self-loops removed in the transformation. Each strong orientation of M can be obtained this way.*

Proof. By Lemmas 1 and 2, we only need to prove that there is a one-to-one correspondence between the strong orientations of G and the ones of M''. Let d_i denote the dummy node of G introduced by the transformation when "splitting" i-th edge $\{x, y\}$. Given an arbitrary orientation of M'', we define an orientation of G in the following way: if (x, y) is the orientation of $\{x, y\}$, then the orientations of $\{x, d_i\}$ and $\{y, d_i\}$ are (x, d_i) and (d_i, y). This mapping is clearly injective.

It is now sufficient to prove that any strong orientation of G is induced by a strong orientation of M'' and vice versa. Let u, w be two nodes of G. We can assume wlog that neither of them is a dummy node: if, say, $u = d_i$ for edge (x, y) of M'', then we have edges $(x, u), (u, y)$ in G and we can replace u with y. Since G is strongly connected, and only has edges between dummy and non-dummy nodes, there exists a directed path $u = v_1, d_1, \ldots, d_{k-1}, v_k = w$ in G, which alternates non-dummy and dummy nodes. By construction of G and the mapping, it follows that $v_1, v_2, \ldots, v_{k-1}, v_k$ is a directed path from u to v in M''. For the converse, let u, w be two nodes of M''. Since M'' is strongly connected, there is a path $u = v_1, \ldots, v_k = w$ in M''. By construction, G has the path $u = v_1, d_1, \ldots, d_{k-1}, v_k$.   □

3   Finding Strong Orientations

In this section we show how to efficiently find all the strong orientations (SOs) of an undirected graph $G = (V, E)$. We assume wlog that $G = (V, E)$ is 2-edge connected: this is a consequence of the following well-known result [21], as otherwise there are no SOs.

Theorem 1 (Robbins' theorem). *A graph G admits a strong orientation iff it is 2-edge connected.*

We introduce the key definitions and properties that will be used to build our algorithm. Using the standard definitions, we call a *cut* of G any bipartition V_1, V_2 of its nodes and we say that an edge $\{x, y\}$ or an arc (x, y) crosses the cut if $x \in V_1$ and $y \in V_2$, or vice versa. We define two kinds of cuts which will help us model the problem, namely ONE-WAY CUT and FORCING CUT (Shown in Fig. 2).

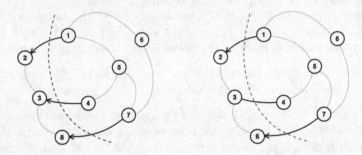

Fig. 2. Two partial orientations of a mixed graph: a ONE-WAY CUT (left) and a FORCING CUT with bound edge $\{3, 4\}$ having $(3, 4)$ as bound direction (right)

Definition 2 (ONE-WAY CUT). *Given a mixed graph $G = (V, E, A)$, we call a cut V_1, V_2 of V a ONE-WAY CUT if the cut is crossed only by arcs, which are all oriented towards V_1 (alternatively, the arcs are all oriented towards V_2).*

We will also exploit another kind of cut that lets us foresee which orientations of which edges will produce a ONE-WAY CUT:

Definition 3 (FORCING CUT). *Given a mixed graph $G = (V, E, A)$, we call a cut V_1, V_2 of V a FORCING CUT if the cut is crossed by exactly one undirected edge, called bound edge, and by one or more arcs that are all oriented towards V_1. We call bound direction the one obtained by orienting the bound edge towards V_2. (The roles of V_1 and V_2 can be interchanged.)*

Note that we cannot have zero arcs in a FORCING CUT of G as otherwise G would not be 2-edge connected.

Lemma 4. *Let $G = (V, E, A)$ be a 2-edge connected mixed graph that has no ONE-WAY CUT. Then any node x reaches any other node y.*

Proof. Let us suppose by contradiction that there exist two nodes x, y such that x does not reach y. Let V_x be the set of nodes that are reachable from x. Since $y \notin V_x$, we have that $V_x, V \setminus V_x$ is a cut of the graph. Moreover, by its definition there can be no edge going from a node of V_x to a node of $V \setminus V_x$, so $V_x, V \setminus V_x$ is a ONE-WAY CUT as the graph is connected. □

The above lemma together with Theorem 3, are crucial to understand the idea behind our approach. For this, we need the following known theorem in [3], that extends Robbins' theorem.

Theorem 2 (Boesch and Tindell). *A mixed graph G has a* SO *if and only if G is strongly connected and 2-edge connected.*

We say that a mixed graph can be completed or extended to a SO if there exists a direction assignment for its edges such that the resulting digraph is a SO. By ensuring that our partial orientation never admits a ONE-WAY CUT, we can ensure the existence of a strongly connected extension using Boesch and Tindell's theorem.

Theorem 3. *A 2-edge connected mixed graph* $G = (V, E, A)$ *can be completed to form a* SO *iff G does not admit a* ONE-WAY CUT.

Proof. If G has a ONE-WAY CUT V_1, V_2, clearly it cannot be extended to a SO. Indeed, as all edges between V_1, V_2 are already oriented, the cut will still be a ONE-WAY CUT in any extension, thus nodes in V_2 will not be reachable by nodes in V_1.

To prove the other implication, note that by Lemma 4 we have that in G any node can reach any other node. Moreover we know by hypothesis that G is 2-edge connected. Boesch and Tindell's theorem implies that such a graph has a SO, proving our result.   □

Finally, we show how the concept of FORCING CUT is important for the completion of an orientation as a SO. In particular, Theorem 4 extends Lemma 2 in [7].

Lemma 5. *Let* $G = (V, E, A)$ *be a 2-edge connected mixed graph, and* V_1, V_2 *a cut of V. Then* V_1, V_2 *can be turned into a* ONE-WAY CUT *by orienting exactly one undirected edge iff* V_1, V_2 *is a* FORCING CUT.

Proof. The proof follows immediately from the definitions of ONE-WAY CUT and FORCING CUT.   □

Theorem 4. *Let* $G = (V, E, A)$ *be a 2-edge connected mixed graph that has no* ONE-WAY CUT, *and* $\{x, y\}$ *an undirected edge in E. Then neither of the orientations* (x, y) *and* (y, x) *of the edge will create a* ONE-WAY CUT *iff* $\{x, y\}$ *is not a bound edge.*

Proof. If $\{x, y\}$ is not a bound edge, both orientations lead to a solution. Indeed, any cut crossed by $\{x, y\}$ is not a FORCING CUT, thus by Lemma 5 any orientation of $\{x, y\}$ will not produce a ONE-WAY CUT. If $\{x, y\}$ is a bound edge, then there is a cut V_1, V_2 of V in which all edges are oriented towards V_1 except for $\{x, y\}$. Orienting $\{x, y\}$ towards V_1 will create a ONE-WAY CUT.   □

3.1   Algorithm Description

The above properties are the guidelines for a simple and efficient algorithm to enumerate the SOs of G. The core idea hinges on bound edges to guarantee that each recursive call either outputs a new SO or yields two calls that will produce at least one new SO each.

Algorithm 1. Finding all strong orientations (SOs)

Input   : Graph $G = (V, E)$.
Output: All SOs of G.
STRONG-ORIENTATIONS$(V, E, \emptyset)$

Function STRONG-ORIENTATIONS(V, E, A)
    $B \leftarrow$ bound edges in mixed graph $G = (V, E, A)$
    $E \leftarrow E \setminus B$
    $A \leftarrow A \cup \{(b, c) : (b, c)$ is the bound direction of $\{b, c\} \in B\}$
    if $E = \emptyset$ **then**
        output SO $\leftarrow \overrightarrow{G} = (V, A)$
    else
        $\{x, y\} \leftarrow$ an arbitrary edge in E
        $E \leftarrow E \setminus \{\{x, y\}\}$
        STRONG-ORIENTATIONS$(V, E, A \cup \{(x, y)\})$
        STRONG-ORIENTATIONS$(V, E, A \cup \{(y, x)\})$

The ideas are detailed in Algorithm 1: it is a recursive approach that consists in incrementally exploring all the possible ways of orienting edges of G that will lead to a solution. In the beginning G is completely undirected, so it will not contain a ONE-WAY CUT. By Theorem 4 we know that the edges that can create a ONE-WAY CUT are exactly all the bound edges; let B be the set of such edges. Each edge in B must be oriented according to its bound direction, as it would otherwise create a ONE-WAY CUT. Note that as a consequence of Boesch and Tindell's theorem [3], if there is at least one SO, then the bound direction does not create a ONE-WAY CUT. For all other edges, we are free to chose any orientation. Thus we orient the edges in B according to their suitable direction, pick an arbitrary edge $\{x, y\}$ (if any), and recur on both possible ways (x, y) and (y, x) of orienting $\{x, y\}$. When there are no more edges that can be oriented we output the current orientation.

It remains to describe how to find the bound edges in B. In any recursive step, our algorithm starts with a mixed graph $G = (V, E, A)$, where A are the edges that have been already directed, and E the ones that have not. We need to find in this graph all the bound edges in E, that is, all the FORCING CUTs of M. As we will show in Lemma 7, these are actually all the undirected edges which are *strong bridges*.

Definition 4 (strong bridge). *Given a mixed graph G, a strong bridge is an edge that, if removed, increases the number of strongly connected components of G.*

Using the algorithm by Italiano et al. [17] we can find all strong bridges in G in $O(|E| + |A|)$ time. The algorithm is intended for directed graphs, but it can also be applied to mixed graphs by considering each undirected edge $\{x, y\}$ as a pair of arcs (x, y), (y, x) with opposite directions, so as to traverse $\{x, y\}$ in both directions: whichever is chosen between (x, y) and (y, x) as a strong bridge, gives the bound direction to $\{x, y\}$. (Note that (x, y) and (y, x) cannot be both chosen as strong bridges.)

3.2   Correctness

As any edge that is not bound can be oriented in both ways and lead to a solution by Theorem 4, we observe the following fact.

Lemma 6. *Let e be an edge that is not bound in G. Then, orienting any bound edge of G in its forced direction does not make e a bound edge.*

Proof. It follows from the observation that any cut involving e has at least two undirected edges, thus orienting the bound edges cannot affect e.

Lemma 7. *Let $\{x, y\}$ be an undirected edge in a strongly connected mixed graph G. Then $\{x, y\}$ is bound iff it is a strong bridge.*

Proof. We will first prove that if $\{x, y\}$ is a bound edge then it is a strong bridge. Indeed, if V_x, V_y is the FORCING CUT of $\{x, y\}$, where all other edges go from V_x to V_y, then removing $\{x, y\}$ makes nodes in V_y unable to reach nodes in V_x, increasing the number of strongly connected components of G, thus $\{x, y\}$ is a strong bridge.

Suppose now that $\{x, y\}$ is a strong bridge. Let V_x and V_y be the set of nodes reachable from respectively x and y without using the edge $\{x, y\}$. Since $\{x, y\}$ is a strong bridge, either $V_x \neq V$ or $V_y \neq V$. Let V_1 be the set, chosen between V_x and V_y, satisfying the latter disequality. Let $V_2 = V \setminus V_1$ be the complement set, which is nonempty, and consider the cut V_1, V_2: all the arcs in this cut (except $\{x, y\}$) must be oriented towards V_1, as otherwise V_1 would be larger. Hence, V_1, V_2 is a FORCING CUT for $\{x, y\}$ because V_1 has no outgoing edges to V_2 other than $\{x, y\}$ itself. $\qquad\qquad\square$

Theorem 5. *Given a 2-edge connected graph $G = (V, E)$, our algorithm correctly outputs all the strong orientations of G exactly once.*

Proof. A 2-edge connected mixed graph can be completed to form a SO iff it does not admit a ONE-WAY CUT by Theorem 3. Hence, we prove by induction on $|E|$ that, if $G' = (V, E, A)$ is a mixed graph with no ONE-WAY CUT, our algorithm outputs all the SOs of G' once.

Base case for $|E| = 0$. Then G' is completely oriented and with no ONE-WAY CUT, so by Lemma 4 it is strongly connected. Moreover it has exactly one SO $\overrightarrow{G'} = (V, A)$, which we output.

Inductive step for $|E| > 0$. We can identify all the bound edges in G' and their bound directions by Lemma 7, using the algorithm in [17]. Orienting bound edges in their bound direction does not alter the set of SOs of G', since there is no SO that has a bound edge in the other direction: as each bound edge belongs to a FORCING CUT, orienting that edge otherwise would create a ONE-WAY CUT by Lemma 5. Also, orienting a bound edge in its bound direction cannot create a new bound edge by Lemma 6. We can thus consider G' as having no bound edges, without loss of generality. If G' has no more undirected edges, we fall back to the base case. Otherwise, given an undirected edge e of G', we know that orienting it either way does not produce any ONE-WAY CUT by Theorem 4. Any SO must have e in either one direction or the other. Let G'_1 and G'_2 be the graphs obtained by orienting e in each way, respectively. Since both G'_1 and G'_2 have a smaller number of undirected edges than G', we know by inductive hypothesis that our algorithm terminates, outputting all the SOs of G'_1 and G'_2 once. Any SO of G' is a SO of either G'_1 and G'_2, and the latter have no intersection as they differ on the orientation of e. Hence, the algorithm produces all the SOs of G' once. □

3.3   Analysis

We now analyze the time and space cost of our algorithm on the graph $G = (V, E)$, with $|V| = n$ and $|E| = m$ assuming wlog that it is connected. We remark that each recursion node which is not a leaf has at least two children, and that every leaf of the computation tree outputs a distinct solution. This gives us a computation tree with no unary nodes[1] and α leaves, where α is the number of solutions. It follows that the total number of recursion nodes is bounded by $2 \cdot \alpha$ and thus the amortized cost per solution of the algorithm is bounded by the cost of a single recursion node.

Consider the structure of Algorithm 1. We show how every step takes $O(m)$ time. Computing bound edges is done in $O(m)$ time by finding the strong bridges and selecting the undirected ones; moreover, the algorithm by Italiano et al. [17] is applied to a directed graph where each undirected edge is represented by two directed arcs, thus finding a strong bridge will immediately give us the bound direction of the corresponding bound edge, making the assignment of bound directions clearly $O(m)$ time. All other steps involve updating or scanning sets of size $O(m)$, which trivially take $O(m)$ time each. The total cost is $O(m \cdot \alpha)$, or equivalently $O(m)$ amortized cost per solution. We remark that this cost is optimal for merely printing each SO.

Finally we show that the space cost is bounded by $O(m)$ as well: indeed, the working space of a single recursion node is $O(m)$, but the information that needs to be passed on to child recursive calls, other than the input, is simply

---

[1] This is crucial, as the presence of unary nodes is the reason behind the $O(m^2)$ cost of the approach based on [7], mentioned in the introduction.

the partial orientation of the graph. If stored as the difference with the partial orientation in the parent node, the space requirement of a root-to-leaf path (and thus of the whole algorithm) is always $O(m)$. Thus the following holds:

Theorem 6. *Given a 2-edge connected graph $G = (V, E)$, Algorithm 1 outputs all the strong orientations of G exactly once, in $O(m)$ amortized time, using $O(m)$ total space.*

We observe that the delay of Algorithm 1 is bounded by the sum of the costs along a leaf-to-root path and a root-to-leaf path. Since the cost of each recursion node is $O(m)$, and the depth of the computation tree is at most m, we obtain $O(m^2)$ delay. We will now show how the delay can be reduced to $O(m)$ using the Output Queue Method by Uno [27], which suitably accumulates solutions that arrives at an irregular pace to output them in a regular fashion, using a queue of bounded size. The method depends on two parameters: T^*, the maximum cumulative cost in a root-to-leaf path of the recursion tree, and $\bar{T}$, an upper bound on the amortized cost per solution in any subtree of the computation. In our case, the former is $O(m^2)$ as discussed above, and the second is $\Theta(m)$, as each k-size subtree of our binary recursion tree has $\Theta(k)$ leaves (i.e. solutions since there are no unary nodes), and a node takes $O(m)$ time. As a result, using a queue of $O(T^*/\bar{T}) = O(m)$ solutions, we can output each solution with delay $O(\bar{T}) = O(m)$. This takes $O(T^* + \bar{T}) = O(m^2)$ preprocessing time and $O(m \cdot T^*/\bar{T}) = O(m^2)$ space.

Theorem 7. *Given a 2-edge connected graph $G = (V, E)$, there exists an algorithm that outputs all the strong orientations of G exactly once, with $O(m)$ delay, using $O(m^2)$ preprocessing time and total space.*

4   Conclusions

In this paper we considered the problem of finding all feasible ways of directing a connected road network, also known as the one-way street problem, and reduced the latter to the problem of listing all the strong orientations in an undirected connected graph. The bounds are optimal if one wants to print each strong orientation. A referee suggests the interesting open problem of enumerating totally cyclic orientations in 3-edge connected graphs [14] in constant amortized time by listing only the edges that get flipped from orientation to orientation.

References

1. Arkin, E.M., Hassin, R.: A note on orientations of mixed graphs. Discrete Appl. Math. **116**(3), 271–278 (2002)
2. Ben-Ameur, W., Glorieux, A., Neto, J.: On the most imbalanced orientation of a graph. In: Xu, D., Du, D., Du, D. (eds.) COCOON 2015. LNCS, vol. 9198, pp. 16–29. Springer, Heidelberg (2015)

3. Boesch, F., Tindell, R.: Robbins' theorem for mixed multigraphs. Am. Math. Monthly **87**(9), 716–719 (1980)
4. Bollobás, B.: Modern Graph Theory. Springer-Verlag, New York (1998)
5. Rainer, E., Feldbacher, K., Klinz, B., Woeginger, G.J.: Minimum-cost strong network orientation problems: classification, complexity, and algorithms. Networks **33**(1), 57–70 (1999)
6. Chartrand, G., Harary, F., Schultz, M., Wall, C.E.: Forced orientation number of a graph. Congressus Numerantium, pp. 183–192 (1994)
7. Chung, F.R.K., Garey, M.R., Tarjan, R.E.: Strongly connected orientations of mixed multigraphs. Networks **15**(4), 477–484 (1985)
8. Chvátal, V., Thomassen, C.: Distances in orientations of graphs. J. Comb. Theory Ser. B **24**(1), 61–75 (1978)
9. Conte, A., Grossi, R., Marino, A., Rizzi, R.: Enumerating cyclic orientations of a graph. In: Lipták, Z., Smyth, W.F. (eds.) IWOCA 2015. LNCS, vol. 9538, pp. 88–99. Springer, Heidelberg (2016)
10. Conte, A., Grossi, R., Marino, A., Rizzi, R.: Listing acyclic orientations of graphs with single, multiple sources. In: Proceedings LATIN (2011) Observation of strains: Theoretical Informatics - 12th Latin American Symposium, Ensenada, 11-15 April 2016, pp. 319–333 (2016)
11. Dankelmann, P., Oellermann, O.R., Jian-Liang, W.: Minimum average distance of strong orientations of graphs. Discrete Appl. Math. **143**(1–3), 204–212 (2004)
12. Fomin, E.V., Matamala, M., Prisner, E., Rapaport, I.: At-free graphs: linear bounds for the oriented diameter. Discrete Appl. Math. **141**(1), 135–148 (2004)
13. Fomin, F.V., Matamala, M.: Complexity of approximating the oriented diameter of chordal graphs. J. Graph Theory **45**(4), 255–269 (2004)
14. Fukuda, K., Prodon, A., Sakuma, T.: Combinatorics and computer science notes on acyclic orientations and the shelling lemma. Theoret. Comput. Sci. **263**(1), 9–16 (2001)
15. Gutin, G.: Minimizing and maximizing the diameter in orientations of graphs. Graphs Comb. **10**(2), 225–230 (1994)
16. Hassin, R., Megiddo, N.: On orientations, shortest paths. Linear Algebra Appl. **114, 115**, 589–602 (1989). Special Issue Dedicated to Alan J. Hoffman
17. Italiano, G.F., Laura, L., Santaroni, F.: Finding strong bridges and strong articulation points in linear time. Theoret. Comput. Sci. **447**, 74–84 (2012)
18. Koh, K.M., Tay, E.G.: Optimal orientations of graphs and digraphs: a survey. Graphs Comb. **18**(4), 745–756 (2002)
19. Kurz, S., Lätsch, M.: Bounds for the minimum oriented diameter. Discrete Math. Theoret. Comput. Sci. **14**(1), 109–140 (2012)
20. Plesník, J.: On the sum of all distances in a graph or digraph. J. Graph Theory **8**(1), 1–21 (1984)
21. Robbins, H.E.: A theorem on graphs, with an application to a problem of traffic control. The American Mathematical Monthly **46**(5), 281–283 (1939)
22. Roberts, F.S.: Graph theory and its applications to problems of society. NSF-CBSM Monograph No. 29. SIAM Publications (1978)
23. Roberts, F.S., Xu, Y.: On the optimal strongly connected orientations of city street graphs. II: two East-West avenues or North-South streets. Networks **19**(2), 221–233 (1989)
24. Roberts, F.S., Xu, Y.: On the optimal strongly connected orientations of city street graphs I: large grids. SIAM J. Discrete Math. **1**(2), 199–222 (1988)

25. Roberts, F.S., Xu, Y.: On the optimal strongly connected orientations of city street graphs. III: three East-West avenues or North-South streets. Networks **22**(2), 109–143 (1992)
26. Roberts, F.S., Xu, Y.: On the optimal strongly connected orientations of city street graphs IV: four East-West avenues or North-South streets. Discrete Appl. Math. **49**(1), 331–356 (1994)
27. Uno, T.: Two general methods to reduce delay and change of enumeration algorithms: NII Technical report NII-2003-004E, Tokyo (2003)

Evangelism in Social Networks

Gennaro Cordasco[2]([⊠]), Luisa Gargano[1], Adele A. Rescigno[1], and Ugo Vaccaro[1]

[1] Department of Informatics, University of Salerno, Fisciano, Italy
{lgargano,arescigno,uvaccaro}@unisa.it
[2] Department of Psychology, Second University of Naples, Caserta, Italy
gennaro.cordasco@unina2.it

Abstract. We consider a population of interconnected individuals that, with respect to a piece information, at each time instant can be subdivided into three (time-dependent) categories: *agnostic, influenced*, and *evangelists*. A dynamical process of information diffusion evolves among the individuals of the population according to the following rules. Initially, all individuals are agnostic. Then, a set of people is chosen from the outside and convinced to start evangelizing, i.e., to start spreading the information. When a number of evangelists, greater than a given threshold, communicate with an node v, the node v becomes influenced, whereas, as soon as the individual v is contacted by a sufficiently *much larger* number of evangelists, it is itself converted into an evangelist and consequently it starts spreading the information. The question is: How to choose a bounded cardinality initial set of evangelists so as to maximize the final number of *influenced* individuals? We prove that the problem is hard to solve, even in an approximate sense, and we present exact polynomial time algorithms for trees and complete graphs. For general graphs, we derive exact algorithms parameterized with respect to neighborhood diversity. We also study the problem when the objective is to select a minimum number of evangelists able of influencing the whole network.

1 Introduction

The Context. Customer Evangelism [26] occurs when a customer *actively* tries to convince other customers to buy or use a particular brand. Fathered by Apple marketing *guru* Guy Kawasaki in the 90's [22], the idea of consumer evangelism has found a new and more powerful incarnation in modern communications media. Social networks like Twitter, Facebook and Pinterest have indeed immensely empowered properly motivated individuals towards brand advocacy and proselytism. We plan to abstract a few algorithmic problems out of this scenario, and provide efficient solutions for some of them.

The Problem. Our model posits an interconnected population consisting of individuals that, with respect to a piece of information and/or an opinion, at each time instant can be subdivided into three time-dependent categories: *agnostic, influenced*, and *evangelists*. Initially, all individuals are agnostic. Then, a set

© Springer International Publishing Switzerland 2016
V. Mäkinen et al. (Eds.): IWOCA 2016, LNCS 9843, pp. 96–108, 2016.
DOI: 10.1007/978-3-319-44543-4_8

of people is chosen and converted into evangelists, that is, convinced to start spreading the information. When a sufficiently large number of evangelists communicate with an node v, the node v becomes influenced; as soon as the individual v has in his neighborood a *much larger* number of evangelists, it is converted to an evangelist and *only then* it starts spreading the information itself. Our model can be seen also as an idealization of diffusion processes studied in the area of memetics. A *meme* [14] is an idea, behavior, or fashion that spreads from person to person within a culture. It is apparent that not every meme learned by a person spreads among the individuals of a population. We are making here the reasonable hypothesis that individuals indeed acquire a meme when it has been heard of from a few friends, but people start spreading the same meme only when they believe it is popular, fashionable, or important, i.e., when it has been communicated to them by *large* number of friends. This is not too far from what has been experimentally observed about how memes evolve and spread within Facebook [2].

A bit more concretely, we are given a graph $G = (V, E)$, abstracting a social network, where the node set V corresponds to people and the edge set to relationships among them. We denote by $N_G(v)$ the neighborhood of node $v \in V$ and by $d_G(v) = |N_G(v)|$ the degree of v in G, we avoid the subscript G whenever the graph is clear from the context. Moreover, let $t_I : V \to \{0, 1, 2, \ldots\}$ and $t_E : V \to \{0, 1, 2, \ldots\}$ be two functions assigning integer thresholds to the nodes in G such that $0 \leq t_I(v) \leq t_E(v) \leq d(v) + 1$, for each $v \in V$.

An evangelization process in G, starting at a subset of nodes $S \subseteq V$, is characterized by two sequences of node subsets $\mathsf{Evg}[S, 0] \subseteq \mathsf{Evg}[S, 1] \subseteq \ldots \subseteq \mathsf{Evg}[S, \tau] \subseteq \ldots \subseteq V$, and $\mathsf{Inf}[S, 0] \subseteq \mathsf{Inf}[S, 1] \subseteq \ldots \subseteq \mathsf{Inf}[S, \tau] \subseteq \ldots \subseteq V$, where for each $\tau = 0, 1, \ldots$, it holds that $\mathsf{Evg}[S, \tau] \subseteq \mathsf{Inf}[S, \tau]$. The process is formally described by the following dynamics: $\mathsf{Evg}[S, 0] = \mathsf{Inf}[S, 0] = S$, and for each $\tau \geq 1$

$$\mathsf{Evg}[S, \tau] = \mathsf{Evg}[S, \tau-1] \cup \Big\{ u : \big|N(u) \cap \mathsf{Evg}[S, \tau-1]\big| \geq t_E(u) \Big\},$$

$$\mathsf{Inf}[S, \tau] = \mathsf{Inf}[S, \tau-1] \cup \Big\{ u : \big|N(u) \cap \mathsf{Evg}[S, \tau-1]\big| \geq t_I(u) \Big\}.$$

In words, a node v becomes influenced if the number of his evangelist neighbors is greater than or equal to its influence threshold $t_I(v)$, and v becomes an evangelist if the number of evangelists among his neighbors reaches its evangelization threshold $t_E(v) \geq t_I(v)$. The process terminates when $\mathsf{Evg}[S, \rho + 1] = \mathsf{Evg}[S, \rho]$ for some $\rho \geq 0$. We denote by $\mathsf{Evg}[S] = \mathsf{Evg}[S, \rho]$ and $\mathsf{Inf}[S] = \mathsf{Inf}[S, \rho + 1]$ the final sets when the process terminates. The initial set S is also denoted as a *seed set* of the evangelization process. Due to foreseeable difficulties in hiring evangelists, it seems reasonable trying to limit their initial number, and see how the dynamics of the spreading process evolves. Therefore, we state our problem as follows:

MAXIMALLY EVANGELIZING SET (MES).
Instance: A graph $G = (V, E)$, thresholds $t_I, t_E : V \to \{0, 1, 2, \ldots\}$, and a budget β.
Question: Find a seed set $S \subseteq V$, with $|S| \leq \beta$, such that $|\mathsf{Inf}[S]|$ is *maximum*.

What is Already Known and What We Prove. The above algorithmic problem has roots in the broad area of the *spread of influence* in Social Networks (see [6,16] and references quoted therein). In the introduction of this paper we have already highlighted the connections of our model to the general area of viral marketing. There, companies wanting to promote products or behaviors might try initially to target and convince a few individuals which, by word-of-mouth effects, can trigger a cascade of influence in the network, leading to an adoption of the products by a much larger number of individuals. Not unexpectedly, viral marketing has also become an important tool in the communication strategies of politicians [25,29]. Less secular applications of our evangelization process can also be envisioned. Here, we shall limit ourselves to discuss the work that is most directly related to ours, and refer the reader to the authoritative texts [6,16] for a synopsis of the area. The first authors to study spread of influence in networks from an algorithmic point of view were Kempe *et al.* [23]. However, they were mostly interested in networks with randomly chosen thresholds. Chen [5] studied the following minimization problem: given an unweighted graph G and fixed thresholds $t(v)$, for each vertex v in G, find a set of minimum size that eventually influences all (or a fixed fraction of) the nodes of G. He proved a strong inapproximability result that makes unlikely the existence of an algorithm with approximation factor better than $O(2^{\log^{1-\epsilon}|V|})$. Chen's result stimulated a series of papers, e.g., [1,3,4,7–13,21,27,28] that isolated interesting cases in which the problem (and variants thereof) become tractable.

All of the above quoted papers considered the basic model in which *any* node, as soon as it is influenced by its neighbors, it also immediately starts spreading influence. The more refined model put forward in this paper, that differenti-ate among active spreaders (evangelists) and plain informed (influenced) nodes, appears to be new, to the best of our knowledge. We would like to point out that we obtain an interesting information diffusion model already in the particular case in which $t_I(v) = 1$, for each node v. In fact, in this case nodes in the sets $\mathsf{Inf}[S, \tau]$ would correspond to people that have simply *heard* about a piece of information, while people in the sets $\mathsf{Evg}[S, \tau]$ would correspond to people who are *actively* spreading that same piece of information.

In Sect. 2, we first prove that the MES problem is hard to solve, even in the approximate sense. Subsequently, we design exact algorithms parameterized with respect to neighborhood diversity (and, as a byproduct, by vertex cover). In Sect. 3, we present exact polynomial time algorithms for trees and complete graphs. Finally, we also study the issue when the objective is to select a minimum number of evangelists capable of influencing the whole network. We refer to this problem as the PERFECT EVANGELIC SET (PES) problem. Namely, given a graph $G = (V, E)$, node thresholds $t_I(v)$ and $t_E(v)$ for each $v \in V$, the PES problem asks for a *minimum size* set $S \subseteq V$ such that $\mathsf{Inf}[S] = V$. In Sect. 4 we study this latter problem in dense graphs.

2   The Complexity of MES

The MES problem includes the INFLUENCE MAXIMIZATION (IM) problem [23], that is known to be NP-hard to approximate within a ratio of $n^{1-\epsilon}$, for any $\epsilon > 0$. In our terminology, the IM problem takes in input a graph G with a threshold function $t : V \to \{0, 1, 2, \ldots\}$ and a budget β, and asks for a subset S of β nodes of G such that $|\mathsf{Evg}[S]|$ is maximum. An instance of the IM problem corresponds to the MES instance consisting of G, β, and threshold functions t_E, t_I, with $t_I(v) = t_E(v) = t(v)$, for each $v \in V$. We can prove that the MES problem remains hard even if the influence threshold t_I is equal to 1, for each node $v \in V$. The proof is quite standard and it is omitted here.

Theorem 1. *It is NP-hard to approximate the MES problem within a ratio of $n^{1-\epsilon}$ for any $\epsilon > 0$ even when $t_I(v) = 1$, for each node $v \in V$.*

2.1   Parameterized Complexity

A parameterized computational problem with input size n and parameter t is called *fixed parameter tractable (FPT)* if it can be solved in time $f(t) \cdot n^c$, where f is a function depending on t only, and c is a constant [15]. In this section we study the effect of some parameters on the computational complexity of the MES problem. As usual, we consider the decision version (α, β)-MES of the problem. It takes in input a graph $G = (V, E)$, node thresholds $t_I : V \to \{0, 1, 2, \ldots\}$ and $t_E : V \to \{0, 1, 2, \ldots\}$, and integer bounds $\alpha, \beta \in \mathbb{N}$, and asks if there exists a seed set $S \subseteq V$ such that $|S| \leq \beta$ and $|\mathsf{Inf}[S]| \geq \alpha$.

We notice that by conveniently choosing the thresholds t_E and t_I, the MES problem specializes in problems whose parameterized complexity is well known. When $t_I(v) = t_E(v)$ for each $v \in V$ and $\alpha = |V|$, the problem becomes the *target set selection* [5]. This problem is $W[2]$-hard[1] with respect to the solution size β [27], it is XP when parameterized with respect to the treewidth [4], and is $W[1]$-hard with respect to the parameters treewidth, cluster vertex deletion number and pathwidth [4,8]. Moreover, the target set selection problem becomes fixed-parameter tractable with respect to the single parameters: Vertex cover number, feedback edge set size, bandwidth [8,27]. In general when $t_I(v) = t_E(v)$ for each $v \in V$, the (α, β)-MES problem has no parameterized approximation algorithm with respect to the parameter β and it is $W[1]$-hard with respect to the combined parameters α and β [3].

In the following, we study the parameterized complexity of the (α, β)-MES problem for the general case $t_I(v) \neq t_E(v)$. We concentrate our attention on two parameters: the well known treewidth and the more recently introduced neighborhood diversity.

Parameterization with Neighborhood Diversity. The *neighborhood diversity* was first introduced in [24]. It has recently received particular attention

---

[1] See [15] for definitions of $W[2]$-hardness, $W[1]$-hardness and the class XP.

[17,19,20] also due to its property of being computable in polynomial time [24]—unlikely other parameters, including treewidth, rankwidth, and vertex cover.

Definition 1. *Given a graph $G = (V, E)$, two nodes $u, v \in V$ have the same type iff $N(v) \setminus \{u\} = N(u) \setminus \{v\}$. The graph G has neighborhood diversity t, if there exists a partition of V into at most t sets, $V_1, V_2, \ldots, V_t$, s.t. all the nodes in V_i have the same type, for $i = 1, \ldots, t$. The family $\mathcal{V} = \{V_1, V_2, \ldots, V_t\}$ is called the* type partition *of G.*

Let $G = (V, E)$ be a graph with type partition $\mathcal{V} = \{V_1, V_2, \ldots, V_t\}$. By Definition 1, each V_i induces either a clique or an independent set in G. For each $V_i, V_j \in \mathcal{V}$, we get that either each node in V_i is a neighbor of each node in V_j or no node in V_i is a neighbor of any node in V_j. Hence, all the nodes in the same V_i have the same neighborhood $N(V_i)$—excluding the nodes in V_i itself.

We present a FPT-algorithm for the MES problem with parameter t. At the end of the evangelization process in G starting at S, we identify the number of evangelists that are neighbors of (all) the nodes in V_i and define

$$N_i(S) = \begin{cases} |\mathsf{Evg}[S] \cap N(V_i)| & \text{if } V_i \text{ is an independent set,} \\ |\mathsf{Evg}[S] \cap (V_i \cup N(V_i))| & \text{if } V_i \text{ is a clique.} \end{cases}$$

It is easy to see that a node $u \in V_i - \mathsf{Evg}[S]$ is influenced if $t_I(u) \leq N_i(S)$.

We can prove the following Lemma.

Lemma 1. *Let S' be a seed set for G. Let $u, v \in V_i$ be s.t. $u \in S'$ and $v \notin S'$, and consider the set $S'' = (S' - \{u\}) \cup \{v\}$. The following properties hold:*
  a) If $t_I(u) \leq N_i(S')$, $t_I(v) \leq N_i(S')$ and $t_E(v) \geq t_E(u)$ then $\mathsf{Inf}[S'] \subseteq \mathsf{Inf}[S'']$;
  b) if $t_I(u) \leq N_i(S')$ and $t_I(v) > N_i(S')$ then $\mathsf{Inf}[S'] \subseteq \mathsf{Inf}[S'']$;
  c) if $t_I(u) > N_i(S')$ and $t_I(v) > N_i(S')$ then $\mathsf{Inf}[S'] = \mathsf{Inf}[S'']$.

We now present our algorithm. We assume that the nodes of G are sorted in order of non–increasing evangelization thresholds and consider all the possible t-ple $(s_1, s_2, \ldots, s_t)$ such that $\sum_{i=1}^{t} s_i = \beta$. For each such $\mathbf{s} = (s_1, s_2, \ldots, s_t)$, we construct the set $S_\mathbf{s}$ in two steps. In the first step we set $S_\mathbf{s} = \cup_{i=1}^{t} S_i$ where S_i is obtained by choosing s_i nodes with the largest evangelization threshold in V_i. In the second step we first consider the evangelization process in G starting at $S_\mathbf{s}$ and then we update each S_i by using the nodes that have not been influenced in the process. In particular, S_i is updated by replacing as many nodes as possible among those that could be influenced (if outside S_i) by nodes that cannot be influenced. The construction of $S_\mathbf{s}$ is detailed in algorithm ME-ND($\mathbf{s}, \mathcal{V}$).

We then consider the evangelization process in G starting at $S_\mathbf{s}$ and get the number $\alpha_\mathbf{s} = |\mathsf{Inf}[S_\mathbf{s}]|$ of influenced nodes at the end of the process. Finally, we determine $\mathbf{s}' = \arg\max_\mathbf{s} \alpha_\mathbf{s}$ and compare α with $\alpha_{\mathbf{s}'}$. If $\alpha_{\mathbf{s}'} \geq \alpha$ then we answer YES to the MES question for G with parameters α and β and $S_{\mathbf{s}'}$ is the desired seed set; otherwise we answer NO.

Theorem 2. *Let t be the neighborhood diversity of G. It is possible to decide the (α, β)-MES question in time $O(nt\, 2^{t \log(\beta+1)})$.*

Proof (Sketch.) We first prove the correctness of our algorithm. For any possible $\mathbf{s} = (s_1, s_2, \ldots, s_t)$, we check if the corresponding set $S_\mathbf{s}$ gets $|\mathsf{Inf}[S_\mathbf{s}]| \geq \alpha$.

For any fixed $\mathbf{s} = (s_1, s_2, \ldots, s_t)$, let $S_\mathbf{s} = \cup_{i=1}^{t} S_i$ be the seed set returned by the algorithm ME-ND. Let now S' be any (α, β)-MES for G such that each $S_i' = S' \cap V_i$ has size $|S_i'| = s_i$, for $i = 1, \ldots, t$. We show that $|\mathsf{Inf}[S_\mathbf{s}]| \geq |\mathsf{Inf}[S']| \geq \alpha$. To this aim, we iteratively tranform each S_i' into S_i by trading a node $u \in S_i' - S_i$ for a node $v \in S_i - S_i'$ without decreasing the number of informed nodes.

Algorithm 1. ME-ND$(\mathbf{s}, \mathcal{V})$

  Input: $\mathbf{s} = (s_1, s_2, \ldots, s_t)$; a type partition $\mathcal{V}$ of G

1 **foreach** $i = 1, \ldots, t$ **do**

2   Let S_i be a the set of s_i nodes of V_i with the largest t()

3 Set $S_\mathbf{s} = \cup_{i=1}^{t} S_i$ and consider the process in G starting at $S_\mathbf{s}$.

4 **foreach** $i = 1, \ldots, t$ **do**   // Update set S_i;

5   **while** $(\exists u \in S_i, t_I(u) \leq N_i(S_\mathbf{s})$ **AND** $\exists v \in V_i - S_i,\ t_I(v) > N_i(S_\mathbf{s}))$ **do**

6        $\overline{u} = \arg\min_{\substack{u \in S_i \\ t_I(u) \leq N_i(S_\mathbf{s})}} \{t_E(u)\}; \qquad S_i = S_i - \{\overline{u}\} \cup \{v\}$

7 **return** $S_\mathbf{s} = \cup_{i=1}^{t} S_i$

If we can choose u, v such that $t_I(u) \leq N_i(S')$ and $t_I(v) \leq N_i(S')$, then by construction of the set S_i we also know that $t_E(v) \geq t_E(u)$ (cfr. line 2 of the algorithm); hence, by Lemma 1 $S'' = (S' - \{u\}) \cup \{v\}$ has $|\mathsf{Inf}[S'']| \geq |\mathsf{Inf}[S']| \geq \alpha$.

If we can choose u, v such that $t_I(v) > N_i(S')$ then by (b) and (c) of Lemma 1 we get $S'' = (S' - \{u\}) \cup \{v\}$ has $|\mathsf{Inf}[S'']| \geq |\mathsf{Inf}[S_i']| \geq \alpha$.

Suppose now that for any choice of u, v it holds $t_I(u) > N_i(S')$ and $t_I(v) \leq N_i(S')$. It is possible to see that the sets S_i (both as initially chosen at line 2 of the algorithm and after each update) maximize the number of evangelized nodes in each V_i and it holds $N_i = N_i(S_\mathbf{s}) \geq N_i(\overline{S})$, for any seed set $\overline{S}$ such $|\overline{S} \cap V_i| = s_i$, for $i = 1, \ldots, t$. Hence,

$$N_i(S_\mathbf{s}) \geq N_i(S'), \qquad \text{for } i = 1, \ldots, t.$$

Notice that, by construction of the sets S_i, it cannot happen that $t_I(u) > N_i(S_\mathbf{s})$ and $t_I(v) \leq N_i(S') \leq N_i(S_\mathbf{s})$ (cfr. lines 5-6 of the algorithm). Therefore, the only remaining case to analyze is when $t_I(u) < N_i(S_\mathbf{s})$ and $t_I(v) \leq N_i(S') \leq N_i(S_\mathbf{s})$ for each $u \in S_i' - S_i$ and $v \in S_i - S_i'$. In such a case, we have

$$\mathsf{Inf}[S'] \cap V_i \subseteq S' \cup \{w \in V_i \mid t_I(w) \leq N_i(S_\mathbf{s})\} \subseteq V_i \cap \mathsf{Inf}[S_\mathbf{s}]$$

Hence, we have $\mathsf{Inf}[S'] \subseteq \mathsf{Inf}[S_\mathbf{s}]$ and we can conclude that $|\mathsf{Inf}[S_\mathbf{s}]| \geq \alpha$.

The correctness follows since for any possible $\mathbf{s} = (s_1, s_2, \ldots, s_t)$, we check if the corresponding set $S_\mathbf{s}$ gets $|\mathsf{Inf}[S_\mathbf{s}]| \geq \alpha$.

Now we evaluate the running time of the algorithm. The number of all the possible t-ple $\mathbf{s} = (s_1, s_2, \ldots, s_t)$ such that $\sum_{i=1}^{t} s_i = \beta$ is $\binom{\beta+t-1}{t-1} < 2^{t \log(\beta+1)}$. Moreover, one needs $O(nt)$ to construct $S_\mathbf{s}$ and $O(nt)$ to obtain $|\mathsf{Inf}[S_\mathbf{s}]|$. Hence, the running time for deciding if a (α, β)-MES for G exists is $O(nt\, 2^{t \log(\beta+1)})$. $\square$

Noticing that the type partition $\mathcal{V}$ can be obtained in polynomial time, one has that the (α, β)-MES problem is in the class FPT when parameterized by the neighborhood diversity t and the solution size β.

Theorem 2 can be used to have FPT linear time algorithms with vertex cover size as parameter. Indeed, graphs of bounded vertex cover have bounded neighborhood diversity—while the opposite is not true since large cliques have neighborhood diversity 1 [20].

Theorem 3. *Given a vertex cover of G of size ℓ, it is possible to decide the (α, β)-MES question in time $O(n(2^\ell + \ell)2^{(2^\ell + \ell)\log \ell})$.*

Parameterization with Treewidth. Roughly speaking, the treewidth measures the "tree-likeness" of a given graph, in particular any tree has treewidth 1. We generalize the results given in [4] for the target set selection problem. We design an algorithm for the PERFECT EVANGELIC SET (PES) problem that runs in $n^{O(w)}$, where w is the treewidth of the input graph. If all the nodes have the same influence threshold we obtain that the problem is FPT.

Definition 2. *A tree decomposition of a graph G is a pair $(\mathcal{T}, \mathcal{X})$, where $\mathcal{X}$ is a family of subsets of $V(G)$, and $\mathcal{T}$ is a tree over $\mathcal{X}$, satisfying the following conditions:*
1. $\cup_{X \in \mathcal{X}} G[X] = G$, and 2. $\forall v \in V(G)$, $\{X \in \mathcal{X} \mid v \in X\}$ is connected in T.

A tree decomposition $(\mathcal{T}, \mathcal{X})$ of a graph G is nice if $\mathcal{T}$ is rooted, binary, each node $X \in \mathcal{X}$ has exactly w vertices, and is of one of the following three types: Leaf node, X is a leaf in $\mathcal{T}$ and consists of w pairwise non-adjacent vertices of G; Replace node, X has one child Y in $\mathcal{T}$, s.t. $X - Y = \{u\}$ and $Y - X = \{v\}$ for $u \neq v$; Join node, X has two children Y and Z in $\mathcal{T}$ with $X = Y = Z$.

The width of T is $\max_{X \in \mathcal{X}} |X| - 1$. The treewidth of G is the minimum width over all tree (nice) decompositions of G.

The algorithm follows a dynamic programming approach computing a table, for each node X of a nice tree decomposition of G, that depends on the pair of thresholds of the vertices in X. Each entry in the table stores the smallest seed set for the subgraph $G[X]$ of G induced by the vertices of the subtree rooted at X. The desired seed set for G is the one corresponding to the root node of the tree decomposition of G. The proof follows the lines of the one in [4] for the target set selection problem (e.g. in the special case $t_E = t_I$), except for the role played by vertices that need to be influenced but not evangelized and by the influence thresholds in computing the entries of the table for each node X. We can prove the following result.

Theorem 4. *In graphs of treewidth w the PES problem can be solved in $n^{O(w)}$ time.*

3   Polynomial Time Algorithms for MES

We show that the MES problem is polynomially solvable on complete graphs and trees. The proof of the following algorithm is omitted for the sake of space.

Theorem 5. *The algorithm MES -K(K, β) solves the MES problem on a clique of size n in O(n) time.*

In case of trees, we give a dynamic programming algorithm that proves Theorem 6.

Algorithm 2. Algorithm MES -K(K, β)

Input: A clique $K = (V, E)$, threshold functions t_I and t_E, budget $\beta \leq |V|$.
Output: S a seed set for K such that $|S| \leq \beta$
1 Order the nodes in V so that $t_E(v_1) \geq t_E(v_2), \ldots, t_E(v_n)$;
2 Set $S = X = \{v_1, v_2, \ldots, v_\beta\}$ and $\eta^* = |\mathsf{Evg}[X]|$
3 **while** ($\exists x \in S$, $t_I(x) \leq \eta^*$ *AND* $\exists y \in V - S$, $t_I(y) > \eta^*$) **do**
   $S = S - \{x\} \cup \{y\}$
4 **return** S

Theorem 6. *The MES problem with bound β can be solved in time $O(\min\{n\Delta^2\beta^3, n^2\beta^3\})$ on any tree with n nodes and maximum degree Δ.*

The rest of this section is devoted to the description and analysis of the algorithm proving Theorem 6. Let $T = (V, E)$ be a tree rooted at any node r and denote by $T(v)$ the subtree rooted at v, for $v \in V$. The algorithm makes a postorder traversal of the input tree T. For each node v, the algorithm solves all possible instances of the MES problem on the subtree $T(v)$, with bound $b \in \{0, 1, \ldots, \beta\}$. Moreover, in order to compute these values one has to consider—for the root node v of $T(v)$—not only the original thresholds $t_I(v)$ and $t_E(v)$ of v, but also the decremented values $t_I(v) - 1$ and $t_E(v) - 1$ which we call the *residual thresholds*. For each node $v \in V$ and integer $b \geq 0$ we define the following quantities:

$NO_v[b]$ is the maximum number of nodes that can be influenced in $T(v)$, (1)
   assuming that at most b of the nodes in $T(v)$ belong to the seed set,
   if v is still agnostic at the end of the evangelization process;

$Inf_v[b]$ is the maximum number of nodes that can be influenced in $T(v)$ (2)
   assuming that at most b of the nodes in $T(v)$ belong to the seed set,
   if, at the end of the process, v is influenced but it is not an evangelist;

$Evg_v[b]$ is the maximum number of nodes that can be influenced in $T(v)$ (3)
   assuming that at most b of the nodes in $T(v)$ belong to the seed set,
   if v is an evangelist at the end of the evangelization process.

Similarly the quantities $\widehat{NO}_v[b]$, $\widehat{Inf}_v[b]$ and $\widehat{Evg}_v[b]$ represent the same quantities as above but considering the decreased thresholds for v (which may reflect the fact that the parent node of v becomes an evangelist before v itself).

We define the above quantities be $-\infty$ if any of the constraints is not satisfiable. For instance, if v is a single node, $b = 0$ and $t_E(v) > 0$ we set[2] $Evg_v[0] = -\infty$.

The maximum number of nodes in T that can be influenced with any seed set of size β can be then obtained by computing

$$\max\{NO_r[\beta], Inf_r[\beta], Evg_r[\beta]\}. \qquad (4)$$

In order to obtain the value in (4), we compute the quantities[3] $NO_v[b]$, $Inf_v[b]$, $Evg_v[b]$, $\widehat{NO}_v[b]$, $\widehat{Inf}_v[b]$ and $\widehat{Evg}_v[b]$ for each $v \in V$ and for each $b = 0, 1, \ldots, \beta$.

We proceed postorder fashion on the tree, so that the computation of the various values for a node v is done after all the values for v's children are known.

For each leaf node ℓ we have the values below. Recall that they refer to the tree $T(\ell)$ consisting of the single node ℓ.

The node ℓ will be not even influenced only if the budget is not sufficient to have ℓ in the seed set (e.g. $b = 0$) while the influence threshold is $t_I(\ell) > 0$. Hence,

$$NO_\ell[b] = \begin{cases} 0 & \text{if } (b = 0 \text{ AND } t_I(\ell) > 0) \\ -\infty & \text{otherwise,} \end{cases} \qquad (5)$$

The node ℓ gets influenced but does not become an evangelist in case the budget is not sufficient to have ℓ in the seed set (e.g. $b = 0$) and the evangelization threshold is $t_E(\ell) > 0$, but the influence threshold is $t_I(\ell) = 0$. Hence,

$$Inf_\ell[b] = \begin{cases} 1 & \text{if } (b = 0 \text{ AND } t_I(\ell) = 0 \text{ AND } t_E(\ell) > 0) \\ -\infty & \text{otherwise.} \end{cases} \qquad (6)$$

The node ℓ becomes evangelist in $T(\ell)$ when either the budget is sufficiently large to have ℓ in the seed set ($b \geq 1$) or its evangelization threshold is $t_E(\ell) = 0$. Hence,

$$Evg_\ell[b] = \begin{cases} 1 & \text{if } (b \geq 1 \text{ OR } t_E(\ell) = 0) \\ -\infty & \text{otherwise.} \end{cases} \qquad (7)$$

The values for $\widehat{NO}_\ell[b]$, $\widehat{Inf}_\ell[b]$ and $\widehat{Evg}_\ell[b]$ are computed similarly by using on ℓ the residual thresholds ($t_I(\ell) - 1$ and $t_E(\ell) - 1$) instead of $t_I(\ell)$ and $t_E(\ell)$.

Lemma 2. *For any internal node v and for any integer $b \in \{0, \ldots, \beta\}$, each of the values $NO_v[b]$, $Inf_v[b]$, $Evg_v[b]$, $\widehat{NO}_v[b]$, $\widehat{Inf}_v[b]$, and $\widehat{Evg}_v[b]$ can be computed in time $O(d^2 b^2)$, where d is the number of children of v in T.*

It follows that the value in (4) can be computed in time $\sum_{v \in V} O(d(v)^2 \beta^2) \times O(\beta) = O(\beta^3) \times \sum_{v \in V} O(d(v)^2) = O(\min\{n\Delta^2 \beta^3, n^2 \beta^3\})$, where Δ is the maximum node degree. Standard backtracking techniques can be used to compute a seed set of size at most β that influences this maximum number of nodes in the same $O(\min\{n\Delta^2 \beta^3, n^2 \beta^3\})$ time. This concludes the proof of Theorem 6.

---

[2] Indeed v should be an evangelist, however the budget is 0 while the threshold is > 0.

[3] For the root node r, the quantities $\widehat{NO}_r[b]$, $\widehat{Inf}_r[b]$ and $\widehat{Evg}_r[b]$ are not required.

4   Dense Graphs

In this section we concentrate on the PERFECT EVANGELIC SET problem in graphs characterized by large minimum degree. In particular, we relate the graph minimum degree to the size of the smallest perfect seed set, e.g., a set $S \subseteq V$ such that $\mathsf{Inf}[S] = V$.

Assuming that $t_I(v) \leq t_I$ and $t_E(v) \leq t_E$, for each $v \in V$, and $t_E + t_I \leq |V| + 2$, the algorithm PES(G, t_E, t_I) selects and returns a set $S \subseteq V$, of size at most $2(t_E - 1)$, that we will prove to be a PES for G whenever the minimum degree of G is $\frac{|V| + t_E + t_I}{2} - 2$.

The construction of the set S returned by the algorithm PES(G, t_E, t_I), immediately implies the fact below.

Fact 1. *1) If $|S| < 2(t_E - 1)$ then each $v \in V - S$ has at least t_I neighbors in S. 2) If $|S| = 2(t_E - 1)$ then the sum of the degrees of the nodes in the subgraph induced by S in G is upper bounded by $[t_I(t_I - 1) - 2] + 2(t_I - 1)[2(t_E - 1) - t_I] = (t_I - 1)(4t_E - t_I - 4) - 2$ if $t_I \geq 2$; it is 0 if $t_I = 1$.*

Theorem 7. *Let $G = (V, E)$ be a graph on n nodes with $t_I(v) \leq t_I$, $t_E(v) \leq t_E$, for each $v \in V$, where $t_E + t_I \leq n + 2$, and $d(v) \geq \frac{n + t_E + t_I}{2} - 2$, for each $v \in V$. The algorithm PES(G, t_E, t_I) returns a PES for G of size at most $2t_E - 2$.*

Proof. Consider the evangelization process in G starting at the set S returned by the algorithm PES(G, t_E, t_I). Let $i \in \{0, 1, \ldots\}$ be a round of the process and $a(i) = |\mathsf{Evg}[S, i] - S|$ be the number of evangelists at round i that not belong to the seed set S. If $V - \mathsf{Inf}[S, i] = \emptyset$ then each node in $V - \mathsf{Evg}[S, i]$ has at least t_I neighbors in $\mathsf{Evg}[S, i]$ and the theorem is proved. Assume then $V - \mathsf{Inf}[S, i] \neq \emptyset$. By 1) of Fact 1, we know that $|S| = 2(t_E - 1)$. Let $\sigma(\mathsf{Evg}[S, i])$ denote the number of edges in the subgraph of G induced by $\mathsf{Evg}[S, i]$. In the following we assume that $t_I \geq 2$. The proof for $t_I = 1$ can be obtained similarly recalling that the value in 2) of Fact 1 is 0 in this case. By 2) of Fact 1 and since each node in $\mathsf{Evg}[S, i] - S$ is connected at most to each other node in $\mathsf{Evg}[S, i] \cup S$, we have that sum of the degrees of the nodes in the subgraph of G induced by $\mathsf{Evg}[S, i]$ is

$$2\sigma(\mathsf{Evg}[S, i]) \leq (t_I - 1)(4t_E - t_I - 4) - 2 + a(i)(a(i) - 1) + 2a(i)[2(t_E - 1)]$$
$$= (t_I - 1)(4t_E - 4 - t_I) - 2 + a(i)^2 + a(i)(4t_E - 5). \tag{8}$$

Recalling that $d(v) \geq \frac{n + t_E + t_I}{2} - 2$ for each $v \in V$, we get that the number $\sigma(\mathsf{Evg}[S, i], V - \mathsf{Evg}[S, i])$ of edges connecting one node in $\mathsf{Evg}[S, i]$ and one in $V - \mathsf{Evg}[S, i]$ is

$$\sigma(\mathsf{Evg}[S, i], V - \mathsf{Evg}[S, i])$$
$$\geq \frac{n + t_E + t_I - 4}{2}[2(t_E - 1) + a(i)] - [(t_I - 1)(4t_E - 4 - t_I) - 2 + a(i)^2 + a(i)(4t_E - 5)]$$

$$= (n + t_E - 3t_I)(t_E - 1) + (t_I - 1)t_I + 2 - a(i)^2 + a(i)\left(\frac{n + t_E + t_I}{2} - 4t_E + 3\right) \tag{9}$$

Algorithm 3. Algorithm $\text{PES}(G, t_E, t_I)$

Input: A graph $G = (V, E)$ having thresholds $t_I(v) \leq t_I$ and $t_E(v) \leq t_E$ for $v \in V$.
Output: S, a perfect seed set for G.

1 Set S as any subset of V such that
2   - $|S| = t_I$ and
3   - at least two nodes in S are independent, if possible [e.g., if G is not a clique],
4 **while** $(|S| < 2(t_E - 1))$ **AND** $(\exists v \in V - S \text{ s.t. } |N(v) \cap S| \leq t_I - 1)$ **do**
   $S = S \cup \{v\}$
5 **return** S

We first determine the minimum value of $a(i)$ that guaranties that at least one node $v \in V - \text{Evg}[S, i]$ becomes an evangelist at round $i + 1$. By contradiction assume that each node in $V - \text{Evg}[S, i]$ has at most $t_E - 1$ neighbors in $\text{Evg}[S, i]$. This assumption implies that $\sigma(\text{Evg}[S, i], V - \text{Evg}[S, i]) \leq (n - 2(t_E - 1) - a(i))(t_E - 1)$.

It is not hard to see that the lower bound in (9) is larger than the above upper bound when $0 \leq a(i) \leq \frac{n + t_E + t_I}{2} - 3t_E + 2$. This leads to a contradiction for such a range of values of $a(i)$. Hence, for each round i for which $0 \leq a(i) \leq \frac{n + t_E + t_I}{2} - 3t_E + 2$ at least one node $v \in V - \text{Evg}[S, i]$ moves from $V - \text{Evg}[S, i]$ to $\text{Evg}[S, i + 1]$ at round $i + 1$.

We show now that if $a(i) = \frac{n + t_E + t_I}{2} - 3t_E + 2$ (i.e., $|\text{Evg}[S, i + 1]| \geq 2(t_E - 1) + \frac{n + t_E + t_I}{2} - 3t_E + 3$) then $|V - \text{Inf}[S, i + 1]| = 0$, thus completing the proof.
Indeed, we have $|V - \text{Evg}[S, i + 1]| \leq n - [2(t_E - 1) + \frac{n + t_E + t_I}{2} - 3t_E + 3] = \frac{n - (t_E + t_I)}{2} + t_E - 1$. This implies that the number of evangelists among the neighbors of any node $v \in V - \text{Evg}[S, i + 1]$ is at least

$$\frac{n + t_E + t_I}{2} - 2 - \frac{n - (t_E + t_I)}{2} - t_E + 2 = t_I.$$

Hence at round $i+1$ each node in $V - \text{Evg}[S, i]$ is influenced and $|V - \text{Inf}[S, i + 1]| = 0$. $\square$

Notice that if $t_E = t_I = 2$, we get the result for Dirac graphs given in [18].

Corollary 1. *Let G be a graph with $d(v) \geq \frac{n}{2}$, for each $v \in V$. The algorithm PES($G, 2, 2$) returns an optimal PES for G of size 2.*

References

1. Ackerman, E., Ben-Zwi, O., Wolfovitz, G.: Combinatorial model and bounds for target set selection. Theor. Comput. Sci. **411**, 4017–4022 (2010)
2. Adamic, L.A., Lento, T.M., Adar, E., Ng, P.C.: Information evolution in social networks. In: Proceedings of the 9th ACM WSDM 2016, pp. 473–482 (2016)
3. Bazgan, C., Chopin, M., Nichterlein, A., Sikora, F.: Parametrized approximability of maximizing the spread of influence in networks. J. Discrete Algorithms **27**, 54–65 (2014)

4. Ben-Zwi, O., Hermelin, D., Lokshtanov, D., Newman, I.: Treewidth governs the complexity of target set selection. Discrete Optim. **8**, 87–96 (2011)
5. Chen, N.: On the approximability of influence in social networks. SIAM J. Discrete Math. **23**, 1400–1415 (2009)
6. Chen, W., Lakshmanan, L.V.S., Castillo, C.: Information and Influence Propagation in Social Networks. Morgan & Claypool, San Rafael (2013)
7. Centeno, C.C., Dourado, M.C., Draque Penso, L., Rautenbach, D., Szwarcfiter, J.L.: Irreversible conversion of graphs. Theor. Comput. Sci. **412**(29), 3693–3700 (2011)
8. Chopin, M., Nichterlein, A., Niedermeier, R., Weller, M.: Constant thresholds can make target set selection tractable. Theor. Comput. Syst. **55**, 61–83 (2014)
9. Chiang, C.-Y., Huang, L.-H., Li, B.-J., Wu, J., Yeh, H.-G.: Some results on the target set selection problem. J. Comb. Optim. **25**, 702–715 (2013)
10. Cicalese, F., Cordasco, G., Gargano, L., Milanič, M., Vaccaro, U.: Latency-bounded target set selection in social networks. Theor. Comput. Sci. **535**, 1–15 (2014)
11. Cicalese, F., Cordasco, G., Gargano, L., Milanič, M., Peters, J.G., Vaccaro, U.: Spread of influence in weighted networks under time and budget constraints. Theor. Comput. Sci. **586**, 40–58 (2015)
12. Coja-Oghlan, A., Feige, U., Krivelevich, M., Reichman, D.: Contagious sets in expanders. In: Proceedings of SODA 2015, pp. 1953–1987 (2015)
13. Cordasco, G., Gargano, L., Mecchia, M., Rescigno, A.A., Vaccaro, U.: A fast and effective heuristic for discovering small target sets in social networks. In: Lu, Z., Kim, D., Wu, W., Li, W., Du, D.-Z. (eds.) Combinatorial Optimization and Applications. LNCS, vol. 9486, pp. 193–208. Springer, Heidelberg (2015)
14. Dawkins, R.: The Selfish Gene. Oxford University Press, Oxford (1989)
15. Downey, R.G., Fellows, M.R.: Parameterized Complexity. Springer, Heidelberg (2012)
16. Easley, D., Kleinberg, J.: Networks, Crowds, and Markets: Reasoning About a Highly Connected World. Cambridge University Press, New York (2010)
17. Fiala, J., Gavenciak, T., Knop, D., Koutecky, M., Kratochvíl, J.: Fixed parameter complexity of distance constrained labeling and uniform channel assignment problems (2015). arXiv:1507.00640
18. Freund, D., Poloczek, M., Reichman, D.: Contagious sets in dense graphs. In: Lipták, Z. (ed.) IWOCA 2015. LNCS, vol. 9538, pp. 185–196. Springer, Heidelberg (2016). doi:10.1007/978-3-319-29516-9_16
19. R. Ganian,Using neighborhood diversity to solve hard problems (2012). arXiv:1201.3091
20. Gargano, L., Rescigno, A.A.: Complexity of conflict-free colorings of graphs. Theor. Comput. Sci. **566**, 39–49 (2015)
21. Gargano, L., Hell, P., Peters, J.G., Vaccaro, U.: Influence diffusion in social networks under time window constraints. Theor. Comput. Sci. **584**, 53–66 (2015)
22. Kawasaki, G.: Selling the Dream: How to Promote Your Product, Company, or Ideas and Make a Difference Using Everyday Evangelism. HarperCollins, New York (1991)
23. Kempe, D., Kleinberg, J.M., Tardos, E.: Maximizing the Spread of Influence through a Social Network. Theor. Comput. **11**, 105–147 (2015)
24. Lampis, M.: Algorithmic meta-theorems for restrictions of treewidth. Algorithmica **64**, 19–37 (2011)
25. Leppaniemi, M., et al.: Targeting young voters in a political campaign: empirical insights into an interactive digital marketing campaign in the 2007 finnish general election. J. Nonprofit Public Sect. Mark. **22**, 14–37 (2010)

26. McConnell, B., Huba, J.: Creating Customer Evangelists: How Loyal Customers Become a Volunteer Sales Force. Lewis Lane Press, USA (2012)
27. Nichterlein, A., Niedermeier, R., Uhlmann, J., Weller, M.: Tractable cases of target set selection. Soc. Netw. Anal. Min. **3**, 1–24 (2012)
28. Reddy, T.V.T., Rangan, C.P.: Variants of spreading messages. J. Graph Algorithms Appl. **15**(5), 683–699 (2011)
29. Tumulty, K.: Obama's Viral Marketing Campaign. TIME Magazine, July 5 2007

Distance Queries in Large-Scale Fully Dynamic Complex Networks

Gianlorenzo D'Angelo[1], Mattia D'Emidio[1(✉)], and Daniele Frigioni[2]

[1] Gran Sasso Science Institute (GSSI), Viale F. Crispi 7, 67100 L'Aquila, Italy
{gianlorenzo.dangelo,mattia.demidio}@gssi.infn.it
[2] Dipartimento di Ingegneria e Scienze dell'Informazione e Matematica,
Universitá degli Studi dell'Aquila, Via Vetoio, 67100 L'Aquila, Italy
daniele.frigioni@univaq.it

Abstract. The 2-hop cover labeling of a graph is a data structure that recently received a lot of attention since it can be exploited to efficiently answer to shortest-path distance queries on large-scale networks. In this paper, we propose the first dynamic algorithm to update 2-hop cover labelings for distance queries under edge removals, and show that: (i) it is efficient in terms of the number of nodes that change their distance toward some other node of the network, as a consequence of an edge removal; (ii) it is able to preserve the *minimality* of the labeling, a desirable property that has impact on both size and query time. In addition, we combine the new method with the unique algorithm in the literature suitable to handle edge additions, thus obtaining the first *fully dynamic* algorithm for updating 2-hop cover labelings for distance queries. We also conduct an extensive experimental study on real and synthetic dynamic networks, to show the scalability and efficiency of our new methods.

1 Introduction

Answering to *shortest-path distance queries* between pairs of nodes of a graph is one of the most fundamental operations on graph data, as it is a building block of some of the most important applications in modern networked systems, such as social networks analysis [16], context-aware search [14], route planning in road networks [1,9], journey planning in transport systems [6], routing and management of resources in computer networks [4].

Distance queries can be easily answered either by using a breadth first search (BFS) on unweighted graphs, or the Dijkstra's algorithm on positively weighted graphs. Unfortunately, networks deriving from real-world applications tend to be huge, yielding unsustainable times to compute shortest paths. For this reason, many smarter methods for efficiently answering distance queries in different application scenarios have been proposed [1,2,5,7,9,13,14], the majority of

Research partially supported by the Italian Ministry of University and Research under the Research Grant 2012C4E3KT PRIN 2012 "AMANDA" (Algorithmics for MAssive and Networked DAta).

© Springer International Publishing Switzerland 2016
V. Mäkinen et al. (Eds.): IWOCA 2016, LNCS 9843, pp. 109–121, 2016.
DOI: 10.1007/978-3-319-44543-4_9

which relies on a preprocessing phase that precomputes data to be exploited for reducing the time required for answering queries. Some of the most efficient of these methods are based on the notion of *2-hop cover labeling* [7]. Among them, the recent *pruned landmark labeling* (PLL) [2] achieves considerable better scalability than other methods in several real-world networks [9].

However, many real-world networks are dynamic and rapidly changing, while most of the above methods have been thought for static networks. In the dynamic case, in order to keep queries correct, the preprocessed data need to be updated after each change. The easiest way is to recompute everything from-scratch. This is in general infeasible in modern large-scale networks since even the fastest methods require too much time. In recent years, some techniques have been developed to solve this issue for classes of dynamic graphs that exhibit a well defined structure, such as road networks [8,10]. However, for *complex* (i.e. with non-trivial topological features) dynamic networks very little has been done.

Recently in [3], a dynamization of PLL has been proposed, which focuses on the *incremental* problem, i.e. on handling *edge additions* only. However, no solution is given w.r.t. the *decremental* problem, i.e. when *edge removals* need to be managed. They motivate this choice by several reasons. First, solving the incremental problem appears to be already quite technically challenging. Second, the authors claim that the decremental problem is even harder, and a solution able to efficiently solving it without making big compromises on performance (e.g. labeling size and hence query time) seems to be very difficult to devise. Finally, the authors claim that removals either never happen in certain kinds of real-world dynamic networks such as interaction networks, and co-author networks, or happen with very low frequency in other kinds of real dynamic networks.

However, there are many prominent scenarios where removals are possible, and often very frequent. Examples are the management of disruptions in both public transportation systems and road-networks, the management of congestions in communication networks, the management of removals of links in network-aware search indices [4,6,12]. More in general, efficiently solving the decremental problem seems to be a crucial building block of all those real-world applications that rely on dynamic large-scale graph-like data, as for instance graph database systems or software modeling tools.

In this paper, we propose the first decremental algorithm for updating 2-hop cover labelings for shortest-path distance query, and show that: (i) it is efficient in terms of the number of nodes that change their distance toward some other node of the network, as a consequence of an edge removal; (ii) it is able to preserve the *minimality* of the labeling, a desirable property that has impact on both size and query time. In addition, we combine the new method with the unique algorithm able to update 2-hop cover labelings in case of incremental graph updates, proposed in [3], thus obtaining the first *fully dynamic* algorithm for updating 2-hop cover labelings. Finally, we propose an extensive experimental study on real and synthetic networks, to demonstrate the scalability and efficiency of our new methods. Our experiments show that: (i) our approaches are orders of magnitude faster than the recomputation from scratch for updating even massively sized labelings; (ii) labelings updated by our algorithms are able to answer to distance queries in large-scale networks in microseconds and their size does not change as graph updates occur. On the one hand our data support our theoretical results in confirming that the decremental problem is much harder to be solved w.r.t. the incremental one. In fact, the time required for updating the labeling in the decremental case is much higher than that required for handling the incremental scenario. On the other hand, our experimental results clearly contradicts the conjecture given in [3] by showing that a practically efficient method for the fully dynamic case can be developed without any compromise on performance.

2   Preliminaries

Given an undirected graph $G = (V, E)$ with $n = |V|$ nodes and $m = |E|$ edges, we denote by $N(v)$ the set of the neighbors of v in G, that is $N(v) = \{u \in V \mid \{u, v\} \in E\}$, and by $d(u, v)$ the distance between nodes u and v, that is the number of edges in a shortest path between u and v. If u and v are not connected, then $d(u, v) = \infty$. A *modification* or *update* to a graph is either the *addition* (also referred as *insertion*) or the *removal* (also referred as *deletion*) of an edge. Nodes additions and removals can be modeled as modifications of multiple edges incident to the same node. We assume that time is described by positive integers and that graph updates occur at each integer t. Symbol G_t denotes the graph after t modifications, where G_0 is the initial graph. Similarly, N_t and d_t denote neighbor and distance functions in G_t, respectively. We omit the parameter t when it is clear by the context.

2-hop Cover Labeling. For each node v of G, the *label* $L(v)$ of v is a set of pairs (u, δ_{uv}), where u is a node in V and $\delta_{uv} = d(u, v)$. The set $\{L(v)\}_{v \in V}$ is referred to as a *labeling*. We use $u \in L(v)$ instead of $(u, \delta_{uv}) \in L(v)$ when the meaning is clear from the context. Labels can be used to answer to a *query* on the distance between two nodes s and t as follows: $\text{QUERY}(s, t, L) = \min\{\delta_{vs} + \delta_{vt} \mid v \in L(s) \wedge v \in L(t)\}$ if $L(s) \cap L(t) \neq \emptyset$, and $\text{QUERY}(s, t, L) = \infty$ otherwise. A labeling L is called a *2-hop cover labeling* of G if, for each pair $s, t \in V$, $L(s) \cap L(t)$ contains at least a node u in a shortest path between s and t (or it is empty if s and t are disconnected). If such a node u exists, it is said to be a *hub* of pair (s, t). It can be easily proven that in a 2-hop cover labeling $\text{QUERY}(s, t, L) = d(s, t)$, for each $s, t \in V$ [7]. For each node $v \in V$, if $L(v)$ is sorted according to the IDs of its nodes, then computing $\text{QUERY}(s, t, L)$ takes $O(|L(s)| + |L(t)|)$ time.

Given a graph G, two nodes s and t in V, and a labeling L of G, a shortest path P from s to t is *induced* by L if for any two nodes u and v in P there exists a hub h of (u, v) such that $h \in P$, or $h = u$, or $h = v$. The set of shortest paths between nodes s and t induced by L is denoted by $\text{PATH}(s, t, L)$. A labeling L is *minimal* if and only if, for each $v \in V$ and for each $(u, \delta_{uv}) \in L(v)$, there exist two nodes s, t such that $\text{QUERY}(s, t, L') \neq d_i(s, t)$, where L' is obtained from L by removing (u, δ_{uv}) in $L(v)$.

In the literature, two algorithms for the computation of a 2-hop cover labeling of a graph are known [2]. The first method is called *naive landmark labeling* and is based on a full computation of n BFSs, starting from all the nodes of the graph. The second method is called *pruned landmark labeling* (PLL) and consists in a tailored pruning of the BFSs of the previous method. For space constraints we omit the details of such algorithms, and refer to [2] for a detailed description.

Even if the two methods exhibit the same worst-case time and space complexity, it has been shown that the latter performs much better in practice [2,9]. In particular, the performance of PLL heavily depends on the ordering of V. It has been experimentally observed that the average label size (and the time complexity of the query algorithm) decreases by several orders of magnitude if the nodes are sorted according to a centrality measure, like e.g. degree or closeness, instead of a random ordering [2]. The best way to select an ordering depends on the graph structure, only recently it has been proposed a sorting algorithm that leads to small label sizes in many graph classes [9]. Note that, given a graph, it has been shown that computing an ordering on the nodes that delivers a labeling of minimum size is NP-Hard [11].

The PLL method guarantees the *well-ordering* property [15] that is, if $v < u$, then u is not in $L(v)$, while v might be in $L(u)$. This property is quite important, as it can be used to prove that the computed labeling is minimal [2]. Minimality is an highly desirable property, given the hardness of finding 2-hop covers of minimum size. In details, PLL builds labeling that are minimal.

Incremental Algorithm. The only known algorithm for the incremental problem has been proposed in [3], and is called ADD in what follows. A peculiarity of ADD is that it does not remove *outdated label entries* from the original labeling, i.e. entries that correspond to distances that have decreased as a consequence of the insertion of an edge. Therefore, they are present also in the updated labeling.

The above observation is motivated by the fact that distances can only decrease due to edge additions and hence they cannot be underestimated because of outdated label entries, i.e. it is possible to correctly answer to distance queries even in presence of such pairs, since the query algorithm searches for the minimum. Moreover, removing outdated entries is avoided also because their detection is computationally expensive. Under this strategy, the minimality of the resulting labeling as a whole can be broken even after a single update, although the set of newly added entries is minimal to answer correct distances. Due to space constraints, we refer to [3] for a thorough description of ADD.

3   Decremental Algorithm

In this section, we present our new algorithm to update a 2-hop cover labeling L of a given graph G under edge removals, called REMOVE. Notice that, in case of removals, outdated label entries cannot be simply ignored, as they might induce underestimated, uncorrect distances. For this reason, the detection and the removal of such entries, unlike the case of edge additions, cannot be avoided. Due to space limitations, the proofs of our results will be given in the full paper.

Let us assume that an edge $\{x, y\}$ is removed from G_{i-1} and that G_i is the resulting graph, for $i \geq 1$. We say that a node v is *affected* by such a removal if there exists a shortest path induced by L between v and any other node u, i.e. a path in PATH(v, u, L) that passes through edge $\{x, y\}$ in G_{i-1}. The set of affected nodes is denoted as *Aff*.

The set of affected nodes is exploited to detect the labels that change as a consequence of the removal of $\{x, y\}$. Indeed, assume that, for a pair of nodes u and v, one of the paths in PATH(u, v, L) passes through edge $\{x, y\}$ in G_{i-1}. Let h be the hub of pair (u, v) corresponding to such path, then one label between (h, δ_{uh}) in $L(u)$ or (h, δ_{vh}) in $L(v)$ might not be correct and hence it must be removed from $L(u)$ or $L(v)$, respectively, and, in this case, a new hub between u and v must be computed (see Fig. 1). This might hold also for the labels of each other node z that contains u (or v) in $L(z)$ and exploit it as a hub from z to v (or u).

Fig. 1. A shortest path between nodes u and v. The solid line represents edge $\{x, y\}$, while dashed lines represent shortest paths. Assume that $h \in L(u) \cap L(v)$, then h is a hub for pair (u, v). If $\{x, y\}$ is removed, then $v, h \in Aff(x)$, $u \in Aff(y)$, and label (h, δ_{uh}) in $L(u)$ is not correct and must be updated.

For each affected node v, $d_{i-1}(v, x) < d_{i-1}(v, y)$ or $d_{i-1}(v, x) > d_{i-1}(v, y)$. Therefore, we divide set *Aff* into two disjoint subsets $Aff(x)$ and $Aff(y)$ that contain the affected nodes closer to x or to y, respectively. Moreover, we observe that if u and v are two affected nodes such that $u \in Aff(x)$ and $v \in Aff(y)$ and a shortest path between u and v passes through $\{x, y\}$, then a hub h of pair (u, v) is also affected and either $h \in Aff(x)$ or $h \in Aff(y)$ (see Fig. 1). The next lemma gives us a way to exploit the set of affected nodes to identify the pairs in L that must be updated.

Lemma 1. *If L is a 2-hop cover labeling of G_{i-1}, then, for each $u, v \in V$, QUERY$(u, v, L) \neq d_i(u, v)$ only if $v \in Aff(x)$ and $u \in Aff(y)$ or $v \in Aff(y)$ and $u \in Aff(x)$.*

By Lemma 1, if L is a 2-hop cover labeling of G_{i-1}, then QUERY$(u, v, L) \neq d_i(u, v)$ only if $v \in Aff(x)$ and $u \in Aff(y)$ (or symmetrically $v \in Aff(y)$ and $u \in Aff(x)$). Moreover, there exists a hub h of pair (u, v) that is also affected and either $h \in Aff(x)$ or $h \in Aff(y)$. In the former case, QUERY$(u, h, L) \neq d_i(u, h)$, while in the latter case QUERY$(v, h, L) \neq \cdot d_i(v, h)$ (Fig. 1 depicts the former case). By repeating this argument, we obtain a pair of affected nodes h', h'' in PATH(u, v, L) such that $(h', \delta_{h'h''}) \in L(h'')$ but $\delta_{h'h''} \neq d_i(h', h'')$. In this case, $(h', \delta_{h'h''})$ must be removed from $L(h'')$ and updated, possibly by finding a new hub for (h', h''), and also for all hubs in the considered path between u and v.

Lemma 2. *To obtain a 2-hop cover labeling of G_i from a 2-hop cover labeling L of G_{i-1} it is enough to update all the pairs (u, δ_{vu}) in label $L(v)$, for all the nodes v and u such that $v \in Aff(x)$ and $u \in Aff(y)$ or $v \in Aff(y)$ and $u \in Aff(x)$.*

Lemma 2 implies that all pairs (u, δ_{vu}) in $L(v)$, for all nodes v and u such that $v \in Aff(x)$ and $u \in Aff(y)$ or $v \in Aff(y)$ and $u \in Aff(x)$ may not be correct in G_i and therefore our approach is to remove them from $L(v)$ and recompute a new hub for (u, v). In fact, note that it is not enough to update (u, δ_{vu}) in $L(v)$ by setting $\delta_{vu} = d_i(u, v)$, but the computation of a new hub might be required.

Our new algorithm REMOVE is made of three phases. In the first phase, we compute sets $Aff(x)$ and $Aff(y)$; in the second phase, we remove the pairs satisfying the condition of Lemma 1; and in the third phase we recompute the missing hubs in order to restore the 2-hop cover labeling according to Lemma 2.

Detecting Affected Nodes. The pseudocode of the algorithm to compute set $Aff(x)$ is given in Algorithm 1, the algorithm to compute $Aff(y)$ is symmetrical and it is not reported. The set of affected nodes is stored in variable $A(x)$. We assume that $A(x)$ is kept sorted according to the node ordering, this could be done e.g. by maintaining $A(x)$ as a heap.

Algorithm 1 mimics a BFS rooted at x but it prunes some of the branches if the reached node is not affected. We use a queue Q to store the nodes to visit and a boolean vector mark to keep track of the visited nodes. At line 3, we initialize $A(x)$, Q, and mark. Then, the pruned BFS is performed at lines 4–11 starting from x. A node v is enqueued in Q only if it is affected. Let v be a node extracted from Q, then v is inserted into $A(x)$ and mark$[v]$ is set to true at line 10. Then, each neighbor u of v is analyzed in order to check whether it is affected and hence must be inserted into Q. Let H be the set of hubs of pair (u, y) in G_{i-1}, then u is inserted into Q if there exists a hub $h \in H$ such that one of the two following conditions holds (see line 9): (i) $h \in A(x)$; (ii) $d_{i-1}(v, y) = d_{i-1}(v, x) + 1$ and either $h = v$ or $h = y$. Condition (i) checks whether a hub h has been already identified as affected. In this case, there exists a shortest path from h to y in PATH(h, y, L) that contains edge $\{x, y\}$ and therefore also one of the shortest paths from u to y in PATH(u, y, L) must contain edge $\{x, y\}$. Note that, since we are performing a BFS starting from x, then the hub h of pair (u, y) is analyzed by the algorithm before u and, if it is affected, then it is inserted in $A(x)$ before u being analyzed from the condition at line 9. Condition (ii) handles the case in which a hub of pair (u, y) is u or y. In this case, u is affected if any shortest path from u to y passes through edge $\{x, y\}$ and therefore we check whether $d_{i-1}(v, y) = d_{i-1}(v, x) + 1$, in fact, in such a case there exists at least a shortest path from u to y passing through edge $\{x, y\}$. If the condition at line 9 is not satisfied, node u is not inserted into Q and then the search is pruned at u.

Lemma 3. *At the end of Algorithm 1, $A(x) = Aff(x)$.*

Removing Affected Hubs. Algorithm 2 removes labels (u, δ_{uv}) from $L(v)$, for all affected nodes u and v that satisfy the condition of Lemma 1. The following lemma follows by simply observing that, by Lemma 3, $A(x) = Aff(x)$.

Algorithm 1. Compute $A(x)$

---

1 **foreach** $v \in V$ **do**
2 $\quad$ mark$[v] \leftarrow$ false;
3 $A \leftarrow \emptyset$; $Q \leftarrow \emptyset$; mark$[x] \leftarrow$ true; Q.Enqueue(x);
4 **while** $Q \neq \emptyset$ **do**
5 $\quad$ $v \leftarrow$ Q.Dequeue();
6 $\quad$ $A(x) \leftarrow A(x) \cup \{v\}$;
7 $\quad$ **foreach** $u \in N_i(v)$: $\neg$mark$[u]$ **do**
8 $\quad\quad$ Let H be the set of hubs of pair (u, y) in G_{i-1};
9 $\quad\quad$ **if** $\exists h \in H$: $h \in A(x) \vee ((h = u \vee h = y) \wedge d_{i-1}(u,y) = d_{i-1}(u,x) + 1)$
    then
10 $\quad\quad\quad$ mark$[u] \leftarrow$ true;
11 $\quad\quad\quad$ Q.Enqueue(u);

Algorithm 2. Remove affected hubs from affected labels

---

1 **foreach** (u, v) : $((v \in A(x) \wedge u \in A(y)) \vee (v \in A(y) \wedge u \in A(x))) \wedge u \in L(v)$ **do**
2 $\quad$ Remove (u, δ_{uv}) from $L(v)$;

Lemma 4. *At the end of Algorithm 2, L does not contain any pair that satisfies the condition of Lemma 1.*

Computing New Hubs. Let us assume w.l.o.g. that $|A(x)| < |A(y)|$. Moreover, we assume that $d_i(x, y) \neq \infty$, since otherwise L is already correct and algorithm REMOVE terminates just after the end of Algorithm 2. Algorithm 3 updates the labels of affected nodes in the case that $d_i(x, y) \neq \infty$. We run Algorithm 3 only on nodes in $A(x)$ and we do not need to run it also on nodes in $A(y)$.

Algorithm 3 performs a BFS rooted at each node $a \in A(x)$. As in Algorithm 1, we use a queue Q to store the nodes to visit and a boolean vector mark to keep track of the visited nodes. We use a vector dist to store the distances between a and any other node. At lines 2–8, we initialize Q, mark, and dist. At lines 10–17, we perform a BFS rooted at each $a \in A(x)$. Let v be the node currently extracted from Q and L be the current label set. If $v \in A(y)$, then the result of a query between a and v that uses the label set might be wrong due to Lemma 1. Therefore, we check whether dist$[v] <$ QUERY(a, v, L) and, in the affirmative case, the path between a and v found by the BFS is shorter than those induced by L and hence we update L either by inserting pair $(v, \text{dist}[v])$ in $L(a)$ or by inserting pair $(a, \text{dist}[v])$ in $L(v)$, depending on the relative ordering of v and a (see lines 12–14). Note that, we do not compute the distances between all the pairs (u, v) such that $v \in A(x)$ and $u \in A(y)$ since in some case a hub h between u and v might have been found in previous iterations of the algorithm.

Theorem 1. *At the end of Algorithm 3, L is a 2-hop cover labeling.*

The next theorem shows that the labeling resulting from Algorithm 3 is minimal. On the one hand, this leads to a reduced query time. On the other

---

Algorithm 3. Compute new hubs

---

1 **foreach** $a \in A(x)$ **do**
2 | Q $\leftarrow \emptyset$; mark$[a] \leftarrow$ true; dist$[a] \leftarrow 0$;
3 | **foreach** $v \in V \setminus \{a\}$ **do**
4 | | mark$[v] \leftarrow$ false;
5 | | dist$[v] \leftarrow \infty$;
6 | **foreach** $v \in N_i(a)$ **do**
7 | | Q.Enqueue(v);
8 | | dist$[v] \leftarrow 1$;
9 | | mark$[v] \leftarrow$ true;
10 | **while** $Q \neq \emptyset$ **do**
11 | | $v \leftarrow$ Q.Dequeue();
12 | | **if** dist$[v] <$ QUERY$(a, v, L) \wedge v \in A(y)$ **then**
13 | | | **if** $v < a$ **then** Insert $(v, \text{dist}[v])$ in $L(a)$;
14 | | | **else** Insert $(a, \text{dist}[v])$ in $L(v)$;
15 | | **foreach** $u \in N_i(v) : \neg\text{mark}[u]$ **do**
16 | | | dist$[u] \leftarrow$ dist$[v] + 1$;
17 | | | Q.Enqueue(u);
18 | | | mark$[u] \leftarrow$ true;

---

hand, it implies that the deletion of a single hub in the labeling would lead to some incorrect queries. However, in Algorithm 1, we exactly identify the hubs that need to be updated in order to keep all the queries correct.

Theorem 2. *If L is a minimal 2-hop cover labeling of G_{i-1} that satisfies the well-ordering property, then at the end of Algorithm 3, L is a minimal 2-hop cover labeling of G_i that satisfies the well-ordering property.*

Note that, in the worst case REMOVE is not better than PLL, which requires $O(nm + n^2)$ time. However, in what follows we show that it is efficient in terms of the number of affected vertices. Given a 2-hop cover labeling L of G, we denote by l the maximum size of the labels in L, i.e. $l = \max_{v \in V} |L(v)|$, by $\hat{A}$ the biggest in size between sets $A(x)$ and $A(y)$, i.e. $\hat{A} = \arg\max\{|A(x)|, |A(y)|\}$, and by $m_{\hat{A}}$ the number of edges incident on the nodes of $\hat{A}$, i.e. $m_{\hat{A}} = \sum_{v \in \hat{A}} |N(v)|$. The next theorem states the complexity of REMOVE as a function of $|\hat{A}|$.

Theorem 3. *Algorithm REMOVE requires $O(m_{\hat{A}} l \log |\hat{A}| + |\hat{A}|(m + n \log |\hat{A}| + nl))$ worst case time.*

Although in the worst case $|\hat{A}| = O(|V|)$, in practice the nodes affected by an edge removal are much less than $|V|$, thus suggesting that REMOVE will behave well in practice. This observation is confirmed by our experimental results, which are shown in the next section.

4   Experimental Evaluation

We implemented, in C++: (i) algorithms ADD and REMOVE; (ii) a combination of them able to handle both edge additions and removals (named FULLY); (iii)

PLL for the from-scratch computation. Concerning the computation of the set of affected nodes within REMOVE, we implemented Algorithm 1 in a way that allows us to avoid to compute the set H of all hubs between pairs (u, y) at line 8. Regarding PLL, we considered its *parallel* version, which exploits multi-cores architectures, while all other algorithms are sequential. As in [3], we sorted nodes in non-increasing order of degree. All code has been compiled with GNU g++ 4.8 (O4 opt. level) under Linux (Kernel 3.13.0–65), and executed on a Quad-core 3.60 GHz Intel Xeon X5687 equipped workstation, with 24 GB of main memory and 12 MB of internal cache.

As input, we used various real-world network instances taken from known repositories for this kind of datasets, such as, e.g., SNAP, PTV, and Konect. In detail, we considered datasets similar to those used in other studies on static and dynamic 2-hop cover based methods [2,3,9]. We tested our algorithms on real-world undirected networks, and on synthetic networks generated via ForestFire model. Graph details on the considered networks are reported in Table 1.

Table 1. Input graphs.

Dataset	Network	$|V|$	$|E|$	AvgDeg
EU-ALL (EUA)	Mail	265 214	365 570	2.77
EPINIONS (EPN)	Social	131 828	841 372	12.76
YOUTUBE (YTB)	Social	1 134 890	2 987 624	5.26
BRIGHTKITE (BKT)	Location-based	58 228	214 078	7.35
GOOGLE (GOO)	Web	875 713	4 322 051	9.87
AS-SKITTER (SKI)	Computer	1 696 415	11 095 298	13.08
DENMARK (DEN)	Road	469 110	545 019	2.32
FLICKRIMG (FLI)	Metadata	105 938	2 316 948	43.74
FORESTFIRE-U (FFU)	Synthetic	1 000 000	7 374 808	14.75
WIKITALK (WTK)	Communication	2 394 385	4 659 565	4.19
FLICKRLINKS (FLL)	Social	1 715 255	15 550 782	18.13
DBPEDIA (DBP)	Miscellaneous	3 966 924	13 820 853	6.97
SIMPWIKI-EN (SWE)	Hyperlink	100 312	826 491	16.5

Executed Tests. We conducted a wide set of experiments which can be logically divided in two categories: *synthetic*, and *real-world* edge operations.

In the synthetic case, for each graph G, we first construct a labeling $L(G)$ by PLL. Then, we generate and perform three different types of *workloads*, called INC, DEC and FUL, respectively. In the INC (DEC, resp.) case, we perform 10 000 randomly chosen edge insertions (deletions, resp.). Note that, edge insertions are performed by selecting pairs of nodes that are not connected in the graph and by inserting and edge between them. By applying i-th operation on G_i, we obtain G_{i+1}. Then, we execute algorithm ADD (REMOVE, resp.) to reflect the change on $L(G_i)$, thus obtaining $L(G_{i+1})$. In parallel, we execute PLL directly on G_{i+1} to compute $L(G_{i+1})$ from-scratch. In the FUL workload case, we perform 10 000 operations of mixed randomly chosen type. For each operation, we either remove or add an edge, run FULLY and compare it with PLL, as above.

In the real-world case, we used a dynamic dataset available on Konect, namely SIMPWIKI-EN, representing a real evolving network. In this scenario, the FUL workload is a realistic sequence of updates, i.e. additions and removals, where the number of additions is around 25 % of the total number of updates within the dataset. In all above tests, we reduce the number of operations whenever dealing with networks of very large size (e.g., to 500 for AS-SKITTER), since PLL can require hundreds of minutes per execution in such cases.

As primary performance metric, after each graph update, we measure (i) the *update time* (UT), i.e. the computational time, in seconds, for updating the labeling by either ADD or REMOVE or FULLY, and (ii) the *building time* (BT), i.e. the computational time, in seconds, for building the labeling from-scratch by PLL. For each graph update, we compute the *speed-up*, i.e. the ratio between UT and BT. As in [3] we also consider space occupancy and query time metrics. In particular, after each graph update, for each considered labeling L, either computed via from-scratch or by dynamic algorithms, we measure: (i) the *labeling size* (LS), i.e. the space required to store L, in bytes, on average; (ii) the *average query time* (QT) i.e. the time needed to answer a distance query, in microseconds, on average, over 100 000 random queries.

The results of the all above experiments are summarized in Table 2 where we report, for each network, average values over all graph operations for all the above measures. We omit the results of the INC workload case, since they are very similar to those of [3]. For completeness' sake, for REMOVE only, we also provide value $|\hat{A}|$, i.e. the average size of set $\hat{A}$. Regarding the speed-up, our results are shown in Fig. 2, where we report the distribution of the speed-up of REMOVE and FULLY via box-plot charts (we highlight minimum, 1st quartile, median, 3rd quartile and maximum values). Due to space constraints, we focus on a subset of the considered networks. Other datasets produce similar outcomes.

Analysis. Our experimental data show that the proposed approach is effective in practice. In fact, we observe (see Table 2) that UT is always by far smaller than BT. In particular, dynamic algorithms are, on average, more than two orders of magnitude faster w.r.t. PLL. Their good performance is even clearer if we focus on Fig. 2, where we can observe that in almost all the instances REMOVE and FULLY are faster than PLL. The median value is always quite high (i.e. more than 100) and there are cases in which both REMOVE and FULLY, are more than 10^8 times faster than PLL. Furthermore, the speed-up seems to increase as the size of network and/or the density increase, which suggests that the proposed methods might scale well to bigger/denser networks. The experiments also support the result of Theorem 3. In fact, the update time of REMOVE decreases whenever the number of affected nodes is small.

Regarding space occupancy, our results show that the use of dynamic algorithms does not induce an increase in labeling size. This confirms the outcome of

Table 2. Experimental results.

Dataset	FUL workload						DEC workload						
	PLL			FULLY			PLL			REMOVE			
	BT	LS	QT	UT	LS	QT	BT	LS	QT	UT	LS	QT	$|\bar{A}|$
Eu-All	20.2	83 MB	9.2	1.98	83 MB	9.7	21.1	83 MB	9.9	3.62	83 MB	9.9	130 556
Epinions	41.9	81 MB	20.3	0.54	81 MB	17.8	41.5	81 MB	20.2	0.99	81 MB	17.5	16 985
YouTube	516	882 MB	22.1	56.16	882 MB	20.6	514	882 MB	22.3	112.43	882 MB	20.7	368 302
Brightkite	9.31	2.5 MB	15.9	0.36	2.5 MB	15.8	9.30	2.5 MB	15.9	0.88	2.5 MB	15.8	13 574
Google	466	663 MB	19.1	27.63	663 MB	17.2	457	663 MB	18.5	73.39	663 MB	17.1	133 688
AS-Skitter	4 930	3.6 GB	53.0	61.99	3.6 GB	67.2	4 930	3.6 GB	55.3	65.05	3.6 GB	53.5	208 709
Denmark	212	1.9 GB	80.7	84.50	2.6 GB	134.0	212	1.9 GB	80.7	92.50	2.6 GB	129.0	46 923
FlickrImg	409	260 MB	70.1	4.76	260 MB	67.2	409	260 MB	69.4	7.41	260 MB	69.5	5 110
ForestFire-U	2 370	2.5 GB	61.7	22.08	2.5 GB	60.5	2 350	2.5 GB	61.8	46.63	2.5 GB	60.3	100 056
WikiTalk	2 720	1.3 GB	35.4	63.6	1.3 GB	16.5	2 840	1.3 GB	30.2	112.0	1.3 GB	16.9	1 167 840
FlickrLinks	6 900	12 GB	50.9	8.9	12 GB	50.9	6 920	12 GB	50.9	16.7	12 GB	52.1	1 003 260
DBPedia	2 760	9.3 GB	21.3	61.1	9.3 GB	21.3	2 760	9.3 GB	17.1	359.0	9.3 GB	17.1	1 785 560
SimpWiki-En	38.3	189 MB	13.4	0.188	190 MB	14.1	41.3	181 MB	15.2	0.814	181 MB	16.0	40 400

our theoretical analysis of Sect. 3 about REMOVE, and that of the experimental evaluation of ADD proposed in [3], where the authors show that, even giving up on the minimality property, the algorithm behaves well in practice w.r.t. space occupancy. Regarding query time, our experiments confirm that dynamic algorithms deliver labelings that exhibit query times comparable to those exhibited by labelings computed from-scratch. This is in agreement with the considerations done w.r.t. average labeling size, which is directly related to query time [9].

In summary, REMOVE can be considered as the first algorithm for updating 2-hop cover labelings that is able to efficiently handle edge removals. In fact, it allows to reflect graph changes on the labeling extremely faster than the fastest available approach, i.e. PLL, and, at the same time, to efficiently answer queries without increasing the labeling size over time. This is clearly a highly desirable behavior in real-world dynamic scenarios. In fact, relying on PLL would imply having wrong answers to distance queries for much longer periods of time.

Regarding FULLY, its performance w.r.t. update time is even better than REMOVE, since it manages additions by the very fast ADD. Moreover, our data show that this is achieved without any degradation w.r.t. labeling size and query time. In fact, even though ADD does not remove outdated labels, and then if used alone it might require the periodic execution of PLL to restore minimality and avoid query time increases, in the FUL experiments we can observe that the average labeling size and query time are always comparable to those of both REMOVE and PLL. This is due to the very effective Algorithm 2, which deletes also outdated label entries that are not removed by ADD.

For the above reasons, FULLY can be considered the first fully dynamic algorithm for updating 2-hop cover labelings, able to deal with both additions and removals, and to answer to distance queries without any compromises on performance thus improving the literature on the matter.

As a last remark, we emphasize that all proposed approaches appear to be extendable to weighted graphs. We plan to investigate this issue as part of our future work.

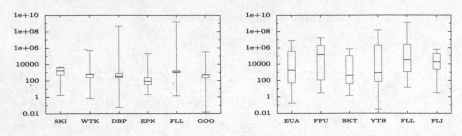

Fig. 2. Speed-up of REMOVE (left) and FULLY (right) against PLL.

References

1. Abraham, I., Delling, D., Goldberg, A.V., Werneck, R.F.: Hierarchical hub labelings for shortest paths. In: Epstein, L., Ferragina, P. (eds.) ESA 2012. LNCS, vol. 7501, pp. 24–35. Springer, Heidelberg (2012)
2. Akiba, T., Iwata, Y., Yoshida, Y.: Fast exact shortest-path distance queries on large networks by pruned landmark labeling. In: Proceedings of International Conference on Management of Data (SIGMOD 2013), pp. 349–360. ACM (2013)
3. Akiba, T., Iwata, Y., Yoshida, Y.: Dynamic and historical shortest-path distance queries on large evolving networks by pruned landmark labeling. In: Proceedings of 23rd International World Wide Web Conference (WWW 2014), pp. 237–248. ACM (2014)
4. Boccaletti, S., Latora, V., Moreno, Y., Chavez, M., Hwang, D.U.: Complex networks: structure and dynamics. Phys. Rep. **424**(4–5), 175–308 (2006)
5. Bruera, F., Cicerone, S., D'Angelo, G., Stefano, G.D., Frigioni, D.: Dynamic multi-level overlay graphs for shortest paths. Math. Comput. Sci. **1**(4), 709–736 (2008)
6. Cionini, A., D'Angelo, G., D'Emidio, M., Frigioni, D., Giannakopoulou, K., Paraskevopoulos, A., Zaroliagis, C.D.: Engineering graph-based models for dynamic timetable information systems. In: ATMOS. OASICS, vol. 42, pp. 46–61. Schloss Dagstuhl (2014)
7. Cohen, E., Halperin, E., Kaplan, H., Zwick, U.: Reachability and distance queries via 2-hop labels. In: Proceedings of the 13th ACM-SIAM Symposium on Discrete Algorithms (SODA 2002), pp. 937–946. ACM/SIAM (2002)
8. D'Angelo, G., D'Emidio, M., Frigioni, D.: Fully dynamic update of arc-flags. Networks **63**(3), 243–259 (2014)
9. Delling, D., Goldberg, A.V., Pajor, T., Werneck, R.F.: Robust distance queries on massive networks. In: Schulz, A.S., Wagner, D. (eds.) ESA 2014. LNCS, vol. 8737, pp. 321–333. Springer, Heidelberg (2014)
10. Delling, D., Goldberg, A.V., Pajor, T., Werneck, R.F.: Customizable route planning. In: Pardalos, P.M., Rebennack, S. (eds.) SEA 2011. LNCS, vol. 6630, pp. 376–387. Springer, Heidelberg (2011)
11. Delling, D., Goldberg, A.V., Savchenko, R., Werneck, R.F.: Hub labels: theory and practice. In: Gudmundsson, J., Katajainen, J. (eds.) SEA 2014. LNCS, vol. 8504, pp. 259–270. Springer, Heidelberg (2014)
12. Delling, D., Italiano, G.F., Pajor, T., Santaroni, F.: Better transit routing by exploiting vehicle GPS data. In: Proceedings of the 7th SIGSPATIAL International Workshop on Computational Transportation Science (IWCTS 2014), pp. 31–40. ACM (2014)

13. Jin, R., Ruan, N., Xiang, Y., Lee, V.E.: A highway-centric labeling approach for answering distance queries on large sparse graphs. In: Proceedings of International Conference on Management of Data (SIGMOD 2012), pp. 445–456. ACM (2012)
14. Potamias, M., Bonchi, F., Castillo, C., Gionis, A.: Fast shortest path distance estimation in large networks. In: Proceedings of 18th ACM Conference on Information and Knowledge Management (CIKM 2009), pp. 867–876. ACM (2009)
15. Qin, Y., Sheng, Q.Z., Zhang, W.E.: SIEF: efficiently answering distance queries for failure prone graphs. In: Proceedings of the 18th International Conference on Extending Database Technology (EDBT 2015), pp. 145–156. OpenProceedings.org (2015)
16. Vieira, M.V., Fonseca, B.M., Damazio, R., Golgher, P.B., de Castro Reis, D., Ribeiro-Neto, B.A.: Efficient search ranking in social networks. In: Proceedings of the 16th Conference on Information and Knowledge Management (CIKM 2007), pp. 563–572. ACM (2007)

Minimax Regret 1-Median Problem in Dynamic Path Networks

Yuya Higashikawa[1,6]([✉]), Siu-Wing Cheng[2], Tsunehiko Kameda[3],
Naoki Katoh[4,6], and Shun Saburi[5]

[1] Department of Information and System Engineering,
Chuo University, Tokyo, Japan
higashikawa.874@g.chuo-u.ac.jp
[2] Department of Computer Science and Engineering,
The Hong Kong University of Science and Technology, Hong Kong, China
scheng@cse.ust.hk
[3] School of Computing Science, Simon Fraser University, Burnaby, Canada
tiko@sfu.ca
[4] Department of Informatics, Kwansei Gakuin University, Sanda, Japan
naoki.katoh@gmail.com
[5] Department of Architecture and Architectural Engineering,
Kyoto University, Kyoto, Japan
as-saburi@archi.kyoto-u.ac.jp
[6] CREST, Japan Science and Technology Agency (JST), Kawaguchi, Japan

Abstract. This paper considers the minimax regret 1-median problem in dynamic path networks. In our model, we are given a dynamic path network consisting of an undirected path with positive edge lengths, uniform positive edge capacity, and nonnegative vertex supplies. Here, each vertex supply is unknown but only an interval of supply is known. A particular assignment of supply to each vertex is called a *scenario*. Given a scenario s and a sink location x in a dynamic path network, let us consider the evacuation time to x of a unit supply given on a vertex by s. The cost of x under s is defined as the sum of evacuation times to x for all supplies given by s, and the *median* under s is defined as a sink location which minimizes this cost. The regret for x under s is defined as the cost of x under s minus the cost of the median under s. Then, the problem is to find a sink location such that the maximum regret for all possible scenarios is minimized. We propose an $O(n^3)$ time algorithm for the minimax regret 1-median problem in dynamic path networks with uniform capacity, where n is the number of vertices in the network.

Keywords: Minimax regret · Sink location · Dynamic flow · Evacuation planning

1 Introduction

The Tohoku-Pacific Ocean Earthquake happened in Japan on March 11, 2011, and many people failed to evacuate and lost their lives due to severe attack by tsunamis.

N. Katoh—JSPS Grant-in-Aid for Scientific Research(A)(25240004).

© Springer International Publishing Switzerland 2016
V. Mäkinen et al. (Eds.): IWOCA 2016, LNCS 9843, pp. 122–134, 2016.
DOI: 10.1007/978-3-319-44543-4_10

From the viewpoint of disaster prevention from city planning and evacuation planning, it has now become extremely important to establish effective evacuation planning systems against large scale disasters in Japan. In particular, arrangements of tsunami evacuation buildings in large Japanese cities near the coast has become an urgent issue. To determine appropriate tsunami evacuation buildings, we need to consider where evacuation buildings are located and how to partition a large area into small regions so that one evacuation building is designated in each region. This produces several theoretical issues to be considered. Among them, this paper focuses on the location problem of the evacuation building assuming that we fix the region such that all evacuees in the region are planned to evacuate to this building. In this paper, we consider the simplest case for which the region consists of a single road.

In order to represent the evacuation, we consider the *dynamic* setting in graph networks, which was first introduced by Ford et al. [11]. In a graph network under the dynamic setting, each vertex is given supply and each edge is given length and capacity which limits the rate of the flow into the edge per unit time. We call such networks under the dynamic setting *dynamic networks*. Unlike in static networks, the time required to move supply from one vertex to a sink can be increased due to congestion caused by the capacity constraints, which require supplies to wait at vertices until supplies preceding them have left. In this paper, we consider the flow on dynamic networks as continuous, that is, each input value is given as a real number, and supply, flow and time are defined continuously. Then each supply can be regarded as fluid, and edge capacity is defined as the maximum amount of supply which can enter an edge per unit time. The *1-sink location problem in dynamic networks* is defined as the problem which requires to find the optimal location of a sink in a given dynamic network so that all supplies are sent to the sink as quickly as possible.

In order to evaluate an evacuation, we can naturally consider two types of criteria: *completion time criterion* and *total time criterion*. In this paper we adopt the latter one (for the former one, refer to [12, 15, 17, 18]). We here define a *unit* as an infinitesimally small portion of supply. Given a sink location x in a dynamic network, let us consider an evacuation to x starting at time 0 and define the *evacuation time* of a unit to x as the time when the unit reaches x in the evacuation. The total time for the evacuation to x is defined as the sum of evacuation times over all infinitesimal units to x. Then, the minimum total time for all possible evacuations to x could be the criterion for the optimality of sink location, which we adopt. Given a dynamic network, we define the *1-median problem* as the problem which requires to find a sink location minimizing the minimum total time, and the optimal solution is called the *median*.

Although the above criterion is reasonable for the sink location, it may not be practical since the number of evacuees in an area may vary depending on the time (e.g., in an office area in a big city, there are many people during the daytime on weekdays while there are much less people on weekends or during the night time). So, in order to take into account the uncertainty of population distribution, we consider the *maximum regret* for a sink location as another evaluation criterion assuming that for each vertex, we only know an interval of vertex supply. Then, the *minimax regret 1-median problem in dynamic path networks* is formulated as follows. A particular assignment of supply to each vertex is called a *scenario*. Here, for a sink location x and a scenario s, we denote the minimum total time by $\Phi^s(x)$. Also let m^s denote the median under s. The problem can be understood as a 2-person Stackelberg game as follows. The first player picks a sink location x and the second player chooses a scenario s that maximizes the *regret* defined as $\Phi^s(x) - \Phi^s(m^s)$. The objective of the first player is to choose x that minimizes the maximum regret.

Related to the minimax regret facility location in graph networks, especially for trees, some efficient algorithms have been presented by [2,3,5–7,9]. For dynamic networks, Cheng et al. [8] first studied the *minimax regret 1-center problem* in path networks, which requires to find a sink location in a path that minimizes the maximum regret where the completion time criterion is adopted instead of the total time one. They presented an $O(n \log^2 n)$ time algorithm. Higashikawa et al. [13] improved the time bound by [8] to $O(n \log n)$, and also Wang [19] independently achieved the same time bound of $O(n \log n)$ with better space complexity. Very recently, Bhattacharya et al. [4] have improved the time bound to $O(n)$. The above problem was extended to the multiple sink location version by Arumugam et al. [1]. For the minimax regret k-center problem in dynamic path networks with uniform capacity, they presented an $O(kn^3 \log n)$ time algorithm, and this time bound was improved to $O(kn^3)$ recently [12]. On the other hand, for dynamic tree networks, only the minimax regret 1-center problem was solved in $O(n^2 \log^2 n)$ time [14,16].

This paper first considers the minimax regret median problem in dynamic networks while all the above works for dynamic networks treated center problems. In this paper, we address the minimax regret 1-median problem in dynamic path networks with uniform capacity and present an $O(n^3)$ time algorithm.

2   Preliminaries

2.1   Dynamic Path Networks Under Uncertain Supplies

Let $P = (V, E)$ be an undirected path with ordered vertices $V = \{v_1, v_2, \ldots, v_n\}$ and edges $E = \{e_1, e_2, \ldots, e_{n-1}\}$ where $e_i = (v_i, v_{i+1})$ for $i \in \{1, \ldots, n-1\}$. Let $\mathcal{N} = (P, l, w, c, \tau)$ be a dynamic network with the underlying path graph P; l is a function that associates each edge e_i with positive length l_i, w is a function that associates each vertex v_i with positive weight w_i, amount of supply at v_i; c is the capacity, a positive constant representing the amount of supply which can enter an edge per unit time; τ is a positive constant representing the time required for a flow to travel a unit distance. In our model, instead of the weight function w on vertices, we are given the weight interval function W that associates each vertex $v_i \in V$ with an interval of supply $W_i = [w_i^-, w_i^+]$ with $0 < w_i^- \leq w_i^+$. We call such a network $\mathcal{N} = (P, l, W, c, \tau)$ with path structures *a dynamic path network under uncertain supplies*.

In the following, we write $p \in P$ to indicate that a point is a vertex of P or lies on one of the edges of P. For any point $p \in P$, we abuse this notation by also letting p denote the distance from v_1 to p. Informally we can regard P as being embedded on a real line with $v_1 = 0$. For two points $p, q \in P$ satisfying $p < q$, let $[p, q]$ (resp. $[p, q)$, $(p, q]$ and (p, q)) denote an interval in P consisting of all points $x \in P$ such that $p \leq x \leq q$ (resp. $p \leq x < q$, $p < x \leq q$ and $p < x < q$).

2.2   Scenarios

Let $\mathcal{S}$ denote the Cartesian product of all W_i for $i \in \{1, \ldots, n\}$:

$$\mathcal{S} = \prod_{i=1}^{n} W_i. \tag{1}$$

An element of $\mathcal{S}$, i.e., a particular assignment of weight to each vertex, is called a *scenario*. Given a scenario $s \in \mathcal{S}$, we denote by w_i^s the weight of a vertex v_i under s.

2.3   Evacuation on a Dynamic Path Network

In our model, the supply is defined continuously. We define a unit as an infinitesimally small portion of supply. Given a sink location $x \in P$ and a scenario $s \in \mathcal{S}$, without loss of generality, an evacuation to x under s is assumed to satisfy the following assumptions. When a unit arrives at a vertex v on its way to x, it has to wait for the departure if there are already some units waiting for leaving v. All units waiting at v for leaving v are processed in the first-come first-served manner. We show the details below.

As shown in Fig. 1, let us consider a path with only three vertices, say v_1, v_2 and x, and two edges. Supplies of w_1 and w_2 are given at vertices v_1 and v_2, respectively, and a sink is at a vertex x, Both edges have capacity c. For a unit of supply, it takes $\tau(v_2 - v_1)$ time to cross the edge (v_1, v_2) and $\tau(x - v_2)$ time to cross the edge (v_2, x). Suppose that all supplies start evacuating to x at time 0. First consider the evacuation of supply given at v_2. Since the amount of supply which can enter an edge in unit time is c, the last unit of v_2 leaves v_2 at time w_2/c and reaches x at time $\tau(x - v_2) + w_2/c$.

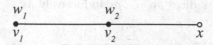

Fig. 1. Illustration of a path with three vertices and two edges

Next consider the evacuation of supply given at v_1. At time $\tau(v_2 - v_1)$, the first unit of v_1 reaches v_2. Then one of the following three situation occurs.

(1) No congestion: If $\tau(v_2 - v_1) > w_2/c$, there is no supply waiting at v_2 when the first unit of v_1 reaches v_2. In this case, every unit from v_1 continues through to x without waiting, say v_1 gets *no congestion* at v_2. Then the last unit of v_1 reaches x at time

$$\tau(x - v_1) + \frac{w_1}{c}. \tag{2}$$

(2) Congestion: If $\tau(v_2 - v_1) < w_2/c$, there is some supply waiting at v_2 when the first unit of v_1 reaches v_2. In this case, every unit from v_1 has to wait at v_2, say v_1 gets *congestion* at v_2. Then the last unit of v_1 reaches x at time

$$\tau(x - v_2) + \frac{w_1 + w_2}{c}. \tag{3}$$

(3) Touching: If $\tau(v_2 - v_1) = w_2/c$, the last unit of v_2 just leaves v_2 when the first unit of v_1 reaches v_2, say v_1 gets *touching* at v_2. Then the last unit of v_1 reaches x at time

$$\tau(x - v_1) + \frac{w_1}{c} = \tau(x - v_2) + \frac{w_1 + w_2}{c}. \tag{4}$$

Note that if v_1 gets congestion or touching at v_2, we can transform the input so that supply of w_1 is moved from v_1 to v_2, which never changes the time when each unit reaches x.

2.4   Total Evacuation Time

For a given $x \in P$ and $s \in \mathcal{S}$, let us consider an evacuation to x under s starting at time 0 and define the evacuation time of a unit to x under s as the time when the unit reaches x. Let $\Phi^s(x)$ denote the sum of evacuation times over all infinitesimal units to x under s. Also let $\Phi_L^s(x)$ (resp. $\Phi_R^s(x)$) denote the sum of evacuation times to x under s for all units on $[v_1, x)$ (resp. $(x, v_n]$). Then, $\Phi^s(x)$ is obviously the sum of $\Phi_L^s(x)$ and $\Phi_R^s(x)$, i.e.,

$$\Phi^s(x) = \Phi_L^s(x) + \Phi_R^s(x). \tag{5}$$

Without loss of generality, we assume $\Phi_L^s(v_1) = 0$ and $\Phi_R^s(v_n) = 0$.

We will show the formula of $\Phi^s(x)$ that has been proved in [15,17]. Suppose that x is located in an open interval (v_h, v_{h+1}) with $1 \leq h \leq n - 1$, i.e., $x \in e_h$. We here show only the formula of $\Phi_L^s(x)$ (the case of $\Phi_R^s(x)$ is symmetric). First, let us define the vertex indices $\rho_1, \ldots, \rho_k$ inductively as

$$\rho_i = \max \left\{ \mathrm{argmax} \left\{ \tau(v_h - v_j) + \frac{\sum_{l=\rho_{i-1}+1}^{j} w_l^s}{c} \; \middle| \; j \in \{\rho_{i-1} + 1, \ldots, h\} \right\} \right\}, \tag{6}$$

where $\rho_0 = 0$. Obviously $\rho_k = h$ holds. We then call a set of all units on $[v_{\rho_{i-1}+1}, v_{\rho_i}]$ the *i-th left cluster* for x (the *i-th right cluster* for x is symmetrically defined), and a vertex v_{ρ_i} the *head* of *i*-th cluster (see Fig. 2). Also, for $i \in \{1, \ldots, k\}$, we define σ_i as $\sigma_i = \sum_{l=\rho_{i-1}+1}^{\rho_i} w_l^s$, which is called the weight of *i*-th cluster.

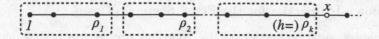

Fig. 2. Illustration of left clusters for x where i-th cluster is headed by a vertex v_{ρ_i}

The definition of (6) can be interpreted as follows. Looking at the i-th left cluster, we find that v_{ρ_i-1} gets congestion or touching at v_{ρ_i}. Therefore transforming the input so that supply of w_{ρ_i-1} is moved from v_{ρ_i-1} to v_{ρ_i} and v_{ρ_i-1} is removed never changes the evacuation time of each unit to x, which implies that $\Phi_L^s(x)$ is never changed. After that, v_{ρ_i-2} also gets congestion or touching at v_{ρ_i}, thus w_{ρ_i-2} can be moved from v_{ρ_i-2} to v_{ρ_i} and v_{ρ_i-2} can be removed without changing $\Phi_L^s(x)$. In the similar manner, even if we transform the input so that all supplies on $[v_{\rho_{i-1}+1}, v_{\rho_i})$ are moved to v_{ρ_i} for every $i \in \{1, \ldots, k\}$, $\Phi_L^s(x)$ is never changed. In the following, we call such a transformation the *left-clustering* for x (the *right-clustering* for x is symmetrically defined). Note that after left-clustering for x, heads of left clusters only remain in the left side of x and $v_{\rho_{i-1}}$ gets no congestion at v_{ρ_i} for any $i \in \{2, \ldots, k\}$, that is,

$$\tau(v_{\rho_i} - v_{\rho_{i-1}}) > \frac{\sigma_i}{c}. \tag{7}$$

In other words, the first unit of each head can reach x without any stop on its way to x. Thus, as in [15,17], $\Phi_L^s(x)$ can be written as

$$\Phi_L^s(x) = \sum_{i=1}^k \left(\sigma_i \tau(x - v_{\rho_i}) + \frac{\sigma_i^2}{2c} \right). \tag{8}$$

2.5 Minimax Regret Formulation

For a scenario $s \in \mathcal{S}$, let m^s be a point in P that minimizes $\Phi^s(x)$ over $x \in P$, called the *median* under s. We now define the *regret* for x under s as

$$R^s(x) = \Phi^s(x) - \Phi^s(m^s). \tag{9}$$

Moreover, we also define the *maximum regret* for x as

$$R_{\max}(x) = \max\{R^s(x) \mid s \in \mathcal{S}\}. \tag{10}$$

If $\hat{s} = \operatorname{argmax}\{R^s(x) \mid s \in \mathcal{S}\}$, we call $\hat{s}$ the *worst case scenario* for x. The goal is to find a point $x^* \in P$, called the *minimax regret median*, that minimizes $R_{\max}(x)$ over $x \in P$, i.e., the objective is to

$$\text{minimize } \{R_{\max}(x) \mid x \in P\}. \tag{11}$$

2.6   Known Properties for the Fixed Scenario Case

We here show some properties on the 1-median problem in a dynamic path network $\mathcal{N} = (P = (V, E), l, w^s, c, \tau)$ when a scenario $s \in \mathcal{S}$ is given, which were basically presented in [15,17]. We first introduce the following two lemmas.

Lemma 1 [15,17]. *For a scenario $s \in \mathcal{S}$, m^s is at a vertex in V.*

Lemma 2 [15,17]. *For a scenario $s \in \mathcal{S}$, all $\Phi^s(v_i)$ over $i \in \{1, \ldots, n\}$ can be computed in $O(n)$ time in total.*

We then can see a corollary of these lemmas.

Corollary 1 [15,17]. *For a scenario $s \in \mathcal{S}$, m^s and $\Phi^s(m^s)$ can be computed in $O(n)$ time.*

Now let us look at the formula of (8). Even if x is moving on an edge e_i (not including endpoints v_i and v_{i+1}), the formation of left clusters for x does not change over $x \in e_i$. Therefore, $\Phi_L^s(x)$ is a linear function of $x \in e_i$, and symmetrically, $\Phi_R^s(x)$ is also a linear function. For $i \in \{1, \ldots, n-1\}$, letting a_i^s and b_i^s be the values such that for $x \in e_i$,

$$\Phi^s(x) = a_i^s x + b_i^s, \tag{12}$$

we can derive the following lemma from [15,17].

Lemma 3. *For a scenario $s \in \mathcal{S}$, all a_i^s and b_i^s over $i \in \{1, \ldots, n-1\}$ can be computed in $O(n)$ time in total.*

3   Properties of Worst Case Scenarios

In this section, we show the important properties which worst case scenarios have. In our problem, a main difficulty lies in evaluating $R^s(x)$ over $s \in \mathcal{S}$ to compute $R_{\max}(x)$ even for a fixed x since the size of $\mathcal{S}$ is infinite. We thus aim to find a scenario set with a finite size (in particular, a polynomial size) which includes a worst case scenario for any $x \in P$. In order to do this, we introduce a new concept, the *gap* between two points $x, y \in P$ under a scenario $s \in \mathcal{S}$, defined by

$$\Gamma^s(x, y) = \Phi^s(x) - \Phi^s(y). \tag{13}$$

By Lemma 1 and the definition of (9), we have

$$R^s(x) = \max\{\Gamma^s(x, y) \mid y \in V\}, \tag{14}$$

and by (10) and (14),

$$R_{\max}(x) = \max\{\max\{\Gamma^s(x, y) \mid y \in V\} \mid s \in \mathcal{S}\}$$
$$= \max\{\max\{\Gamma^s(x, y) \mid s \in \mathcal{S}\} \mid y \in V\}. \tag{15}$$

From (15), if we can compute $\max\{\Gamma^s(x,y) \mid s \in \mathcal{S}\}$ for a fixed pair $\langle x, y \rangle \in P \times V$, $R_{\max}(x)$ can also be computed by repeating the same maximization over $y \in V$. We call a scenario that maximizes $\Gamma^s(x, y)$ for a fixed $\langle x, y \rangle$ a worst case scenario for $\langle x, y \rangle$. In the following, we show a scenario set of size $O(n)$ that includes a worst case scenario for a fixed $\langle x, y \rangle$, which implies a scenario set of size $O(n^2)$ that includes a worst case scenario for a fixed x.

3.1   Bipartite Scenario

We first introduce the concept of the *bipartite scenario*, which was originally introduced as the *dominant scenario* in [8,13]. Let us consider a scenario $s \in \mathcal{S}$. A scenario s is said to be *left-bipartite* (resp. *right-bipartite*) if $w_j^s = w_j^+$ (resp. w_j^-) over $j \in \{1, \ldots, i\}$ and $w_j^s = w_j^-$ (resp. w_j^+) over $j \in \{i+1, \ldots, n\}$ for some $i \in \{1, \ldots, n-1\}$. Obviously the number of such scenarios is $O(n)$. The authors of [8,13] treated the minimax regret 1-center problem in dynamic path networks, which requires to find a sink location in a path that minimizes the maximum regret similarly defined as (10) where the completion time criterion is adopted instead of the total time one. They proved that for any point in an input path, at least one worst case scenario is left-bipartite or right-bipartite.

3.2   Pseudo-bipartite Scenario

We here introduce the concept of the *pseudo-bipartite scenario*. A scenario s is said to be *left-pseudo-bipartite* (resp. *right-pseudo-bipartite*) if $w_j^s = w_j^+$ (resp. w_j^-) over $j \in \{1, \ldots, i-1\}$ and $w_j^s = w_j^-$ (resp. w_j^+) over $j \in \{i+1, \ldots, n\}$ for some $i \in \{2, \ldots, n-1\}$. In this definition, we do not care about the weight of a vertex v_i, called the *intermediate vertex*. Given a pseudo-bipartite scenario with the intermediate vertex v_i, we call intervals $[v_1, v_i)$ and $(v_i, v_n]$ the *left part* and the *right part*, respectively. Let $\mathcal{S}_L$ (resp. $\mathcal{S}_R$) denote a set of all left-pseudo-bipartite scenarios (resp. right-pseudo-bipartite scenarios). We then prove the following lemma (the proof is omitted).

Lemma 4. *Given a pair $\langle x, y \rangle \in P \times V$ satisfying $y < x$ (resp. $x < y$), there exists a worst case scenario for $\langle x, y \rangle$ belonging to $\mathcal{S}_L$ (resp. $\mathcal{S}_R$) such that x is in the right part (resp. the left part), and y is in the left part (resp. the right part) or at the intermediate vertex.*

3.3   Critical Pseudo-bipartite Scenario

By Lemma 4, we studied the property of a worst case scenario for a fixed $\langle x, y \rangle \in P \times V$, however the sizes of $\mathcal{S}_L$ and $\mathcal{S}_R$ are still infinite since the weight of the intermediate vertex in a pseudo-bipartite scenario is not fixed. In the rest of this section, we focus on the weight of the intermediate vertex in a pseudo-bipartite scenario which is worst for $\langle x, y \rangle$.

Given a pair $\langle x, y \rangle \in P \times V$ satisfying $y < x$, let us consider a scenario $s \in \mathcal{S}_L$ such that the intermediate vertex is v_i and $y \leq v_i < x$. Suppose that the weight of v_i is set as the minimum, i.e., $w_i^s = w_i^-$. Performing the right-clustering for y under s (mentioned in Sect. 2.4), we will get right clusters for y such that the head of l-th cluster is ρ_l for $l \in \{1, \ldots, k\}$ satisfying $\rho_k < \ldots < \rho_1$, and the weight of l-th cluster is σ_l. Suppose that the intermediate vertex v_i belongs to the j-th cluster.

Let us increase the weight of v_i, little by little, without changing the weight of any other vertex. Let $s(w)$ be a scenario in $\mathcal{S}_L$ such that the intermediate vertex is v_i whose weight is $w \in [w_i^-, w_i^+]$. Note that $v_{\rho_{j-1}}$ gets no congestion at v_{ρ_j} under s, i.e., $s(w_i^-)$. When the weight of v_i reaches some value ω, $v_{\rho_{j-1}}$ may get touching at v_{ρ_j}. If so, the following equality holds:

$$\tau(v_{\rho_{j-1}} - v_{\rho_j}) = \frac{\sigma_j + (\omega - w_i^-)}{c}. \tag{16}$$

Here $\sigma_j + (\omega - w_i^-)$ corresponds to the weight of j-th cluster under $s(\omega)$. At that moment, by the definition of (6), the $(j-1)$-th cluster is immediately merged to the j-th cluster. We then call $s(\omega)$ a *critical left-pseudo-bipartite scenario* for y. Note that such critical scenarios may occur several times while increasing the weight of v_i from w_i^- to w_i^+. Also, $s(w_i^-)$ and $s(w_i^+)$ are assumed to be critical left-pseudo-bipartite scenarios for y even if any merge does not occur at those moments. *Critical right-pseudo-bipartite scenarios* for y are symmetrically defined. Let $\mathcal{S}_y$ denote a set of all critical left-pseudo-bipartite scenarios and critical right-pseudo-bipartite scenarios for y, and $\mathcal{S}^* = \bigcup_{y \in V} \mathcal{S}_y$. We will show two lemmas (the proof of Lemma 5 is omitted).

Lemma 5. *Given a pair $\langle x, y \rangle \in P \times V$, there exists a worst case scenario for $\langle x, y \rangle$ belonging to $\mathcal{S}_y$.*

Lemma 6. *For a vertex $y \in V$, the size of $\mathcal{S}_y$ is $O(n)$, and all scenarios in $\mathcal{S}_y$ can be computed in $O(n)$ time.*

Proof. We first prove that the number of critical left-pseudo-bipartite scenarios for y is $O(n)$ (the case of critical right-pseudo-bipartite scenarios is symmetric). Suppose that $y = v_j$. For $i \in \{j+1, \ldots, n\}$ and $w \in [w_i^-, w_i^+]$, let $s(i, w)$ be a scenario in $\mathcal{S}_L$ such that the intermediate vertex is v_i whose weight is w. Here, let us define the total ordering between two scenarios $s(i, w)$ and $s(i', w')$: $s(i, w) \prec s(i', w')$ holds if and only if (a) $i < i'$ or (b) $i = i'$ and $w < w'$. For $i \in \{j+1, \ldots, n\}$, we also define p_i and q_i as follows. Let p_i be the number of critical left-pseudo-bipartite scenarios for y such that the intermediate vertex is v_i (including $s(i, w_i^-)$ and $s(i, w_i^+)$). Let q_i be, under a scenario $s(i, w_i^+)$, the number of right clusters for y that follow a cluster including v_i.

Let us consider computing all critical left-pseudo-bipartite scenarios for y in ascending order, and suppose that the weight of v_i now increases from w_i^- to w_i^+. Here let c be the right cluster for y including v_i. While the weight increases, since $p_i - 2$ critical left-pseudo-bipartite scenarios for y occur (except $s(i, w_i^-)$ and $s(i, w_i^+)$), and at each such scenario, one or more clusters are merged into c, at least $p_i - 2$ clusters are merged into c in total. We thus have $q_i \leq q_{i-1} - (p_i - 2)$ for $i \in \{j+1, \ldots, n\}$, i.e.,

$$p_i \leq q_{i-1} - q_i + 2. \tag{17}$$

Note that the total number of critical left-pseudo-bipartite scenarios for y is exactly $1 + \sum_{i=j+1}^{n}(p_i - 1)$. By (17), we have

$$\sum_{i=j+1}^{n}(p_i - 1) \leq \sum_{i=j+1}^{n}(q_{i-1} - q_i + 1) = q_j - q_n + (n - j), \tag{18}$$

which is $O(n)$ since $q_j \leq n - j$ and $q_n = 0$.

In the rest of the proof, we show that all critical left-pseudo-bipartite scenarios for $y = v_j$ can be computed in $O(n)$ time. Recall that all critical left-pseudo-bipartite scenarios for y are computed in ascending order. The algorithm first gets $s(j+1, w_{j+1}^-)$, and performs the right clustering for y under $s(j+1, w_{j+1}^-)$. As claimed in [15,17], it is easy to see that the right clustering for a fixed y can be obtained in $O(n)$ time.

Suppose that for particular $i \in \{j+1, \ldots, n\}$ and $\omega \in [w_i^-, w_i^+]$, $s(i, \omega)$ is critical for y, and the algorithm has already obtained $s(i, \omega)$ and the right clusters for y. We then show how to compute the subsequent critical left-pseudo-bipartite scenario. Let c_y be a right cluster for y including v_i and c_y' be a right cluster for y immediately following c_y. Also, let ρ_y (resp. ρ_y') be the index of a vertex that corresponds to the head of c_y (resp. c_y'), and σ_y (resp. σ_y') be the weight of c_y (resp. c_y').

There are two cases: [Case 1] $\omega < w_i^+$; [Case 2] $\omega = w_i^+$. For Case 2, we notice that $s(i+1, w_{i+1}^-)$ is equivalent to $s(i, w_i^+)$. Therefore, this case immediately results in Case 1 by letting i be $i+1$ and ω be w_{i+1}^- (although a right cluster for y including v_{i+1} may be c_y', not c_y). We thus consider only Case 1 in the following.

The algorithm will compute the subsequent critical left-pseudo-bipartite scenario $s(i, \omega')$ where ω' satisfies $\omega < \omega' \leq w_i^+$. In order to compute ω', the algorithm test if there exists $w \in (\omega, w_i^+]$ such that

$$\tau(v_{\rho_y'} - v_{\rho_y}) = \frac{\sigma_y + (w - \omega)}{c}, \tag{19}$$

which is similar to (16). If yes, for such w, the algorithm returns $\omega' = w$ and updates the right clusters for y by merging c_y' into c_y. Otherwise, $\omega' = w_i^+$ is just returned. Such testing and updating are done in $O(1)$ time. Since the number of critical left-pseudo-bipartite scenarios for y is $O(n)$ and each of those is computed in $O(1)$ time, we completes the proof. □

By (15), we have a corollary of Lemma 5.

Corollary 2. *Given a point $x \in P$, there exists a worst case scenario for x belonging to $\mathcal{S}^*$.*

Also, a corollary of Lemma 6 immediately follows.

Corollary 3. *The size of $\mathcal{S}^*$ is $O(n^2)$, and all scenarios in $\mathcal{S}^*$ can be computed in $O(n^2)$ time.*

4   Algorithm

In this section, we show an algorithm that computes the minimax regret median, which minimizes $R_{\max}(x)$ over $x \in P$. The algorithm basically consists of two phases:

[**Phase 1**] Compute $R_{\max}(v_i)$ over $i \in \{1, \ldots, n\}$, and
[**Phase 2**] Compute $\min\{R_{\max}(x) \mid x \in e_i\}$ over $i \in \{1, \ldots, n-1\}$.
After these, the algorithm evaluates all the $2n - 1$ values obtained and finds the minimax regret median in $O(n)$ time.

By Corollary 2, we only have to consider scenarios in $\mathcal{S}^*$ to compute $R_{\max}(x)$ for any $x \in P$. Therefore, the algorithm computes all scenarios in $\mathcal{S}^*$ in advance, which can be done in $O(n^2)$ time by Corollary 3. Subsequently, it computes all the values $\Phi^s(m^s)$ over $s \in \mathcal{S}^*$ for Phase 1 and Phase 2. By Corollaries 1 and 3, this can be done in $O(n^3)$ time in total.

First let us see details in Phase 1. For a fixed scenario $s \in \mathcal{S}^*$, since all $\Phi^s(v_i)$ over $i \in \{1, \ldots, n\}$ can be computed in $O(n)$ time by Lemma 2 and $\Phi^s(m^s)$ has already been computed before Phase 1, all $R^s(v_i)$ over $i \in \{1, \ldots, n\}$ can also be computed in $O(n)$ time (refer to (9)). After the algorithm obtains $R^s(v_1), \ldots, R^s(v_n)$ over $s \in \mathcal{S}^*$ in $O(n^3)$ time, for each $i \in \{1, \ldots, n\}$, $R^s(v_i)$ over $s \in \mathcal{S}^*$ are evaluated to obtain $R_{\max}(v_i)$. Thus, it is easy to see that Phase 1 can be done in $O(n^3)$ time in total.

We next focus on Phase 2. As mentioned at the end of Sect. 2.6, for a fixed scenario $s \in \mathcal{S}^*$, $\Phi^s(x)$ is a linear function of $x \in e_i$ for each $i \in \{1, \ldots, n-1\}$ (not including v_i and v_{i+1}). Therefore, $R^s(x)$ is also linear for $x \in e_i$ for each i. Referring to (12), a function $R^s(x)$ on an edge e_i is written as

$$R^s(x) = a_i^s x + b_i^s - \Phi^s(m^s). \tag{20}$$

Recall that $\Phi^s(m^s)$ has already been computed. Then, by Lemma 3, $R^s(x)$ on e_i over $i \in \{1, \ldots, n-1\}$ can be computed in $O(n)$ time. After the algorithm does the same computation over $s \in \mathcal{S}^*$ in $O(n^3)$ time, on each edge e_i, we have $O(n^2)$ linear functions $R^s(x)$ over $s \in \mathcal{S}^*$. By the definition of (10), $\min\{R_{\max}(x) \mid x \in e_i\}$ can be obtained by solving a linear programming problem in two dimensions with $O(n^2)$ constraints, i.e.,

$$\begin{aligned}
\text{minimize} \quad & y \\
\text{subject to} \quad & a_i^s x + b_i^s - \Phi^s(m^s) \leq y, \quad \forall s \in \mathcal{S}^* \\
& v_i \leq x \leq v_{i+1}.
\end{aligned}$$

This problem can be solved in $O(n^2)$ time by [10]. Repeating the same operations over $i \in \{1, \ldots, n-1\}$, Phase 2 is completed in $O(n^3)$ time.

Theorem 1. *The minimax regret 1-median problem in dynamic path networks with uniform capacity can be solved in $O(n^3)$ time.*

5   Conclusion

In this paper, we address the minimax regret 1-median problem in dynamic path networks with uniform capacity and present an $O(n^3)$ time algorithm. Additionally, this is the first work that treats the minimax regret facility location problem in dynamic networks where the total time criterion is adopted. Two natural questions immediately follow. The first one is whether we can reduce the number of scenarios to be considered. The other one is whether we can extend the problem to the k-median version with $k \geq 2$, or the problem in more general networks.

References

1. Arumugam, G.P., Augustine, J., Golin, M.J., Srikanthan, P.: A polynomial time algorithm for minimax-regret evacuation on a dynamic path. CoRR abs/1404.5448. arXiv:1404.5448
2. Averbakh, I., Berman, O.: Algorithms for the robust 1-center problem on a tree. Eur. J. Oper. Res. **123**(2), 292–302 (2000)
3. Bhattacharya, B., Kameda, T.: A linear time algorithm for computing minmax regret 1-median on a tree. In: Gudmundsson, J., Mestre, J., Viglas, T. (eds.) COCOON 2012. LNCS, vol. 7434, pp. 1–12. Springer, Heidelberg (2012)
4. Bhattacharya, B., Kameda, T.: Improved algorithms for computing minmax regret 1-sink and 2-sink on path network. In: Zhang, Z., Wu, L., Xu, W., Du, D.-Z. (eds.) COCOA 2014. LNCS, vol. 8881, pp. 146–160. Springer, Heidelberg (2014)
5. Bhattacharya, B., Kameda, T., Song, Z.: A linear time algorithm for computing minmax regret 1-median on a tree network. Algorithmica **62**, 1–20 (2013)
6. Brodal, G.S., Georgiadis, L., Katriel, I.: An $O(n \log n)$ version of the Averbakh-Berman algorithm for the robust median of a tree. Oper. Res. Lett. **36**(1), 14–18 (2008)
7. Chen, B., Lin, C.: Minmax-regret robust 1-median location on a tree. Networks **31**(2), 93–103 (1998)
8. Cheng, S.-W., Higashikawa, Y., Katoh, N., Ni, G., Su, B., Xu, Y.: Minimax regret 1-sink location problems in dynamic path networks. In: Chan, T.-H.H., Lau, L.C., Trevisan, L. (eds.) TAMC 2013. LNCS, vol. 7876, pp. 121–132. Springer, Heidelberg (2013)
9. Conde, E.: A note on the minmax regret centdian location on trees. Oper. Res. Lett. **36**(2), 271–275 (2008)
10. Dyer, M.E.: Linear time algorithms for two- and three-variable linear programs. SIAM J. Comput. **13**(1), 31–45 (1984)
11. Ford Jr., L.R., Fulkerson, D.R.: Constructing maximal dynamic flows from static flows. Oper. Res. **6**, 419–433 (1958)
12. Higashikawa, Y.: Studies on the Space Exploration and the Sink Location under Incomplete Information towards Applications to Evacuation Planning. Doctoral Dissertation, Kyoto University (2014)
13. Higashikawa, Y., Augustine, J., Cheng, S.W., Katoh, N., Ni, G., Su, B., Xu, Y.: Minimax regret 1-sink location problem in dynamic path networks. Theor. Comput. Sci. **588**, 24–36 (2015). doi:10.1016/j.tcs.2014.02.010

14. Higashikawa, Y., Golin, M.J., Katoh, N.: Minimax regret sink location problem in dynamic tree networks with uniform capacity. In: Pal, S.P., Sadakane, K. (eds.) WALCOM 2014. LNCS, vol. 8344, pp. 125–137. Springer, Heidelberg (2014)
15. Higashikawa, Y., Golin, M.J., Katoh, N.: Multiple sink location problems in dynamic path networks. In: Gu, Q., Hell, P., Yang, B. (eds.) AAIM 2014. LNCS, vol. 8546, pp. 149–161. Springer, Heidelberg (2014)
16. Higashikawa, Y., Golin, M.J., Katoh, N.: Minimax regret sink location problem in dynamic tree networks with uniform capacity. J. Graph Algorithms Appl. **18**(4), 539–555 (2014)
17. Higashikawa, Y., Golin, M.J., Katoh, N.: Multiple sink location problems in dynamic path networks. Theor. Comput. Sci. **607**(Part 1), 2–15 (2015). doi:10.1016/j.tcs.2015.05.053
18. Mamada, S., Uno, T., Makino, K., Fujishige, S.: An $O(n \log^2 n)$ algorithm for the optimal sink location problem in dynamic tree networks. Discrete Appl. Math. **154**(16), 2387–2401 (2006)
19. Wang, H.: Minmax regret 1-facility location on uncertain path networks. In: Cai, L., Cheng, S.-W., Lam, T.-W. (eds.) Algorithms and Computation. LNCS, vol. 8283, pp. 733–743. Springer, Heidelberg (2013)

Enumeration

On Maximal Chain Subgraphs
and Covers of Bipartite Graphs

Tiziana Calamoneri[1], Mattia Gastaldello[1,2($\boxtimes$)], Arnaud Mary[2],
Marie-France Sagot[2], and Blerina Sinaimeri[2]

[1] Sapienza University of Rome, via Salaria 113, 00198 Roma, Italy
gastaldello@di.uniroma1.it
[2] INRIA and Université de Lyon, Université Lyon 1,
LBBE, CNRS UMR558, Villeurbanne, France

Abstract. In this paper, we address three related problems. One is the
enumeration of all the maximal *edge induced* chain subgraphs of a bipartite graph, for which we provide a polynomial delay algorithm. We give
bounds on the number of maximal chain subgraphs for a bipartite graph
and use them to establish the input-sensitive complexity of the enumeration problem. The second problem we treat is the one of finding the
minimum number of chain subgraphs needed to cover all the edges a
bipartite graph. For this we provide an exact exponential algorithm with
a non trivial complexity. Finally, we approach the problem of enumerating all minimal chain subgraph covers of a bipartite graph and show
that it can be solved in quasi-polynomial time.

Keywords: Chain subgraph cover problem · Enumeration algorithms ·
Exact exponential algorithms

1 Introduction

Enumerating (listing) the subgraphs of a given graph plays an important role
in analysing its structural properties. It thus is a central issue in many areas,
notably in data mining and computational biology.

In this paper, we address the problem of enumerating without repetitions all
maximal *edge induced* chain subgraphs of a bipartite graph. These are graphs
that do not contain a $2K_2$ as induced subgraph. From now on, we will refer to
them as *chain subgraphs* for short when there is no ambiguity.

Bipartite graphs arise naturally in many applications, such as biology as
will be mentioned later in the introduction, since they enable to model the
relations between two different classes of objects. The problem of enumerating
in bipartite graphs all subgraphs with certain properties has thus already been
considered in the literature. These concern for instance maximal bicliques for
which polynomial delay enumeration algorithms in bipartite [6,11] as well as
in general graphs [5,11] were provided. In the case of maximal *induced* chain
subgraphs, their enumeration can be done in output polynomial time as it can

© Springer International Publishing Switzerland 2016
V. Mäkinen et al. (Eds.): IWOCA 2016, LNCS 9843, pp. 137–150, 2016.
DOI: 10.1007/978-3-319-44543-4_11

be reduced to the enumeration of a particular case of the minimal hitting set problem [7] (where the sets in the family are of cardinality 4). However, the existence of a polynomial delay algorithm for this problem remains open. To the best of our knowledge, nothing is known so far about the problem of enumerating maximal *edge induced* chain subgraphs in bipartite graphs.

In this paper, we propose a polynomial delay algorithm to enumerate all maximal chain subgraphs of a bipartite graph. We also provide an analysis of the time complexity of this algorithm in terms of input size. In order to do this, we prove some upper bounds on the maximum number of maximal chain subgraphs of a bipartite graph G with n nodes and m edges. This is also of intrinsic interest as combinatorial bounds on the maximum number of specific subgraphs in a graph are difficult to obtain and have received a lot of attention (see for *e.g.* [8,12]).

We then address a second related problem called the *minimum chain subgraph cover* problem. This asks to determine, for a given graph G, the minimum number of chain subgraphs that cover all the edges of G. This has already been investigated in the literature as it is related to other well-known problems such as maximum induced matching (see *e.g.* [3,4]). For bipartite graphs, the problem was shown to be NP-hard [14].

Calling m the number of edges in the graph, we provide an exact exponential algorithm which runs in time $O^*((2 + \varepsilon)^m)$, for every $\varepsilon > 0$ by combining our results on the enumeration of maximal chain subgraphs with the inclusion-exclusion technique [1] (by O^* we denote standard big O notation but omitting polynomial factors). Notice that, since a chain subgraph cover is a family of subsets of edges, the existence of an algorithm whose complexity is close to 2^m is not obvious. Indeed, the basic search space would have size 2^{2^m}, which corresponds to all families of subsets of edges of a graph on m edges.

Finally, we approach the problem of enumerating all minimal covers by chain subgraphs. To this purpose, we provide a total output quasi-polynomial time algorithm to enumerate all *minimal* covers by maximal chain subgraphs of a bipartite graph. To do so, we prove that this can be polynomially reduced to the enumeration of the minimal set covers of a hypergraph.

Besides their theoretical interest, the problems of finding one minimum chain subgraph cover and of enumerating all such covers have also a direct application in biology. Nor *et al.* [13] showed that a minimum chain subgraph cover of such a bipartite graph provides a good model for identifying the minimum genetic architecture enabling to explain one type of manipulation, called *cytoplasmic incompatibility*, by some parasite bacteria on their hosts. This phenomenon results in the death of embryos produced in crosses between males carrying the infection and uninfected females. The observed cytoplasmic compatibility relationships, can be then represented by a bipartite graph with males and females in different classes. Moreover, as different minimum (resp. minimal) covers may correspond to solutions that differ in terms of their biological interpretation, the capacity to enumerate all such minimal chain covers becomes crucial.

The remainder of the paper is organised as follows. In Sect. 2, we give some definitions and preliminary results that will be used throughout the paper. Section 3 then provides a polynomial delay algorithm to enumerate all maximal chain subgraphs in a bipartite graph G with n nodes and m edges, and Sect. 4 presents an upper bound on their maximum number. We use the latter to further establish the input-sensitive complexity of the enumeration algorithm. In Sect. 5, we detail the exact algorithm for finding the minimum size of a minimum chain cover in bipartite graphs, and in Sect. 6 we exploit the connection of this problem with the minimal set cover of a hypergraph to show that it is possible to enumerate in quasi-polynomial time all minimal covers by maximal chain subgraphs of a bipartite graph. Finally, we conclude with some open problems.

2 Preliminaries

Throughout the paper, we assume that the reader is familiar with the standard graph terminology, as contained for instance in [2]. We consider finite undirected graphs without loops or multiple edges. For each of the graph problems in this paper, we let n denote the number of nodes and m the number of edges of the input graph.

Given a bipartite graph $G = (U \cup W, E)$ and a node $u \in U$, we denote by $N_G(u)$ the set of nodes adjacent to u in G and by $E_G(u)$ the *set of edges incident to u in G*. Moreover, given $U' \subseteq U$ and $W' \subseteq W$, we denote by $G[U', W']$ the *subgraph of G induced by $U' \cup W'$*. A node $u \in U$ such that $N_G(u) = W$ is called a *universal node*.

A bipartite graph is a *chain graph* if it does not contain a $2K_2$ as an induced subgraph. Equivalently, a bipartite graph is a chain graph if and only if for each two nodes v_1 and v_2 both in U (resp. in W), it holds that either $N_G(v_1) \subseteq N_G(v_2)$ or $N_G(v_2) \subseteq N_G(v_1)$. Given a chain subgraph $C = (X \cup Y, F)$ of G, we say that a permutation π of the nodes of U is a *neighbourhood ordering* of G if $N_C(u_{\pi(1)}) \subseteq N_C(u_{\pi(2)}) \subseteq \ldots \subseteq N_C(u_{\pi(|U|)})$. Observe that if $X \subset U$, the sets $N_C(u_{\pi(1)}), \ldots, N_C(u_{\pi(l)})$ for some integer $l \leq |U|$, may be empty and, in the case C is connected, $l = |U| - |X|$. By the *largest neighbourhood of C*, we mean the neighbourhood of a node x in X for which the set $N_C(x) \subseteq Y$ has maximum cardinality. A set $Y' \subseteq Y$ is a *maximal neighborhood* of G, if there exists $x \in X$ such that $N_G(x) = Y'$ and there does not exist a node $x' \in X$ such that $N_G(x) \subset N_G(x')$. Two nodes x, x' such that $N_C(x) = N_C(x')$ are called *twins*.

In this paper, we always consider *edge induced* chain subgraphs of a graph G. Hence, here a chain subgraph C of G is a set of edges $E(C) \subseteq E(G)$ and in that case its set of nodes will be constituted by all the nodes of G incident to at least one edge in C (sometimes abusing notation, we more simply write $C \setminus E(D)$ with D a subgraph of G, $e \in C$, $C \subseteq E(D)$ or equivalently $C \subseteq D$ to say that C is a subgraph of D).

A *maximal chain subgraph* C of a given bipartite graph G is a connected chain subgraph such that no superset of $E(C)$ is a chain subgraph. We denote by $\mathscr{C}(G)$ the set of all maximal chain subgraphs in G.

A set of chain subgraphs $C_1, \ldots, C_k$ is a *cover* for G if $\cup_{1 \leq i \leq k} E(C_i) = E(G)$. Observe that, given any cover of G by chain subgraphs $C = \{C_1, \ldots C_k\}$, there exists another cover of same size $C' = \{C'_1, \ldots C'_k\}$ whose chain subgraphs are all maximal; more precisely, for each $i = 1, \ldots, k$, C'_i is a maximal chain subgraph of G and C'_i admits C_i as subgraph. In order to avoid redundancies, from now on, although not explicitly highlighted, we will restrict our attention to the covers by maximal chain subgraphs.

We denote by $\mathcal{S}(G)$ the set of all minimal chain covers of a bipartite graph G.

An enumeration algorithm is said to be *output polynomial* or *total polynomial* if the total running time is polynomial in the size of the input and the output. It is said to be *polynomial delay* if the time between the output of any one solution and the next one is bounded by a polynomial function of the input size [10].

3   Enumerating All Maximal Chain Subgraphs

In this section, we provide a polynomial delay algorithm for enumerating all the maximal chain subgraphs of a given bipartite graph. We start by proving the following result.

Proposition 1. *Let* $C = (X \cup Y, F)$ *be a chain subgraph of* $G = (U \cup W, E)$, *with* $X \subseteq U$, $Y \subseteq W$ *and* $F \subseteq E$, *and let* $x \in X$ *be a node with largest neighbourhood in* C. *Then* C *is a maximal chain subgraph of* G *if and only if both the following conditions hold:*

(i) $N_C(x) = N_G(x)$ *is a maximal neighbourhood of* G, *i.e. there does not exist a node* $x' \in X$ *such that* $N_G(x) \subset N_G(x')$.

(ii) $C \setminus E_G(x)$ *is a maximal chain subgraph of* $G[U \setminus \{x\}, N_G(x)]$.

Proof. ($\Rightarrow$) Let $C = (X \cup Y, F)$ be a maximal chain subgraph of $G = (U \cup W, E)$. To prove that *(i)* holds, suppose by contradiction that $N_C(x)$ is not a maximal neighbourhood of G, i.e. there exists $x' \in U$ with $N_C(x) \subset N_G(x')$ (possibly $x' = x$). Since $N_C(x)$ is the largest neighbourhood of C, for all $z \in X$, we have $N_C(z) \subseteq N_C(x) \subset N_G(x')$, so we can then add to C all the edges incident to x' and still obtain a chain subgraph thereby contradicting the maximality of C. To prove that *(ii)* holds, first observe that $N_G(x) = Y$ (otherwise we would violate (i) with $x' = x$). By contradiction, assume that $C \setminus E_G(x)$ is not maximal in $G[U \setminus \{x\}, N_G(x)]$. Then, there exists a chain subgraph C' such that $C \setminus E_G(x) \subset C' \subseteq G[U \setminus \{x\}, N_G(x)]$. By adding to each one of the previous graphs the edges in $E_G(x)$, we have that the strict inclusion is preserved because the added edges were not present in any one of the three graphs. Since C' with the addition of $E_G(x)$ is still a chain subgraph with $N_G(x)$ as its largest neighbourhood, we reach a contradiction with the hypothesis that C is maximal in G.

($\Leftarrow$) We show that if both *(i)* and *(ii)* hold, then the chain subgraph C of G is maximal. Suppose by contradiction that C is not maximal in G, and let C' be a chain subgraph of G such that $C \subset C'$. Let x be the node with the largest neighbourhood in C. It follows that $N_C(x) \subseteq N_{C'}(x)$. As *(i)* holds, we have that $N_G(x) = N_C(x) \subseteq N_{C'}(x) \subseteq N_G(x)$ from which we derive that $N_{C'}(x) = N_G(x)$, and that $C' \subseteq G[U, N_G(x)]$ since $N_{C'}(x)$ is a maximal neighbourhood of G, hence the largest neighbourhood of C' (and C by the hypothesis). This implies also that C and C' differ in some node different from x, *i.e.* $C \setminus E_G(x) \subset C' \setminus E_G(x) \subseteq G[U \setminus \{x\}, N_G(x)]$. Notice that $C' \setminus E_G(x)$ is still a chain subgraph because we simply removed node x and all its incident edges. We then get a contradiction with *(ii)*. $\qquad\square$

Proposition 1 leads us to the design of Algorithm 1 which efficiently enumerates all maximal chain subgraphs of G. It exploits the fact that, in each maximal chain subgraph, a node u whose neighbourhood is largest is also maximal in G (part *(i)* of Proposition 1) and this holds recursively in the chain subgraph obtained by removing node u and restricting the graph to $N_C(u)$ (part *(ii)* of Proposition 1). To compute the maximal neighbourhood nodes, the algorithm uses a function, `computeCandidates`, that, given sets U and W, returns for each maximal neighbourhood $Y \subset W$, a unique node u, called *candidate*, for which $N_G(u) = Y$. This means that in case of twin nodes, the function `computeCandidates` extracts only one representative node according to some fixed order on the nodes (*e.g.* the node with the smallest label according to the lexicographical order). If the graph has no edges, the function returns the empty set.

Proposition 2 (Correctness). *Algorithm 1 correctly enumerates all the maximal chain subgraphs of the input graph G without repetitions.*

Proof. Let $G = (U \cup W, E)$ be a bipartite graph. We prove the correctness of Algorithm 1 by induction on $|U|$, *i.e.* we show that all the solutions are output, without repetitions.

When $|U| = 1$, let u be the only node in U. We have that $N_G(u)$ is the only neighbourhood in W, and line 3 returns $\{u\}$ as unique candidate. In line 9, the algorithm reduces the graph of interest. In line 10, the whole $E_G(u)$ is added to the current chain subgraph C. Then the function is recursively recalled, with $U' = \emptyset$ so the condition at line 4 is true and C is printed; it is in fact the only chain subgraph of G, it is trivially maximal and there are no repetitions. Correctness then follows when $|U| = 1$.

Assume now that $|U| = k$ with $k > 1$. As inductive hypothesis, let the algorithm work correctly when $|U| = k - 1$.

For each candidate u, the algorithm recursively recalls the same function on a reduced graph and, by the inductive hypothesis, outputs all chain subgraphs of this reduced subgraph without repetitions. By Proposition 1, if we add to each one of these chain subgraphs the node u and all the edges incident to u in $G[U, W]$, we get a different maximal chain subgraph of G since each maximal chain subgraph has one and only one maximal neighborhood and the function `computeCandidates` returns only one representative node. Recall that in the case of twin nodes the algorithm will always consider the nodes in a precise order and so no repetition occurs. Moreover, iterating this process for all candidates guarantees that all maximal chain subgraphs are enumerated and no one is missed. $\qquad\square$

Algorithm 1. Enumerate All Maximal Chain Subgraphs

   Input: A bipartite graph $G = (U \cup W, E)$
   Output: All maximal chain subgraphs of G
1 $C \longleftarrow \emptyset$;   /* C is the set of edges of the current chain subgraph */

2 enumerateMaximalChain(U, W, C)
3      $Candidates \longleftarrow$ computeCandidates(U, W)

4   **if** $Candidates == \emptyset$ **then**
5   print(C);
6   return;
7   **end**
8   **for** $u \in Candidates$ **do**
9         $U' \longleftarrow U \setminus \{u\}$; $W' \longleftarrow W \cap N_G(u)$;   /* reduced graph */
10       $F(u) \longleftarrow \{$edges of $E_G(u)$ incident to some node in $W'\}$
11   enumerateMaximalChain$(U', W', C \cup F(u))$;
12   **end**

Let $G = (U \cup W, E)$ be a bipartite graph, with $n = |U| + |W|$ and $m = |E|$. Before proving the time complexity of Algorithm 1, we observe that the running time of the function ComputeCandidates is $O(nm)$. Indeed, if we assume that the adjacency lists of the graph are ordered, for each node $u_i \in U$, it requires only time proportional to $i \cdot deg(u_i) \leq n \cdot deg(u_i)$ to check whether the neighbourhood of u_i either is included, or includes the neighbourhood of u_j, for each $j < i$.

Proposition 3 (Time Complexity and Polynomial Delay). *Let $G = (U \cup W, E)$ be a bipartite graph. The total running time of Algorithm 1 is $O(|\mathscr{C}(G)|n^2m)$ where $|\mathscr{C}(G)|$ is the number of maximal chains subgraph of G. Moreover, the solutions are enumerated in polynomial time delay $O(n^2m)$.*

Proof. Represent the computation of Algorithm 1 as a tree of the recursion calls of enumerateMaximalChain, each node of which stores the current graph on which the recursion is called at line 11. Of course, the root stores G and on each leaf the condition $Candidates = \emptyset$ is true and a new solution is output. Observe that each leaf contains a feasible solution, and that no repetitions occur in view of Proposition 2, so the number of leaves is exactly $|\mathscr{C}(G)|$.

Since at each call the size of U is reduced by one, the tree height is necessarily bounded by $|U| = O(n)$; moreover, on each tree node, $O(nm)$ time is spent for running function ComputeCandidates.

It follows that, since the algorithm explores the tree in DFS fashion starting from the root, between two solutions the running time is at most $O(n^2m)$ and the total running time is $O(|\mathscr{C}(G)|n^2m)$.   □

4 Upper Bounds on the Number of Maximal Chain Subgraphs

In this section, we give two upper bounds on the maximum number of maximal chain subgraphs of a bipartite graph G with n nodes and m edges. The first bound is given in terms of n while the second depends on m. These bounds are of independent interest, however we will use them in two directions. First, they will allow us to determine the (input-sensitive) complexity of Algorithm 1. Indeed, in Proposition 3, we proved that the total running time of Algorithm 1 is of the form $O(D(n) \times |\mathscr{C}(G)|)$, where $D(n)$ is the delay of the algorithm and $|\mathscr{C}(G)|$ is the number of maximal chain subgraphs of G. Thus, a bound on $|\mathscr{C}(G)|$ leads to a bound on the running time of Algorithm 1 depending on the size of the input. Second, the bound on $|\mathscr{C}(G)|$ in terms of edges allows us to compute the time complexity of an exact exponential algorithm for the minimum chain subgraph cover problem in Sect. 5.

4.1 Bound in Terms of Nodes

The following lemma claims that a given permutation is the neighbourhood ordering of at most one maximal chain subgraph.

Lemma 1. Let C_1 and C_2 be two maximal chain subgraphs of $G = (U \cup W, E)$ and let π_1 (resp. π_2) be a neighbourhood ordering of C_1 (resp. C_2). Then, $\pi_1 = \pi_2 \Longrightarrow C_1 = C_2$.

Proof. The proof proceeds by induction on the number of nodes of U.

If $|U| = 1$ then G has only one maximal chain subgraph and the result trivially holds.

Assume now that $|U| > 1$. By Proposition 1, we have that $N_{C_1}(u_{\pi(|U|)}) = N_G(u_{\pi(|U|)}) = N_{C_2}(u_{\pi(|U|)})$. Using again Proposition 1, we obtain that $C_1' := C_1[U \setminus \{u_{\pi(|U|)}\}, N_G(u_{\pi(|U|)})]$ and $C_2' := C_2[U \setminus \{u_{\pi(|U|)}\}, N_G(u_{\pi(|U|)})]$ are maximal chain subgraphs of the graph defined as $G[U \setminus \{u_{\pi(|U|)}\}, N_G(u_{\pi(|U|)})]$. Applying the inductive hypothesis with the permutations restricted to the $|U| - 1$ elements, we have that $C_1' = C_2'$. Finally, since $N_{C_1}(u_{\pi(|U|)}) = N_{C_2}(u_{\pi(|U|)})$, we conclude that $C_1 = C_2$. □

As a corollary, the maximum number of chain subgraphs of a graph $G = (U \cup W, E)$ is bounded by $|U|!$. Since the same reasoning can be applied on W, we have that $|\mathscr{C}(G)| \leq |W|!$ and hence:

$$|\mathscr{C}(G)| \leq \min(|U|, |W|)! \leq \frac{n}{2}!$$

This bound is tight as shown by the following family of graphs that reaches it.

Consider the *antimatching graph with n nodes* $A_n = (U \cup W, E)$ defined as the complement of an $n/2$ edge perfect matching, *i.e.*:

$$U := \{u_1, \ldots, u_{n/2}\}, \quad W := \{w_1, \ldots, w_{n/2}\},$$
$$E := \{(u_i, w_j) \in U \times W : i \neq j\}$$

It is not difficult to convince oneself that the maximal chain subgraphs of A_n are exactly $(n/2)!$ and that a different permutation corresponds to each of them. In particular, for each permutation π of the nodes of U, the corresponding maximal chain subgraph C_π of A_n can be defined by means of the set of neighbourhoods as follows:

$$N_{C_\pi}(u_i) := \{w_k s.t. \pi^{-1}(k) < \pi^{-1}(i)\}.$$

The so-defined graph C_π is a chain subgraph since all the neighbourhoods form a chain of inclusions. Moreover, it is maximal since if we added to the neighbourhood of u_i any one of the missing edges (u_i, w_j) with $\pi^{-1}(j) \geq \pi^{-1}(i)$, we would introduce a $2K_2$ with the existing edge (u_j, w_i) as (u_j, w_j) and (u_i, w_i) are not in E.

4.2   Bound in Terms of Edges

Let $T(m)$ be the maximum number of maximal chain subgraphs over all bipartite graphs with m edges. We prove that $T(m) \leq 2^{\sqrt{m} \log(m)}$.

Lemma 2. *Let $G = (U \cup W, E)$ be a bipartite graph. Then $|\mathscr{C}(G)| \leq |U| \cdot T(m - |W|)$.*

Proof. In view of how the algorithm works and of Proposition 1, at the beginning, there at most $|U|$ candidates. For each candidate x, we can build as many chain subgraphs as there are in $G[U \setminus \{x\}, N_G(x)]$. We claim that this latter graph has at most $m - |W|$ edges. Indeed, in order to construct $G[U \setminus \{x\}, N_G(x)]$, we remove from G exactly $|E_G(x)|$ edges when deleting x from U, and $|W| - |N_G(x)|$ nodes (each one connected to at least a different edge as G is connected) when reducing W to $N_G(x)$. Observing that $|E_G(x)| = |N_G(x)|$, in total we remove $|W|$ edges. The proof follows from the fact that the number of chain subgraphs of $G[U \setminus \{x\}, N_G(x)]$ is bounded by $T(m - |W|)$.   □

Theorem 1. *Let $G = (U \cup W, E)$ be a bipartite graph with n nodes and m edges; then $|\mathscr{C}(G)| \leq 2^{\sqrt{m} \log m}$, i.e. $T(m) \leq 2^{\sqrt{m} \log m}$.*

Proof. Assume w.l.o.g that $|U| \leq |W|$. The proof is by induction on m. Note that for $m = 1$ the theorem holds trivially.

Applying the inductive hypothesis and Lemma 2, we have:

$$|\mathscr{C}(G)| \leq |U| T(m - |W|) \leq \frac{n}{2} 2^{\left(\sqrt{m - \frac{1}{2}n} \log(m - \frac{1}{2}n)\right)}.$$

Since the function $x \mapsto x 2^{\sqrt{m-x} \log(m-x)}$ is decreasing in the interval $[\sqrt{m}, m-1]$, the maximum of $\frac{n}{2} 2^{\sqrt{m-\frac{n}{2}} \log(m-\frac{n}{2})}$ is reached when $n/2$ is minimum. Note that trivially for a bipartite graph we have $n/2 > \sqrt{m}$. Hence,

$$|\mathscr{C}(G)| \le \sqrt{m}\, 2^{\sqrt{m-\sqrt{m}} \log(m-\sqrt{m})}$$

Let $A := \sqrt{m} - \sqrt{m - \sqrt{m}}$ and $B := \frac{m-\sqrt{m}}{m}$. We then have:

$$|\mathscr{C}(G)| \le \sqrt{m}\, 2^{(\sqrt{m}-A) \log(mB)}$$
$$= 2^{\sqrt{m}\log m} \times \sqrt{m}\, 2^{\log B(\sqrt{m}-A) - A \log(m)}$$

Let us show that $Z := \sqrt{m}\, 2^{\log B(\sqrt{m}-A) - A \log m} \le 1$ by showing that $\log Z \le 0$:

$$\log Z = \log \sqrt{m} + \log B(\sqrt{m} - A) - A \log(m)$$
$$= \log \sqrt{m}(1 - 2A) + \log B(\sqrt{m} - A)$$
$$\le 0$$

considering that $B < 1$ and $1/2 < A \le 1$ since:

$$A = \frac{1}{1 + \sqrt{B}} = \frac{1}{1 + \sqrt{1 - \frac{1}{\sqrt{m}}}}$$

□

Corollary 1. *The (input-sensitive) complexity of Algorithm 1 is bounded by* $O^*(2^{\sqrt{m} log(m)})$.

5   Minimum Chain Subgraph Cover

In this section, we show how to find in polynomial space the minimum size of a chain subgraph cover in time $O^*((2 + \epsilon)^m)$, for every $\varepsilon > 0$. Since a chain subgraph cover is a family of subsets of edges, the existence of an algorithm whose complexity is close to 2^m is not obvious. Indeed the basic search space has size 2^{2^m}, as it corresponds to a family of subsets of edges. To obtain this result, we exploit Algorithm 1, the bound obtained in Theorem 1 and the inclusion/exclusion method [1,8] that has already been successfully applied to exact exponential algorithms for many partitioning and covering problems.

We first express the problem as an inclusion-exclusion formula over the subsets of edges of G.

Proposition 4. *[1] Let $c_k(G)$ be the number of chain subgraph covers of size k of a graph G. We have that:*

$$c_k(G) = \sum_{A \subseteq E} (-1)^{|A|} a(A)^k$$

where $a(A)$ denotes the number of maximal chain subgraphs not intersecting A.

Exploring this result brings to the exact algorithm as described in the proof of the next theorem.

Theorem 2. *Given a bipartite graph G with m edges, for all $k \in \mathbb{N}^*$ and for all $\varepsilon > 0$, $c_k(G)$ can be computed in time $O^*((2 + \epsilon)^m)$.*

Proof. Let $G = (U \cup W, E)$ be a bipartite graph, $k \in \mathbb{N}^*$ and $\varepsilon > 0$. Using the formula of Proposition 4, c_k can be computed in time $\sum_{i=0}^{m} \binom{m}{i} C(i)$, where $C(i)$ is the time complexity needed to compute $a(A)$, $|A| = i$.

Notice that to compute $a(A)$ for a given $A \subseteq E$, one can naively compute all maximal chain subgraphs of $G' = (U \cup W, E \setminus A)$ and, for each of them, check whether it is maximal in G. Using this fact, and Corollary 1, $C(i)$ can be determined in time $O(n^2 m 2^{\sqrt{m-i} \log(m-i)})$.

Thus we have that $c_k(G)$ can be computed in time $\sum_{i=0}^{m} \binom{m}{i} n^2 m 2^{\sqrt{m-i} \log(m-i)}$.
Observe now that since $2^{\sqrt{m-i} \log(m-i)} = o((1+\varepsilon)^m)$, there exists a constant n_ε such that for all $m > n_\varepsilon$, $2^{\sqrt{m-i} \log(m-i)} < (1+\varepsilon)^m$.

Recalling that G is connected and thus $m \geq n$, we then have:

$$\sum_{i=0}^{m} \binom{m}{i} n^2 m 2^{\sqrt{m-i} \log(m-i)} = n^2 m \left(\sum_{i=0}^{m-n_\varepsilon-1} \binom{m}{i} 2^{\sqrt{m-i} \log(m-i)} + \sum_{i=m-n_\varepsilon}^{m} \binom{m}{i} 2^{\sqrt{m-i} \log(m-i)} \right)$$

$$\leq n^2 m \left(\sum_{i=0}^{m-n_\varepsilon-1} \binom{m}{i} (1+\varepsilon)^{m-i} + n_\varepsilon m^{n_\varepsilon} 2^{\sqrt{n_\varepsilon} \log(n_\varepsilon)} \right)$$

$$\leq n^2 m \left(\sum_{i=0}^{m} \binom{m}{i} (1+\varepsilon)^{m-i} + n_\varepsilon m^{n_\varepsilon} 2^{\sqrt{n_\varepsilon} \log(n_\varepsilon)} \right)$$

$$\leq n^2 m (2+\varepsilon)^m + n^2 n_\varepsilon m^{1+n_\varepsilon} 2^{\sqrt{n_\varepsilon} \log(n_\varepsilon)}$$

$$= O^*((2+\varepsilon)^m).$$

We conclude, by observing that the size of a minimum chain cover is given by the smallest value of k for which $c_k(G) \neq 0$. $\square$

6   Enumeration of Minimal Chain Subgraph Covers

In this section, we prove that the enumeration of all minimal chain subgraph covers can be polynomially reduced to the enumeration of the minimal set covers of a hypergraph. This reduction implies that there is a quasi-polynomial time algorithm to enumerate all minimal chain subgraph covers. Indeed, the result in [9] implies that all the minimal set covers of a hypergraph can be enumerated in time $N^{\log N}$ where N is the sum of the input size (*i.e.* $n + m$) and of the output size (*i.e.* the number of minimal set covers).

Let $G = (U \cup W, E)$ be a bipartite graph, $\mathscr{C} = \mathscr{C}(G)$ the set of all its maximal chain subgraphs, and $\mathcal{S} = \mathcal{S}(G)$ the set of its minimal chain subgraph covers. Notice that the minimal chain subgraph covers of G are the minimal set covers of the hypergraph $\mathcal{H} := (V, \mathcal{E})$ where $V = E$ and $\mathcal{E} = \mathscr{C}$. Unfortunately, the size of $\mathcal{H}$ might be exponential in the size of G plus the size of S. Indeed not every maximal chain subgraph in $\mathscr{C}$ will necessarily be part of some minimal chain subgraph cover. To obtain a quasi-polynomial time algorithm to enumerate all minimal chain subgraph covers, we need to enumerate only those maximal chain subgraphs that belong to a minimal chain subgraph cover.

Given an edge $e \in E$, let $\mathscr{C}_e$ be the set of all maximal chain subgraphs of G containing e and $\mathcal{M}_e$ the set of all edges $e' \in E$ inducing a $2K_2$ in G together with e.

We call an edge $e \in E$ *non-essential* if there exists another edge $e' \in E$ such that $\mathscr{C}_{e'} \subset \mathscr{C}_e$. An edge which is not non-essential is said to be *essential*. Note that for every non-essential edge e, there exists an essential edge e_1 such that $\mathscr{C}_{e_1} \subset \mathscr{C}_e$. Indeed, by applying iteratively the definition of a non-essential edge, we obtain a list of inclusions $\mathscr{C}_e \supset \mathscr{C}_{e_1} \supset \mathscr{C}_{e_2} \ldots$, where no $\mathscr{C}_{e_i}$ is repeated as the inclusions are strict. The last element of the list will correspond to an essential edge.

The following lemma claims that if a maximal chain subgraph C contains at least one essential edge, then it belongs to at least one minimal chain subgraph cover.

Lemma 3. *Let C be a maximal chain subgraph of a bipartite graph $G = (U \cup W, E)$. Then C belongs to a minimal chain subgraph cover of G if and only if C contains an essential edge.*

Proof. ($\Rightarrow$) Let C belong to a minimal chain subgraph cover M and assume that C contains no essential edge. Given $e \in C$, e therefore being non-essential, there exists an essential edge e' such that $\mathscr{C}_{e'} \subset \mathscr{C}_e$. Moreover, $e' \notin C$. As M is a cover, there exists $C' \in M$ such that $e' \in C'$. Thus, $C' \neq C$, $C' \in \mathscr{C}_{e'} \subset \mathscr{C}_e$, hence $e \in C'$. Since for every edge $e \in C$, there exists $C' \in M$ containing it, we have that $M \setminus \{C\}$ is a cover, contradicting the minimality of M.

($\Leftarrow$) Assume C contains an essential edge e. Let $\mathscr{C}' = \{D \in \mathscr{C}(G) : e \notin D\}$. Note that $\mathscr{C}' = \mathscr{C} \setminus \mathscr{C}_e$. We show that $\mathscr{C}' \cup \{C\}$ is a cover. Suppose on the contrary that there exists $e' \in E \setminus E(C)$ and e' is not covered by $\mathscr{C}'$ and thus $\mathscr{C}_{e'} \cap \mathscr{C}' = \emptyset$. This implies that $\mathscr{C}_{e'} \subseteq \mathscr{C} \setminus \mathscr{C}' = \mathscr{C}_e$ and as e is essential, we obtain $\mathscr{C}_{e'} = \mathscr{C}_e$ from which we deduce that $e' \in C$. Thus, $M = \mathscr{C}' \cup \{C\}$ is a cover and clearly it contains a minimal one. Finally, we conclude by observing that, since by construction C is the only chain subgraph of M that contains e, it belongs to any minimal cover contained in M. $\square$

It follows that the set of maximal chain subgraphs that can contribute to a minimal chain cover is $\tilde{\mathscr{C}} = \cup \mathscr{C}_e$ where the index e of the union operation runs over all the essential edges of G. In the following, we show how to detect essential edges. This problem then consists in detecting all the couples e_1, e_2 such that $\mathscr{C}_{e_1} \subseteq \mathscr{C}_{e_2}$ before enumerating all useful maximal chain subgraphs. The following lemma holds.

Lemma 4. *Let C be a maximal chain subgraph of a bipartite graph $G = (U \cup W, E)$ and let $e \in E$ be such that for all $e' \in E(C)$, it holds that $e \notin \mathcal{M}_{e'}$. Then $e \in C$.*

Using this lemma we can now prove the following result.

Theorem 3. *Given a bipartite graph $G = (U \cup W, E)$, for any two edges $e, e' \in E$, $\mathcal{C}_e \subseteq \mathcal{C}_{e'}$ if and only if $\mathcal{M}_e \supseteq \mathcal{M}_{e'}$.*

Proof. ($\Rightarrow$) Given two edges $e, e' \in E$, suppose that $\mathcal{C}_e \subseteq \mathcal{C}_{e'}$, and assume on the contrary that there exists $f \in \mathcal{M}_{e'}$ and $f \notin \mathcal{M}_e$. Then there exists a maximal chain C' containing e and f (as they do not form a $2K_2$ in G) but not e' ($f \in \mathcal{M}_{e'}$). Hence, $C' \in \mathcal{C}_e$ but $C' \notin \mathcal{C}_{e'}$, contradicting the assumption that $\mathcal{C}_e \subseteq \mathcal{C}_{e'}$.

($\Leftarrow$) Suppose now $\mathcal{M}_e \supseteq \mathcal{M}_{e'}$. Let $C \in \mathcal{C}_e$. By definition, none of the edges of $\mathcal{M}_e$ appears in C. Hence, e' does not form a $2K_2$ with any edge in C in the graph G (as $\mathcal{M}_e \supseteq \mathcal{M}_{e'}$). By Lemma 4 $e' \in C$. Thus, $\mathcal{C}_e \subseteq \mathcal{C}_{e'}$.   □

Notice that, given an edge $e = (u, w) \in E$, $u \in U$ and $w \in W$, it is easy to determine the set $\mathcal{M}_e$. We just need to start from E and delete all edges that are incident either to u or to w, as well as all edges at distance 2 from e (that is all edges $e' = (u', w')$ such that either u' is adjacent to w or w' is adjacent to u). Checking whether $\mathcal{M}_e \supseteq \mathcal{M}_{e'}$ is also easy: it suffices to sort the edges in each set in lexicographic order, and then the inclusion of each pair can be checked in linear time in their size, that is in $O(m)$. It is thus possible to enumerate in polynomial delay only those maximal chain subgraphs that contain at least one essential edge by modifying Algorithm 1. Due to space limits, we do not detail the algorithm here. Finally, we are now able to state the main result of this section.

Theorem 4. *Given a bipartite graph $G = (U \cup W, E)$, one can enumerate all its minimal chain subgraph covers, i.e. all the elements in $\mathcal{S}$, in time $O(|\mathcal{S}|^{\log(|\mathcal{S}|)+2})$.*

Proof. We first construct the hypergraph $\mathcal{H} = (V, \mathcal{E})$ where $V := E'$ is the set of essential edges of G and $\mathcal{E} := \mathcal{C}_{ess}$ is the set of maximal chain subgraphs of G that contain at least one essential edge. This takes time $O(n^2 m |\mathcal{C}_{ess}|)$. Applying then the algorithm given in [9], one can enumerate all minimal set covers of $\mathcal{H}$ (i.e. all minimal chain subgraph covers) in time $O((|\mathcal{H}| + |\mathcal{S}|)^{\log(|\mathcal{H}|+|\mathcal{S}|)}) = O((|\mathcal{C}_{ess}| + |\mathcal{S}|)^{\log(|\mathcal{C}_{ess}|+|\mathcal{S}|)})$. The total running time is thus $O(n^2 m |\mathcal{C}_{ess}| + (|\mathcal{C}_{ess}| + |\mathcal{S}|)^{\log(|\mathcal{C}_{ess}|+|\mathcal{S}|)})$. Notice now that since by Lemma 3, every maximal chain subgraph in $\mathcal{C}_{ess}$ belongs to at least one minimal chain subgraph cover, we have that $|\mathcal{C}_{ess}| \leq m|\mathcal{S}|$. Finally, we obtain that the total running time is $O(n^2 m^2 |\mathcal{S}| + (|\mathcal{S}| + |\mathcal{S}|)^{\log(|\mathcal{S}|+|\mathcal{S}|)}) = O(|\mathcal{S}|^{\log(|\mathcal{S}|)+2})$.

7   Conclusion

In this paper, we studied different problems related to maximal chain subgraphs and chain subgraph covers in bipartite graphs. This work raises many questions.

First, it remains an open problem whether it is possible to enumerate the minimal chain covers of a graph in polynomial delay. Indeed, our problem is more constrained than an arbitrary instance of the set cover of a hypergraph. A future goal is to better exploit the connections between these two problems. Second, it would be interesting to determine the exact value of $T(m)$. We conjecture that a tighter bound may be $\left(\frac{1+\sqrt{1+4m}}{2} \right)!$. Finally, it is worth exploring the different nature of the problems considered here in the case where we deal with an hereditary property (induced chain subgraphs) instead of a non-hereditary one (edge induced chain subgraphs). In particular, it remains unknown whether enumerating maximal induced subgraphs can be done in polynomial delay.

Acknowledgments. T. Calamoneri is supported in part by the Italian Ministry of Education, University, and Research (MIUR) under PRIN 2012C4E3KT national research project "AMANDA - Algorithmics for MAssive and Networked DAta" and in part by Sapienza University of Rome project "Graph Algorithms for Phylogenetics: A Promising Approach". M. Gastaldello is supported by the Università Italo-Francese project "Algorithms and Models for the solution of difficult problems in biology". A. Mary is supported by the ANR project GraphEN "Enumération dans les graphes et les hypergraphes: algorithmes et complexité", ANR-15-CE40-0009.

References

1. Björklund, A., Husfeldt, T., Koivisto, M.: Set partitioning via inclusion-exclusion. SIAM J. Comput. **39**(2), 546–563 (2009)
2. Bollobás, B.: Modern graph theory. Graduate Texts in Mathematics. Springer-Verlag, Heidelberg (1998)
3. Brandstädt, A., Eschen, E.M., Sritharan, R.: The induced matching and chain subgraph cover problems for convex bipartite graphs. Theor. Comput. Sci. **381**(1), 260–265 (2007)
4. Chang-Wu, Y., Gen-Huey, C., Tze-Heng, M.: On the complexity of the k-chain subgraph cover problem. Theor. Comput. Sci. **205**(1), 85–98 (1998)
5. Dias, V.M.F., de Figueiredo, C.M.H., Szwarcfiter, J.L.: Generating bicliques of a graph in lexicographic order. Theor. Comput. Sci. **337**(1–3), 240–248 (2005)
6. Dias, V.M.H., de Figueiredo, C.M.H., Szwarcfiter, J.L.: On the generation of bicliques of a graph. Discrete Appl. Math. **155**(14), 1826–1832 (2007)
7. Eiter, T., Gottlob, G.: Identifying the minimal transversals of a hypergraph and related problems. SIAM J. Comput. **24**(6), 1278–1304 (1995)
8. Fedor, V.: Fomin and Dieter Kratsch. Exact Exponential Algorithms. Springer-Verlag New York Inc, New York, NY, USA (2010)
9. Fredman, M.L., Khachiyan, L.: On the complexity of dualization of monotone disjunctive normal forms. J. Algorithms **21**(3), 618–628 (1996)
10. Johnson, D.S., Yannakakis, M., Papadimitriou, C.H.: On generating all maximal independent sets. Inf. Process. Lett. **27**(3), 119–123 (1988)
11. Makino, K., Uno, T.: New algorithms for enumerating all maximal cliques. In: Hagerup, T., Katajainen, J. (eds.) SWAT 2004. LNCS, vol. 3111, pp. 260–272. Springer, Heidelberg (2004)

12. Moon, J.W., Moser, L.: On cliques in graphs. Isr. J. Math. **3**(1), 23–28 (1965)
13. Nor, I., Engelstädter, J., Duron, O., Reuter, M., Sagot, M.-F., Charlat, S.: On the genetic architecture of cytoplasmic incompatibility: inference from phenotypic data. Am. Nat. **182**(1), E15–E24 (2013)
14. Yannakakis, M.: The complexity of the partial order dimension problem. SIAM J. Algebraic Discrete Methods **3**(3), 351–358 (1982)

Weighted de Bruijn Graphs
for the Menage Problem and Its Generalizations

Max A. Alekseyev[✉]

The George Washington University, Washington, DC, USA
maxal@gwu.edu

Abstract. We address the problem of enumeration of seating arrangements of married couples around a circular table such that no spouses sit next to each other and no k consecutive persons are of the same gender. While the case of $k = 2$ corresponds to the classical *problème des ménages* with a well-studied solution, no closed-form expression for the number of seating arrangements is known when $k \geq 3$.

We propose a novel approach for this type of problems based on enumeration of walks in certain algebraically weighted de Bruijn graphs. Our approach leads to new expressions for the menage numbers and their exponential generating function and allows one to efficiently compute the number of seating arrangements in general cases, which we illustrate in detail for the ternary case of $k = 3$.

1 Introduction

The famous *menage problem* (*problème des ménages*) asks for the number M_n of seating arrangements of n married couples of opposite sex around a circular table such that

1. no spouses sit next to each other;
2. females and males alternate.

The problem was formulated by Edouard Lucas in 1891 [6]. A complete solution was first obtained by Touchard in 1934 [9].

Let us call a couple seating next to each other *close*. The restriction of the menage problem can be equivalently stated as

1. there are no close couples;
2. no $k = 2$ consecutive people are of the same sex.

This reformulation allows us to generalize the menage problem to other values of k, such as $k = 3$ which we refer to as the *ternary menage problem*. The ternary menage problem was posed by Hugo Pfoertner in 2006 as the sequence A114939 in the OEIS [8], for which he then managed to correctly compute only the first three terms.

The work is supported by the National Science Foundation under grant No. IIS-1462107.

© Springer International Publishing Switzerland 2016
V. Mäkinen et al. (Eds.): IWOCA 2016, LNCS 9843, pp. 151–162, 2016.
DOI: 10.1007/978-3-319-44543-4_12

In this work, we propose a novel approach for the generalized menage problem based on the transfer-matrix method [7, Section 4.7] applied to certain algebraically weighted de Bruijn graphs. We illustrate our approach on the classical case $k = 2$, where we obtain new formulae for the menage numbers M_n and their exponential generating function (EGF). While an explicit expression (in terms of the modified Bessel functions) for the EGF was earlier derived by Wyman and Moser [10], they admitted it be "quite complicated". In contrast, our expression (and its derivation) is much simpler and can be stated in terms of a certain power series or the exponential integral function. We further apply our approach for the ternary case $k = 3$, which apparently has not been addressed in the literature before. While the resulting formulae in this case are not that simple, they provide an efficient method for computing the corresponding number of seating arrangements, which we used to compute many new terms for A114939 and related sequences in the OEIS.

2   Classical Approaches for Menage Problem

To the best of our knowledge, there exist three major approaches for solving the menage problem, which we briefly discuss below.

Ladies First. A straightforward approach to the menage problem is first to seat all ladies (in $2 \cdot n!$ ways) and then to seat all gentlemen, obeying the close couple restriction. This way the problem reduces to enumerating placements of non-attacking rooks on a square board like the one shown in Fig. 1. Using the rook theory [7, Section 2.3], this leads to the *Touchard formula*:

$$M_n = 2 \cdot n! \cdot \sum_{k=0}^{n} (-1)^k \frac{2n}{2n-k} \binom{2n-k}{k} (n-k)! . \tag{1}$$

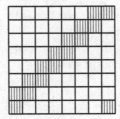

Fig. 1. A board corresponding to the menage problem with $n = 8$ couples. For a fixed seating arrangement of ladies, the seating arrangements of gentlemen are in one-to-one correspondence with the placements of n non-attacking rooks at non-shaded cells.

Hamiltonian Cycles in Crown Graphs. The seating arrangements satisfying the menage problem correspond to directed Hamiltonian cycles in the *crown graph* on $2n$ vertices obtained from the complete bipartite graph $K_{n,n}$ with removal of a perfect matching. Here males/females represent the partite sets of $K_{n,n}$ with every male vertex connected to every female vertex, except for the spouses (corresponding to the removed perfect matching). For odd integers n, crown graphs on $2n$ vertices represent circulant graphs, where Hamiltonian cycles can be systematically enumerated [4].

Non-Sexist Inclusion-Exclusion. Bogart and Doyle [2] suggested to compute M_n with the inclusion-exclusion principle (e.g., see [7, Section 2.1]) as the number of alternating male-female seating arrangements that have no close couples. To do so, they computed the number $W_{n,j}$ of alternating male-female seating arrangements of n couples with j fixed couples being close as

$$W_{n,j} = 2 \cdot \frac{2n}{2n-j} \binom{2n-j}{j} \cdot j! \cdot (n-j)!^2 , \qquad (2)$$

where:

- the factor 2 accounts for two ways to reserve alternating seats for males and females;
- $\frac{2n}{2n-j} \binom{2n-j}{j}$ is the number of ways to select $2j$ seats for the j close couples;
- $j!$ is the number of seating arrangements of the j close couples at the $2j$ selected seats;
- $(n-j)!^2 = (n-j)! \cdot (n-j)!$ is the number of ways to seat females and males from the other $n-j$ couples.

The inclusion-exclusion principle then implies that

$$M_n = \sum_{j=0}^{n} (-1)^j \cdot \binom{n}{j} \cdot W_{n,j}$$

$$= 2 \cdot \sum_{j=0}^{n} (-1)^j \cdot \binom{n}{j} \cdot \frac{2n}{2n-j} \binom{2n-j}{j} \cdot j! \cdot (n-j)!^2 , \qquad (3)$$

which trivially simplifies to (1).

The aforementioned approaches for the menage problem do not seem to easily extend to the ternary case, since there is no nice male-female alternating structure anymore. In particular, the ladies-first approach does not reduce the problem to a uniform board and there is no obvious reduction to a Hamiltonian cycle problem. The (non-sexist) inclusion-exclusion approach is most prominent, but it is unclear what should be in place of $\frac{2n}{2n-j} \binom{2n-j}{j}$. In order to generalize the solution to the menage problem to the ternary case, we suggest to look at this problem at a different angle as described below.

3   De Bruijn Graph Approach for Menage Problem

So far, a seating arrangement in the menage problem was viewed as a cyclic (clockwise) sequence of females (f_i) and males (m_j):

$$f_{i_1} \rightarrow m_{j_1} \rightarrow f_{i_2} \rightarrow m_{j_2} \rightarrow \cdots \rightarrow f_{i_n} \rightarrow m_{j_n} \rightarrow f_{i_1} .$$

However, it can also be viewed as a cyclic sequence of *pairs* of people sitting next to each other:

$$(f_{i_1}, m_{j_1}) \to (m_{j_1}, f_{i_2}) \to (f_{i_2}, m_{j_2}) \to \cdots \to (f_{i_n}, m_{j_n}) \to (m_{j_n}, f_{i_1}) \to (f_{i_1}, m_{j_1}).$$

A similar idea was used by Nicolaas de Bruijn [3] to construct a shortest sequence, which contains every subsequence of length ℓ (called ℓ-*mer*) over a given alphabet. He introduced directed graphs, now named after him, whose nodes represent $(\ell - 1)$-mers and arcs represent ℓ-mers (the arc corresponding to an ℓ-mer s connects the prefix of s with the suffix of s).

We employ de Bruijn graphs for $\ell = 3$ for solving the menage problem. However, in contrast to conventional *unweighted* de Bruijn graphs, we introduce algebraic weights to account for (i) the balance between females and males; and (ii) the number of close couples.

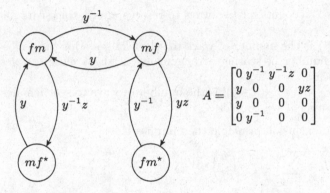

Fig. 2. The weighted de Bruijn graph for the menage problem and its adjacency matrix A.

The (weighted) de Bruijn graph for the menage problem and its adjacency matrix A are shown in Fig. 2. This graph has 4 nodes labeled fm (for clockwise adjacent female–male pair), mf (clockwise adjacent male–female pair), and their starred variants indicating close couples. There is an arc connecting every pair of nodes uv and vw (at most one of which may be starred) for $u, v, w \in \{f, m\}$. Each such arc has an algebraic weight $y^p z^q$ with $p = \pm 1$ and $q \in \{0, 1\}$ such that the degree of indeterminate y accounts for the males-females balance, while the degree of indeterminate z accounts for the number of close couples. Namely, $p = 1$ whenever $w = m$ (and $p = -1$ whenever $w = f$), while $q = 1$ whenever vw is starred.

Any seating arrangement corresponds to a cyclic sequence of nodes fm and mf, some of which may be starred to indicate close couples. Such sequence with j close couples corresponds to a walk of length $2n$ and algebraic weight $y^0 z^j$. The transfer-matrix method [7, Section 4.7] implies that the number of such walks equals $[y^0 z^j]$ $\text{tr}(A^{2n})$, i.e., the coefficient of $y^0 z^j$ in the trace of matrix A^{2n}. This leads to a new formula for $W_{n,j}$:

$$W_{n,j} = [y^0 z^j] \ \text{tr}(A^{2n}) \cdot j! \cdot (n-j)!^2 \,,$$

where the factors $j!$ and $(n-j)!^2$ bear the same meaning as in (2). Similarly to (3), we then obtain

$$M_n = n! \cdot \sum_{j=0}^{n} (-1)^j \cdot (n-j)! \cdot [y^0 z^j] \ \text{tr}(A^{2n}). \qquad (4)$$

Comparison of (4) and (1) suggests the following identity, which we will prove explicitly:

Lemma 1. *For the matrix A defined in Fig. 2 and any integers $n > 1$, $j \geq 0$,*

$$[y^0 z^j] \ \text{tr}(A^{2n}) = 2 \cdot \frac{2n}{2n-j} \cdot \binom{2n-j}{j}.$$

Proof. The eigenvalues of A are $\frac{1 \pm \sqrt{1+4z}}{2}$, each of multiplicity 2.[1] It follows that

$$[y^0 z^j] \ \text{tr}(A^{2n}) = 2 \cdot [z^j] \left(\left(\frac{1 + \sqrt{1+4z}}{2} \right)^{2n} + \left(\frac{1 - \sqrt{1+4z}}{2} \right)^{2n} \right).$$

We further remark that $\frac{1-\sqrt{1+4z}}{2} = -zC(-z)$ and $\frac{1+\sqrt{1+4z}}{2} = C(-z)^{-1}$, where $C(x) = \frac{1-\sqrt{1-4x}}{2x}$ is the ordinary generating function for Catalan numbers.

Since $j \leq n$ and $n > 1$, we have $[z^j] \ (-zC(-z))^{2n} = 0$. On the other hand, since $[x^k] \ C(x)^m = \frac{m}{2k+m} \binom{2k+m}{m}$ (e.g., see [5, formula (5.70)]), we have

$$C(-z)^{-2n} = (-1)^j \frac{-2n}{2j-2n} \binom{2j-2n}{j} = \frac{2n}{2n-2j} \binom{2n-j-1}{j}$$

$$= \frac{2n}{2n-2j} \binom{2n-j-1}{2n-2j-1} = \frac{2n}{2n-j} \binom{2n-j}{j},$$

which concludes the proof. $\qquad \square$

Lemma 1 proves that our formula (4) implies the Touchard formula (1). In the next section, we show that it also implies another (apparently new) formula for M_n. But most importantly, the matrix formula (4) can be generalized for the ternary menage problem as we show in Sect. 5.

---

[1] We remark that A^2 does not depend on y, so it is not surprising that the eigenvalues of A do not depend on y either.

4   New Formulae for Menage Numbers and Their EGF

We find it convenient to define the *series Laplace transform* $\mathcal{L}_{x,y}$ of a function $f(x)$ as the conventional Laplace transform of $f(yt)$ (as a function of t) evaluated at 1, i.e.,

$$\mathcal{L}_{x,y}[f] = \int_0^\infty f(yt) \cdot e^{-t} \cdot dt.$$

It can be easily seen that $\mathcal{L}_{x,y}[x^k] = k! \cdot y^k$ for any integer $k \geq 0$ and thus for a power series $f(x)$, we have

$$\mathcal{L}_{x,y}[f] = \sum_{k=0}^\infty k! \cdot y^k \cdot [x^k] \, f(x).$$

In particular,

$$\sum_{k=0}^\infty k! \cdot [x^k] \, f(x) = \mathcal{L}_{x,1}[f] = \int_0^\infty f(t) \cdot e^{-t} \cdot dt.$$

Lemma 2. *Let U, V be same-size square matrices that do not depend on indeterminate z. Then for any integer $n \geq 0$,*

$$\sum_{j=0}^n (-1)^j \cdot (n-j)! \cdot [z^j] \, \mathrm{tr}((U + V \cdot z)^n) = \int_0^\infty \mathrm{tr}((U \cdot t - V)^n) \cdot e^{-t} \cdot dt.$$

Proof. We have

$$[z^j] \, \mathrm{tr}((U + V \cdot z)^n) = [z^{n-j}] \, \mathrm{tr}((U \cdot z + V)^n) = (-1)^j \cdot [z^{n-j}] \, \mathrm{tr}((U \cdot z - V)^n).$$

Hence

$$\sum_{j=0}^n (-1)^j \cdot (n-j)! \cdot [z^j] \, \mathrm{tr}((U + V \cdot z)^n) = \sum_{j=0}^n (n-j)! \cdot [z^{n-j}] \, \mathrm{tr}((U \cdot z - V)^n)$$

$$= \mathcal{L}_{z,1}[\mathrm{tr}((U \cdot z - V)^n)],$$

which concludes the proof.   □

We are now ready to derive new closed-form expressions for the menage numbers M_n and their exponential generating function.

Theorem 1. *For all integers $n > 1$,*

$$M_n = 2 \cdot n! \cdot \int_0^\infty \left(\left(\frac{t - 2 + \sqrt{t^2 - 4t}}{2} \right)^n + \left(\frac{t - 2 - \sqrt{t^2 - 4t}}{2} \right)^n \right) \cdot e^{-t} \cdot dt. \quad (5)$$

Furthermore,

$$\sum_{n=0}^{\infty} M_n \frac{x^n}{n!} = -1 - 2x + 2 \cdot \int_0^{\infty} \frac{x^2 - 1}{xt - (x+1)^2} \cdot e^{-t} \cdot dt \qquad (6)$$

$$= -1 + 2x + 2 \cdot \frac{1-x}{1+x} \cdot \sum_{k=0}^{\infty} \frac{k! \cdot x^k}{(1+x)^{2k}} \qquad (7)$$

$$= -1 + 2x + 2 \cdot \frac{1-x^2}{x} \cdot e^{-\frac{(x+1)^2}{x}} \cdot \mathrm{Ei}\left(\frac{(x+1)^2}{x}\right), \qquad (8)$$

where $\mathrm{Ei}(t)$ *is the exponential integral.*

Proof. For the matrix A defined in Fig. 2, we have $A^2 = U + V \cdot z$, where

$$U = \begin{bmatrix} 1 & 0 & 0 & 0 \\ 0 & 1 & 0 & 0 \\ 0 & 1 & 0 & 0 \\ 1 & 0 & 0 & 0 \end{bmatrix} \quad \text{and} \quad V = \begin{bmatrix} 1 & 0 & 0 & 1 \\ 0 & 1 & 1 & 0 \\ 0 & 0 & 1 & 0 \\ 0 & 0 & 0 & 1 \end{bmatrix}.$$

Then Lemma 2 and formula (4) imply

$$M_n = n! \cdot \int_0^{\infty} \mathrm{tr}((U \cdot t - V)^n) \cdot e^{-t} \cdot dt.$$

Since the eigenvalues of the matrix $U \cdot t - V$ are $\frac{t - 2 \pm \sqrt{t^2 - 4t}}{2}$, each of multiplicity 2, we obtain formula (5).

To derive (6) from (5), we notice that

$$\sum_{n=0}^{\infty} \left(\left(\frac{t - 2 + \sqrt{t^2 - 4t}}{2} \right)^n + \left(\frac{t - 2 - \sqrt{t^2 - 4t}}{2} \right)^n \right) \cdot x^n = 1 + \frac{x^2 - 1}{xt - (x+1)^2}$$

and take special care of the initial values $M_0 = 1$ and $M_1 = 0$. Expanding the last expression as a power series in t, we get

$$\frac{x^2 - 1}{xt - (x+1)^2} = \frac{1 - x}{1 + x} \cdot \frac{1}{1 - \frac{x}{(x+1)^2} t} = \frac{1 - x}{1 + x} \cdot \sum_{k=0}^{\infty} \frac{x^k}{(x+1)^{2k}} \cdot t^k.$$

Applying the series Laplace transform $\mathcal{L}_{t,1}$, we obtain (7). Expression (8) now follows from (7), since $\sum_{k=0}^{\infty} k! \cdot y^k = \frac{e^{-1/y}}{y} \cdot \mathrm{Ei}\left(\frac{1}{y}\right)$ (e.g., see [1, formula (1.1.7)]). $\qquad \square$

5   De Bruijn Graph Approach for Ternary Menage Problem

In contrast to the menage problem, in the ternary case two females or two males can sit next to each other. Hence, the de Bruijn graph in this case can be

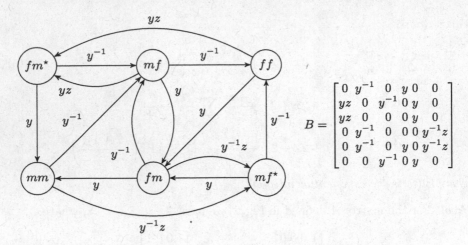

Fig. 3. The weighted de Bruijn graph for the ternary menage problem and its adjacency matrix B.

obtained from the de Bruijn graph for the menage problem by adding two more nodes labeled ff and mm, connected to the other nodes following the same rules (Fig. 3).

Let T_n be the number of seating arrangements of n couples in the ternary menage problem.

Theorem 2. *For $n > 1$, the number T_n can be computed in the following ways:*

$$T_n = n! \cdot \sum_{j=0}^{n} (-1)^j \cdot (n-j)! \cdot [y^0 z^j] \ \mathrm{tr}(B^{2n}),$$ (9)

where B is defined in Fig. 3; or

$$T_n = n! \cdot \int_0^\infty [y^n] \ \mathrm{tr}(B_2^n) \cdot e^{-t} \cdot dt, \quad B_2 = \begin{bmatrix} -y & yt & t & 0 & yt & -y \\ -y & y(t-1) & 0 & y^2(t-1) & yt & -y \\ 0 & y(t-1) & 0 & y^2(t-1) & 0 & -y \\ -y & 0 & t-1 & 0 & y(t-1) & 0 \\ -y & yt & t-1 & 0 & y(t-1) & -y \\ -y & yt & 0 & y^2 t & yt & -y \end{bmatrix}; \quad (10)$$

or

$$T_n = n! \cdot \int_0^\infty [x^n y^n] \ \frac{a(xy,t) + b(xy,t) \cdot (x + xy^2)}{c(xy,t) + d(xy,t) \cdot (x + xy^2)} \cdot e^{-t} \cdot dt,$$ (11)

where

$a(p,t) = -2\,p^5 t^3 + 2\,p^4 t^4 + 4\,p^5 t^2 - 8\,p^4 t^3 - 2\,p^5 t + 12\,p^4 t^2 - 8\,p^4 t + 6\,p^3 t - 4\,p^2 t^2$
$\qquad + 16\,p^2 - 10\,pt + 20\,p + 6\,,$

$b(p,t) = -p^2 t(2 + p - t)(p - 3\,t + 6)\,,$

$c(p,t) = p^6 t^2 - 2\,p^5 t^3 + p^4 t^4 + 4\,p^5 t^2 - 4\,p^4 t^3 - 2\,p^5 t + 6\,p^4 t^2 - 4\,p^4 t + 2\,p^3 t - p^2 t^2$
$\qquad + 4\,p^2 - 2\,pt + 4\,p + 1\,,$

$d(p,t) = -p^2 t(2 + p - t)^2\,.$

Proof. Formula (9) is similar to (4) and follows directly from the definition of the de Bruijn graph in Fig. 3.

To avoid dealing with negative powers, we notice that $[y^0 z^j]\ \mathrm{tr}(B^{2n}) = [y^{2n} z^j]\ \mathrm{tr}((yB)^{2n})$. Furthermore, the matrix $(yB)^2$ has entries that are polynomial in y^2 and z with the degree in z being at most 1, that is $(yB)^2 = U + V \cdot z$, where matrices U, V do not depend on z. Since the specified matrix B_2 equals $U \cdot t - V$ where y^2 is replaced with y, formula (10) easily follows from (9) and Lemma 2.

According to [7, Corollary 4.7.3],

$$\sum_{n=0}^{\infty} \mathrm{tr}(B_2^n) \cdot x^n = 6 - \frac{xQ'(x)}{Q(x)},$$

where $Q(x) = \det(I - x \cdot B_2)$ and I is the 6×6 identity matrix. Direct computation shows that $Q(x) = c(xy, t) + d(xy, t) \cdot (x + xy^2)$ and $6Q(x) - xQ'(x) = a(xy, t) + b(xy, t) \cdot (x + xy^2)$, implying that

$$\sum_{n=0}^{\infty} \mathrm{tr}(B_2^n) \cdot x^n = \frac{a(xy, t) + b(xy, t) \cdot (x + xy^2)}{c(xy, t) + d(xy, t) \cdot (x + xy^2)}.$$

Substitution of this expression into (10) yields (11). □

While formulae (9) and (10) provide an efficient way for computing T_n for a given integer $n > 1$, the special form of the rational function in (11) further enables us to obtain a closed-form expression for the EGF for the numbers T_n.

Lemma 3. *Let $a(z)$, $b(z)$, $c(z)$, $d(z)$ be polynomials such that $[z^0] c(z) = 1$ (i.e., $c(z)$ is invertible as a series in z). Then for any integer $n \geq 0$,*

$$[x^n y^n] \frac{a(xy) + b(xy) \cdot (x + xy^2)}{c(xy) + d(xy) \cdot (x + xy^2)} = [p^n] \left(\frac{a(p) \cdot d(p) - b(p) \cdot c(p)}{d(p) \cdot \sqrt{c(p)^2 - 4 \cdot p^2 \cdot d(p)^2}} + \frac{b(p)}{d(p)} \right).$$

Proof. Let $p = xy$. Then $\frac{a(xy) + b(xy) \cdot (x + xy^2)}{c(xy) + d(xy) \cdot (x + xy^2)} = \frac{a(p) + b(p) \cdot (x + py)}{c(p) + d(p) \cdot (x + py)}$. In the series expansion of this function, we will discard all terms with distinct degrees in x and y, while in the remaining terms (with equal degrees in x and y), we will replace xy with p to obtain a univariate power series in p. We start with the following expansion:

$$\frac{a(p) + b(p) \cdot (x + py)}{c(p) + d(p) \cdot (x + py)} = \frac{a(p) + b(p) \cdot (x + py)}{c(p)} \sum_{k=0}^{\infty} \left(\frac{-d(p)}{c(p)} \right)^k (x + py)^k$$

$$= \frac{a(p)}{c(p)} \sum_{k=0}^{\infty} \left(\frac{-d(p)}{c(p)} \right)^k (x + py)^k + \frac{b(p)}{c(p)} \sum_{k=0}^{\infty} \left(\frac{-d(p)}{c(p)} \right)^k (x + py)^{k+1}.$$

Here from each power of $x + py$ we extract the term with the equal degrees in x and y and replace it with the corresponding power of p. This yields

$$\frac{a(p)}{c(p)} \sum_{k=0}^{\infty} \left(\frac{-d(p)}{c(p)}\right)^{2k} \binom{2k}{k} p^{2k} + \frac{b(p)}{c(p)} \sum_{k=0}^{\infty} \left(\frac{-d(p)}{c(p)}\right)^{2k+1} \binom{2k+2}{k+1} p^{2k+2}$$

$$= \frac{a(p)}{c(p)} \cdot f\left(\left(\frac{d(p)}{c(p)} p\right)^2\right) - \frac{b(p)}{d(p)} \cdot \left(f\left(\left(\frac{d(p)}{c(p)} p\right)^2\right) - 1\right)$$

$$= \frac{a(p) \cdot d(p) - b(p) \cdot c(p)}{c(p) \cdot d(p)} \cdot f\left(\left(\frac{d(p)}{c(p)} \cdot p\right)^2\right) + \frac{b(p)}{d(p)}$$

$$= \frac{a(p) \cdot d(p) - b(p) \cdot c(p)}{d(p) \cdot \sqrt{c(p)^2 - 4 \cdot p^2 \cdot d(p)^2}} + \frac{b(p)}{d(p)},$$

where $f(z) = (1 - 4z)^{-1/2} = \sum_{k=0}^{\infty} \binom{2k}{k} \cdot z^k$. By construction, the coefficient of $x^n y^n$ in the original expression equals the coefficient of p^n in the last expression. $\square$

Theorem 3. *Let polynomials $a(p,t)$, $b(p,t)$, $c(p,t)$, and $d(p,t)$ be defined as in Theorem 2. Then for all $n > 1$,*

$$T_n = n! \cdot \int_0^{\infty} [p^n] \left(\frac{a(p,t) \cdot d(p,t) - b(p,t) \cdot c(p,t)}{d(p,t) \cdot \sqrt{c(p,t)^2 - 4 \cdot p^2 \cdot d(p,t)^2}} + \frac{b(p,t)}{d(p,t)}\right) \cdot e^{-t} \cdot dt. \quad (12)$$

Correspondingly, the exponential generating function for T_n equals

$$\sum_{n=0}^{\infty} T_n \frac{x^n}{n!} = -2 + 2 \cdot x - 2 \cdot x \cdot e^{-x-2} \cdot \mathrm{Ei}(x + 2)$$

$$+ \int_0^{\infty} \frac{(t^3 x^2 + (-2x^3 - 4x^2 - x)t^2 + (x^4 + 4x^3 + 7x^2 + 4x + 3)t - 6x^2 - 9x - 6) \cdot e^{-t}}{(t - (x+2))\sqrt{(t^2 x^2 - tx^3 - 2tx^2 - 3xt + 4x^2 + 4x + 1)}(t^2 x^2 - tx^3 - 2tx^2 + xt + 1)} \, dt.$$

Proof. Formula (12) directly follows from (11) and Lemma 3. Multiplying (12) by $\frac{x^n}{n!}$ and summing over n, we further get

$$\sum_{n=0}^{\infty} T_n \frac{x^n}{n!} = -5 + 2x$$

$$+ \int_0^{\infty} \left(\frac{a(x,t) \cdot d(x,t) - b(x,t) \cdot c(x,t)}{d(x,t) \cdot \sqrt{c(x,t)^2 - 4 \cdot p^2 \cdot d(x,t)^2}} + \frac{b(x,t)}{d(x,t)}\right) \cdot e^{-t} \cdot dt, \quad (13)$$

where the terms $-5 + 2x$ take care of the initial values $T_0 = 1$ and $T_1 = 0$. While we are not aware if it is possible to simplify the integral of the term involving a square root, below we evaluate the integral of the rational term.

Expansion of $\frac{b(x,t)}{d(x,t)}$ as a power series in t yields

$$\frac{b(x,t)}{d(x,t)} = \frac{x - 3t + 6}{x - t + 2} = 3 - \frac{2x}{2+x} \cdot \frac{1}{1 - \frac{t}{2+x}} = 3 - \frac{2x}{2+x} \cdot \sum_{k=0}^{\infty} \frac{t^k}{(2+x)^k}.$$

Similarly to the proof of Theorem 1, this further allows us to evaluate the integral

$$\int_0^\infty \frac{b(x,t)}{d(x,t)} \cdot e^{-t} \cdot dt = 3 - 2 \cdot x \cdot e^{-x-2} \cdot \text{Ei}(x+2).$$

Plugging this expression into (13) completes the proof.   □

6 Computing Numerical Values

The Online Encyclopedia of Integer Sequences [8] contains a number of sequences related to the menage problem and its ternary variant:

Sequence	Terms for $n = 1, 2, 3, \ldots$	OEIS index
M_n	$0, 0, 12, 96, 3120, 115200, 5836320, 382072320, \ldots$	A059375
$M_n/2n!$	$0, 0, 1, 2, 13, 80, 579, 4738, 43387, 439792, \ldots$	A000179
$M_n/2n$	$0, 0, 2, 12, 312, 9600, 416880, 23879520, \ldots$	A094047
T_n	$0, 8, 84, 3456, 219120, 19281600, 2324085120, \ldots$	A258338
$T_n/4n$	$0, 1, 7, 216, 10956, 803400, 83003040, \ldots$	A114939

While the Touchard formula (1) can be used to efficiently compute the menage numbers M_n and associated sequences, our formula (9) enables the same for the numbers T_n. In particular, we have computed many terms of sequences A114939 and A258338 in the OEIS.

We also remark that the formula (12) provides another way to compute T_n by extracting the coefficient of x^p (which is a polynomial in t) and applying the transform $\mathcal{L}_{t,1}$ (i.e., replacing every t^k with $k!$).

References

1. Bleistein, N., Handelsman, R.A.: Asymptotic Expansions of Integrals. Dover Books on Mathematics, Revised edn. Dover Publications, New York (2010)
2. Bogart, K.P., Doyle, P.G.: Non-sexist solution of the ménage problem. Am. Math. Monthly **93**, 514–519 (1986)
3. de Bruijn, N.G.: A combinatorial problem. Proceedings of the Section of Sciences of the Koninklijke Nederlandse Akademie van Wetenschappen te Amsterdam **49**(7), 758–764 (1946)
4. Golin, M.J., Leung, Y.C.: Unhooking circulant graphs: a combinatorial method for counting spanning trees and other parameters. In: Hromkovič, J., Nagl, M., Westfechtel, B. (eds.) WG 2004. LNCS, vol. 3353, pp. 296–307. Springer, Heidelberg (2004)
5. Graham, R.L., Knuth, D.E., Patashnik, O.: Concrete Mathematics: A Foundation for Computer Science, 2nd edn. Addison-Wesley, Reading, MA (1994)
6. Lucas, E.: Théorie des Nombres. Gauthier-Villars, Paris (1891)

7. Stanley, R.P.: Enumerative Combinatorics, vol. 1. Cambridge University Press, New York, NY (1997)
8. The OEIS Foundation: The On-Line Encyclopedia of Integer Sequences. Published electronically at http://oeis.org (2016)
9. Touchard, J.: Sur un probléme de permutations. C. R. Acad. Sci. Paris **198**, 631–633 (1934)
10. Wyman, M., Moser, L.: On the problème des ménages. Can. J. Math. **10**, 468–480 (1958)

Reconfiguration of Steiner Trees
in an Unweighted Graph

Haruka Mizuta[1,2($\boxtimes$)], Takehiro Ito[1,3], and Xiao Zhou[1]

[1] Graduate School of Information Sciences, Tohoku University,
Aoba-yama 6-6-05, Sendai 980-8579, Japan
haruka.mizuta.s4@dc.tohoku.ac.jp, {takehiro,zhou}@ecei.tohoku.ac.jp
[2] JST, ERATO, Kawarabayashi Large Graph Project,
c/o Global Research Center for Big Data Mathematics, NII,
2-1-2 Hitotsubashi, Chiyoda-ku, Tokyo 101-8430, Japan
[3] CREST, JST, 4-1-8 Honcho, Kawaguchi, Saitama 332-0012, Japan

Abstract. We study a reconfiguration problem for Steiner trees in an unweighted graph, which determines whether there exists a sequence of Steiner trees that transforms a given Steiner tree into another one by exchanging a single edge at a time. In this paper, we show that the problem is PSPACE-complete even for split graphs (and hence for chordal graphs), while solvable in linear time for interval graphs.

1 Introduction

The STEINER TREE problem on graphs is one of the most well-known NP-complete problems [3]. For an unweighted graph G and a vertex subset $S \subseteq V(G)$, a *Steiner tree* for S is a subtree of G which includes all vertices in S; each vertex in S is called a *terminal*. For example, Fig. 1 illustrates five Steiner trees of the same graph G for the same terminal set S. Given an unweighted graph G, a terminal set $S \subseteq V(G)$, and an integer $k \geq 0$, the STEINER TREE problem is to determine whether there exists a Steiner tree T of G for S such that T consists of at most k edges. This problem is known to be NP-complete [3].

The concept of Steiner trees has several applications such as network routing and VLSI design. In the network routing problem, a graph represents a computer network such that each terminal corresponds to a user or a server, each non-terminal to a router, and each edge to a communication link. Then, we wish to find a routing which connects all users and servers to provide the service; thus, a Steiner tree of the graph represents such a routing.

1.1 Our Problem

However, the network routing problem could be considered in more "dynamic" situations: In order to temporarily remove routers for maintenance, we sometimes need to change the current routing (i.e., Steiner tree) into another one.

JSPS KAKENHI Grant Numbers 15H00849 and 16K00004 (T. Ito), and 16K00003 (X. Zhou).

© Springer International Publishing Switzerland 2016
V. Mäkinen et al. (Eds.): IWOCA 2016, LNCS 9843, pp. 163–175, 2016.
DOI: 10.1007/978-3-319-44543-4_13

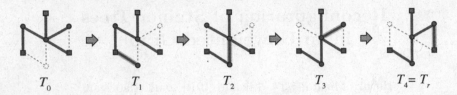

Fig. 1. A sequence $\langle T_0, T_1, \ldots, T_4 \rangle$ of Steiner trees of the same graph G for the same terminal set S, where the terminals are depicted by squares, non-terminals by circles, the edges in Steiner trees by thick lines.

To minimize disruption, this transformation needs to be done by switching communication links one by one, while keeping the connectivity among all users and servers to provide the service even during the transformation.

In this paper, we thus study the following problem: Suppose that we are given two Steiner trees of a graph G for a terminal set $S \subseteq V(G)$ (e.g., the leftmost and rightmost ones in Fig. 1), and we are asked whether we can transform one into the other via Steiner trees for S such that each intermediate Steiner tree can be obtained from the previous one by exchanging a single edge, that is, two consecutive Steiner trees T and T' in the transformation satisfy both $|E(T) \setminus E(T')| = 1$ and $|E(T') \setminus E(T)| = 1$. We call this decision problem the STEINER TREE RECONFIGURATION problem. For the particular instance of Fig. 1, the answer is yes as illustrated in the figure.

1.2 Known and Related Results

Similar problems have been extensively studied under the reconfiguration framework [6], which arises when we wish to find a step-by-step transformation between two feasible solutions of a combinatorial (search) problem such that all intermediate solutions are also feasible. The reconfiguration framework has been applied to several well-studied problems, including SATISFIABILITY [4,11], INDEPENDENT SET [2,6,9], VERTEX COVER [6,7,12], CLIQUE [6,8], and so on. (See also a survey [5].)

Ito et al. [6] studied the SPANNING TREE RECONFIGURATION problem, which can be seen as STEINER TREE RECONFIGURATION when restricted to the case where all vertices in a given graph are terminals. They showed that any instance of SPANNING TREE RECONFIGURATION is a yes-instance, that is, there always exists a desired transformation between two spanning trees in any graph.

1.3 Our Contribution

In this paper, we study STEINER TREE RECONFIGURATION from the viewpoint of graph classes and paint an interesting picture of the boundary between intractability and polynomial-time solvability (See Fig. 2.)

More specifically, we prove that the problem is PSPACE-complete even for split graphs, while is solvable in linear time for interval graphs. To do so,

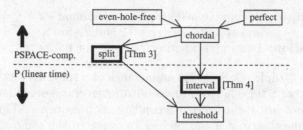

Fig. 2. Our results, where each arrow represents the inclusion relationship between graph classes: $A \rightarrow B$ represents that a graph class B is properly included in a graph class A [1].

we first give a sufficient condition and a necessary condition for the existence of a desired transformation between two Steiner trees; we emphasize that these conditions hold for any graph. We then show that our necessary condition is indeed a necessary and sufficient condition for interval graphs.

2   Preliminaries

In this section, we first define some basic terms and notation. Then, we introduce a sufficient condition and a necessary condition for the existence of a desired transformation between two Steiner trees.

2.1   Definitions

In this paper, we assume without loss of generality that graphs are simple and connected. For a graph G, we denote by $V(G)$ and $E(G)$ the vertex set and edge set of G, respectively. For a vertex subset $V' \subseteq V(G)$, we denote by $G[V']$ the subgraph of G induced by V'. We simply write $G \setminus V' = G[V(G) \setminus V']$.

For a graph G and a terminal set $S \subseteq V(G)$, a subtree T of G is a *Steiner tree* for S if $S \subseteq V(T)$ holds. For convenience, although T is not a rooted tree, we call each degree-1 vertex of T a *leaf* of T. We say that a leaf v_f of T is *free* if it is a non-terminal, that is, $v_f \in V(T) \setminus S$. Thus, $T \setminus \{v_f\}$ is also a Steiner tree for S, and hence a Steiner tree having a free leaf is not minimal.

For a graph G and a terminal set S, we say that two Steiner trees T and T' for S are *adjacent* if both $|E(T) \setminus E(T')| = 1$ and $|E(T') \setminus E(T)| = 1$ hold; we write $T \leftrightarrow T'$ in this case. For two Steiner trees T_p and T_q for S, a sequence $\langle T_0, T_1, \ldots, T_\ell \rangle$ of Steiner trees for S is called a *reconfiguration sequence* between T_p and T_q if $T_0 = T_p$, $T_\ell = T_q$, and $T_{i-1} \leftrightarrow T_i$ holds for each $i \in \{1, 2, \ldots, \ell\}$. Note that any reconfiguration sequence is *reversible*, that is, $\langle T_\ell, T_{\ell-1}, \ldots, T_0 \rangle$ is a reconfiguration sequence between T_q and T_p. We write $T_p \leadsto T_q$ if there is a reconfiguration sequence between T_p and T_q. Then, the STEINER TREE RECONFIGURATION problem is defined as follows:

Input: An unweighted graph G, a terminal set $S \subseteq V(G)$,
and two Steiner trees T_0 and T_r for S
Question: Determine whether $T_0 \leftrightsquigarrow T_r$ or not.

We denote by a 4-tuple (G, S, T_0, T_r) an instance of STEINER TREE RECONFIGU-
RATION. Note that STEINER TREE RECONFIGURATION is a decision problem and
hence it does not ask for an actual reconfiguration sequence. Throughout the
paper, we assume that $S \neq \emptyset$ and Steiner trees are of the same size; otherwise
such an instance is trivial.

2.2 Sufficient Condition and Necessary Condition

In this subsection, we give a sufficient condition and a necessary condition for
the existence of a reconfiguration sequence between two Steiner trees. These
conditions will play important roles in this paper to prove our results, and we
emphasize that they hold for any graph.

We first give a sufficient condition, as follows.

Theorem 1. *Let* (G, S, T_0, T_r) *be an instance of* STEINER TREE RECONFIGU-
RATION. *If* $V(T_0) = V(T_r)$, *then it is a* yes-*instance.*

Proof. Suppose that $V(T_0) = V(T_r)$ holds. Then, we have $G[V(T_0)] = G[V(T_r)]$.
Therefore, both T_0 and T_r form spanning trees of $G[V(T_0)] = G[V(T_r)]$. It is
known that any two spanning trees are reconfigurable each other by exchanging
a single edge at a time [6, Proposition 1], and hence the theorem follows.   □

Theorem 1 says that any two Steiner trees are reconfigurable each other as
long as they consist of the same vertex set. On the other hand, since we can
exchange only a single edge at a time, two adjacent Steiner trees having different
vertex sets satisfy a special property, as in the following proposition.

Proposition 1. *Suppose that* $T \leftrightarrow T'$ *holds for two Steiner trees* T *and* T' *of a
graph* G *with a terminal set* S. *If* $V(T) \neq V(T')$, *then*

– $V(T) \setminus V(T')$ *contains exactly one vertex* v_f, *and* v_f *is a free leaf in* T; *and*
– $V(T') \setminus V(T)$ *contains exactly one vertex* v'_f, *and* v'_f *is a free leaf in* T'.

Proof. Suppose for a contradiction that $V(T) \setminus V(T')$ contains more than one
vertex. (The proof is symmetric for the case where $V(T') \setminus V(T)$ contains more
than one vertex.) Since $S \neq \emptyset$ and $V(T) \setminus V(T')$ contains no terminal, we know
that $V(T) \cap V(T') \neq \emptyset$. Then, since T is connected, T has at least two edges inci-
dent to vertices in $V(T) \setminus V(T')$ and hence $|E(T) \setminus E(T')| \geq 2$. This contradicts
the assumption that $T \leftrightarrow T'$.

In this way, we have verified that $V(T) \setminus V(T')$ contains exactly one vertex
v_f, and hence it is a leaf in T. Since both T and T' are Steiner trees for S, we
know $V(T) \triangle V(T') = (V(T) \setminus V(T')) \cup (V(T') \setminus V(T)) \subseteq V(G) \setminus S$. Thus, v_f is
free.   □

We now give a sufficient condition for no-instance; by taking a contrapositive, this yields a necessary condition for yes-instance.

Theorem 2. *Let* (G, S, T_0, T_r) *be an instance of* STEINER TREE RECONFIGU-RATION. *Then, it is a* no-*instance if the following conditions* (*a*) *and* (*b*) *hold*:

(a) $V(T_0) \neq V(T_r)$; *and*
(b) *at least one of* $G[V(T_0)]$ *and* $G[V(T_r)]$ *has no Steiner tree for* S *with a free leaf.*

Proof. Suppose for a contradiction that (G, S, T_0, T_r) is a yes-instance even though it satisfies both Conditions (a) and (b). Then, there exists a reconfiguration sequence $\mathcal{T}$ between T_0 and T_r. Let T_{i+1} be the first Steiner tree in $\mathcal{T}$ such that $V(T_{i+1}) \neq V(T_0)$; such a Steiner tree exists since $V(T_0) \neq V(T_r)$. Then, the Steiner tree T_i in $\mathcal{T}$ satisfies $T_i \leftrightarrow T_{i+1}$ and $V(T_i) = V(T_0)$. By Proposition 1, $V(T_i) \setminus V(T_{i+1})$ contains exactly one vertex v_f which is a free leaf in T_i. Since $V(T_i) = V(T_0)$, we can conclude that $G[V(T_0)]$ has a Steiner tree T_i for S with a free leaf v_f. By the symmetric arguments, $G[V(T_r)]$ has a Steiner tree for S with a free leaf, too. This contradicts the assumption that Condition (b) holds.   □

3   PSPACE-Completeness for Split Graphs

In this section, we show the computational hardness of STEINER TREE RECON-FIGURATION. A graph is a *split graph* if it can be partitioned into a clique and an independent set.

Theorem 3. STEINER TREE RECONFIGURATION *is* PSPACE-*complete for split graphs.*

We prove Theorem 3 in the remainder of this section. Observe that the prob-lem can be solved in (most conveniently, nondeterministic [13]) polynomial space, and hence it is in PSPACE. Therefore, we show that the problem is PSPACE-hard for split graphs, by giving a polynomial-time reduction from VERTEX COVER RECONFIGURATION [14].

3.1   Reduction

Recall that a *vertex cover* C of a graph G is a vertex subset of G which contains at least one of the two endpoints of every edge in G. We say that two vertex covers C and C' of G are *adjacent* if $|C \setminus C'| = |C' \setminus C| = 1$, that is, C' can be obtained from C by exchanging a single vertex; we write $C \leftrightarrow C'$ in this case. (This adjacency relation is sometimes called the TJ *rule.*) Given two vertex covers C_0 and C_r of a graph G, the VERTEX COVER RECONFIGURATION problem (under the TJ rule) is to determine whether there exists a sequence $\langle C_0, C_1, \ldots, C_\ell \rangle$ of vertex covers of G such that $C_\ell = C_r$ and $C_{i-1} \leftrightarrow C_i$ for all $i \in \{1, 2, \ldots, \ell\}$. This problem is known to be PSPACE-complete even for planar graphs with bounded bandwidth [14, Theorem 9]. We denote by a triple (G, C_0, C_r) an instance of

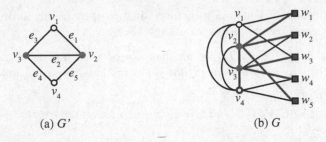

(a) G' (b) G

Fig. 3. (a) Given a graph G' for VERTEX COVER RECONFIGURATION with a vertex cover $C = \{v_2, v_3\}$, and (b) its corresponding split graph G for STEINER TREE RECONFIGU-RATION with a Steiner tree corresponding to C having a free leaf v_4.

VERTEX COVER RECONFIGURATION, and assume without loss of generality that neither $C_0 = V(G)$ nor $C_r = V(G)$ holds; otherwise it is a trivial instance.

Given an instance (G', C_0, C_r) of VERTEX COVER RECONFIGURATION, we now construct a corresponding instance (G, S, T_0, T_r) of STEINER TREE RECON-FIGURATION, as follows. (See Fig. 3 as an example.) Let $A = V(G')$. For the edge set $E(G') = \{e_1, e_2, \ldots, e_{|E(G')|}\}$, we define a new vertex set $B = \{w_1, w_2, \ldots, w_{|E(G')|}\}$ in which each vertex w_i corresponds to the edge $e_i \in E(G')$. Then, G is defined to be a graph such that

(1) $V(G) = A \cup B$;
(2) $G[A]$ forms a clique, and $G[B]$ forms an independent set; and
(3) for each edge $e_i = v_j v_k \in E(G')$, there are two edges $w_i v_j$ and $w_i v_k$ in G.

Let $S = B$, and let T_0' and T_r' be any spanning trees in $G[C_0 \cup B]$ and $G[C_r \cup B]$, respectively; we will prove in Lemma 1 that both T_0' and T_r' are Steiner trees for S. Since $C_0 \neq V(G') = A$ and $G[A]$ is a clique, we can choose a vertex $v_f^0 \in A \setminus C_0$ and join it to any vertex in $C_0 = V(T_0') \cap A$; let T_0 be the resulting tree, then v_f^0 is a free leaf in T_0. Similarly, since $C_r \neq V(G') = A$, we can obtain a Steiner tree T_r having a free leaf $v_f^r \in A \setminus C_r$. This completes the construction of the corresponding instance (G, S, T_0, T_r) of STEINER TREE RECONFIGURATION. Clearly, this construction can be done in polynomial time.

3.2   Correctness of the Reduction

To prove the correctness of our reduction, we first show that both T_0 and T_r (in particular, T_0' and T_r') are Steiner trees for S, as in the following lemma.

Lemma 1. *A vertex subset $C \subseteq V(G')$ forms a vertex cover of G' if and only if G has a Steiner tree T such that $V(T) = C \cup B$.*

Proof. We first prove the only-if direction. Suppose that $C \subseteq V(G')$ forms a vertex cover of G'. Then, for each edge $e_i = v_j v_k \in E(G')$, we have $v_j \in C$ or $v_k \in C$. Since $G[C]$ forms a clique and G has two edges $w_i v_j$ and $w_i v_k$ for the vertex $w_i \in V(G)$ corresponding to $e_i \in E(G')$, we thus know that $G[C \cup B]$ is connected. By taking any spanning tree of $G[C \cup B]$, we can obtain a Steiner tree T such that $V(T) = C \cup B$.

We then prove the if direction. Suppose that G has a Steiner tree T such that $V(T) = C \cup B$. Since B forms an independent set of G, every vertex $w_i \in B$ must be adjacent to at least one vertex $v_j \in C$ in T. By the construction of the graph G, the edge e_i in G' (corresponding to $w_i \in V(G)$) is incident to $v_j \in V(G')$. Thus, C forms a vertex cover of G'.   □

The following lemma completes the proof of Theorem 3, whose proof is omitted from this extended abstract.

Lemma 2. *An instance* (G', C_0, C_r) *of* VERTEX COVER RECONFIGURATION *is a* yes-*instance if and only if the corresponding instance* (G, S, T_0, T_r) *of* STEINER TREE RECONFIGURATION *is a* yes-*instance.*

4   Algorithm for Interval Graphs

A graph G with $V(G) = \{v_1, v_2, \ldots, v_n\}$ is an *interval graph* if there exists a set $\mathcal{I}$ of (closed) intervals $I_1, I_2, \ldots, I_n$ such that $v_i v_j \in E(G)$ if and only if $I_i \cap I_j \neq \emptyset$ for each $i, j \in \{1, 2, \ldots, n\}$. We call the set $\mathcal{I}$ of intervals an *interval representation* of the graph. For a given graph G, it can be determined in linear time whether G is an interval graph, and if so we can obtain an interval representation of G in linear time [10].

In this section, we prove that STEINER TREE RECONFIGURATION is solvable in linear time for interval graphs. The key is the following theorem, whose proof will be given in the remainder of this section.

Theorem 4. *Let* (G, S, T_0, T_r) *be an instance of* STEINER TREE RECONFIGURATION *such that* G *is an interval graph. Then, it is a* yes-*instance if and only if the following conditions* (a) *or* (b) *hold:*

(a) $V(T_0) = V(T_r)$; *or*
(b) *each of* $G[V(T_0)]$ *and* $G[V(T_r)]$ *has a Steiner tree for* S *with a free leaf.*

Then, we have the following corollary.

Corollary 1. STEINER TREE RECONFIGURATION *can be solved in linear time for interval graphs.*

Proof. It suffices to show that Conditions (a) and (b) of Theorem 4 can be checked in linear time. We can clearly check Condition (a) in linear time. Thus, we show that Condition (b) can be checked in linear time, as follows.

Notice that, for a non-terminal vertex $v \in V(T_0) \setminus S$, if the induced graph $G[V(T_0) \setminus \{v\}]$ is connected, then any spanning tree T of $G[V(T_0) \setminus \{v\}]$ is a Steiner tree for S; by adding the non-terminal vertex v to T as a leaf, we can obtain a Steiner tree with a free leaf. The same holds for T_r, too.

We now check in linear time whether such a non-terminal vertex $v \in V(T_0) \setminus S$ exists or not. Since $G[V(T_0)]$ is an interval graph, we first obtain its interval representation in linear time [10]. Then, by traversing the interval representation from left to right, we can enumerate all cut-vertices in $G[V(T_0)]$ in linear time, and hence the existence of a desired non-terminal vertex $v \in V(T_0) \setminus S$ can be checked in linear time. (The same is applied to T_r, too.)   □

We give a proof of Theorem 4 in the remainder of this section. The only-if direction is immediate from Theorem 2 (by taking a contrapositive). In addition, when Condition (a) holds, the if direction is also immediate from Theorem 1. Therefore, it suffices to prove that (G, S, T_0, T_r) is a yes-instance if both $V(T_0) \neq V(T_r)$ and Condition (b) hold.

Let (G, S, T_0, T_r) be a given instance of STEINER TREE RECONFIGURATION such that G is an interval graph, $V(T_0) \neq V(T_r)$, and Condition (b) of Theorem 4 holds. Then, $G[V(T_0)]$ has a Steiner tree for S with a free leaf, and by Theorem 1 there exists a reconfiguration sequence between T_0 and the Steiner tree with a free leaf; the same holds for T_r. Therefore, we assume without loss of generality that two given Steiner trees T_0 and T_r have free leaves. We will construct a reconfiguration sequence between T_0 and T_r.

Let $\mathcal{I}$ be an interval representation of G. For an interval $I_i \in \mathcal{I}$, we denote by $l(I_i)$ and $r(I_i)$ the left and right coordinates of I_i, respectively; we sometimes call the values $l(I_i)$ and $r(I_i)$ the l-value and r-value of I_i, respectively. We may assume without loss of generality that all l-values and r-values are distinct. For notational convenience, we sometimes identify a vertex $v_i \in V(G)$ with its corresponding interval $I_i \in \mathcal{I}$, and simply write $l(v_i) = l(I_i)$ and $r(v_i) = r(I_i)$. We say that a path P in G is r-increasing if the r-values of the vertices along P are increasing. Let s_{left} be the terminal in S which has the minimum l-value, that is, $l(s_{\text{left}}) = \min\{l(v) : v \in S\}$, while let s_{right} be the terminal in S which has the maximum r-value, that is, $r(s_{\text{right}}) = \max\{r(v) : v \in S\}$. Note that $s_{\text{left}} = s_{\text{right}}$ may hold. By the definition of s_{right}, all vertices v with $r(s_{\text{right}}) < r(v)$ are non-terminals. Because we can greedily exchange edges incident to such non-terminals, we simply ignore them (or if some of the vertices are required for the connectedness of G, we can shorten their right coordinates with keeping the reconfigurability); the details are omitted from this extended abstract. Then, we say that a Steiner tree F for S is in *standard form* if

– the unique path P in F from s_{left} to s_{right} is r-increasing; and
– every terminal in $S \setminus V(P)$ is a leaf in F which is adjacent to some vertex in P.

(See Fig. 4(c) as an example.)

Lemma 3. *For any Steiner tree T of an interval graph G, there exists a Steiner tree F of G such that F is in standard form, all free leaves in T are free leaves in F, $V(F) = V(T)$, and $T \rightsquigarrow F$.*

Proof. Let V_{free} be the set of all free leaves in T, and let T' be the subtree of T obtained by deleting the vertices in V_{free}. (See Fig. 4(a) in which T' is illustrated by the thick dotted lines.) We first prove the existence of a Steiner tree F' in standard form for S such that $V(F') = V(T')$.

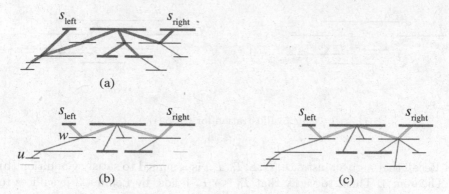

Fig. 4. (a) Steiner tree T of an interval graph G, (b) Steiner tree F' of $G[V(T')]$ in a standard form, and (c) Steiner tree F of G in a standard form. In the figure, graphs are illustrated by their interval representations; each terminal in S is depicted by thick (red) line, and each non-terminal by thin (black) line. Steiner trees are depicted by dotted lines on the interval representations. In (b) and (c), the thick (green) dotted lines represent the paths from s_{left} to s_{right} (Color figure online).

Consider the induced subgraph $G[V(T')]$ of G. (See Fig. 4(b).) Since T' is connected, $G[V(T')]$ is also connected. Therefore, we can greedily find an r-increasing path P in $G[V(T')]$ from s_{left} to s_{right}. By the choice of s_{left} and s_{right}, every terminal s in $S \setminus V(P)$ intersects with at least one vertex in P; we arbitrarily choose such a vertex in P, and connect s with it.

To finish the construction of F', we now claim that every vertex in $V(T') \setminus S$ is either on P or has a path to a vertex w in P which consists of only non-terminal vertices except for w. (See the vertex u in Fig. 4(b) as an example for the latter case.) Then, the terminals in $S \setminus V(P)$ remain leaves in F', as required in standard form. Suppose for a contradiction that a vertex u in $V(T') \setminus S$ does not have such a path. If both $l(u) < r(s_{\text{right}})$ and $l(s_{\text{left}}) < r(u)$ hold, then u intersects with some vertex in P. Thus, u must satisfy either $r(s_{\text{right}}) < l(u)$ or $r(u) < l(s_{\text{left}})$. Consider the case where $r(u) < l(s_{\text{left}})$ holds; the other case is symmetric. Then, since $G[V(T')]$ is connected but u has no desired path to any vertex in P, there must exist a terminal $s \in S$ such that $l(s) < l(s_{\text{left}})$; this contradicts the definition of s_{left}.

In this way, there exists a Steiner tree F' in standard form such that $V(F') = V(T')$. Then, since $G[V(T)]$ is connected and every vertex u with $l(u) < r(s_{\text{right}})$ and $l(s_{\text{left}}) < r(u)$ intersects with a vertex in P, we can add the vertices in V_{free} to F' as leaves so that the terminals in $S \setminus V(P)$ remain leaves in F'; let F be the resulting tree. (See Fig. 4(b) and (c).)

Therefore, we have verified the existence of a Steiner tree F in standard form such that $V(F) = V(T)$ and all free leaves in T are free leaves also in F. Then, since $V(F) = V(T)$ holds, Theorem 1 yields that $T \rightsquigarrow F$. $\qquad\square$

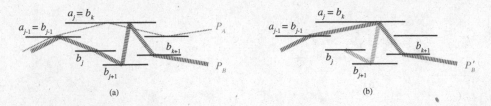

Fig. 5. Illustration for Case (i).

Recall that a given instance (G, S, T_0, T_r) is assumed to satisfy Condition (b) of Theorem 4. Then, to verify that $T_0 \leftrightsquigarrow T_r$ holds, by Lemma 3 it suffices to construct a reconfiguration sequence between two Steiner trees T_0' and T_r' such that $V(T_0') = V(T_0)$, $V(T_r') = V(T_r)$, both T_0' and T_r' are in standard form and have free leaves. Thus, the following lemma completes the proof of Theorem 4.

Lemma 4. *Let T_A and T_B be any two Steiner trees of G for S such that $|V(T_A)| = |V(T_B)|$, and both T_A and T_B are in standard form and have free leaves. Then, $T_A \leftrightsquigarrow T_B$.*

Proof. Let $P_A = (a_1, a_2, \ldots, a_{\ell_A})$ and $P_B = (b_1, b_2, \ldots, b_{\ell_B})$ be the paths from s_{left} to s_{right} in T_A and T_B, respectively; and hence $a_1 = b_1 = s_{\text{left}}$ and $a_{\ell_A} = b_{\ell_B} = s_{\text{right}}$. We prove the lemma by induction on the number of vertices in $V(P_A) \triangle V(P_B)$.

First, consider the case where $V(P_A) \triangle V(P_B) = \emptyset$. Since both T_A and T_B are in standard form, we know $P_A = P_B$. Furthermore, all terminals in $S \setminus V(P_A)$ are leaves and adjacent to vertices in P_A. Therefore, by greedily exchanging the edges in $E(T_A) \triangle E(T_B)$, we can obtain a reconfiguration sequence between T_A and T_B. We thus have $T_A \leftrightsquigarrow T_B$.

Second, consider the case where $V(P_A) \triangle V(P_B) \neq \emptyset$. Let j be the first index such that $a_j \neq b_j$. (See Fig. 5(a).) Since both P_A and P_B are r-increasing, a_j and b_j intersect with each other and hence $a_j b_j \in E(G)$. Assume without loss of generality that $r(b_j) < r(a_j)$ holds, as illustrated in Fig. 5(a). (The other case is symmetric.) Then, we have $a_j b_{j+1} \in E(G)$. We deal with this case according to the following three sub-cases.

Case (i): a_j appears in P_B. (See Fig. 5.)

Let k be the index such that $b_k = a_j$. Since $r(b_j) < r(a_j) = r(b_k)$ and P_B is r-increasing, we know that $k > j$ holds. Therefore, we simply exchange the edge $b_{j-1} b_j \in E(P_B)$ with the edge $b_{j-1} b_k \in E(G) \setminus E(T_B)$, and obtain a Steiner tree T_B' for S with the path $P_B' = (b_1, b_2, \ldots, b_{j-1}, b_k, b_{k+1}, \ldots, b_{\ell_B})$ from $b_1 = s_{\text{left}}$ to $b_{\ell_B} = s_{\text{right}}$. (See Figs. 5(a) and (b).) Then, P_B' is r-increasing. Since $b_{j-1} b_j \in E(P_B)$, neither b_{j-1} nor b_j is a free leaf in T_B. Thus, free leaves in T_B remain free leaves also in T_B'. If needed, we can transform T_B' into a Steiner tree T_B'' in standard form with keeping the free leaves, as in the proof of Lemma 3. Then, since $|V(P_A) \triangle V(P_B')| < |V(P_A) \triangle V(P_B)|$, we can apply the induction hypothesis to T_A and T_B''. Therefore, we have $T_B \leftrightarrow T_B' \leftrightsquigarrow T_B'' \leftrightsquigarrow T_A$.

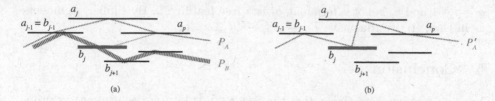

(a)   (b)

Fig. 6. Illustration for Case (ii).

Case (ii): b_j is a terminal in S. (See Fig. 6.)

Since P_A is r-increasing and we have assumed without loss of generality that $r(b_j) < r(a_j)$ holds, b_j does not appear in P_A. Then, since T_A is in standard form, b_j must be a leaf in T_A which is adjacent to a vertex a_p in P_A for some index p. If $p \neq j$, then we first exchange the edge $a_p b_j \in E(T_A)$ with the edge $a_j b_j \in E(G) \setminus E(T_A)$. We then exchange the edge $a_{j-1} a_j \in E(P_A)$ with the edge $a_{j-1} b_j \in E(G) \setminus E(T_A)$, and obtain a Steiner tree T_A' for S with the path $P_A' = (a_1, a_2, \ldots, a_{j-1}, b_j, a_j, a_{j+1}, \ldots, a_{\ell_A})$ from $a_1 = s_{\text{left}}$ to $a_{\ell_A} = s_{\text{right}}$. (See Figs. 6(a) and (b).) Note that, since $r(a_{j-1}) = r(b_{j-1}) < r(b_j) < r(a_j)$ holds, P_A' is r-increasing. Since b_j is a terminal, it is not a free leaf. In addition, since $a_{j-1} a_j \in E(P_A)$, neither a_{j-1} nor a_j is a free leaf in T_A. Thus, free leaves in T_A remain free leaves also in T_A'. Then, by similar arguments as in Case (i), we thus have $T_A \leftrightsquigarrow T_A' \leftrightsquigarrow T_B$.

Case (iii): b_j is not a terminal in S. (See Fig. 7.)

If a_j appears in P_B, then we apply Case (i) above. We now consider the case where a_j does not appear in P_B. Let b_q be any vertex in P_B such that $l(b_q) < r(a_j) < r(b_q)$; by the definitions of s_{left} and s_{right}, such a vertex b_q always exists. Recall that $a_j b_{j+1} \in E(G)$ holds, and hence we know that $q \geq j + 1$. If $a_j \notin V(T_B)$, then we exchange an arbitrary chosen edge $e_f \in E(T_B)$ incident to a free leaf with the edge $b_q a_j \in E(G) \setminus E(T_B)$. Otherwise, we pick the first edge on the path in T_B from a_j to a vertex in P_B, and exchange it with the edge $b_q a_j \in E(G) \setminus E(T_B)$. We then exchange the edge $b_{j-1} b_j \in E(P_B)$ with the edge $b_{j-1} a_j$, and obtain a Steiner tree T_B' for S with the path $P_B' = (b_1, b_2, \ldots, b_{j-1}, a_j, b_q, b_{q+1}, \ldots, b_{\ell_B})$ from $b_1 = s_{\text{left}}$ to $b_{\ell_B} = s_{\text{right}}$. (See Figs. 7(a) and (b).) By the choice of b_q, P_B' is r-increasing. In addition, since

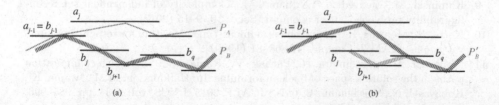

(a)   (b)

Fig. 7. Illustration for Case (iii).

$q \geq j + 1$ and b_j is not a terminal, b_j is a free leaf in T'_B. By similar arguments as in Case (i), we thus have $T_B \leftrightsquigarrow T'_B \leftrightsquigarrow T_A$.   □

5 Conclusion

In this paper, we have shown that the STEINER TREE RECONFIGURATION problem is PSPACE-complete even for split graphs (and hence for chordal graphs), while solvable in linear time for interval graphs. Thus, as illustrated in Fig. 2, we have clarified a boundary on the graph classes lying between intractability and tractability, because the structure of split graphs (resp., chordal graphs) can be seen as a star-like (resp., tree-like) structure of cliques, while that of interval graphs can be seen as a path-like structure of cliques.

Acknowledgment. We are grateful to Tatsuhiko Hatanaka for valuable discussions with him.

References

1. Brandstädt, A., Le, V.B., Spinrad, J.P.: Graph Classes: A Survey, SIAM (1999)
2. Demaine, E.D., Demaine, M.L., Fox-Epstein, E., Hoang, D.A., Ito, T., Ono, H., Otachi, Y., Uehara, R., Yamada, T.: Linear-time algorithm for sliding tokens on trees. Theor. Comput. Sci. **600**, 132–142 (2015)
3. Garey, M.R., Johnson, D.S.: Computers and Intractability: A Guide to the Theory of NP-Completeness. Freeman, San Francisco (1979)
4. Gopalan, P., Kolaitis, P.G., Maneva, E.N., Papadimitriou, C.H.: The connectivity of Boolean satisfiability: computational and structural dichotomies. SIAM J. Comput. **38**, 2330–2355 (2009)
5. van den Heuvel, J.: The complexity of change. Surveys in Combinatorics 2013, London Mathematical Society Lecture Notes Series 409 (2013)
6. Ito, T., Demaine, E.D., Harvey, N.J.A., Papadimitriou, C.H., Sideri, M., Uehara, R., Uno, Y.: On the complexity of reconfiguration problems. Theor. Comput. Sci. **412**, 1054–1065 (2011)
7. Ito, T., Nooka, H., Zhou, X.: Reconfiguration of vertex covers in a graph. IEICE Trans. Inf. Syst. **E99–D**, 598–606 (2016)
8. Ito, T., Ono, H., Otachi, Y.: Reconfiguration of cliques in a graph. In: Jain, R., Jain, S., Stephan, F. (eds.) TAMC 2015. LNCS, vol. 9076, pp. 212–223. Springer, Heidelberg (2015)
9. Kamiński, M., Medvedev, P., Milanič, M.: Complexity of independent set reconfigurability problems. Theor. Comput. Sci. **439**, 9–15 (2012)
10. Korte, N., Möhring, R.: An incremental linear-time algorithm for recognizing interval graphs. SIAM J. Comput. **18**, 68–81 (1989)
11. Mouawad, A.E., Nishimura, N., Pathak, V., Raman, V.: Shortest reconfiguration paths in the solution space of boolean formulas. In: Halldórsson, M.M., Iwama, K., Kobayashi, N., Speckmann, B. (eds.) ICALP 2015. LNCS, vol. 9134, pp. 985–996. Springer, Heidelberg (2015)

12. Mouawad, A.E., Nishimura, N., Raman, V.: Vertex cover reconfiguration and beyond. In: Ahn, H.-K., Shin, C.-S. (eds.) ISAAC 2014. LNCS, vol. 8889, pp. 452–463. Springer, Heidelberg (2014)
13. Savitch, W.J.: Relationships between nondeterministic and deterministic tape complexities. J. Comput. Syst. Sci. **4**, 177–192 (1970)
14. van der Zanden, T.C.: Parameterized complexity of graph constraint logic. In: Proceedings of IPEC 2015, LIPIcs 43, pp. 282–293 (2015)

Online Algorithms

Weighted Online Problems with Advice

Joan Boyar, Lene M. Favrholdt$^{(\boxtimes)}$, Christian Kudahl,
and Jesper W. Mikkelsen

Department of Mathematics and Computer Science,
University of Southern Denmark, Odense, Denmark
{joan,lenem,kudahl,jesperwm}@imada.sdu.dk

Abstract. Recently, the first online complexity class, AOC, was introduced. The class consists of many online problems where each request
must be either accepted or rejected, and the aim is to either minimize or
maximize the number of accepted requests, while maintaining a feasible
solution. All AOC-complete problems (including Independent Set, Vertex
Cover, Dominating Set, and Set Cover) have essentially the same advice
complexity. In this paper, we study weighted versions of problems in
AOC, i.e., each request comes with a weight and the aim is to either minimize or maximize the total weight of the accepted requests. In contrast
to the unweighted versions, we show that there is a significant difference
in the advice complexity of complete minimization and maximization
problems. We also show that our algorithmic techniques for dealing with
weighted requests can be extended to work for non-complete AOC problems such as Matching (giving better results than what follow from the
general AOC results) and even non-AOC problems such as scheduling.

1 Introduction

An online problem is an optimization problem for which the input is divided into
small pieces, usually called requests, arriving sequentially. An online algorithm
must serve each request, irrevocably, without any knowledge of possible future
requests. The quality of online algorithms is traditionally measured using the
competitive ratio [10,14], which is essentially the worst case ratio of the online
performance to the performance of an optimal offline algorithm, i.e., an algorithm
that knows the whole input sequence from the beginning and has unlimited
computational power.

For some online problems such as Independent Set or Vertex Cover, the best
possible competitive ratio is linear in the sequence length. This gives rise to the
question of what would happen, if the algorithm knew *something* about future
requests. Sometimes a semi-online setting is studied where it is assumed that the
algorithm has some specific knowledge such as the value of an optimal solution.
The extra knowledge may also be more problem specific such as an access graph
for paging. In contrast to problem specific approaches, advice complexity [3,7,9]

This work was partially supported by the Villum Foundation and the Danish Council
for Independent Research, Natural Sciences, grant DFF-1323-00247.

© Springer International Publishing Switzerland 2016
V. Mäkinen et al. (Eds.): IWOCA 2016, LNCS 9843, pp. 179–190, 2016.
DOI: 10.1007/978-3-319-44543-4_14

is a quantitative and standardized way of relaxing the online constraint. The main idea of advice complexity is to provide an online algorithm, ALG, with some partial knowledge of the future in the form of advice bits provided by a trusted oracle which has unlimited computational power and knows the entire request sequence. Informally, the advice complextity of an algorithm, ALG, is the maximum number of advice bits read by ALG for input sequences of a given length, and the advice complexity of a problem is the advice complexity of the best possible algorithm for the problem. Upper bounds on the advice complexity for a problem can sometimes lead to (or come from) semi-online algorithms, and lower bounds can show that such algorithms do not exist. Since its introduction, advice complexity has been a very active area of research. Lower and upper bounds on advice complexity have been obtained for a large number of online problems; a recent list can be found in [15].

Recently in [5], the first complexity class for online problems, AOC, was introduced. The class consists of online problems that can be described in the following way: The input is a sequence of requests and each request must either be accepted or rejected. The set of accepted requests is called the solution. For each request sequence, there is at least one feasible solution. The class contains minimization as well as maximization problems. For a minimization problem, the goal is to accept as few requests as possible, while maintaining a feasible solution, and for maximization problems, the aim is to accept as many requests as possible. For minimization problems, any super set of a feasible solution is also a solution, and for maximization problems, any subset of a feasible solution is also a feasible solution.

In this paper, we consider a generalization of the problems in the class AOC in which each request comes with a weight. The goal is now to either minimize or maximize the total weight of the accepted requests. We separately consider the classes of maximization and minimization problems. For AOC-complete maximization problems, we get advice complexity results quite similar to those for the unweighted versions of the problems, but for AOC-complete minimization problems, the results are a lot more negative, so this gives a complexity class containing harder problems than AOC. This is in contrast to unweighted AOC-complete problems, where minimization and maximization problems are equally hard in terms of advice complexity. Recently, differences between (unweighted) AOC minimization and maximization problems were found with respect to online bounded analysis [4] and min- and max-induced subgraph problems [11].

Our upper bound techniques are also useful for non-complete AOC problems such as Matching as well as non-AOC problems such as Scheduling.

Previous results. For any AOC-complete problem, $\Theta(n/c)$ advice bits are necessary and sufficient to obtain a competitive ratio of c. More specifically, for competitive ratio c, the advice complexity is $B(n, c) \pm O(\log n)$, where

$$B(n, c) = \log \left(1 + \frac{(c-1)^{c-1}}{c^c}\right) n, \tag{1}$$

and $an/c \le B(n,c) \le n/c$, $a = 1/(e\ln(2)) \approx 0.53$. This is an upper bound on the advice complexity of all problems in AOC. In [5], a list of problems including Independent Set, Vertex Cover, Dominating Set, and Set Cover were proven AOC-complete.

The paper [1] studies a semi-online version of scheduling where it is allowed to keep several parallel schedules and choose the best schedule in the end. The scheduling problem considered is makespan minimization on m identical machines. Using $(1/\varepsilon)^{O(\log(1/\varepsilon))}$ parallel schedules, a $(4/3 + \varepsilon)$-competitive algorithm is obtained. Moreover, a $(1+\varepsilon)$-competitive algorithm using $(m/\varepsilon)^{O(\log(1/\varepsilon)/\varepsilon)}$ parallel schedules is given along with an almost matching lower bound. Note that keeping s different schedules until the end corresponds to working with s different online algorithms. Thus, this particular semi-online model easily translates to the advice model, the advice being which of the s algorithms to run. In this way, the results of [1] correspond to a $(4/3 + \varepsilon)$-competitive algorithm using $O(\log^2(1/\varepsilon))$ advice bits and a $(1 + \varepsilon)$-competitive algorithm using $O(\log(m/\varepsilon) \cdot \log(1/\varepsilon)/\varepsilon)$ advice bits. In particular, note that this algorithm uses constant advice in the size of the input and only logarithmic advice in the number of machines.

In [13], scheduling on identical machines with a more general type of objective function (including makespan, minimizing the ℓ_p-norm, and machine covering) was studied. The paper considers the advice-with-request model where a fixed number of advice bits are provided along with each request. The main result is a $(1 + \varepsilon)$-competitive algorithm that uses $O((1/\varepsilon) \cdot \log(1/\varepsilon))$ advice bits per request, totaling $O((n/\varepsilon) \cdot \log(1/\varepsilon))$ bits of advice for the entire sequence.

Our results. We prove that adding arbitrary weights, AOC-complete minimization problems become a lot harder than AOC-complete maximization problems:

- For AOC-complete maximization problems, the weighted version is not significantly harder than the unweighted version: For any maximization problem in AOC (this includes, e.g., Independent Set), the c-competitive algorithm given in [5] for the unweighted version of the problem can be converted into a $(1+\varepsilon)c$-competitive algorithm for the weighted version using only $O((\log^2 n)/\varepsilon)$ additional advice bits. Thus, a $(1 + \varepsilon)c$-competitive algorithm using at most $B(n,c) + O((\log^2 n)/\varepsilon)$ bits of advice is obtained.

  For non-complete AOC problems, better trade-offs between the competitive ratio and number of advice bits can be obtained. We show that any c-competitive algorithm for an AOC maximization problem, P, using b advice bits can be converted into a $O(c \cdot \log n)$-competitive algorithm for the weighted version of P using $b + O(\log n)$ advice bits. For Weighted Matching, this implies a $O(\log n)$-competitive algorithm reading $O(\log n)$ bits of advice. We show that this is best possible in the following sense: For a set of weighted AOC problems including Matching, Independent Set, and Clique, no algorithm reading $o(\log n)$ bits of advice can have a competitive ratio bounded by any function of n. Furthermore, any $O(1)$-competitive algorithm for Weighted Matching must read $\Omega(n)$ advice bits.

- For all minimization problems known to be AOC-complete (this includes, e.g., Vertex Cover, Dominating Set, and Set Cover), $n - O(\log n)$ bits of advice are required to obtain a competitive ratio bounded by a function of n. This should be contrasted with the fact that n bits of advice trivially yields a strictly 1-competitive algorithm.

If the largest weight $w_{\max}$ cannot be arbitrarily larger than the smallest weight $w_{\min}$, the c-competitive algorithm given in [5] for the unweighted version can be converted into a $c(1 + \varepsilon)$-competitive algorithm for the weighted versions using $B(n, c) + O(\log^2 n + \log(\log(w_{\max}/w_{\min})/\varepsilon))$ advice bits in total.

Our main upper bound technique is a simple exponential classification scheme that can be used to sparsify the set of possible weights. This technique can also be used for problems outside of AOC. For example, for *scheduling on related machines*, we show that for many important objective functions (including makespan minimization and minimizing the ℓ_p-norm), there exist $(1 + \varepsilon)$-competitive algorithms reading $O((\log^2 n)/\varepsilon)$ bits of advice. For scheduling on m *unrelated* machines where m is constant, we get a similar result, but with $O((\log n)^{m+1}/\varepsilon^m)$ advice bits. Finally, for unrelated machines, where the goal is to *maximize* a seminorm (as in machine covering), we show that there is a $(1 + \varepsilon)$-competitive algorithm reading $O((\log n)^{m+1}/\varepsilon^m)$ bits of advice.

For scheduling on related and unrelated machines, our results are the first non-trivial upper bounds on the advice complexity. For the case of makespan minimization on identical machines, the algorithm of [1] is strictly better than ours. However, for minimizing the ℓ_p-norm or maximizing a seminorm on identical machines, we exponentially improve the previous best upper bound [13] (which was linear in n).

The missing proofs and definitions can be found in the full paper [6].

2   Preliminaries

Throughout the paper, we let n denote the number of requests in the input. We let $\mathbb{R}_+$ denote the set containing 0 and all positive real numbers. We let log denote the binary logarithm $\log_2$. For $k \geq 1$, $[k] = \{1, 2, \ldots, k\}$. For any bit string y, let $|y|_0$ and $|y|_1$ denote the number of zeros and the number of ones, respectively, in y. We write $x \sqsubseteq y$ if for all indices, i, $x_i = 1 \Rightarrow y_i = 1$.

In this paper, we use the "advice-on-tape" model [3]. Before the first request arrives, the oracle, which knows the entire request sequence, prepares an *advice tape*, an infinite binary string. The algorithm ALG may, at any point, read some bits from the advice tape. The *advice complexity* of ALG is the maximum number of bits read by ALG for any input sequence of at most a given length. OPT is an optimal offline algorithm.

Advice complexity is combined with competitive analysis to determine how many bits of advice are necessary and sufficient to achieve a given competitive ratio.

Definition 1 (Competitive ratio [10,14] and advice complexity [3]). *The input to an online problem, P, is a request sequence $\sigma = \langle r_1, \ldots, r_n \rangle$. An online algorithm with advice, ALG, computes the output $y = \langle y_1, \ldots, y_n \rangle$, where y_i*

is computed from $\varphi, r_1, \ldots, r_i$, where φ is the content of the advice tape. Each possible output for P is associated with a cost/profit. For a request sequence σ, $\text{ALG}(\sigma)$ $(\text{OPT}(\sigma))$ denotes the cost/profit of the output computed by ALG (OPT) when serving σ.

If P is a maximization (minimization) problem, then ALG is $c(n)$-competitive if there exists a constant, α, such that, for all $n \in \mathbb{N}$, $\text{OPT}(\sigma) \leq c(n) \cdot \text{ALG}(\sigma) + \alpha$, $(\text{ALG}(\sigma) \leq c(n) \cdot \text{OPT}(\sigma) + \alpha)$, for all request sequences, σ, of length at most n. If the relevant inequality holds with $\alpha = 0$, we say that ALG is strictly $c(n)$-competitive.

The advice complexity, $b(n)$, of an algorithm, ALG, is the largest number of bits of φ read by ALG over all possible request sequences of length at most n. The advice complexity of a problem, P, is a function, $f(n, c)$, $c \geq 1$, such that the smallest possible advice complexity of a strictly c-competitive online algorithm for P is $f(n, c)$.

We only consider deterministic online algorithms (with advice). Note that both the advice read and the competitive ratio may depend on n, but, for ease of notation, we often write b and c instead of $b(n)$ and $c(n)$. Also, with this definition, $c \geq 1$, for both minimization and maximization problems.

In this paper, we consider the complexity class AOC from [5].

Definition 2 (AOC [5]). *A problem, P, is in AOC (Asymmetric Online Covering) if it can be defined as follows: The input to an instance of P consists of a sequence of n requests, $\sigma = \langle r_1, \ldots, r_n \rangle$, and possibly one final dummy request. An algorithm for P computes a binary output string, $y = y_1 \ldots y_n \in \{0, 1\}^n$, where $y_i = f(r_1, \ldots, r_i)$ for some function f.*

For minimization (maximization) problems, the score function, s, maps a pair, (σ, y), of input and output to a cost (profit) in $\mathbb{N} \cup \{\infty\}$ $(\mathbb{N} \cup \{-\infty\})$. For an input, σ, and an output, y, y is feasible if $s(\sigma, y) \in \mathbb{N}$. Otherwise, y is infeasible. There must exist at least one feasible output. Let $S_{\min}(\sigma)$ $(S_{\max}(\sigma))$ be the set of those outputs that minimize (maximize) s for a given input σ.

If P is a minimization problem, then for every input, σ, the following must hold:

1. *For a feasible output, y, $s(\sigma, y) = |y|_1$.*
2. *An output, y, is feasible if there exists a $y' \in S_{\min}(\sigma)$ such that $y' \sqsubseteq y$. If there is no such y', the output may or may not be feasible.*

If P is a maximization problem, then for every input, σ, the following must hold:

1. *For a feasible output, y, $s(\sigma, y) = |y|_0$.*
2. *An output, y, is feasible if there exists a $y' \in S_{\max}(\sigma)$ such that $y' \sqsubseteq y$. If there is no such y', the output may or may not be feasible.*

Recall that no problem in AOC requires more than $B(n, c) + O(\log n)$ bits of advice (see Eq. (1) for the definition of $B(n, c)$). The problems in AOC requiring the most advice are AOC-complete [5]:

Definition 3 (AOC-complete [5]). *A problem* $P \in AOC$ *is* AOC-complete *if for all* $c > 1$, *any c-competitive algorithm for* P *must read at least* $B(n,c) - O(\log n)$ *bits of advice.*

In [5], an abstract guessing game, MINASGK (Minimum Asymmetric String Guessing with Known History), was introduced and shown to be AOC-complete. The MINASGK-problem itself is very artificial, but it is well-suited as the starting point of reductions. All other minimization problems known to be AOC-complete have been shown to be so via reductions from MINASGK.

The input for MINASGK is a secret string $x = x_1 x_2 \ldots x_n \in \{0,1\}^n$ given in n rounds. In round $i \in [n]$, the online algorithm must answer $y_i \in \{0,1\}$. Immediately after answering, the correct answer x_i for round i is revealed to the algorithm. If the algorithm answers $y_i = 1$, it incurs a cost of 1. If the algorithm answers $y_i = 0$, then it incurs no cost if $x_i = 0$, but if $x_i = 1$, then the output of the algorithm is declared to be infeasible (and the algorithm incurs a cost of ∞). The objective is to minimize the total cost incurred. Note that the optimal solution has cost $|x|_1$.

The problem MINASGK is based on the *binary string guessing* problem [2,9]. Binary string guessing is similar to asymmetric string guessing, except that any wrong guess (0 instead of 1 or 1 instead of 0) gives a cost of 1.

In Theorem 1, we show a very strong lower bound for a weighted version of MINASGK. In Theorem 2, via reductions, we show that this lower bound implies similar strong lower bounds for a weighted version of other AOC-complete minimization problems.

We now formally define weighted versions of the problems in AOC.

Definition 4 (Weighted AOC). *Let* P *be a problem in* AOC. *We define the weighted version of* P, *denoted* P_w, *as follows: A* P_w-*input* $\sigma = \langle \{r_1, w_1\}, \{r_2, w_2\}, \ldots, \{r_n, w_n\} \rangle$ *consists of* n P-*requests,* $r_1, \ldots, r_n$, *each of which has a weight* $w_i \in \mathbb{R}_+$. *The* P-*request* r_i *and its weight* w_i *are revealed simultaneously. An output* $y = y_1 \ldots y_n \in \{0,1\}^n$ *is feasible for the input* σ *if and only if* y *is feasible for the* P-*input* $\langle r_1, \ldots, r_n \rangle$. *The cost (profit) of an infeasible solution is* ∞ ($-\infty$).

If P *is a minimization problem, then the cost of a feasible* P_w-*output* y *for an input* σ *is*

$$s(\sigma, y) = \sum_{i=1}^{n} w_i y_i$$

If P *is a maximization problem, then the profit of a feasible* P_w-*output* y *for an input* σ *is*

$$s(\sigma, y) = \sum_{i=1}^{n} w_i (1 - y_i)$$

3 Weighted Versions of **AOC**-Complete Minimization Problems

In the weighted version of MINASGK, MINASGK$_W$, each request is a weight for the current request and the value 0 or 1 of the previous request. Producing a feasible solution requires *accepting* (answering 1 to) all requests with value 1, and the cost of a feasible solution is the sum of all weights for requests which are accepted.

We start with a negative result for MINASGK$_W$ and then use it to obtain similar results for the weighted online version of Vertex Cover, Set Cover, Dominating Set, and Cycle Finding.

Theorem 1. *For* MINASGK$_W$, *the competitive ratio of any algorithm with less than n bits of advice is not bounded by any function of n.*

Proof (sketch). Let ALG be any algorithm for MINASGK$_W$ reading at most $n-1$ bits of advice. We show how an adversary can construct input sequences where the cost of ALG is arbitrarily larger than that of OPT. To this end, we describe a way to assign weights to the requests in MINASGK$_W$ such that if ALG makes a single mistake (either guessing 0 when the correct answer is 1 or vice versa), its competitive ratio is unbounded. We use a large number $a > 1$, which we allow to depend on n. All weights are from the interval $[1, a]$ (note that they are not necessarily integers). Let $x = x_1, \ldots, x_n$ be an arbitrary input string with at least one 1, and set $w_1 = a^{1/2}$. For $i > 1$, we set $w_i = w_{i-1} \cdot a^{(-2^{-i})}$ if $x_{i-1} = 0$ and we set $w_i = w_{i-1} \cdot a^{(2^{-i})}$ if $x_{i-1} = 1$. Since the weights are only a function of previous requests, they do not reveal any information to ALG about future requests. This finishes the description of the hard input instance.

Since ALG reads strictly less than n bits of advice, it will make at least one mistake on at least one input string x. We claim that this implies that the competitive ratio of ALG is not bounded by any function of n. Indeed, if ALG guesses 0 for a request, but the correct answer is 1, the solution is infeasible and ALG gets a cost of ∞. We now consider the case where ALG guesses 1 for a request j, but the correct answer is $x_j = 0$. By our choice of weights, for every $j' \neq j$ such that $x_j = 1$, the weight $w_{j'}$ of $x_{j'}$ is smaller than w_j by at least a (multiplicative) factor of $a^{(2^{-n})}$. Thus, the total cost of OPT, which is the sum of weights $w_{j'}$ for which $x_{j'} = 1$, is at least a factor of $a^{(2^{-n})}/n$ smaller than the weight w_j alone. This proves the lower bound since a can be arbitrarily large. $\square$

In order to show that similar lower bounds apply to the weighted versions of all minimization problems known to be AOC-complete, we define a simple type of advice preserving reduction for online problems (Definition 5 in the full paper [6]). These are much less general than those defined by Sprock in his PhD dissertation [16], mainly because we do not allow the amount of advice needed to change by a multiplicative factor. Using these reductions and previous results [5] showing that the underlying AOC problems are AOC-complete, the following result can be shown.

Theorem 2. *For the weighted online versions of Vertex Cover, Cycle Finding, Dominating Set, Set Cover, an algorithm reading less than $n - O(\log n)$ bits of advice cannot have a competitive ratio bounded by any function of n.*

4   Exponential Sparsification

Assume that we are faced with an online problem which we know how to efficiently solve, possibly using advice, in the unweighted version (or when there are only few possible different weights). We use *exponential sparsification*, a simple technique which can be of help when designing algorithms with advice for weighted online problems by reducing the number of different possible weights the algorithm has to handle. The first step is to partition the set of possible weights into intervals of exponentially increasing length, i.e., for some small ε, $0 < \varepsilon < 1$,

$$\mathbb{R}_+ = \bigcup_{k=-\infty}^{\infty} \left[(1+\varepsilon)^k, (1+\varepsilon)^{k+1} \right).$$

How to proceed depends on the problem at hand. We now informally explain the meta-algorithm that we repeatedly use in this paper. Note that if $w_1, w_2 \in \left[(1+\varepsilon)^k, (1+\varepsilon)^{k+1} \right)$ and $w_1 \le w_2$, then $w_1 \le w_2 \le (1+\varepsilon)w_1$. For many online problems, this means that an algorithm can treat all requests whose weights belong to this interval as if they all had weight $(1+\varepsilon)^{k+1}$ with only a small loss in the competitive ratio.

Consider now a set of weights and let $w_{\max}$ denote the largest weight in the set. Let $k_{\max}$ be the integer for which $w_{\max} \in \left[(1+\varepsilon)^{k_{\max}}, (1+\varepsilon)^{k_{\max}+1} \right)$. We say that a request with weight $w \in \left[(1+\varepsilon)^k, (1+\varepsilon)^{k+1} \right)$ is *unimportant* if $k < k_{\max} - \lceil \log_{1+\varepsilon}(n^2) \rceil$. Furthermore, we will often categorize the request as *important* if $k_{\max} - \lceil \log_{1+\varepsilon}(n^2) \rceil \le k < k_{\max}+1$ and as *huge* if $k \ge k_{\max}+1$. Each unimportant request has weight $w \le (1+\varepsilon)^{k+1} \le (1+\varepsilon)^{k_{\max}-\lceil \log_{1+\varepsilon}(n^2) \rceil -1+1} \le w_{\max}/n^2$, so the total sum of the unimportant weights is $O(w_{\max}/n)$. For many weighted online problems, this means that an algorithm can easily serve the requests with unimportant weights. In maximization problems, this is done by rejecting them. In minimization problems, it is done by accepting them. Thus, exponential sparsification (when applicable) essentially reduces the problem of computing a good approximate solution for a problem with n distinct weights to that of computing a good approximate solution with only $O(\log_{1+\varepsilon} n)$ distinct weights.

For a concrete problem, several modifications of this meta-algorithm might be necessary. Often, the most tricky part is how the algorithm can learn $k_{\max}$ without using too much advice. One approach that we often use is the following: The oracle encodes the index i of the first request whose weight is close enough to $(1+\varepsilon)^{k_{\max}}$ that the algorithm only needs a little bit of advice to deduce $k_{\max}$ from the weight of this request. If it is somehow possible for the algorithm to serve all requests prior to i reasonably well, then this approach works well.

Our main application of exponential sparsification is to weighted AOC problems. We begin by considering maximization problems. Note that no assumptions are made about the weights of P_w in Theorem 3.

Theorem 3. *If* $P \in$ *AOC is a maximization problem, then for any $c > 1$ and $0 < \varepsilon \leq 1$, P_w has a strictly $(1 + \varepsilon)c$-competitive algorithm using $B(n, c) + O(\varepsilon^{-1} \log^2 n)$ advice bits.*

Proof (sketch). We split the requests into *classes*, where class k contains all requests with weights in $[(1 + \varepsilon)^k, (1 + \varepsilon)^{k+1})$. The algorithm learns the class $k_{\max}$ of the request with the largest weight which is accepted by OPT, using the approach described above. The unimportant and huge requests are rejected. This is a safe choice, since the huge requests are also rejected by OPT and the total weight of the unimportant requests is insignificant.

In each class with important requests, we use the covering design based c-competitive algorithm for unweighted MAXASG that was used in [5] to obtain the upper bound of $B(n, c) + O(\log n)$ on the advice complexity of AOC problems. This gives a feasible solution, since the accepted requests constitute a subset of those accepted by OPT. We use that the function $B(n, c)$ is linear, which ensures that, in total, these smaller covering designs use roughly the same number of advice bits as one covering design for all n requests. For each smaller covering design, there are an additional $O(\log n)$ advice bits used, which adds up to the $O(\varepsilon^{-1} \log^2 n)$ additional advice bits in the theorem.   □

It may be surprising that adding weights to AOC-complete maximization problems has almost no effect, while adding weights to AOC-complete minimization problems drastically changes the advice complexity. In particular, one might wonder why the technique used in Theorem 3 does not work for minimization problems. The key difference lies in the beginning of the sequence.

For maximization problems, the algorithm can safely reject all requests before the first important one. For minimization problems, this approach does not work, since the algorithm must accept a superset of what OPT accepts in order to ensure that its output is feasible. Thus, rejecting an unimportant request that OPT accepts may result in an infeasible solution. This essentially means that the algorithm is forced into accepting all requests before the first important request arrives. Accepting all unimportant requests is no problem, since they will not contribute significantly to the total cost. However, accepting even a single huge request can give an unbounded contribution to the algorithm's cost. As shown in Theorem 1, it is not possible in general for the algorithm to tell if a request in the beginning of the sequence is unimportant or huge without using a lot of advice.

However, if the ratio of the largest weight, $w_{\max}$, to the smallest weight, $w_{\min}$, is not too large, exponential sparsification is also useful for minimization problems in AOC. Essentially, when this ratio is bounded, it is possible for the algorithm to learn a good approximation of $w_{\max}$ when the first request arrives. This is formalized in Theorem 4, the proof of which is very similar to the proof of Theorem 3.

Theorem 4. *If* $P \in AOC$ *is a minimization problem and* $0 < \varepsilon \le 1$, *then* P_w *with all weights in* $[w_{min}, w_{max}]$ *has a* $(1+\varepsilon)c$-*competitive algorithm with advice complexity at most*

$$B(n,c) + O\left(\varepsilon^{-1} \log^2 n + \log\left(\varepsilon^{-1} \log \frac{w_{max}}{w_{min}}\right)\right).$$

5   Matching and Other Non-Complete AOC Problems

We first provide a general theorem that works for all maximization problems in AOC, giving better results in some cases than that in Theorem 3.

Theorem 5. *Let* $P \in AOC$ *be a maximization problem. If there exists a* c-*competitive* P-*algorithm reading* b *bits of advice, then there exists a* $O(c \cdot \log n)$-*competitive* P_w-*algorithm reading* $O(b + \log n)$ *bits of advice.*

In the online matching problem, edges arrive online, the algorithm must irrevocably accept or reject them as they arrive, and the goal is to maximize the number of edges accepted. The natural greedy algorithm for this problem is well known to be 2-competitive. In terms of advice, the problem is known to be in AOC, but is not AOC-complete [5]. We remark that a version of unweighted online matching with vertex arrivals (incomparable to our weighted matching with edge arrivals) has been studied with advice in [8].

Corollary 1. *For Weighted Matching, there exists a* $O(\log n)$-*competitive algorithm reading* $O(\log n)$ *bits of advice.*

Proof. The result follows from Theorem 5 since there exists a 2-competitive algorithm without advice for (unweighted) Matching.   □

5.1   Lower Bounds

Theorem 3 shows that for weighted maximization problems in AOC, we cannot hope to achieve lower bounds similar to that of Theorem 2. However, we do have the following lower bound for algorithms reading very little advice.

Theorem 6. *For the weighted online versions of Independent Set, Clique, Disjoint Path Allocation, and Matching, an algorithm reading* $o(\log n)$ *bits of advice cannot have a competitive ratio bounded by any function of* n.

Returning to the example of Weighted Matching, we now know that $O(\log n)$ bits suffice to be $O(\log n)$-competitive, and that $o(\log n)$ bits of advice leads to a competitive ratio unbounded by any function of n. Furthermore, using a technique introduced in [12], we prove the following result:

Theorem 7. *An* $O(1)$-*competitive algorithm for Weighted Matching must read at least* $\Omega(n)$ *bits of advice.*

In particular, we cannot achieve a constant competitive ratio using $O(\log n)$ bits of advice for Weighted Matching. We leave it as an open problem to close the gap between $\omega(1)$ and $O(\log n)$ on the competitive ratio of Matching algorithms with advice complexity $O(\log n)$.

6   Scheduling with Sublinear Advice

For the scheduling problems studied, the requests are jobs, each characterized by its size. Each job must be assigned to one of m available machines. If the machines are *identical*, the *load* of a job on any machine is simply its size. If the machines are *related*, each machine has a speed, and the load of a job, J, assigned to a machine with speed s is the size of J divided by s. If the machines are *unrelated*, each job arrives with a vector specifying its load on each machine.

Consider a sequence $\sigma = \langle r_1, \ldots, r_n \rangle$ of n jobs that arrive online. Each job $r_i \in \sigma$ has an associated weight-function $w_i : [m] \to \mathbb{R}_+$. Upon arrival, a job must irrevocably be assigned to one of the m machines. The *load* L_j of a machine $j \in [m]$ is defined as $L_j = \sum_{i \in M_j} w_i(j)$ where M_j is the set of (indices of) jobs scheduled on machine j. The *total load* of a schedule for σ is the vector $\mathbf{L} = (L_1, \ldots, L_m)$. We say that $(L_1, \ldots, L_m) \leq (L'_1, \ldots, L'_m)$ if and only if $L_i \leq L'_i$ for $1 \leq i \leq m$. A scheduling problem of the above type is specified by an *objective function* $f : \mathbb{R}_+^m \to \mathbb{R}_+$ and by specifying if the goal is to minimize or maximize $f(\mathbf{L}) = f(L_1, \ldots, L_m) \in \mathbb{R}_+$. We assume that f is non-decreasing, i.e., $f(\mathbf{L}) \leq f(\mathbf{L}')$ for all $\mathbf{L} \leq \mathbf{L}'$. Some of the classical choices of objective function include:

- Minimizing the ℓ_p-norm $f_p(\mathbf{L}) = f_p(L_1, \ldots, L_m) = \|(L_1, \ldots, L_m)\|_p$ for some $1 \leq p \leq \infty$. That is, for $1 \leq p < \infty$, the goal is to minimize $\left(\sum_{j \in [m]} L_j^p \right)^{1/p}$ and for $p = \infty$, the goal is to minimize the makespan $\max_{j \in [m]} L_j$.
- Maximizing the minimum load $f(\mathbf{L}) = \min_{j \in [m]} L_j$. This is also known as machine covering. Note that this objective function is not a norm, but it is a seminorm.[1]

We begin with a result for *unrelated* machines.

Theorem 8. *Let* P *be a scheduling problem on m unrelated machines where the goal is to minimize an objective function f. Assume that f is a norm. Then, for $0 < \varepsilon \leq 1$, there exists a $(1 + \varepsilon)$-competitive* P*-algorithm reading $O\big(\big(\frac{4}{\varepsilon}\log(n) + 2\big)^m \log(n)\big)$ bits of advice. In particular, if $m = O(1)$ and $\varepsilon = \Omega(1)$, then there exists a $(1 + \varepsilon)$-competitive algorithm reading $O(\mathrm{polylog}(n))$ bits of advice.*

Roughly speaking, we prove Theorem 8 by carefully using exponential sparsification to partition the set of jobs into a sufficiently small number of *types*, and then giving as advice the number of jobs of each type.

The advice complexity of the algorithm for unrelated machines in Theorem 8 depends quite heavily on the number of machines m. In Theorem 9, we show that the dependency on m can be removed when the machines are related.

Theorem 9. *Let* P *be a scheduling problem on m related machines where the goal is to minimize an objective function f. Assume that f is a norm. Then, for $0 < \varepsilon \leq 1$, there exists a $(1 + \varepsilon)$-competitive* P*-algorithm with advice complexity $O\big(\varepsilon^{-1} \log^2 n\big)$.*

---

[1] f is a norm if $f(a\mathbf{v}) = |a| f(\mathbf{v})$, $f(\mathbf{v} + \mathbf{v}) \leq f(\mathbf{v}) + f(\mathbf{v})$, and $f(\mathbf{v}) = 0 \Rightarrow \mathbf{v} = \mathbf{0}$. A seminorm does not require this last condition.

We now consider scheduling problems where the goal is to maximize an objective function f. Recall that we assume that the objective function is non-decreasing. The most notable example is when f is the minimum load.

Theorem 10. *Let* P *be a scheduling problem on* m *unrelated machines where the goal is to maximize an objective function* f. *Assume that* f *is a seminorm. Then, for every* $0 < \varepsilon \leq 1$, *there exists a* $(1 + \varepsilon)$-*competitive* P-*algorithm with advice complexity* $O((\frac{4}{\varepsilon}\log(n) + 2)^m m^2 \log n)$. *In particular, if* $m = O(1)$ *and* $\varepsilon = \Omega(1)$, *the advice complexity is* $O(\text{polylog}(n))$.

References

1. Albers, S., Hellwig, M.: Online makespan minimization with parallel schedules. In: Ravi, R., Gørtz, I.L. (eds.) SWAT 2014. LNCS, vol. 8503, pp. 13–25. Springer, Heidelberg (2014)
2. Böckenhauer, H.-J., Hromkovič, J., Komm, D., Krug, S., Smula, J., Sprock, A.: The string guessing problem as a method to prove lower bounds on the advice complexity. Theor. Comput. Sci. **554**, 95–108 (2014)
3. Böckenhauer, H.-J., Komm, D., Královič, R., Královič, R., Mömke, T.: On the advice complexity of online problems. In: Dong, Y., Du, D.-Z., Ibarra, O. (eds.) ISAAC 2009. LNCS, vol. 5878, pp. 331–340. Springer, Heidelberg (2009)
4. Boyar, J., Epstein, L., Favrholdt, L.M., Larsen, K.S., Levin, A.: Online bounded analysis. In: Kulikov, A.S., Woeginger, G.J. (eds.) CSR 2016. LNCS, vol. 9691, pp. 131–145. Springer, Heidelberg (2016). doi:10.1007/978-3-319-34171-2_10
5. Boyar, J., Favrholdt, L.M., Kudahl, C., Mikkelsen, J.W.: Advice complexity for a class of online problems. In: STACS, vol. 30 of LIPIcs, pp. 116–129 (2015)
6. Boyar, J., Favrholdt, L.M., Kudahl, C., Mikkelsen, J.W.: Weighted online problems with advice (2016). http://arxiv.org/abs/1606.05210
7. Dobrev, S., Královič, R., Pardubská, D.: Measuring the problem-relevant information in input. RAIRO Theor. Inf. Appl. **43**(3), 585–613 (2009)
8. Dürr, C., Konrad, C., Renault, M.P.: On the power of advice and randomization for online bipartite matching. In: ESA, 2016 to appear
9. Emek, Y., Fraigniaud, P., Korman, A., Rosén, A.: Online computation with advice. Theor. Comput. Sci. **412**(24), 2642–2656 (2011)
10. Karlin, A.R., Manasse, M.S., Rudolph, L., Sleator, D.D.: Competitive snoopy caching. Algorithmica **3**, 77–119 (1988)
11. Komm, D., Královič, R., Královič, R., Kudahl, C.: Advice complexity of the online induced subgraph problem. In: MFCS, 2016 to appear
12. Mikkelsen, J.W.: Randomization can be as helpful as a glimpse of the future in online computation. In: ICALP, 2016 to appear
13. Renault, M.P., Rosén, A., van Stee, R.: Online algorithms with advice for bin packing and scheduling problems. Theor. Comput. Sci. **600**, 155–170 (2015)
14. Sleator, D.D., Tarjan, R.E.: Amortized efficiency of list update and paging rules. Commun. ACM **28**(2), 202–208 (1985)
15. Smula, J.: Information Content of Online Problems: Advice versus Determinism and Randomization. Ph.D. thesis, ETH, Zürich (2015)
16. Sprock, A.: Analysis of hard problems in reoptimization and online computation. Ph.D. thesis, ETH, Zürich (2013)

Finding Gapped Palindromes Online

Yuta Fujishige[1]($\boxtimes$), Michitaro Nakamura[2], Shunsuke Inenaga[1], Hideo Bannai[1], and Masayuki Takeda[1]

[1] Department of Informatics, Kyushu University, Fukuoka, Japan
{yuta.fujishige,inenaga,bannai,takeda}@inf.kyushu-u.ac.jp
[2] Department of Physics, Kyushu University, Fukuoka, Japan

Abstract. A string s is said to be a *gapped palindrome* iff $s = xyx^R$ for some strings x, y such that $|x| \geq 1$, $|y| \geq 2$, and x^R denotes the reverse image of x. In this paper we consider two kinds of gapped palindromes, and present efficient *online* algorithms to compute these gapped palindromes occurring in a string. First, we show an online algorithm to find all maximal *g-gapped palindromes* with fixed gap length $g \geq 2$ in a string of length n in $O(n \log \sigma)$ time and $O(n)$ space, where σ is the alphabet size. Second, we show an online algorithm to find all maximal *length-constrained gapped palindromes* with arm length at least $A \geq 1$ and gap length in range $[g_{\min}, g_{\max}]$ in $O(n(\frac{g_{\max} - g_{\min}}{A} + \log \sigma))$ time and $O(n)$ space. We also show that if A is a constant, then there exists a string of length n which contains $\Omega(n(g_{\max} - g_{\min}))$ maximal LCGPs, which implies we cannot hope for a significant speed-up in the worst case.

1 Introduction

A *palindrome* is a string of form xax^R, where x is a string called the left arm, a is either the empty string or a single character, and x^R is the reversed string of x called the right arm. Finding palindromic substrings in a given string w is a classical problem on string processing. The earliest work on this problem dates back to at least 1970's when Manacher [10] proposed an online algorithm to find all prefix palindromes in w in $O(n)$ time, where n is the length of w. Later, Apostolico et al. [1] pointed out that Manacher's algorithm can be used to find all maximal palindromes in w in $O(n)$ time, where a maximal palindrome is a substring palindrome $w[i..j] = w[i..j]^R$ of w whose arms cannot be further extended based on the same center $\frac{i+j}{2}$.

A natural generalisation of palindromes is *gapped palindromes* of form xyx^R, where y is a string of length at least 2 called a *gap*[1]. Finding gapped palindromes has applications in bioinformatics, e.g.; RNA secondary structures called hairpins can be regarded as a kind of gapped palindrome $xy\overline{x}^R$, where $\overline{x}$ represents the complement of x ($\overline{x}$ is obtained by exchanging A with U and exchanging C with G in x). The most basic type of gapped palindromes is *g-gapped palindromes*, where $g \geq 2$ is a pre-defined fixed length of the gaps. For three parameters $g_{\min}$,

---

[1] If y is a single character, then xyx^R is a palindrome of odd length. Thus we here assume y is of length at least 2.

© Springer International Publishing Switzerland 2016
V. Mäkinen et al. (Eds.): IWOCA 2016, LNCS 9843, pp. 191–202, 2016.
DOI: 10.1007/978-3-319-44543-4_15

$g_{\max}$, and A such that $2 \leq g_{\min} \leq g_{\max}$ and $A \geq 1$, Kolpakov and Kucherov [8] introduced *length-constrained gapped palindromes* (*LCGPs*) which has arms of length at least A and gaps of length in range $[g_{\min}, g_{\max}]$. This is a natural generalisation of g-gapped palindromes with $g_{\min} = g_{\max} = g$ and $A = 1$.

In this paper, we consider the problems of finding these gapped palindromes in a string in an *online manner*. Namely, our input is a growing string to which new characters can be appended, and each character of the string arrives one by one, from left to right. Let n be the length of the final string w. We propose:

(1) An online algorithm to compute all maximal g-gapped palindromes in w in $O(n \log \sigma)$ time and $O(n)$ space, where σ is the alphabet size. This algorithm can be modified to output only *distinct* maximal g-gapped palindromes in an online manner, in the same complexity.
(2) An online algorithm to compute all maximal LCGPs in w in $O(n(m+\log \sigma))$ time and $O(n)$ space, where $m = \max\{\frac{g_{\max}-g_{\min}}{A}, 1\}$.

Formal definitions of the maximality of these gapped palindromes will be given in Sects. 3 and 4, respectively.

We remark that using a slightly modified version of Solution (1), it is trivial to obtain an $O(n(g_{\max} - g_{\min} + \log \sigma))$-time solution for finding all maximal LCGPs, by simply testing gap lengths $g_{\min}, g_{\min}+1, \ldots, g_{\max}$ separately. Hence, in the case where A is not a constant and $\log \sigma$ is not a dominating term, then Solution (2) speeds up this trivial method by a factor of A. On the other hand, in the case where A is a constant, then we show that there exists a string of length n which contains $\Omega(nm)$ maximal LCGPs, meaning that we cannot hope significant speed-up in the worst case.

Solution (2) is based on Solution (1) and is quite different from the offline solution by Kolpakov and Kucherov [8]. To our knowledge, these are the first efficient online algorithms that compute *any kind* of gapped palindromes.

Related work. A number of efficient *offline* algorithms for computing various kinds of gapped palindromes have been proposed in the literature.

Let w be an input string w of length n over the integer alphabet. There exists a folklore $O(n)$-time algorithm (see e.g. [6]) which finds all maximal g-gapped palindromes for a given fixed gap length g; the suffix tree [4,12] of string $w^R \# w\$$ and a constant-time LCA data structure [2] over the suffix tree are constructed during preprocessing, and then computing each maximal g-gapped palindrome reduces to an outward longest common extension (LCE) query, which can be answered by an LCA query on the tree. Our algorithm for computing all maximal g-gapped palindromes can be regarded as an online version of this algorithm.

Kolpakov and Kucherov [8] proposed an $O(n + L)$-time offline algorithm to find all maximal LCGPs, where L is the number of outputs. Their algorithm consists of the following two steps: In the first step, it computes all (not necessarily outward maximal) LCGPs whose arms are of length exactly A. Let (i,j) be the pair of the ending position i and the beginning position j of the left and right arms of each of the above LCGPs, respectively. In the second step, for each LCGP computed above, the algorithm performs an outward LCE query from

i and j, using the same suffix-tree based data structure as for the maximal g-gapped palindromes above. However, each time a new character is appended to the growing string, the LCE value from the same pair of positions may increase, and it is impossible to know beforehand when the growth of the LCE value for each pair of positions stops. Thus, it seems difficult to apply Kolpakov and Kucherov's solution to our online setting.

There exist efficient offline solutions for finding other kinds of gapped palindromes. Kolpakov and Kucherov [8] also proposed an $O(n)$-time[2] offline algorithm to compute all maximal *long-armed palindromes* (those whose arms are longer than their gap) in a given string w of length n. Kolpakov and Kucherov's algorithm uses a variant of Lempel-Ziv factorisation called the reversed LZ factorisation of strings. Let $f_1, \ldots, f_k$ be the reversed LZ factorisation of w. Then, for each pair f_i of adjacent factors, their algorithm focuses on positions $\frac{|f_i|}{2^k}$ for every $1 \leq k \leq \lceil \frac{|f_i|}{2} \rceil$ in f_i. This implies that the length of each f_i needs to be precomputed. However, in the online setting, the length of the last factor that is a suffix of the current string can extend each time a new character is appended. It is therefore unclear whether we can extend their solution to the online scenario.

Very recently, Gawrychowski et al. [5] considered a generalisation of long-armed palindromes called α-*gapped palindromes*; For a parameter $\alpha > 1$, a gapped palindrome xyx^R is said to be an α-gapped palindrome iff $|xy| \leq \alpha |y|$. Gawrychowski et al. [5] proposed an $O(\alpha n)$-time offline algorithm which computes all maximal α-gapped palindromes in an input string w of length n. This algorithm requires a preprocessing of the input w for integer $c \geq 2$ such that the occurrences of a substring of length 2^k (called a basic factor therein) in another substring of length $c2^k$ can be computed efficiently. Thus, it seems difficult to apply their result to the online setting.

2 Preliminaries

2.1 Strings

Let Σ be an ordered alphabet of size σ. An element of Σ^* is called a *string*. The length of string w is denoted by $|w|$. The empty string is denoted by ε. For any non-empty string w, $w[i]$ denotes the character at position i of w for $1 \leq i \leq |w|$, and $w[i..j]$ denotes the substring of w that begins at position i and ends at position j in w for $1 \leq i \leq j \leq |w|$. For convenience, let $w[i..j] = \varepsilon$ for $i > j$. For $0 \leq i \leq |w| + 1$, $w[1..i]$ and $w[i..|w|]$ are called a *prefix* and a *suffix* of w, respectively. Let w^R denotes the *reversed image* of w, namely, $w^R = x[|x|] \cdots x[1]$. For instance, if $w = \mathtt{desserts}$, then $w^R = \mathtt{stressed}$. For any strings x and y, let $lcp(x, y)$ denote the length of the longest common prefix of x and y.

---

[2] Originally, Kolpakov and Kucherov [8] stated their algorithm works in $O(n+S)$ time, where S is the number of outputs. It follows from a recent work by Gawrychowski et al. [5] that $S = O(n)$.

2.2   Gapped Palindromes

A string p is said to be a *gapped palindrome* iff $p = xyx^R$ for some non-empty strings x, y with $|y| > 1$. The intervals $[1, |x|]$, $[|y| + 1, |xy|]$, and $[|xy| + 1, |xyx|]$ in p are called the *left arm*, *gap*, and *right arm* of gapped palindrome $p = xyx^R$. Note that in general the choice of arms and gap are not unique for the same string p. For instance, if $p = $ abccbba, then we can take $x = $ ab and $y = $ ccb, or $x = $ a and $y = $ bccbb.

A gapped palindrome xyx^R is said to be a *length-constrained palindrome* (*LCGP*) iff $|x| \geq A$ and $g_{min} \leq |y| \leq g_{max}$ for some fixed integer parameters $A \geq 1$ and $1 < g_{min} \leq g_{max}$. A gapped palindrome xyx^R is said to be a *g-gapped* palindrome iff $|y| = g$ for some fixed integer $g > 1$. Note that any g-gapped palindrome is a special case of a length-constrained palindrome with $g_{min} = g_{max} = g$ and $A = 1$.

An occurrence of a gapped palindrome $p = xyx^R$ in a string w is identified by a triple (i, j, a) such that a denotes the length of each arm, and i, j denote the ending and beginning positions of the left and right arms of p, respectively. Namely, $w[i - a + 1..i] = x$, $w[i + 1..j - 1] = y$, and $w[j..j + a - 1] = x^R$. The *center* of an occurrence (i, j, a) of a gapped palindrome in w is $\frac{i+j}{2}$.

2.3   Suffix Trees and LCE Queries

The suffix tree of a string w, denoted $STree(w)$, is a path-compressed trie which represents all suffixes of w. More formally, $STree(w)$ is an edge-labelled rooted tree such that (1) Every internal node is branching; (2) The out-going edges of every internal node begin with mutually distinct characters; (3) Each edge is labelled by a non-empty substring of w; (4) For each suffix s of w, there is a unique path from the root which spells out s (the path possibly ends on an edge). It follows from the definition of $STree(w)$ that if $n = |w|$ then the number of nodes and edges in $STree(w)$ is $O(n)$. By representing every edge label x by a pair (i, j) of integers such that $x = w[i..j]$, $STree(w)$ can be represented with $O(n)$ space.

For any node v of $STree(w)$, let $str(v)$ denotes the substring of w that is obtained by concatenating the edge labels in the path from the root to v. Each node v stores the length $|str(v)|$ of the string it represents. For each non-root node v, let $slink(v) = (v, u)$ be a reversed edge called the *suffix link* of v, such that $str(u) = str(v)[2..|str(v)|]$. It is well-known that $STree(w)$ with the suffix links of all nodes can be constructed online in $O(n \log \sigma)$ time and $O(n)$ space [11].

The *locus* of a substring x of w in $STree(w)$ is the ending point of the path P_x that spells out x from the root. If the ending point of P_x lies on an edge label, then the locus is represented by triple $\langle u, s, t \rangle$ such that u is the deepest node in the path P_x and s, t are positions of w with $str(u)w[s..t] = x$.

Given an ordered pair (i, j) of positions in a string w of length n, a *reversed longest common extension query* $rlce_w(i, j)$ returns $lcp((w[1..i])^R, w[j..n])$. Computing $rlce_w(i, j)$ reduces to the lowest common ancestor (LCA) problem on $STree(w')$, where $w' = w^R\#w$ and $\#$ is a special delimiter which does not occur in w. Let $v_{i,j}$ be the LCA of the two leaves which represent the suffixes $w'[n - i + 1..2n + 1]$ and $w'[n + j + 1..2n + 1]$. Then, we have that $|str(v_{i,j})| = rlce_w(i,j)$. Using an LCA data structure (e.g. [2]), we can answer $rlce_w(i,j)$ query for any pair (i,j) of positions in $O(1)$ time after an $O(n)$-time preprocessing on $STree(w')$.

3 Online Algorithms to Compute All Maximal g-gapped Palindromes

An occurrence (i,j,a) of a g-gapped palindrome xyx^R in a string w is said to be *maximal*, if the arms x, x^R cannot be extended outward, i.e., if $w[b-1] \neq w[e+1]$, $b = 1$, or $e = n$, where $b = i - a + 1$ and $e = j + a - 1$[3].

Example 1. Consider string `aabaacabbcaabb` and let $g = 3$. This string has 3-gapped maximal palindromes $(1,5,1) = \text{a} \cdot \text{aba} \cdot \text{a}$, $(6,10,4) = \text{baac} \cdot \text{ab}$ $\text{b} \cdot \text{caab}$, $(7,11,1) = \text{a} \cdot \text{bbc} \cdot \text{a}$, and $(9,13,2) = \text{bb} \cdot \text{caa} \cdot \text{bb}$.

3.1 Computing all Maximal g-gapped Palindromes Online

In this subsection, we propose online algorithms to compute all maximal g-gapped palindromes in a string w of length n, where $g > 1$ is a given fixed integer parameter (since $g = 1$ gives odd palindromes, we set $g > 1$).

As was mentioned in Sect. 1, there exists an *offline* algorithm which computes all g-gapped maximal palindromes in $O(n)$ time and space for an input string w of length n over an integer alphabet. However, in our scenario the input string w is given online, and we wish to process each character from left to right. In the sequel, we will show our online algorithm which can deal with this setting.

For each $k = 1, \ldots, n$, our algorithm maintains the *longest* g-gapped suffix palindrome of $w[1..k]$ (if it exists). For each g-gapped palindrome to compute, we maintain two variables i, j ($i < j < k$) that represent the ending position of the left arm and the beginning position of the right arm of g-gapped palindrome, respectively. Assume $(i, j, a_{i,j})$ is the longest g-gapped suffix palindrome of $w[1..k]$, where the gap of length g is $w[i + 1..j - 1]$, $j = i + g + 1$ and $j + a_{i,j} - 1 = k$. In case there are no g-gapped suffix palindromes of $w[1..k]$, then let $a_{i,j} = 0$, $i = k - g$ and $j = k + 1$. Depending on the next character $w[k + 1]$, we have two cases:

1. If $w[i - a_{i,j}] = w[k + 1]$, then there exists a longer g-gapped palindrome centered at $\frac{i+j}{2}$. We then naïvely extend the arm length by $a_{i,j} \leftarrow a_{i,j} + 1$, and proceed to the forthcoming character by updating $k \leftarrow k + 1$.

---

[3] Since the gap length is fixed to g and since it simplifies the description of the algorithm, here we do not consider inward maximality of the arms. However, it is easy to modify our algorithm so that it outputs all g-gapped palindromes that are both outward and inward maximal with the same efficiency.

2. If $w[i - a_{i,j}] \neq w[k+1]$, then it appears that $(i, j, a_{i,j})$ is the longest g-gapped maximal palindrome ending at position k, and hence we output it. We then shift the gap to the right by updating $i \leftarrow i + 1$ and $j \leftarrow j + 1$. There are two-sub cases.

   (a) If $j > k+1$, then it appears that there is no g-gapped suffix palindrome of $w[1..k+1]$. We therefore update $k \leftarrow k+1$ and proceed to the forthcoming character, with the current values of i and j.

   (b) If $j \leq k+1$, then we compute $a_{i,j}$ (we will later describe how to efficiently compute it for updated i and j). There are two sub-cases:

      i. If $j + a_{i,j} - 1 = k + 1$, then $(i, j, a_{i,j})$ is the longest g-gapped suffix palindrome of $w[1..k + 1]$. We proceed to the forthcoming character by updating $k \leftarrow k + 1$.

      ii. If $j + a_{i,j} - 1 < k + 1$, then $(i, j, a_{i,j})$ is the maximal g-gapped palindrome with the gap beginning at position $i + 1$, and hence we output it. We then shift the gap to the right by updating $i \leftarrow i + 1$ and $j \leftarrow j + 1$, and go to either Case 2a or Case 2b depending on the value of j.

In order to efficiently compute $a_{i,j}$ of Case 2 above in our online scenario, we utilize the following results:

Theorem 1 ([7]). *There exists an $O(n \log \sigma)$-time $O(n)$-space algorithm to maintain the suffix tree with suffix links for a bidirectionally growing string to which new characters can be prepended and appended, where n is the length of the final string.*

Theorem 2 ([3]). *There exists a linear-space algorithm for a rooted tree that supports the following operations and query in $O(1)$ worst-case time: which supports the following operations and query in $O(1)$ worst-case time: (1) Insert a new node; (2) Delete an existing node; (3) LCA query for any pair of nodes in the current tree.*

We are ready to show the main result of this section:

Theorem 3. *For a growing string to which new characters are appended, we can compute all maximal g-gapped palindromes in an online manner, in $O(n \log \sigma)$ time and $O(n)$ space, where n is the length of the final string.*

Proof. The correctness immediately follows from the above arguments.

The time complexity is shown as follows. In the sequel, we consider the amortised time cost for each $k = 1, \ldots, n$. For each k that falls into Case 1, it clearly takes $O(1)$ time. For each k that falls into Case 2b, we output several maximal g-gapped palindromes. It takes $O(1)$ time to output the longest maximal g-gapped palindrome. The key is how to compute the arm lengths $a_{i,j}$ of shorter maximal g-gapped palindromes. For this sake we maintain $STree(w'_k)$ where $w'_k = (w[1..k])^R \# w[1..k]$, where $\#$ is a special delimiter which does not appear elsewhere in w'_k (see also Fig. 1 for an example).

Note that computing $a_{i,j}$ is equivalent to computing $rlce_{w[1..k]}(i, j)$, and thus

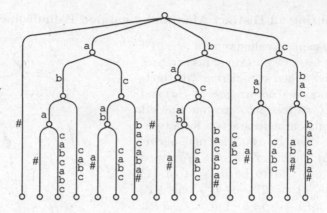

Fig. 1. $STree(w_k')$ with $w[1..k] =$ abacabcabc and $w_k' =$ cbacbacaba#abacabcabc. The label strings after # are omitted for simplicity.

is equivalent to computing $|str(v_{i,j})|$, where $v_{i,j}$ is the LCA of the nodes of $STree(w_k')$ which represent the suffixes $w_k'[k-i+1..2k+1]$ and $w_k'[k+j+1..2k+1]$ of w_k'. Since # is unique in w_k', the suffix $w_k'[k-i+1..2k+1]$ is always represented by a leaf of $STree(w_k')$ and hence can easily be accessed in $O(1)$ time. However, notice that the other suffix $w_k'[k+j+1..2k+1]$ is not represented by a node when the path that spells out $w_k'[k+j+1..2k+1]$ from the root ends on an edge (this can happen when $w_k'[k+j+1..2k+1] = w[j..k]$ is a prefix of another suffix of $w[1..k]$). Consider such a case, and let $\langle u_j, s_j, t_j \rangle$ be the locus for the suffix $w_k'[k+j+1..2k+1]$. Since u_j is the nearest ancestor to the locus, we can use u_j for the LCA query instead of the locus for $w_k'[k+j+1..2k+1]$.

What remains is how to quickly find the loci for increasing j. For this we can use a similar technique to Ukkonen's online suffix tree construction algorithm [11]: Assume that the locus $\langle u_j, s_j, t_j \rangle$ for the suffix $w_k'[k+j+1..2k+1] = w[j..k]$ in $STree(w_k')$ is given. To find the locus for $\langle u_{j+1}, s_{j+1}, t_{j+1} \rangle$ for the next suffix $w_k'[k+j+2..2k+1] = w[j+1..k]$, we first follow the suffix link of u_j and arrive at $z = slink(u_j)$. We then traverse the path from z which spells out $w_k'[s_{j+1}..t_{j+1}]$. The last piece of this path gives the locus $\langle u_{j+1}, s_{j+1}, t_{j+1} \rangle$ (see also Fig. 2).

Using a similar analysis to [11], the cost to find this locus is amortised to $O(\log \sigma)$. Since the total number of outputs (maximal g-gapped palindromes) is linear in n, the amortised cost per output is $O(\log \sigma)$. The cost to update $STree(w_k')$ to $STree(w_{k+1}')$ is amortised to $O(\log \sigma)$ by Theorem 1. Each LCA query can be answered in $O(1)$ time by Theorem 2. Hence, the total time complexity is $O(n \log \sigma)$. The total space requirement is clearly $O(n)$. This completes the proof. □

3.2   Computing all Distinct Maximal g-gapped Palindromes Online

Consider a g-gapped palindrome $p = xyx^R$ which has at least two maximal occurrences in a string w. When considering "distinctness" of two maximal occurrences (i, j, a) and (i', j', a) of p, we take into account the left and right neighbouring characters for a technical reason. Namely, two maximal occurrences (i, j, a) and (i', j', a) of a g-gapped palindromes are said to be *distinct* iff (1) $w[b-1] \neq w[b'-1]$ or (2) $w[e+1] \neq w[e'+1]$, where $b = i-a+1$, $e = j+a-1$, $b' = i'-a+1$, and $e' = j'+a-1$.

Our online algorithm of Sect. 3.1 can be modified to output all distinct maximal g-gapped palindromes in an online manner.

For any string w, let $lusuf(w)$ denote the longest suffix of w which appears at least twice in w (we assume that the empty string ε appears $|w|+1$ times in w so $lusuf(w)$ always exists). We make use of the following simple observation:

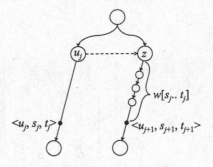

Fig. 2. Illustration of how to find the locus $\langle u_{j+1}, s_{j+1}, t_{j+1} \rangle$ of the next suffix $w'_k[k + j + 2..2k + 1] = w[j+1..k]$ using the suffix link of u_j, where $\langle u_j, s_j, t_j \rangle$ is the locus of the previous suffix $w'_k[k + j + 1..2k + 1] = w[j..k]$. The cost for walking down from node z to the locus for $\langle u_{j+1}, s_{j+1}, t_{j+1} \rangle$ is $O(\log \sigma)$ amortised.

Observation 1. *Let (i, j, a) be an occurrence of a maximal g-gapped palindrome xyx^R in a string w, and let $c_\ell = w[i - a]$ and $c_r = w[j + a]$. Then, it is the first (i.e. left-most) maximal occurrence of xyx^R in w iff $|c_\ell xyx^R c_r| = j-i+2a+1 > |lusuf(w[1..j + a - 1])|$.*

Theorem 4. *For a growing string to which new characters are appended, we can compute all distinct maximal g-gapped palindromes in an online manner, in $O(n \log \sigma)$ time and $O(n)$ space, where n is the length of the final string.*

Proof. On top of $STree(w'_k)$ used in Theorem 3, we build another suffix tree $STree(w[1..k])$ for increasing $k = 1, \ldots, n$ using Ukkonen's online algorithm [11]. For each k, Ukkonen's algorithm maintains an invariant called the *active point* which indicates the locus of $lusuf(w[1..k])$. When we process the kth character $w[k]$, we store $|lusuf(w[1..h])|$ for all $1 \leq h \leq k$. Let $(i, j, a_{i,j})$ be an occurrence of a maximal g-maximal found at the k-th stage of the algorithm where we have processed $w[1..k]$. Then, we can determine in $O(1)$ time whether or not it is the first maximal occurrence of the g-gapped palindrome using Observation 1 (recall that the right mismatched position $j + a_{i,j}$ never exceeds k and hence we know $|lusuf(w[1..j+a_{i,j}])|$). Since Ukkonen's online algorithm works in $O(n \log \sigma)$ time and $O(n)$ space, the theorem holds.   □

We note that a similar technique was used by Kosolobov et al. [9] in their online algorithm to find all distinct palindromes (without gaps) in a given string.

4   Online Algorithms to Compute all Maximal LCGPs

An occurrence (i, j, a) of an LCGP in a string w of length n is said to be *outward-maximal* iff $w[i - a] \neq w[j + a]$, $i - a + 1 = 1$, or $j + a - 1 = n$, and it is said to be *inward-maximal* iff $w[i + 1] \neq w[j - 1]$. It is said to be *maximal* iff it is both outward-maximal and inward-maximal[4].

Example 2. Consider string aabaacabbcaabb and let $g_{\min} = 1$, $g_{\max} = 4$, and $A = 2$. All the maximal LCGPs in this string are $(2, 4, 2) = $ aa $\cdot$ b $\cdot$ aa, $(4, 7, 2) = $ ba $\cdot$ ac $\cdot$ ab, $(6, 10, 4) = $ baac $\cdot$ abb $\cdot$ caab, and $(9, 13, 2) = $ bb $\cdot$ caa $\cdot$ bb.

4.1   Computing all Maximal LCGPs Online

In this section, we present an online algorithm to compute all maximal LCGPs of a given string w. This algorithm works in $O(n(m + \log \sigma))$ time and $O(n)$ space, where $n = |w|$ and $m = \max\{\frac{g_{\max} - g_{\min}}{A}, 1\}$.

Let $d = \frac{g_{\max} - g_{\min}}{2}$. For ease of explanation, we assume that $d \mod A = 0$ and we will describe our algorithm for this case. However, the algorithm can easily be extended to a general case with $d \mod A \neq 0$, retaining the same efficiency.

For each $k = 1, \ldots, n$ in increasing order, we maintain a pair (i, j) of positions such that $j - i = g_{\min} + 1$ and the longest inward-maximal suffix LCGP of $w[1..k]$ is centered at $\frac{i+j}{2}$ (if it exists). If it does not exist, then let $i = k - g_{\max}$ and $j = k - g_{\max} + g_{\min} + 1$. For $1 \leq l \leq \frac{d}{A}$, we consider the positions $i - l \cdot A$ and $j + l \cdot A$ in $w[1..k]$, called *sampled* positions. The following simple lemma suggests how we can use these sampled positions for efficient computation of LCGPs.

Lemma 1. *Let (i', j', a') be any maximal LCGP whose center is $\frac{i+j}{2}$ (i.e., $\frac{i'+j'}{2} = \frac{i+j}{2}$). Then, there exists l ($1 \leq l \leq \frac{d}{A}$) such that $j + l \cdot A \in [j', j' + a' - 1]$ and $i - l \cdot A \in [i' - a' + 1, i']$. Moreover, for each such l, (i', j', a') is the unique maximal LCGP satisfying the above conditions.*

Proof. The existence of l is clear from the fact that the arms of LCGPs must be at least A long (see also Fig. 3). By definition, the arms of two different maximal LCGPs with the same center cannot overlap. Thus, for each l, there exists at most one LCGP whose left and right arms contain sampled positions $i - l \cdot A$ and $j + l \cdot A$, respectively. This completes the proof. $\qquad\square$

Let l ($1 \leq l \leq \frac{d}{A}$) be the smallest integer such that $i - l \cdot A$ (resp. $j + l \cdot A$) is contained in the left arm (resp. the right arm) of the longest suffix inward-maximal LCGP of $w[1..k]$ that is centered at $\frac{i+j}{2}$, and let a_l be the length of the arm of this LCGP. Also, let i_l, j_l be the ending position of the left arm and the beginning position of the right arm of this LCGP, respectively. Note $\frac{i_l + j_l}{2} = \frac{i+j}{2}$ and $j_l + a_l - 1 = k$. Depending on the next character $w[k+1]$, we have two cases:

----

[4] Since the gap length varies in range $[g_{\min}, g_{\max}]$, we here consider both outward and inward maximality of the arms.

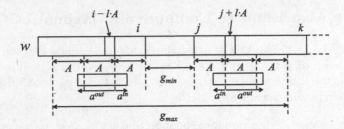

Fig. 3. Illustration for Lemma 1. Since any LCGP centered at $\frac{i+j}{2}$ with gap length in range $[g_{\min}, g_{\max}]$ contains a pair $(i - l \cdot A, j + l \cdot A)$ of sampled positions for some l, we can compute it by two LCEs from the sampled positions.

1. If $w[i_l - a_l] = w[k+1]$, then $(i_l, j_l, a_l + 1)$ is the longest suffix inward-maximal LCGP of $w[1..k+1]$ centered at $\frac{i+j}{2}$. Thus, we naïvely extend the arm length outward by $a_l \leftarrow a_l + 1$, and proceed to the forthcoming character by updating $k \leftarrow k+1$.
2. If $w[i_l - a_l] \neq w[k+1]$, then it appears that (i_l, j_l, a_l) is a maximal LCGP centered at $\frac{i+j}{2}$ and ending at position k, and hence we output it. To compute other maximal LCGPs centered at $\frac{i+j}{2}$, we do the following: We update $l \leftarrow l+1$, and consider a pair $(i - l \cdot A, j + l \cdot A)$ of the sampled positions and compute the outward LCE $a_l^{out} = rlce_{w[1..k+1]}(i - l \cdot A, j + l \cdot A)$ and the inward LCE $a_l^{in} = rlce_{w[1..k+1]}(j + l \cdot A - 1, i - l \cdot A + 1)$ from these sampled positions (see also Fig. 3). There are three sub-cases depending on the LCE values:
   (a) If $a_l^{out} + a_l^{in} < A$ or $a_l^{in} > l \cdot A$, then there is no maximal LCGP with gap length in range $[g_{\min}, g_{\max}]$ that is centered at $\frac{i+j}{2}$ and contains the sampled positions $i - l \cdot A$ and $j + l \cdot A$. We update $l \leftarrow l+1$, and go to one of the following sub-cases.
      i. If $l \leq \frac{d}{A}$, then we compute the outward and inward LCEs from the pair of sampled positions with l.
      ii. If $l > \frac{d}{A}$, then there is no suffix gapped palindrome of $w[1..k]$ that is centered at $\frac{i+j}{2}$ and has a gap length in range $[g_{\min}, g_{\max}]$. We therefore update $i \leftarrow i+1$, $j \leftarrow j+1$, $l \leftarrow 1$, $k \leftarrow k+1$ and proceed to the forthcoming character.
   (b) If $a_l^{out} + a_l^{in} \geq A$, $a_l^{in} \leq l \cdot A$, and $j + l \cdot A + a_l^{out} \leq k$, then $(i_l, j_{'l}, a_l)$ is a maximal LCGP centered at $\frac{i+j}{2}$ where $i_l = i - l \cdot A + a_l^{in}$, $j + l \cdot A + a_l^{out}$, and $a_l = a_l^{out} + a_l^{in}$. We output it and update $l \leftarrow l + 1 + \lfloor \frac{a_l^{out}}{A} \rfloor$ (this is to skip the subsequent sampled positions which are also contained in the same LCGP due to Lemma 1).
      i. If $l \leq \frac{d}{A}$, then we compute the outward and inward LCEs from the pair of sampled positions with l.
      ii. If $l > \frac{d}{A}$, then there is no inward-maximal suffix gapped palindrome of $w[1..k]$ that is centered at $\frac{i+j}{2}$ and has a gap length in range $[g_{\min}, g_{\max}]$. We therefore update $i \leftarrow i+1$, $j \leftarrow j+1$, $l \leftarrow 1$, $k \leftarrow k+1$ and proceed to the forthcoming character.

(c) If $a_l^{out} + a_l^{in} \geq A$, $a_l^{in} \leq l \cdot A$, and $j + l \cdot A + a_l^{out} = k+1$, then (i_l, j_l, a_l) is an inward-maximal gapped suffix palindrome of $w[1..k+1]$ with gap length in range $[g_{min}, g_{max}]$. Moreover, since we have processed l in increasing order, it is guaranteed that (i_l, j_l, a_l) is the longest such one. Hence, we proceed to the next character by updating $k \leftarrow k + 1$.

Theorem 5. *For a growing string to which new characters are appended, we can compute all LCGPs in an online manner, in $O(n(m + \log \sigma))$ time and $O(n)$ space, where n is the length of the final string and $m = \max\{\frac{g_{max} - g_{min}}{A}, 1\}$.*

Proof. The correctness should be clear from the above arguments.

For each $k = 1, \ldots, n$, we consider a fixed center $\frac{i+j}{2}$ and compute all LCGPs with this center. We perform at most $\frac{2d}{A}$ LCE queries for each k, as there are $\frac{d}{A}$ sampled positions for each k. Since each LCE query can be answered in $O(1)$ time as in the proof of Theorem 3, the total time cost of the LCE queries for all $k = 1, \ldots, n$ is $O(\frac{d}{A}n) = O(mn)$. We use additional $O(n \log \sigma)$ time to maintain the suffix tree augmented with the dynamic LCA data structure for bidirectionally growing string $w_k' = (w[1..k])^R \# w[1..k]$. Thus the total time complexity is $O(n(m + \log \sigma))$.

The total space requirement is dominated by the suffix tree and the dynamic LCA data structure, and hence is $O(n)$. □

4.2 Optimality of our Algorithm

The following corollary is immediate from Theorem 5.

Corollary 1. *For constant parameters g_{min}, g_{max}, A and a constant-size alphabet, we can compute all maximal LCGPs in a string of length n in an online manner, in optimal $O(n)$ time and space.*

We can show that even for non-constant gap constraints g_{min} and g_{max}, the running-time of our algorithm is optimal in the worst case. For any string w, let L_w denote the number of all maximal LCGPs in w w.r.t. given parameters g_{min}, g_{max}, and A. It immediately follows from Lemma 1 that L_w is upper-bounded by the total number of sampled positions in w. Hence $L_w = O(mn)$, where $n = |w|$ and $m = \max\{\frac{g_{max} - g_{min}}{A}, 1\}$. It is also true that there is an instance w for which $L_w = \Omega(mn)$ if A is a constant: For example, consider string $z = (abc)^{\frac{n}{3}}$. This string z contains maximal gapped palindromes of form $a(bc(abc)^p)a$ with arm a, $b(c(abc)^p a)b$ with arm b, and $c((abc)^p ab)c$ with arm c for all $0 \leq p \leq \frac{n}{3} - 2$. Thus, for $A = 1$ and for any $2 \leq g_{min} \leq g_{max}$, the string z contains $L_z = \Theta((g_{max} - g_{min})n) = \Theta(\frac{g_{max} - g_{min}}{A}n) = \Theta(mn)$ maximal LCGPs. Hence the running time $O(m(n + \log \sigma))$ of our algorithm is optimal in the worst case, for a constant-size alphabet.

5 Conclusions

In this paper, we presented an online algorithm which finds all maximal g-gapped palindromes occurring in a string w of length n in $O(n \log \sigma)$ time, where σ is the alphabet size. We also showed that the above online algorithm can be extended to find more general length-constrained gapped palindromes (LCGPs) occurring in w in $O(n(\frac{g_{\min}-g_{\max}}{A} + \log\sigma))$ time, for given parameters $2 \leq g_{\min} \leq g_{\max}$ and $A \geq 1$. We also showed that if A is a constant, then there exists a string which contains $\Omega((g_{\min}-g_{\max})n)$ maximal LCGPs. This implies that for a constant-size alphabet the running time of our algorithm is optimal in the worst case.

To our knowledge, the proposed methods are the first online algorithms to find any kind of gapped palindromes in strings. Therefore, there remain many open problems. In particular, we are interested in the following:

- Is there a string of length n which contains $\Omega(\frac{g_{\min}-g_{\max}}{A}n)$ maximal LCGPs for *non-constant A*?
- Can we reduce the $n\frac{g_{\min}-g_{\max}}{A}$ factor to L_w in the $O(n(\frac{g_{\min}-g_{\max}}{A} + \log\sigma))$-time algorithm for finding all maximal LCGPs, thereby obtaining an optimal algorithm?
- Can the *maximal α-gapped palindromes* [5] of a given string be computed online efficiently?

References

1. Apostolico, A., Breslauer, D., Galil, Z.: Parallel detection of all palindromes in a string. Theor. Comput. Sci. **141**(1&2), 163–173 (1995)
2. Bender, M.A., Farach-Colton, M.: The LCA problem revisited. In: Gonnet, G.H., Viola, A. (eds.) LATIN 2000. LNCS, vol. 1776, pp. 88–94. Springer, Heidelberg (2000)
3. Cole, R., Hariharan, R.: Dynamic LCA queries on trees. SIAM J. Comput. **34**(4), 894–923 (2005)
4. Farach-Colton, M., Ferragina, P., Muthukrishnan., S.: On the sorting-complexity of suffix tree construction. J. ACM **47**(6), 987–1011 (2000)
5. Gawrychowski, P., Tomohiro, I., Inenaga, S., Köppl, D., Manea, F.: Efficiently finding all maximal α-gapped repeats. In: STACS 2016 (to appear, 2016). http://arxiv.org/abs/1509.09237
6. Gusfield, D.: Algorithms on Strings, Trees, and Sequences. Cambridge University Press, New York (1997)
7. Inenaga, S.: Bidirectional construction of suffix trees. Nord. J. Comput. **10**(1), 52 (2003)
8. Kolpakov, R., Kucherov, G.: Searching for gapped palindromes. Theor. Comput. Sci. **410**(51), 5365–5373 (2009)
9. Kosolobov, D., Rubinchik, M., Shur, A.M.: Finding distinct subpalindromes online. In: PSC 2013, pp. 63–69 (2013)
10. Manacher, G.K.: A new linear-time on-line algorithm for finding the smallest initial palindrome of a string. J. ACM **22**(3), 346–351 (1975)
11. Ukkonen, E.: On-line construction of suffix trees. Algorithmica **14**(3), 249–260 (1995)
12. Weiner, P.: Linear pattern matching algorithms. In: 14th Annual Symposium on Switching and Automata Theory, pp. 1–11 (1973)

Advice Complexity of the Online Search Problem

Jhoirene Clemente[1]([⊠]), Juraj Hromkovič[2], Dennis Komm[2],
and Christian Kudahl[3]

[1] Department of Computer Science, University of the Philippines Diliman,
Quezon City, Philippines
jbclemente@up.edu.ph
[2] Department of Computer Science, ETH Zürich, Zürich, Switzerland
{juraj.hromkovic,dennis.komm}@inf.ethz.ch
[3] Department of Mathematics and Computer Science,
University of Southern Denmark, Odense, Denmark
kudahl@imada.sdu.dk

Abstract. The online search problem is a fundamental problem in finance. The numerous direct applications include searching for optimal prices for commodity trading and trading foreign currencies. In this paper, we analyze the advice complexity of this problem. In particular, we are interested in identifying the minimum amount of information needed in order to achieve a certain competitive ratio. We design an algorithm that reads b bits of advice and achieves a competitive ratio of $(M/m)^{1/(2^b+1)}$ where M and m are the maximum and minimum price in the input. We also give a matching lower bound. Furthermore, we compare the power of advice and randomization for this problem.

1 Introduction

We study the online search problem (abbreviated ONLINE SEARCH), which is formulated as an online (profit) maximization problem. For such problems, the input arrives gradually in consecutive time steps. Each piece of input is called a *request*. After a request is given, an online algorithm (also called the *online player*) has to produce a definite piece of the output, called an *answer*. Each answer is thus computed without any knowledge about further requests [6]. The goal is to produce an output with a *profit* that is as large as possible. In ONLINE SEARCH, the online player searches for the maximum price of a certain asset that unfolds sequentially. Suppose the player, in this context a trader, would like to transfer its assets from, say, USD to CHF in one transaction. Each day (formally, each time step), the trader receives a quotation of the current exchange rate and

J. Clemente—Supported by ERDT Scholarship. Sandwich program funded by PCIEERD-BCDA.
J. Hromkovič—Supported by SNF grant 200021-146372.
C. Kudahl—Supported by the Villum Foundation and the Stibo-Foundation.

© Springer International Publishing Switzerland 2016
V. Mäkinen et al. (Eds.): IWOCA 2016, LNCS 9843, pp. 203–212, 2016.
DOI: 10.1007/978-3-319-44543-4_16

decides whether to trade on the same day or to wait. The trading duration is finite, and it may be known or unknown to the trader. Formally, we define ONLINE SEARCH as follows.

Definition 1 (Online Search Problem). *Let $\sigma = (p_1, p_2, \ldots, p_n)$, with $0 < m \leq p_i \leq M$ for all $1 \leq i \leq n$, be a sequence of prices that arrives in an online fashion. Here, M and m are upper and lower bounds on the prices, respectively. For each day i, price p_i is revealed, and the online player has to choose whether to trade on the same day or to wait for the new price quotation on the next day. If the player trades on day i, its profit is p_i. If the player did not trade for the first $n - 1$ days, it must accept p_n. The player's goal is to maximize the obtained price (i. e., its profit).*

We assume that the parameters m and M for the price range are fixed and known to the online algorithm in advance. The duration of the trading period n is finite and may or may not be known to the online algorithm. We do not take into account sampling costs in the profit, i. e., the price for each day is freely given by the market to the trader. However, some direct applications of ONLINE SEARCH may do require to consider the sampling costs. For instance, obtaining prices of a certain product may induce some cost, either in the form of time or money, from the player. For a study of such more involved cost variants, we refer the reader to Xu et al. [19], where the authors considered the accumulated sampling cost while maximizing the player's profit.

1.1 Competitive Analysis and Advice Complexity

Competitive analysis was introduced by Sleator and Tarjan in 1985 [18] to analyze the solution quality of online algorithms. The measure used in the analysis is called the *competitive ratio*, which can be obtained by comparing the profit of the online algorithm to the one of an optimal offline solution. The term "offline" is used when the whole input sequence is known in advance. Note that it is generally not possible for an online algorithm to compute the optimal offline solution in advance, because parts of the output have to be specified before the whole input is known. It is merely taken into account to analyze the profit that can hypothetically be obtained if the whole input is known in advance. The competitive ratio of an online algorithm is formally defined as follows.

Definition 2 (Competitive Ratio). *Let Π be an online maximization problem, let* ALG *be an online algorithm for Π, and let $c > 1$.* ALG *is said to be c-competitive if, for every instance I of Π, we have*

$$c \cdot \text{profit}(\text{ALG}(I)) \geq \text{profit}(\text{OPT}(I)),$$

where $\text{profit}(\text{ALG}(I))$ is the profit of ALG *on input I, and $\text{profit}(\text{OPT}(I))$ denotes the optimal offline profit.*

In this paper, we study the *advice complexity* of ONLINE SEARCH. More specifically, we ask about the additional information both sufficient and necessary in order to improve the obtainable competitive ratio. In a way, this approach can be seen as measuring the *information content* of the problem at hand [13].

This tool, which was introduced by Dobrev et al. in 2008 [9] and then revised by Böckenhauer et al. [4], Hromkovič et al. [13], and Emek et al. [11], is a complementary tool to analyze online problems. In order to study the information that is needed in order to outperform purely deterministic (or randomized) online algorithms, we introduce a trusted source, referred to as an *oracle*, which sees the whole input in advance and may write binary information on a so-called *advice tape*. These *advice bits* are allowed to be any function of the entire input. The algorithm, which is called an *online algorithm with advice* in this setting, may then use the advice to compute the output for the given input. The approach is quantitative and problem-independent. In other words, the information supplied can be arbitrary (as long as it is computable). This is in particular interesting to give lower bounds for many other measurements or relaxations of online problems. More specifically, hardness results in advice complexity give useful negative results about various semi-online approaches. If it is for example shown that $O(\log_2 n)$ bits of advice do not help any online algorithm to achieve a better competitive ratio, this gives a negative answer to questions of the form: Would it help the algorithm to know the length of the input? Would it help the algorithm to know the number of requests of a certain type?

Many prominent online problems have been studied in this framework, including paging [4,9], the k-server problem [4,11,12,17], metrical task systems [11], and the online knapsack problem [5]. Negative results on the advice complexity can be transferred by a special kind of reduction [2,8,11]. Moreover, advice complexity has a close and non-trivial relation to randomization [1,3,14,16]. We now define online algorithms with advice formally.

Definition 3 (Advice Complexity). *Let $x_1, \ldots, x_n$ be the input for an online problem Π. An online algorithm with advice, ALG, for Π computes the output sequence $y_1, \ldots, y_n$, where y_i is allowed to depend on $x_1, \ldots, x_{i-1}$ as well as on an advice string ϕ. The advice, ϕ, is written in binary on an infinite tape and is allowed to depend on the request sequence $x_1, \ldots, x_n$. The advice complexity of ALG is the largest number of advice bits it reads from ϕ over all inputs of length at most n.*

Our paper is devoted to both creating online algorithms with advice for ONLINE SEARCH that achieve a certain output quality while using a certain number of advice bits, and to show that such algorithms cannot exist if the advice complexity is below some certain threshold.

Most of the work in advice complexity theory considers problems where at least n advice bits are required for an algorithm to be optimal. Here, we study a problem where only $\log_2 n$ bits give an optimal algorithm. We investigate how this problem behaves when the number of advice bits is in the interval $[1, \log_2 n]$. For the ease of presentation, we assume that $\log_2 n$ is integer.

2 Related Work

The search problem in an offline setting, i. e., where the set of prices is known in advance, can easily be solved optimally in time $O(n)$. However, for a lot of online environments such as stock trading and foreign exchange, decisions should be made even though there is no knowledge of the future prices of the currencies. These problems are intrinsically online.

The most common approaches are Bayesian. These approaches rely on a prior distribution of prices where the online algorithm computes a certain *reservation price* based on the distribution. The trader accepts any price that is larger than or equal to the reservation price. If this certain price is not met, the player has to trade on the last day (according to Definition 1). Throughout this paper, ALG[p] denotes the algorithm that accepts the first price it sees that is at least p.

Since the prior distribution of prices is not necessarily known in advance, El-Yaniv et al. [10] proposed to measure the quality of online trading algorithms using competitive analysis. Moreover, for some assets, the goal is not just to increase the profit but to minimize the loss by considering the possible worst-case scenarios in the market. Competitive analysis in financial problems such as ONLINE SEARCH can provide a guaranteed performance measure for the trader's profit. The best deterministic online algorithm with respect to competitive analysis is ALG[$\sqrt{Mm}$], i. e., the algorithm that accepts the first price it sees that is at least $\sqrt{Mm}$ (or it accepts p_n if no such price is ever seen). This algorithm has a competitive ratio of $\sqrt{M/m}$, which is provably the best competitive ratio any deterministic online algorithm without advice can achieve [6].

Boyar et al. [7] studied how the problem behaves when applying a variety of difference performance measures (and not just competitive ratio).

3 Advice for the Online Search Problem

In this section, we explore the advice complexity of ONLINE SEARCH. We start by studying how much advice is necessary and sufficient in order to obtain an optimal output. After that, we study general c-competitiveness.

3.1 Advice for Optimality

It is possible for an algorithm to be optimal using $\log_2 n$ bits of advice if n is known in advance by simply encoding the day where the largest price is offered. If n is not known in advance, it has to be encoded with a self-delimiting encoding, for example, by writing the length of $\log_2 n$ in unary followed by $\log_2 n$. This requires $2 \log_2 n$ bits [4].

Moreover, optimality can also be achieved by encoding the value of $p_{\max}$ using $O(\log_2(M/m))$ bits, but since M and m can be arbitrarily large, this may be very expensive. We now give a complementing lower bound.

Theorem 1. *At least $\log_2 n$ bits of advice are necessary to obtain an optimal solution for* ONLINE SEARCH. *This holds even if n is known to the algorithm.*

Proof. We use that an algorithm with b advice bits can be viewed as dealing with the best of 2^b algorithms without advice, for the particular instance chosen. First, we generate a set of request sequences $\mathcal{S}$. Then, we show that, for $\mathcal{S}$, there is no set of $n-1$ or fewer deterministic algorithms, which can ensure that at least one algorithm always gets the optimal solution.

We construct the set $\mathcal{S}$ in such a way that each request has a unique optimal solution. The construction is as follows. Let $\mathcal{S} = \{\sigma_1, \sigma_2, \ldots, \sigma_n\}$, such that

$$\sigma_i = (\underbrace{m + \delta, m + 2\delta, \ldots, m + i\delta}_{i}, \underbrace{m, \ldots, m}_{n-i}),$$

where $\delta = (M - m)/n$. Each σ_i is thus a sequence of n prices that follow an increasing order until day i. Then the price drops to the minimum m for the remaining $n - i$ days. The optimal solution for each σ_i clearly is to trade on day i and obtain a profit of $m + i\delta$. From the construction, it is impossible for any deterministic online algorithm to distinguish the request sequence σ_i from any other sequence of requests σ_j, for $j > i$, until the price for day $i + 1$ is offered. This is due to the fact that the set of requests $\{\sigma_i, \sigma_{i+1}, \ldots, \sigma_n\}$ have the same prices offered from day 1 up to day i. Since we have n such input instances with different optimal solutions, and fewer than n algorithms, there is one algorithm that gets chosen for at least one the above instances. Clearly, this algorithm cannot be optimal for both these instances. Thus, any online algorithm with advice needs $\log_2 n$ bits of advice to identify the actual input from these n possible cases.   □

Note that if it is required that the prices are integral, this construction still works by picking m and M such that δ is an integer.

3.2   Advice for c-Competitiveness

Next, we investigate the advice complexity of ONLINE SEARCH if we have less than $\log_2 n$ advice bits. This means we study a tradeoff between the number b of advice bits supplied and the competitive ratio c obtainable. Recall that, without advice bits, the optimal trader strategy is to use ALG$[p]$, where the reservation price is $p = \sqrt{Mm}$.

Before we present the upper bounds for online algorithms with advice for ONLINE SEARCH that achieve c-competitiveness, we give a simple intuition behind our strategy. We can think of it as having 2^b deterministic algorithms with different reservation prices. The computation of each reservation price p_i is obtained by computing the solution of the following equation.

$$\frac{p_1}{m} = \frac{p_2}{p_1} = \ldots = \frac{p_{2^i}}{p_{2^i - 1}} = \ldots = \frac{M}{p_{2^b}}$$

Theorem 2. *For every $b > 0$, there exists an online algorithm with advice for* ONLINE SEARCH *which reads b bits of advice and achieves a competitive ratio of at most $(M/m)^{\frac{1}{2^b+1}}$. This holds even if n is unknown.*

Proof. We describe an algorithm ALG with advice which reads b bits of advice and achieves the claimed competitive ratio. First, the oracle simulates the algorithms

$$\text{ALG}\left[m^{\frac{2^b+1-i}{2^b+1}}M^{\frac{i}{2^b+1}}\right]$$

for $i = 1, \ldots, 2^b$. Let A denote the set of these algorithms. Then, it writes the value of i for the algorithm that achieves the best competitive ratio. We argue that at least one of the algorithms gets a competitive ratio of at most

$$\left(\frac{M}{m}\right)^{\frac{1}{2^b+1}}.$$

We have three cases for $p_{\max}$. The first case is when $p_{\max} < m^{\frac{2^b}{2^b+1}}M^{\frac{1}{2^b+1}}$. Here, each algorithm in A will get the price offered on the last day, which is at least m. The competitive ratio for ALG is at most

$$\frac{m^{\frac{2^b}{2^b+1}}M^{\frac{1}{2^b+1}}}{m} = \left(\frac{M}{m}\right)^{\frac{1}{2^b+1}}.$$

The second case is when $p_{\max} \geq m^{\frac{1}{2^b+1}}M^{\frac{2^b}{2^b+1}}$. In this case,

$$\text{ALG}\left[m^{\frac{1}{2^b+1}}M^{\frac{2^b}{2^b+1}}\right]$$

gets a price of at least $m^{\frac{1}{2^b+1}}M^{\frac{2^b}{2^b+1}}$. Since OPT gets at most M, the competitive ratio for ALG is again at most

$$\frac{M}{m^{\frac{1}{2^b+1}}M^{\frac{2^b}{2^b+1}}} = \left(\frac{M}{m}\right)^{\frac{1}{2^b+1}}.$$

The last case is when $m^{\frac{2^b+1-i}{2^b+1}}M^{\frac{i}{2^b+1}} \leq p_{\max} < m^{\frac{2^b-i}{2^b+1}}M^{\frac{i+1}{2^b+1}}$ for some $i < 2^b$. In this case,

$$\text{ALG}\left[m^{\frac{2^b+1-i}{2^b+1}}M^{\frac{i}{2^b+1}}\right]$$

gets at least its reservation price. Thus, also here, the competitive ratio for ALG is at most

$$\frac{m^{\frac{2^b-i}{2^b+1}}M^{\frac{i+1}{2^b+1}}}{m^{\frac{2^b+1-i}{2^b+1}}M^{\frac{i}{2^b+1}}} = \left(\frac{M}{m}\right)^{\frac{1}{2^b+1}}.$$

All in all, we have shown that, in each case, ALG obtains a competitive ratio of at most $(M/m)^{\frac{1}{2^b+1}}$ as we claimed. $\square$

We now present a matching lower bound.

Theorem 3. *Let* ALG *be an algorithm with advice for* ONLINE SEARCH *which reads* $b < \log_2 n$ *bits of advice. The competitive ratio of* ALG *is at least* $(M/m)^{\frac{1}{2^b+1}}$.

Proof. For any given $b < \log_2 n$, let ALG be an algorithm with advice that reads at most b bits of advice. Again, we view this advice as 2^b deterministic online algorithms. We now give a class of request sequences that ensure that each of them gets a competitive ratio of at least

$$\left(\frac{M}{m}\right)^{\frac{1}{2^b+1}}.$$

Consider the sequence $(p_1, p_2, \ldots, p_{2^b})$ with

$$p_i = m^{\frac{2^b+1-i}{2^b+1}} M^{\frac{i}{2^b+1}}.$$

The adversary simulates all 2^b algorithms on this sequence. We consider two cases. If a request p_i is rejected by all algorithms, it requests $p_1, p_2, \ldots, p_i$ followed by requests that are all equal to m. For the first case, assume that there exists a request p_i, which is rejected by all 2^b algorithms. The remaining requests are all m. This means that ALG gets a price of at most p_{i-1} (the largest request that was not p_i) while OPT gets a price of p_i. Note that, if the first request is rejected, ALG gets a price of at most $m = p_0$. In this case, the competitive ratio for ALG is at least

$$\frac{p_i}{p_{i-1}} = \frac{m^{\frac{2^b+1-i}{2^b+1}} M^{\frac{i}{2^b+1}}}{m^{\frac{2^b+2-i}{2^b+1}} M^{\frac{i-1}{2^b+1}}} = \left(\frac{M}{m}\right)^{\frac{1}{2^b+1}}.$$

Thus, ALG cannot obtain a competitive ratio which is better than $(M/m)^{\frac{1}{2^b+1}}$ if a request is rejected by all the algorithms.

Next, we consider the second case. Here, every request in σ is accepted by some algorithm. Since there are 2^b requests in σ, it follows that all algorithms accept a price that is at most p_{2^b}. Since $2^b < n$, the adversary can still make a request. The final request is then M. The competitive ratio for ALG is therefore bounded from below by

$$\frac{M}{p_{2^b}} = \frac{M}{m^{\frac{1}{2^b+1}} M^{\frac{2^b}{2^b+1}}} = \left(\frac{M}{m}\right)^{\frac{1}{2^b+1}}.$$

In both cases, ALG has a competitive ratio of at least $(M/m)^{\frac{1}{2^b+1}}$ as claimed by the theorem. □

4   Advice and Randomization

Randomization is often used to improve the competitive ratio of online algorithms (in expectation). Here, the online player is allowed to base some of its answers on a random source. An oblivious adversary knows the algorithm, but not the outcome of the random decisions. To provide an improvement over the lower bound of deterministic online algorithms for ONLINE SEARCH, El-Yaniv et al. [10] provided an upper bound by presenting a randomized algorithm with an expected competitive ratio of $\log_2(M/m)$. Lorenz et al. [15] provided an asymptotically matching lower bound of $(\log_2(M/m))/2$ for randomized online algorithms for ONLINE SEARCH.

In this section, we compare the power of advice to the ability of an online algorithm to access random bits for ONLINE SEARCH. The competitive ratio of online algorithms with advice (with an increasing number of advice bits) is shown in Fig. 1. We fixed a fluctuation ratio M/m, and we highlighted the competitive ratio of the best deterministic algorithm, i. e., $(M/m)^{\frac{1}{2}}$, and the corresponding upper (i. e., $\log_2(M/m)$) and lower (i. e., $\log_2(M/m)/2$) bounds of randomized algorithms for ONLINE SEARCH.

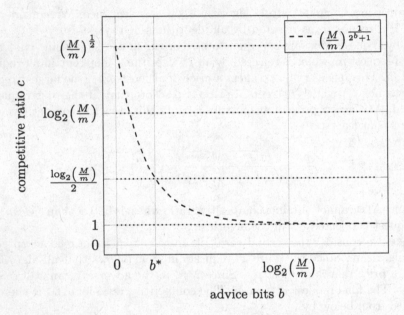

Fig. 1. Plot comparing the competitive ratio of the online algorithm with advice with respect to the lower bound for deterministic and randomized algorithms.

It is interesting to point out that, with the number of advice bits greater than

$$b^* = \log_2\left(\frac{\log_2(M/m)}{\log_2\left(\frac{\log_2(M/m)}{2}\right)} - 1\right),$$

our online algorithm for ONLINE SEARCH outperforms the lower bound of randomized online algorithms. And as we increase the number of advice bits, the better the competitive ratio we get. In the plot shown in Fig. 1, we considered a fluctuation ratio $M/m = n$. Note that the competitive ratio is asymptotic to 1, but it is actually possible to get an optimal solution with $\log_2 n$ advice bits.

5 Conclusion and Future Work

We studied the advice complexity of ONLINE SEARCH and determined upper and lower bounds on the advice complexity to achieve both optimality and c-competitiveness. We presented a tight lower bound of $\log_2 n$ for the number of advice needed by any online algorithm to obtain optimal solutions, as shown in Theorem 1. We also provided a strategy with b bits of advice and achieved a tight bound of $(M/m)^{\frac{1}{2^b+1}}$ for the competitive ratio as shown in Theorems 2 and 3.

We compared the power of advice and randomization in terms of competitive ratio. The comparison of the competitive ratio is shown in Fig. 1.

For future work, it would be interesting to extend the results to the ONE-WAY TRADING problem with advice. It is known that ONLINE SEARCH and the ONE-WAY TRADING are closely related. In fact, they are equivalent in the sense that, for every randomized algorithm for ONLINE SEARCH, there exists an equivalent deterministic algorithm for ONE-WAY TRADING [10]. Although randomization significantly improved the competitive ratio of algorithms for ONLINE SEARCH, it can be shown that it cannot help to improve the competitive ratio of algorithms for ONE-WAY TRADING. It would be interesting to investigate the tradeoff between advice and competitive ratio in ONE-WAY TRADING.

References

1. Barhum, K., Böckenhauer, H.-J., Forišek, M., Gebauer, H., Hromkovič, J., Krug, S., Smula, J., Steffen, B.: On the power of advice and randomization for the disjoint path allocation problem. In: Geffert, V., Preneel, B., Rovan, B., Štuller, J., Tjoa, A.M. (eds.) SOFSEM 2014. LNCS, vol. 8327, pp. 89–101. Springer, Heidelberg (2014)
2. Böckenhauer, H.-J., Hromkovič, J., Komm, D., Krug, S., Smula, J., Sprock, A.:The string guessing problem as a method to prove lower bounds on the advice complexity.Theor. Comput. Sci. 554, 95–108 (2014). Elsevier Science Publishers
3. Böckenhauer, H.-J., Komm, D., Královič, R., Královič, R.: On the advice complexity of the k-server problem. In: Du, D.-Z., Zhang, G. (eds.) ICALP 2011. LNCS, vol. 6755, pp. 207–218. Springer, Heidelberg (2013)
4. Böckenhauer, H.-J., Komm, D., Královič, R., Královič, R., Mömke, T.: On the advice complexity of the online problem. In: Dong, Y., Du, D.-Z., Ibarra, O. (eds.) ISAAC 2009. LNCS, vol. 5878, pp. 331–340. Springer, Heidelberg (2013)
5. Böckenhauer, H.-J., Komm, D., Královič, R., Rossmanith, P.: The online knapsack problem: advice and randomization. Theor. Comput. Sci. 527, 61–72 (2014)
6. Borodin, A., El-Yaniv, R.: Online Computation and Competitive Analysis. Cambridge University Press, New York (1998)

7. Boyar, J., Larsen, K.S., Maiti, A.: A comparison of performance measures via online search. In: Snoeyink, J., Lu, P., Su, K., Wang, L. (eds.) FAW-AAIM 2012. LNCS, vol. 7285, pp. 303–314. Springer, Heidelberg (2012)
8. Boyar, J., Favrholdt, L.M., Kudahl, C., Mikkelsen, J.W.: Advice complexity for a class of online problems. In: Proceedings of the 32nd Symposium on Theoretical Aspects of Computer Science (STACS 2015). Leibniz International Proceedings in Informatics, vol. 30, pp. 116–129. Schloss Dagstuhl – Leibniz-Zentrum für Informatik (2015)
9. Dobrev, S., Královič, R., Pardubská, D.: How much information about the future is needed? In: Geffert, V., Karhumäki, J., Bertoni, A., Preneel, B., Návrat, P., Bieliková, M. (eds.) SOFSEM 2008. LNCS, vol. 4910, pp. 247–258. Springer, Heidelberg (2008)
10. El-Yaniv, R., Fiat, A., Karp, R., Turpin, G.: Optimal search and one-way trading online algorithms. Algorithmica **30**, 101–139 (2001)
11. Emek, Y., Fraigniaud, P., Korman, A., Rosén, A.: Online computation with advice. Theor. Comput. Sci. **412**(24), 2642–2656 (2011)
12. Gupta, S., Kamali, S., López-Ortiz, A.: On advice complexity of the k-server problem under sparse metrics. In: Moscibroda, T., Rescigno, A.A. (eds.) SIROCCO 2013. LNCS, vol. 8179, pp. 55–67. Springer, Heidelberg (2013)
13. Hromkovič, J., Královič, R., Královič, R.: Information complexity of online problems. In: Hliněný, P., Kučera, A. (eds.) MFCS 2010. LNCS, vol. 6281, pp. 24–36. Springer, Heidelberg (2010)
14. Komm, D., Královič, R.: Advice complexity and barely random algorithms. Theor. Inform. Appl. (RAIRO) **45**(2), 249–267 (2011)
15. Lorenz, J., Panagiotou, K., Steger, A.: Optimal algorithms for k-search with application in option pricing. Algorithmica **55**(2), 311–328 (2009)
16. Mikkelsen, J.: Randomization can be as helpful as a glimpse of the future in online computation. CoRR, abs/1511.05886 (2015)
17. Renault, M.P., Rosén, A.: On Online Algorithms with Advice for the k-Server Problem. In: Solis-Oba, R., Persiano, G. (eds.) WAOA 2011. LNCS, vol. 7164, pp. 198–210. Springer, Heidelberg (2012)
18. Sleator, D.D., Tarjan, R.E.: Amortized efficiency of list update and paging rules. Commun. ACM **28**(2), 202–208 (1985)
19. Xu, Y., Zhang, W., Zheng, F.: Optimal algorithms for the online time series search problem. Theoret. Comput. Sci. **412**(3), 192–197 (2011)

Packed Compact Tries: A Fast and Efficient Data Structure for Online String Processing

Takuya Takagi[1]([✉]), Shunsuke Inenaga[2], Kunihiko Sadakane[3], and Hiroki Arimura[1]

[1] Graduate School of IST, Hokkaido University, Sapporo, Japan
{tkg,arim}@ist.hokudai.ac.jp
[2] Department of Informatics, Kyushu University, Fukuoka, Japan
inenaga@inf.kyushu-u.ac.jp
[3] Graduate School of Information Science and Technology,
University of Tokyo, Tokyo, Japan
sada@mist.i.u-tokyo.ac.jp

Abstract. We present a new data structure called the *packed compact trie* (*packed c-trie*) which stores a set S of k strings of total length n in $n \log \sigma + O(k \log n)$ bits of space and supports fast pattern matching queries and updates, where σ is the alphabet size. Assume that $\alpha = \log_\sigma n$ letters are packed in a single machine word on the standard word RAM model, and let $f(k, n)$ denote the query and update times of the dynamic predecessor/successor data structure of our choice which stores k integers from universe $[1, n]$ in $O(k \log n)$ bits of space. Then, given a string of length m, our packed c-tries support pattern matching queries and insert/delete operations in $O(\frac{m}{\alpha} f(k, n))$ worst-case time and in $O(\frac{m}{\alpha} + f(k, n))$ expected time. Our experiments show that our packed c-tries are faster than the standard compact tries (a.k.a. Patricia trees) on real data sets. We also discuss applications of our packed c-tries.

1 Introduction

The trie for a set S of strings of total length n is a classical data structure which occupies $O(n \log n + n \log \sigma)$ bits of space and allows for prefix search and insertion/deletion for a given string of length m in $O(m \log \sigma)$ time, where σ is the alphabet size. The *compact trie* for S is a path-compressed trie where the edges in every non-branching path are merged into a single edge [16]. By representing each edge label by a pair of positions in a string in S, the compact trie can be stored in $n \log \sigma + O(k \log n)$ bits of space, where k is the number of strings in S, retaining the same time efficiency for prefix search and insertion/deletion for a given string. Thus, compact tries have widely been used in numerous applications such as dynamic dictionary matching [12], suffix trees [19], sparse suffix trees [15], external string indexes [8], and grammar-based text compression [11].

In this paper, we show how to accelerate prefix search queries and update operations of compact tries on the standard word RAM model with machine word size $w = \log n$, still keeping $n \log \sigma + O(k \log n)$-bit space usage. A basic idea is

© Springer International Publishing Switzerland 2016
V. Mäkinen et al. (Eds.): IWOCA 2016, LNCS 9843, pp. 213–225, 2016.
DOI: 10.1007/978-3-319-44543-4_17

to use the *packed string matching* approach [5], where $\alpha = \log_\sigma n$ consecutive letters are packed in a single word and can be manipulated in $O(1)$ time. In this setting, we can read a given pattern P of length m in $O(\frac{m}{\alpha})$ time, but, during the traversal of P over a compact trie, there can be at most m branching nodes. Thus, a naïve implementation of a compact trie takes $O(\frac{m}{\log_\sigma n} + m \log \sigma) = O(m \log \sigma)$ time even in the packed matching setting.

To overcome the above difficulty, we propose how to quickly process long non-branching paths using bit manipulations, and how to quickly process dense branching subtrees using fast predecessor/successor queries and dictionary lookups. As a result, we obtain a new compact trie called the *packed compact trie* (*packed c-trie*) for a dynamic set S of strings with the following efficiency:

Theorem 1 (main result). *Let $f(k, n)$ be the query/update times of an arbitrary dynamic predecessor/successor data structure using $O(k \log n)$ bits of space for a dynamic set of k integers from the universe $[1, n]$. Our packed c-trie stores a set S of k strings of total length n in $n \log \sigma + O(k \log n)$ bits of space and supports prefix search and insertion/deletion for a given string of length m in $O(\frac{m}{\alpha} f(k, n))$ worst-case time or in $O(\frac{m}{\alpha} + f(k, n))$ expected time.*

Using Beame and Fich's data structure [3] or Willard's y-fast trie [20] as the dynamic predecessor/successor data structure, we obtain the following corollary:

Corollary 1. *There exists a packed c-trie for a dynamic set S of strings which uses $n \log \sigma + O(k \log n)$ bits of space, and supports prefix search and insert/delete operations for a given string of length m in $O(\frac{m}{\alpha} \cdot \frac{\log \log k \log \log n}{\log \log \log n})$ worst-case time or in $O(\frac{m}{\alpha} + \log \log n)$ expected time.*

Unlike most other (compact) tries, our packed c-trie does *not* maintain a dictionary or a search structure for the children of each node. Instead, we partition our c-trie into $\lceil h/\alpha \rceil$ levels, where h is the length of the longest string in S. Then each subtree of height α, called a *micro c-trie*, maintains a predecessor/successor dictionary that processes prefix search inside the micro c-trie. A reduction from prefix search to predecessor/successor queries was already considered in an earlier work by Cole et al. [6], however, their data structure is static. On the other hand, our micro c-tries are dynamic. A similar technique to our packed c-trie was used in the linked dynamic uncompacted trie by Jansson et al. [14].

Our experiments show that our packed c-tries are faster than Patricia trees for both construction and prefix search in almost all data sets we tested.

We show that our packed c-tries can be applied to efficient online construction of *evenly sparse suffix trees* [15], *word suffix trees* [13] and its extension [17]. Also, packed c-tries can be used for online computation of the *LZ-Double factorization* [11] (*LZDF*), a state-of-the-art online grammar-based text compressor.

Related Work. Belazzougui et al. [4] proposed a *randomized* compact trie called the *signed dynamic z-fast trie*, which stores a dynamic set S of k strings in $n \log \sigma + O(k \log n)$ bits of space. Given a string of length m, the signed dynamic z-fast trie supports prefix search in $O(\frac{m}{\alpha} + \log m)$ worst-case time *only with high probability*, and supports insert/delete operations in $O(\frac{m}{\alpha} + \log m)$ expected time *only with high probability*.[1] On the other hand, our packed c-trie always return the correct answer for prefix search, and always insert/delete a given string correctly, in the bounds stated in Theorem 1 and Corollary 1.

Andersson and Thorup [2] proposed the *exponential search tree* which uses $n \log \sigma + O(k \log n)$ bits of space, and supports prefix search and insert/delete operations in $O(m + \sqrt{\frac{\log k}{\log \log k}})$ worst-case time. Each node v of the exponential search tree stores a constant-time look-up dictionary for some children of v and a dynamic predecessor/successor data structure for the other children of v. This implies that given a string of length m, at most m nodes in the search path for the string must be processed one by one, and hence packing $\alpha = \log_\sigma n$ letters in a single word does not seem to speed-up the exponential search tree.

Fischer and Gawrychowski's *wexponential search tree* [9] proposed uses $n \log \sigma + O(k \log n)$ bits of space, and supports prefix search and insert/delete operations in $O(m + \frac{(\log \log \sigma)^2}{\log \log \log \sigma})$ worst-case time. When $\sigma = \mathrm{polylog}(n)$, our packed c-trie achieves $O(m \frac{\log \sigma \log \log k \log \log n}{\log n \log \log \log n}) = O(m \frac{(\log \log n)^2}{\log n \log \log \log n}) = O(o(1)m)$ worst-case time, while the wexponential search tree requires $O(m + \frac{(\log \log \log n)^2}{\log \log \log \log n})$ time[2].

2   Preliminaries

Let Σ be the alphabet of size σ. An element of Σ^* is called a string. For any string X of length n, $|X|$ denotes its length, namely $|X| = n$. We denote the empty string by ε. For any $1 \le i \le n$, $X[i]$ denotes the ith character of X. For any $1 \le i \le j \le |X|$, $X[i..j]$ denotes the substring $X[i] \cdots X[j]$. For convenience, $X[i..j] = \varepsilon$ for $i > j$. For any strings X, Y, $LCP(X, Y)$ denotes the longest common prefix of X and Y.

Throughout this paper, the base of the logarithms will be 2, unless otherwise stated. For any integers $i \le j$, $[i, j]$ denotes the interval $\{i, i+1, \ldots, j\}$. Our model of computation is the standard word RAM of word size $w = \log n$ bits. For simplicity, we assume that w is a multiple of $\log \sigma$, so $\alpha = \log_\sigma n$ letters are packed in a single word. Since we can read w bits in constant time, we can read and process α consecutive letters in constant time.

Let $S = \{X_1, \ldots, X_k\}$ be a set of k non-empty strings of total length n. We consider dynamic data structures for S allowing for fast prefix searches of given patterns over strings in S, and fast insertion/deletion of strings to/from S.

Suppose S is prefix-free. The *trie* of S is a tree s.t. each edge is labeled by a single letter, the labels of the out edges of each node are distinct, and for each $X_i \in S$ there is a unique leaf ℓ_i s.t. the path from the root to ℓ_i spells out X_i.

---

[1] The $O(\log m)$ expected bound for insertion/deletion stated in [4] assumes that the prefix search for the string has already been performed.

[2] For sufficiently long patterns of length $m = \Theta(n)$, our packed c-trie achieves worst-case *sublinear* $o(n)$ time while the wexponential search tree requires $O(n)$ time.

The *compact trie* $\mathcal{T}_S$ of S is a path-compressed trie obtained by contracting non-branching paths into single edges. Namely, in $\mathcal{T}_S$, each edge is labeled by a non-empty substring of T, each internal node has at least two children, the out-going edges from each node begin with distinct letters, and each edge label x is encoded by a triple $\langle i, a, b \rangle$ such that $x = X_i[a..b]$ for some $1 \leq i \leq k$ and $1 \leq a \leq b \leq |X_i|$. The *length* of an edge e, denoted $|e|$, is the length of its label string. Let $root(\mathcal{T}_S)$ denote the root of the compact trie $\mathcal{T}_S$. For any node v, let $parent(v)$ denotes its parent. For convenience, let $\perp$ be an auxiliary node s.t. $parent(root(\mathcal{T}_S)) = \perp$. We assume the edge from $\perp$ to $root(\mathcal{T}_S)$ is labeled by an arbitrary letter. For any node v, let $str(v)$ denotes the string obtained by concatenating the edge labels from the root to v. Each node v stores $|str(v)|$.

Let s be a prefix of any string in S. Let v be the shallowest node of $\mathcal{T}_S$ such that s is a suffix of $str(v)$ (notice s can be equal to $str(v)$), and let $u = parent(v)$. The *locus* of string s in $\mathcal{T}_S$ is a pair $\phi = (e, h)$, where e is the edge from u to v and h is the offset from u, namely, $h = |s| - |str(u)|$.[3] We extend the str function to locus ϕ, so that $str(\phi) = s$. The *string depth* of locus ϕ is $d(\phi) = |str(\phi)|$. A string P is *recognized* by $\mathcal{T}_S$ iff there is a locus ϕ with $str(\phi) = P$.

We consider the following query and operations on dynamic compact tries.

LPS(ϕ, P): Given a locus in $\mathcal{T}_S$ and a pattern string P, it returns the locus $\hat{\phi}$ of string $str(\phi)Q$ in $\mathcal{T}_S$, where Q is the longest prefix of P for which $str(\phi)Q$ is recognized by $\mathcal{T}_S$. When $\phi = ((\perp, root(\mathcal{T}_S)), 1)$, then the query is known as the *longest prefix search* for the pattern P in the compact trie.

Insert(ϕ, X): Given a locus ϕ in $\mathcal{T}_S$ and a string X, it inserts a new leaf which corresponds to a new string $str(\phi)X \in S$ into the compact trie, from the given locus ϕ. When there is no node at the locus $\hat{\phi} = \mathsf{LPS}(\phi, X)$, then a new node is created at $\hat{\phi}$ as the parent of the leaf. When $\phi = ((\perp, root(\mathcal{T}_S)), 1)$, then this is standard insertion of string X to $\mathcal{T}_S$.

Delete(X_i): Given a string $X_i \in S$, it deletes the leaf ℓ_i. If the out-degree of the parent v of ℓ_i becomes 1 after the deletion of ℓ_i, then the in-coming and out-going edges of v are merged into a single edge, and v is also deleted.

For a dynamic set $I \subseteq [1, n]$ of k integers of $w = \log n$ bits each, *dynamic predecessor data structures* (e.g., [3,4,21]) efficiently support predecessor query $\mathsf{Pred}(X) = \max(\{Y \in I \mid Y \leq X\} \cup \{0\})$, successor query $\mathsf{Succ}(X) = \min(\{Y \in I \mid Y \leq X\} \cup \{n + 1\})$, and insert/delete operations for I. Let $f(k, n)$ be the time complexity of for predecessor/successor queries and insert/delete operations of an arbitrary dynamic predecessor/successor data structure which occupies $O(k \log n)$ bits of space. Beame and Fich's data structure [3] achieves $f(k, n) = O(\frac{(\log \log k)(\log \log n)}{\log \log \log n})$ worst-case time, while Willard's Y-fast trie [20] achieves $f(k, n) = O(\log \log n)$ expected time.

---

[3] In the literature the locus is represented by (u, c, h) where c is the first letter of the label of e. Since our packed c-trie does not maintain a search structure for branches, we represent the locus directly on e.

3 Packed Dynamic Compact Tries

This section presents our new dynamic compact tries called the *packed dynamic compact tries (packed c-tries)* for a dynamic set $S = \{X_1, \ldots, X_k\}$ of k strings of total length n, which achieves the main result in Theorem 1. In the sequel, a string $X \in \Sigma^*$ is called *short* if $|X| \le \alpha = \log_\sigma n$, and is called *long* if $|X| > \alpha$.

Micro Dynamic Compact Tries for Short Strings. In this subsection, we present our data structure storing short strings. Our input is a dynamic set $S = \{X_1, \ldots, X_k\}$ of k strings of total length n, such that $|X_i| \le \alpha = \log_\sigma n$ for every $1 \le i \le k$. Hence it holds that $k \le \sigma^\alpha = n$. For simplicity, we assume for now that $|X_i| = \alpha$ for every $1 \le i \le k$. The general case where S contains strings shorter than α will be explained later in Remark 1.

The dynamic data structure for short strings, called a *micro c-trie* and denoted $\mathcal{MT}_S$, consists of the following: (i) A dynamic compact trie of height exactly α storing the set S. Let $\mathcal{N}$ be the set of internal nodes, and let $\mathcal{L} = \{\ell_1, \ldots, \ell_k\}$ be the set of k leaves such that ℓ_i corresponds to X_i for $1 \le i \le k$. Since every internal node is branching, $|\mathcal{N}| \le k - 1$. Every node v of $\mathcal{MT}_S$ corresponds to the string $str(v)$ of $\log n$ bits. Overall, this compact trie requires $n \log \sigma + O(k \log n)$ bits of space (including S). (ii) A dynamic predecessor/successor data structure $\mathcal{D}$ which stores the set $S = \{X_1, \ldots, X_k\}$ of strings in $O(k \log n)$ bits of space, where each X_i is regarded as a $\log n$-bit integer. $\mathcal{D}$ supports predecessor/successor queries and insert/delete operations in $f(k, n)$ time each. Clearly $\mathcal{MT}_S$ requires $n \log \sigma + O(k \log n)$ bits of total space.

The next lemma shows how to support in $O(1)$ time LCP queries for strings represented by two given nodes on the dynamic micro c-trie $\mathcal{MT}_S$. This is related to the *labeling scheme* (e.g., see [1]) which assigns a short label to each node so that later, given the labels of two nodes, the label of the LCA of the nodes can be answered in $O(1)$ time. Although the static tree is considered in the labeling scheme, our micro c-trie is dynamic. Also, our algorithm is much simpler than applying the dynamic LCA data structure [7] to our micro c-tries.

Lemma 1. *For any nodes u and v of the dynamic micro c-trie $\mathcal{MT}_S$, we can compute $LCP(str(u), str(v))$ in $O(1)$ time.*

Proof. We pad $str(u)$ and/or $str(v)$ with an arbitrary letter c so they become α long each, namely, let $P = str(u)c^{\alpha - |str(u)|}$ and $Q = str(v)c^{\alpha - |str(v)|}$. We compute the most significant bit (msb) of the XOR of the bit representations of P and Q. Let b the bit position of the msb, and let $z = (b-1)/\log \sigma$. W.l.o.g. assume $|str(u)| \le |str(v)|$. (1) If $z < str(u)$, then $str(u)[1, z] = LCP(str(u), str(v))$. In this case, there exists a branching node y such that $str(y) = str(u)[1, z]$, and hence $LCP(str(u), str(v)) = str(y)$. (2) If $z \ge str(u)$, then $str(u) = LCP(str(u), str(v))$, and hence $str(u) = LCP(str(u), str(v))$.

Since each of P and Q is stored in a single machine word, we can compute the XOR of P and Q in $O(1)$ time. The msb can be computed in $O(1)$ time using the technique of Fredman and Willard [10]. This completes the proof. □

On micro c-tries, prefix searches and insertion operations can be started not only from the root but from *any* node. This is necessary for online sparse suffix tree construction based on Ukkonen's algorithm [18], since during the suffix link traversal we have to insert new leaves from non-root internal nodes.

Theorem 2. *The micro c-trie $\mathcal{MT}_S$ supports* LPS(ϕ, X) *queries in $O(f(k, n))$ time.*

Proof. Let P be the prefix of $str(\phi)X$ of length α, i.e., $P = str(\phi)X[1..\alpha - d(phi)]$. The case where P is represented by a leaf is easy, and thus, in what follows we focus on the case where P is not represented by a leaf.

First, we compute the string depth $d = d(\phi) \in [0, \alpha]$. Observe that $d = \max\{|LCP(P, \mathsf{Pred}(P))|, |LCP(P, \mathsf{Succ}(P))|\}$. Given P, we compute $\mathsf{Pred}(P)$ and $\mathsf{Succ}(P)$ in $O(f(k, n))$ time. Then, we can compute $|LCP(P, \mathsf{Pred}(P))|$ in $O(1)$ time by computing the msb of the XOR of the bit representations of P and $\mathsf{Pred}(P)$, as in Lemma 1. $|LCP(P, \mathsf{Succ}(P))|$ can be computed analogously, and thus, $d = d(\phi)$ can be computed in $O(f(k, n))$ time.

Second, we locate $e = (u, v)$. See also Fig. 1. Let $Z = P[1, d]$. Let $LB = Zc_1^{\alpha - |Z|}$ and $UB = Zc_\sigma^{\alpha - |Z|}$ be the lexicographically least and greatest strings of length α with prefix Z, respectively. To locate u in $\mathcal{MT}_S$, we find the leftmost and rightmost leaves X_L and X_R below ϕ by $X_L = \mathsf{Succ}(LB)$ and $X_R = \mathsf{Pred}(UB)$. Then, the longer one of $LCP(X_{L-1}, X_L)$ and $LCP(X_R, X_{R+1})$ corresponds to the origin node u of e, and $LCP(X_L, X_R)$ corresponds to the destination node v of e. These LCPs can be computed in $O(1)$ time by Lemma 1. What remains is how to access the nodes u and v representing these strings. In so doing, let \$ be a special character that does not appear in any strings in S. For each string Y represented by an internal node of $\mathcal{MT}_S$, we pad \$ at the end of Y so its length becomes exactly α, namely, we obtain $Y\$^{\alpha - |Y|}$. We insert this padded string into a dynamic dictionary dedicated only for internal nodes (here we use a predecessor/successor data structure). Now, given a string represented by an internal node, we can access the corresponding node in $O(f(k, n))$ time. Finally we obtain $\phi = ((u, v), d - |str(u)|)$ in overall $O(f(k, n))$ time. □

It follows from the proof of Theorem 2 that a dynamic predecessor/successor data structure is enough to support pattern matching queries on our dynamic micro c-tire. This implies that we do not have to store (the triples for) the edge labels in the micro c-trie. This observation is important when we consider delete operations on the set S, as we will see in the next lemma.

Lemma 2. *The micro c-trie $\mathcal{MT}_S$ supports* Insert(ϕ, X) *and* Delete(X) *operations in $O(f(k, n))$ time. We assume that $d(\phi) + |X| \leq \alpha$ so that the height of the micro compact trie will always be kept within α.*

Proof. We show how to support Insert(ϕ, X) in $O(f(k, n))$ time. Initially $S = \emptyset$, the micro compact trie $\mathcal{MT}_S$ consists only of $root(\mathcal{MT}_S)$, and predecessor/successor dictionary $\mathcal{D}$ contains no elements. When the first string X is inserted to S, then we create a leaf below the root and insert X to $\mathcal{D}$.

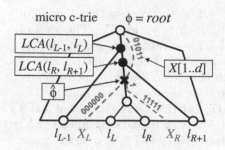

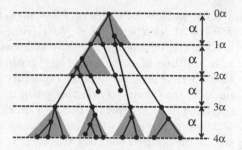

Fig. 1. Given the initial locus ϕ (which is on the root in this figure) and query pattern $P = 01011010110$, the algorithm of Theorem 2 answers the LPS(ϕ, P) query on the micro c-trie as in this figure. The answer to the query is the locus $\hat{\phi}$ for $P[1..5] = 01011$.

Fig. 2. Micro-trie decomposition: The packed c-trie is decomposed into a number of micro c-tries (gray rectangles) each of which is of height $\alpha = \log_\sigma n$. Each micro-trie is equipped with a dynamic predecessor/successor data structure.

Suppose that the data structure maintains a string set S with $|S| \geq 1$. To insert a string X from the given locus ϕ, we first conduct the LPS(ϕ, X) query of Theorem 2, and let $\hat{\phi} = (e, h)$ be the answer to the query. If $h = |e|$, then we simply insert a new leaf ℓ from the destination node of e. Otherwise, we split e at $\hat{\phi}$ and create a new node v there as the parent of the new leaf, such that $str(v) = str(\hat{\phi})$. The rest is the same as in the former case. After the new leaf is inserted, we insert $str(\phi)X$ to $\mathcal{D}$ in $O(f(k, n))$ time.

We consider Delete(X). Recall that each edge of the micro c-trie does *not* store the triple representing its string label. Thanks to this property, we need not consider updates of the labels of the edges in the path from the root to the deleted leaf (which usually becomes problematic in compact tries). Thus, we can support Delete(X) in a similar way to Insert(ϕ, X), in $O(f(k, n))$ time. □

Remark 1. When $d(\phi) + |X| < \alpha$, then we can support Insert(ϕ, X) and LPS(ϕ, X) as follows. When inserting X, we pad X with a special letter \$ which does not appear in S. Namely, we perform Insert(ϕ, X) operation with $X' = X\$^{\alpha - d(\phi) - |X|}$. When computing LPS$(\phi, X)$, we pad X with another special letter $\# \neq \$$ which does not appear in S. Namely, we perform LPS(ϕ, X'') query with $X' = X\#^{\alpha - d(\phi) - |X|}$. This gives us the correct locus for LPS(ϕ, X).

Packed Dynamic Compact Tries for Long Strings. In this subsection, we present the *packed dynamic compact trie (packed c-trie)* $\mathcal{PT}_S$ for a set S of variable-length strings of length at most $O(2^w) = O(n)$.

Micro Trie Decomposition. We decompose $\mathcal{PT}_S$ into a number of micro c-tries. See also Fig. 2. Let $h > \alpha$ be the length of the longest string in S.

We categorize the nodes of $\mathcal{PT}_S$ into $\lceil h/\alpha \rceil + 1$ levels: We say that a node of $\mathcal{PT}_S$ is at level i $(0 \le i \le \lceil h/\alpha \rceil)$ iff $|str(v)| \in [i\alpha, (i+1)\alpha - 1]$. The level of a node v is denoted by $level(v)$. A locus ϕ of $\mathcal{PT}_S$ is called a *boundary* iff $d(\phi)$ is a multiple of α. Consider any path from $root(\mathcal{PT}_S)$ to a leaf, and assume that there is no node at some boundary $k\alpha$ on this path. We create an auxiliary node at that boundary on this path, iff there is at least one non-auxiliary (i.e., original) node at level $i-1$ or $i+1$ on this path. Let $\mathcal{BN}$ denote the set of nodes at the boundaries, called the *boundary nodes*. For each boundary node $v \in \mathcal{BN}$, we create a micro compact trie $\mathcal{MT}$ whose root $root(\mathcal{MT})$ is v, internal nodes are all descendants u of v with $level(u) = level(v)$, and leaves are all boundary descendants ℓ of v with $level(\ell) = level(v) + 1$. Notice that each boundary node is the root of a micro c-trie at its level and is also a leaf of a micro c-trie at the previous level. An edge is said to be a *long edge* iff its label is at least α long. We store the label of each long edge by a triple of integers. Recall that, on the other hand, we do not store (encodings) of the edge labels in the micro c-tries.

Lemma 3. *The packed c-trie $\mathcal{PT}_S$ for a prefix-free set S of k strings requires $n \log \sigma + O(k \log n)$ bits of space.*

Proof. Firstly, we show the number of auxiliary boundary nodes in $\mathcal{PT}_S$. At most 2 auxiliary boundary nodes are created on each *original* edge of $\mathcal{PT}_S$. Since there are at most $2k - 2$ original edges, the total number of auxiliary boundary nodes is at most $4k - 4$.

Since there are at most $2k - 1$ original nodes in $\mathcal{PT}_S$, the total number of nodes in $\mathcal{PT}_S$ is at most $6k - 5$. Clearly, the total number of *short* strings of length at most α maintained by the micro c-tries is no more than the number of all nodes in $\mathcal{PT}_S$. The number of long edges in $\mathcal{PT}_S$ is no more than the number of its nodes. Overall, the total space of $\mathcal{PT}_S$ is $n \log \sigma + O(k \log n)$ bits.   □

For any locus ϕ on $\mathcal{PT}_S$, $ld(\phi)$ denotes the local string depth of ϕ in the micro c-trie $\mathcal{MT}$ that contains ϕ. Namely, if $root(\mathcal{MT}) = v$, the parent of u in $\mathcal{PT}_S$ is u, and $e = (u, v)$, then $ld(\phi) = d(\phi) - d((e, |e|))$. Prefix search queries and insert/delete operations can be supported by our packed c-trie, as follows.

Lemma 4. *The packed c-trie $\mathcal{PT}_S$ supports $\mathsf{LPS}(\phi, P)$ query in $O(\frac{m}{\alpha} f(k, n))$ worst-case time, where $m = |P| > \alpha$.*

Proof. If $m + ld(\phi) \le \alpha$, the bound immediately follows from Theorem 2. Assume $m + ld(\phi) > \alpha$, and let $q = \alpha - ld(\phi) + 1$. We factorize P into $h + 1$ blocks as $p_0 = P[1, q-1]$, $p_1 = P[q, q+\alpha-1]$, ..., $p_{h-1} = P[q+(h-1)\alpha, q+h\alpha-1]$, and $p_h = P[q+h\alpha, m]$, where $1 \le |p_0| \le \alpha$, $|p_i| = \alpha$ for $1 \le i \le h-1$, and $1 \le |p_h| \le \alpha$. Each block can be computed in $O(1)$ time by standard bit operations. If there is a mismatch in p_0, we are done. Otherwise, for each i in increasing order from 1 to h, we perform $\mathsf{LPS}(\gamma, p_i)$ query from the root γ of the corresponding micro c-trie at each level of the corresponding path starting from ϕ. This continues until we find either the first mismatch for some i or complete matches for all

i's. Each LPS query with each micro c-trie takes $O(f(k, n))$ time by Theorem 2. Since $h = O(\frac{m}{\alpha})$, it takes $O(\frac{m}{\alpha} f(k, n))$ total time.   □

Lemma 5. *The packed c-trie $\mathcal{PT}_S$ supports* Insert(ϕ, X) *and* Delete(X_i) *operations in* $O(\frac{m}{\alpha} f(k, n))$ *worst-case time, where $m = |X| > \alpha$.*

Proof. Insert(ϕ, X): we first perform LPS(ϕ, X) in $O(\frac{m}{\alpha} f(k, n))$ time (Lemma 4). Let $x_0, \ldots, x_h$ be the factorization of X w.r.t. ϕ, and let x_j be the block of the factorization containing the first mismatch. Then, we conduct Insert(γ, x_j) operation on the corresponding micro c-trie, where γ is its root. It takes $O(f(k, n))$ time (Lemma 2). If $j = h$ (x_j is the last block in the factorization of X), then we are done. Otherwise, we create a new edge with label $x'_j x_{j+1} \cdots x_k$, where x'_j is the suffix of X_j which begins at the mismatched position, leading to the new leaf ℓ. We create a new boundary node if necessary. These operations take $O(1)$ time each. Hence, Insert(ϕ, X) takes $O(\frac{m}{\alpha} f(k, n))$ total time.

Delete(X_i): Let Q be the path from the root r of $\mathcal{PT}_S$ to leaf ℓ_i. If ℓ_i is a child of the root of $\mathcal{PT}_S$, then we simply delete the single edge in Q. Otherwise, for each sub-path of Q that belongs to a micro c-trie, we perform Delete operation of Lemma 2 in this micro c-trie. Since the path Q spans at most $\frac{m}{\alpha}$ micro c-tries, the delete operations on these micro c-tries take $O(\frac{m}{\alpha} f(k, n))$ total time. For each long edge in Q whose label refers to X_i, let $\langle i, a, b \rangle$ be the triple representing the label. We replace the triple with $\langle i', a', b' \rangle$, where $X_{i'}$ is the predecessor of X_i in S and $X_{i'}[a'..b'] = X[a..b]$ (if X_i does not have a predecessor, then we can use the successor of S instead). We can find $X_{i'}$ as follows. First, we compute $\phi = \mathsf{LPS}(r, X_i) = LCA(\ell_{i'}, \ell_i)$. Then, we can find $\ell_{i'}$ by traversing the right-most path from ϕ that is to the left of the sub-path of Q from ϕ to ℓ_i. This can be done in $O(\frac{m}{\alpha} f(k, n))$ time. The positions a' and b' in $X_{i'}$ can be computed by simple arithmetics, since we know the total length of the labels in the path from ϕ to $\ell_{i'}$. Since the path Q contains less than $\frac{m}{\alpha}$ long edges, the triples for all long edges in Q can be updated in $O(\frac{m}{\alpha})$ time.   □

Speeding-up with Hashing. By augmenting each micro c-trie with a hash table storing the short strings, we achieve a good expected performance, as follows:

Lemma 6. *The packed c-trie $\mathcal{PT}_S$ augmented with hashing supports* LPS(ϕ, X) *query,* Insert(ϕ, X) *and* Delete(X) *operations in* $O(\frac{m}{\alpha} + f(k, n))$ *expected time.*

4   Applications to Online String Processing

Sparse Suffix Trees. The *suffix tree* [19] of a string T of length n is a compact trie which stores all n suffixes of T. A *sparse suffix tree* for a set $K \subseteq [1, n]$ of *sampled positions* of T is a compact trie which stores only the subset $S = \{T[i..n] \mid i \in K\}$ of the suffixes of T beginning at the sampled positions in K. It is known that if the set K of sampled positions satisfy some properties (e.g., every r positions for some fixed $r > 1$ or the positions immediately after the word delimiters), the sparse suffix tree can be constructed in an online manner in $O(n \log \sigma)$ time and $n \log \sigma + O(n \log n)$ bits of space [13,15,17].

Packed c-tries can speed up online construction and pattern matching for these sparse suffix trees: Here each input string X to Insert is given as a pair (i, j) of positions in T s.t. $X = T[i..j]$. As Lemma 7 states, Insert operation in such a case can be processed more quickly than in Lemma 4.

Lemma 7. *Given a pair (i, j) of positions in T s.t. $X = T[i..j]$, we can support* Insert(ϕ, X) *in $O(\frac{q}{\alpha} f(k, n))$ worst-case time or $O(\frac{q}{\alpha} + f(k, n))$ expected time, where q is the length of the longest prefix of X that can be spelled out from ϕ.*

Theorem 3. *Using packed c-tries, we can construct in an online manner the sparse suffix trees of [13, 15, 17] for a given text T of length n in $O((\frac{n}{\alpha} + k) f(k, n))$ worst-case time or in $O(\frac{n}{\alpha} + k f(k, n))$ expected time with $n \log \sigma + O(k \log n)$ bits of space, where k is the number of sampled positions. At any moment during the construction, pattern matching queries take $O(\frac{m}{\alpha} f(k, n))$ worst-case time or in $O(\frac{m}{\alpha} + f(k, n))$ expected time, where m is the the pattern length.*

LZ-Double Factorization. *LZ-Double factorization (LZDF)* [11] is a generalization of Lempel-Ziv 78 factorization [22]. The ith factor $g_i = g_{i_1} g_{i_2}$ of the LZDF of a string T of length n is the concatenation of previous factors g_{i_1} and g_{i_2} s.t. g_{i_1} is the longest prefix of $T[1 + \sum_{j=1}^{i-1} |g_j|, n]$ that is a previous factor (one of $\{g_1, \ldots, g_{i-1}\} \cup \Sigma$), and g_{i_2} is the longest prefix of $T[1 + |g_{i_1}| + \sum_{j=1}^{i-1} |g_j|, n]$ that is a previous factor. Goto et al. [11] proposed a Patricia-tree based algorithm which computes the LZDF of a given string T of length n in $O(k(M + \min\{k, M\} \log \sigma))$ worst-case time[4] with $O(k \log n) = O(n \log \sigma)$ bits of space[5], where k is the number of factors and M is the length of the longest factor. With packed c-tries, we can achieve a good expected performance:

Theorem 4. *Using our packed c-trie, we can compute the LZDF of string T in $O(k(\frac{M}{\alpha} + f(k, n)))$ expected time with $O(n \log \sigma)$ bits of space.*

5   Preliminary Experiments

This section shows some preliminary experimental results which compare our implementations of packed c-tries against that of the classical c-trie (Patricia tree). Table 1 shows the datasets and their statistics used in our experiments, where the first six datasets are from Pizza&Chili Corpus[6], the seventh consists of URLs in uk domain[7], and the eighth consists of all titles from Japanese Wikipedia[8]. The datasets were treated as binary strings.

---

[4] Since $kM \geq n$ always hods, the n term is hidden in the time complexity.

[5] Since all the factors of the LZDF are distinct, $k = O(\frac{n}{\log_\sigma n})$ holds [22].

[6] Pizza&Chili Corpus, http://pizzachili.dcc.uchile.cl.

[7] Laboratory for webalgorithmics, uk-2005.urls.gz, http://law.di.unimi.it/datasets.php.

[8] jawiki, https://dumps.wikimedia.org/jawiki/.

Table 1. Description of the datasets we used in our experiments.

Dataset	Original alhpabet size	Actual alphabet size	Total size (bytes)	Number of strings	Ave. string length (bits)
DNA	4	2	52,428,800	337	1,244,600.59
DBLP	128	2	52,428,800	3,229,589	129.87
english	128	2	52,428,800	9,400,185	44.62
pitches	128	2	52,428,800	93,354	4,492.90
proteins	20	2	52,428,800	186,914	2,243.98
sources	128	2	52,428,800	5,998,228	69.93
urls	128	2	52,010,031	707,658	587.97
jawiki	$\geq 2^{16}$	2	30,414,297	1,643,827	148.02

We tested three implementations of c-tries by the authors: an implementation CT of classical c-tries, and two simplified implementations PCT_{xor} and PCT_{hash} of our packed c-tries in Sect. 3 as a proof-of-concept versions. CT uses *unordered_map* in the C++/STL library to maintain the branching out-going edges of its nodes. For our implementations of packed c-tries, we set $\alpha = 32$. The first implementation PCT_{xor} only uses the XOR-based technique of Theorem 4 to quickly process long edges, while branching out-going edges are processed as in CT. The second implementation PCT_{hash} is a simplified version of our packed c-tries of Lemma 6 using XOR and hashing. Each micro c-trie in PCT_{hash} is equipped with a hash table for α-bit integers. We again used unordered_map in the C++/STL library for hash tables on micro c-tries. For simplicity, each micro c-trie is not equipped with a predecessor/successor data structure.

We compiled all programs with gcc 4.9.3 using -O3 option, and ran all experiments on a PC (2.8 GHz Intel Core i7 processor, register size 64 bits, 16 GB of memory) running on MacOS X 10.10.5, where consecutive $\alpha = 32$ bits of strings were packed into a machine word.

Table 2. Summary of our experimental results.

Dataset	Tree size (# of nodes)			Const. time (msec)			Query time (msec)		
	CT	PCT_{xor}	PCT_{hash}	CT	PCT_{xor}	PCT_{hash}	CT	PCT_{xor}	PCT_{hash}
DNA	674	674	985	**14,494**	15,270	18,596	6,690	7,381	**5,342**
DBLP	1,059,656	1,059,656	1,204,651	16,662	16,987	**14,139**	8,083	8,905	**7,209**
english	448,379	448,379	532,750	17,496	**16,944**	18,197	**9,127**	9,916	10,452
pitches	86,205	86,205	121,943	18,816	16,571	**16,520**	7,022	9,009	**6,053**
proteins	310,392	310,392	437,768	17,957	**15,733**	18,673	8,511	8,851	**6,749**
sources	1,314,571	1,314,571	1,616,872	17,398	**15,929**	16,892	8,111	8,444	**7,852**
urls	1,341,200	1,341,200	1,357,730	14,038	**13,422**	13,585	6,939	6,903	**5,918**
jawiki	2,365,821	2,365,821	3,043,817	9,440	**9,116**	10,107	4,477	4,061	**3,962**

In Table 2, we show our experimental results. First, we consider the first groups of columns for the tree sizes. We observe that the number of nodes of PCT_{hash} increases from both of CT and PCT_{xor}. The gain varies from 101.3 % on urls to 146.1 % on DNA. This comes from the addition of boundary nodes. Next, we consider the second groups of columns for the construction times. We observe that PCT_{xor} is slightly faster than the classical CT in most case. The construction time of PCT_{hash} is slightly faster against CT for DBLP, pitches, sources and urls, and slower for DNA, english, proteins and jawiki. Yet, the construction time of PCT_{hash} per node is faster than CT for all datasets. We, however, do not observe clear advantage of PCT_{hash} over PCT_{xor}. We guess that the inconsistency is due to the balance of utility and the overhead for creating the boundary nodes. Finally, we consider the third groups of columns for query times. In these experiments, we used all strings from the dataset as query patterns, and searched them on each c-trie. The table shows the total times for all the pattern searches. Among all the datasets except english, PCT_{hash} is clearly faster than CT, where the former achieved 5 % to 20 % speed-up over the latter. This indicates that PCT_{hash} is superior to the classic c-tries in prefix searches.

References

1. Alstrup, S., Gavoille, C., Kaplan, H., Rauhe, T.: Nearest common ancestors: a survey and a new distributed algorithm. Theory Comp. Sys. **37**, 441–456 (2002)
2. Andersson, A., Thorup, M.: Dynamic ordered sets with exponential search trees. J. ACM **54**(3), 13 (2007)
3. Beame, P., Fich, F.E.: Optimal bounds for the predecessor problem and related problems. J. Comput. Syst. Sci. **65**(1), 38–72 (2002)
4. Belazzougui, D., Boldi, P., Vigna, S.: Dynamic Z-Fast tries. In: Chavez, E., Lonardi, S. (eds.) SPIRE 2010. LNCS, vol. 6393, pp. 159–172. Springer, Heidelberg (2010)
5. Ben-Kiki, O., Bille, P., Breslauer, D., Gasieniec, L., Grossi, R., Weimann, O.: Optimal packed string matching. In: FSTTCS 2011, pp. 423–432 (2011)
6. Cole, R., Gottlieb, L., Lewenstein, M.: Dictionary matching and indexing with errors and don't cares. In: Proceedings of the STOC 2004, pp. 91–100 (2004)
7. Cole, R., Hariharan, R.: Dynamic LCA queries on trees. SIAM J. Comput. **34**(4), 894–923 (2005)
8. Ferragina, P., Grossi, R.: The string B-tree: a new data structure for string search in external memory and its applications. J. ACM **46**(2), 236–280 (1999)
9. Fischer, J., Gawrychowski, P.: Alphabet-dependent string searching with wexponential search trees. In: Cicalese, F., Porat, E., Vaccaro, U. (eds.) CPM 2015. LNCS, vol. 9133, pp. 160–171. Springer, Heidelberg (2015)
10. Fredman, M.L., Willard, D.E.: Surpassing the information theoretic bound with fusion trees. J. Comput. Syst. Sci. **47**(3), 424–436 (1993)
11. Goto, K., Bannai, H., Inenaga, S., Takeda, M.: LZD factorization: simple and practical online grammar compression with variable-to-fixed encoding. In: Cicalese, F., Porat, E., Vaccaro, U. (eds.) CPM 2015. LNCS, vol. 9133, pp. 219–230. Springer, Heidelberg (2015)
12. Hon, W.-K., Lam, T.-W., Shah, R., Tam, S.-L., Vitter, J.S.: Succinct index for dynamic dictionary matching. In: Dong, Y., Du, D.-Z., Ibarra, O. (eds.) ISAAC 2009. LNCS, vol. 5878, pp. 1034–1043. Springer, Heidelberg (2009)

13. Inenaga, S., Takeda, M.: On-line linear-time construction of word suffix trees. In: Lewenstein, M., Valiente, G. (eds.) CPM 2006. LNCS, vol. 4009, pp. 60–71. Springer, Heidelberg (2006)

14. Jansson, J., Sadakane, K., Sung, W.: Linked dynamic tries with applications to LZ-compression in sublinear time and space. Algorithmica **71**(4), 969–988 (2015)

15. K"arkk"ainen, J., Ukkonen, E.: Sparse suffix trees. In: Cai, J.-Y., Wong, C.K. (eds.) COCOON 1996. LNCS, vol. 1090, pp. 219–230. Springer, Heidelberg (1996)

16. Morrison, D.R.: PATRICIA: practical algorithm to retrieve information coded in alphanumeric. J. ACM **15**(4), 514–534 (1968)

17. Uemura, T., Arimura, H.: Sparse and truncated suffix trees on variable-length codes. In: Giancarlo, R., Manzini, G. (eds.) CPM 2011. LNCS, vol. 6661, pp. 246–260. Springer, Heidelberg (2011)

18. Ukkonen, E.: On-line construction of suffix-trees. Algorithmica **13**(3), 249–260 (1995)

19. Weiner, P.: Linear pattern-matching algorithms. In: Proceedings of 14th IEEE Annual Symposium on Switching and Automata Theory, pp. 1–11 (1973)

20. Willard, D.E.: Log-logarithmic worst-case range queries are possible in space $\Theta(N)$. Inf. Process. Lett. **17**, 81–84 (1983)

21. Willard, D.E.: New trie data sturucture which support very fast search operations. J. Comput. Syst. Sci. **28**, 379–394 (1984)

22. Ziv, J., Lempel, A.: Compression of individual sequences via variable-length coding. IEEE Trans. Inf. Theory **24**(5), 530–536 (1978)

Algorithmic Graph Theory

A Boundary Property for Upper Domination

Hassan AbouEisha[1], Shahid Hussain[1], Vadim Lozin[2(✉)], Jérôme Monnot[3],
Bernard Ries[4], and Viktor Zamaraev[2]

[1] King Abdullah University of Science and Technology, Thuwal, Saudi Arabia
{hassan.aboueisha,shahid.hussain}@kaust.edu.sa
[2] Mathematics Institute, University of Warwick, Coventry CV4 7AL, UK
{V.Lozin,V.Zamaraev}@warwick.ac.uk
[3] CNRS, LAMSADE UMR 7243, University Paris-Dauphine, Paris, France
jerome.monnot@dauphine.fr
[4] Department of Informatics, University of Fribourg, Bd de Pérolles 90,
1700 Fribourg, Switzerland
bernard.ries@unifr.ch

Abstract. An upper dominating set in a graph is a minimal (with respect to set inclusion) dominating set of maximum cardinality. The problem of finding an upper dominating set is generally NP-hard, but can be solved in polynomial time in some restricted graph classes, such as P_4-free graphs or $2K_2$-free graphs. For classes defined by finitely many forbidden induced subgraphs, the boundary separating difficult instances of the problem from polynomially solvable ones consists of the so called *boundary classes*. However, none of such classes has been identified so far for the upper dominating set problem. In the present paper, we discover the first boundary class for this problem.

1 Introduction

In a graph $G = (V, E)$, a *dominating set* is a subset of vertices $D \subseteq V$ such that any vertex outside of D has a neighbour in D. A dominating set D is *minimal* if no proper subset of D is dominating. An *upper dominating set* is a minimal dominating set of maximum cardinality. The UPPER DOMINATING SET problem (i.e. the problem of finding an upper dominating set in a graph) is known to be NP-hard [5]. Moreover, it remains difficult under substantial restrictions, for instance, for triangle-free graphs and the complements of bipartite graphs [1]. On the other hand, in some particular graph classes, the problem can be solved in polynomial time, which is the case for bipartite graphs [6], chordal graphs [10], generalized series-parallel graphs [9], graphs of bounded clique-width [7] and $2K_2$-free graphs [1]. What other restrictions are necessary and sufficient for polynomial-time solvability of the problem? For classes defined by finitely many forbidden induced subgraphs, this question can be answered through the notion of boundary classes. This notion was introduced in [2] to study the MAXIMUM

V. Lozin—The author gratefully acknowledges support from EPSRC, grant EP/L020408/1.

© Springer International Publishing Switzerland 2016
V. Mäkinen et al. (Eds.): IWOCA 2016, LNCS 9843, pp. 229–240, 2016.
DOI: 10.1007/978-3-319-44543-4_18

INDEPENDENT SET problem and was later applied to many other algorithmic graph problems (see e.g. [3,4,13]). However, for the UPPER DOMINATING SET problem no boundary classes have been identified so far. In the present paper, we reveal the first boundary class for this problem.

The organization of the paper is as follows. In Sect. 2, we introduce basic definitions, including the notion of a boundary class, and prove some preliminary results. Section 3 contains the main result of the paper. Finally, in Sect. 4 we discuss an open problem.

2   Preliminaries

We denote by $\mathcal{G}$ the set of all simple graphs, i.e. undirected graphs without loops and multiple edges. The *girth* of a graph $G \in \mathcal{G}$ is the length of a shortest cycle in G. As usual, we denote by K_n, P_n and C_n the complete graph, the chordless path and the chordless cycle with n vertices, respectively. Also, $\overline{G}$ denotes the complement of G. A *star* is a connected graph in which all edges are incident to a same vertex, called the *center* of the star.

Let $G = (V, E)$ be a graph with vertex set V and edge set E, and let u and v be two vertices of G. If u is adjacent to v, we write $uv \in E$ and say that u and v are *neighbours*. The *neighbourhood* of a vertex $v \in V$ is the set of its neighbours; it is denoted by $N(v)$. The *degree* of v is the size of its neighbourhood.

A subgraph of G is *spanning* if it contains all vertices of G, and it is *induced* if two vertices of the subgraph are adjacent if and only if they are adjacent in G. If a graph H is isomorphic to an induced subgraph of a graph G, we say that G contains H. Otherwise we say that G is H-free. Given a set of graphs M, we denote by Free(M) the set of all graphs containing no induced subgraphs from M.

A class of graphs (or graph property) is a set of graphs closed under isomorphism. A class is *hereditary* if it is closed under taking induced subgraphs. It is well-known (and not difficult to see) that a class X is hereditary if and only if $X = $ Free(M) for some set M. If M is a finite set, we say that X is *finitely defined*.

A class of graphs is *monotone* if it is closed under taking subgraphs (not necessarily induced). Clearly, every monotone class is hereditary.

In a graph, a *clique* is a subset of pairwise adjacent vertices, and an *independent set* is a subset of vertices no two of which are adjacent. A graph is *bipartite* if its vertices can be partitioned into two independent sets. It is well-known that a graph is bipartite if and only if it is free of odd cycles, i.e. if and only if it belongs to Free($C_3, C_5, C_7, \ldots$). We say that a graph G is *co-bipartite* if $\overline{G}$ is bipartite. Clearly, a graph is co-bipartite if and only if it belongs to Free($\overline{C}_3, \overline{C}_5, \overline{C}_7, \ldots$).

We complete this part of the section with the following technical lemma, proved in [1], where a *private* neighbour of a vertex $x \in D$ is a vertex $y \notin D$ such that x is the only neighbour of y in D.

Lemma 1. *Let G be a connected graph and D a minimal dominating set in G. If there are vertices in D that have no private neighbour outside of D, then D*

can be transformed in polynomial time into a minimal dominating set D' with $|D'| \leq |D|$ in which every vertex has a private neighbour outside of D'.

2.1 Boundary Classes of Graphs

As we mentioned earlier, the notion of boundary classes of graphs was introduced in [2] to study the MAXIMUM INDEPENDENT SET problem in hereditary classes. Later this notion was applied to some other problems of both algorithmic [3,4,13,17] and combinatorial [14,15,19] nature. Assuming $P \neq NP$, the notion of boundary classes can be defined, with respect to algorithmic graph problems, as follows.

Let Π be an algorithmic graph problem, which is generally NP-hard. We will say that a hereditary class X of graphs is Π-*tough* if the problem is NP-hard for graphs in X and Π-*easy*, otherwise. We define the notion of a boundary class for Π in two steps. First, let us define the notion of a limit class.

Definition 1. *A hereditary class X is a* LIMIT CLASS *for Π if X is the intersection of a sequence $X_1 \supseteq X_2 \supseteq X_3 \supseteq \ldots$ of Π-tough classes, in which case we also say that the sequence* CONVERGES *to X.*

Example. To illustrate the notion of a limit class, let us quote a result from [20] stating that the MAXIMUM INDEPENDENT SET problem is NP-hard for graphs with large girth, i.e. for $(C_3, C_4, \ldots, C_k)$-free graphs for each fixed value of k. With k tending to infinity, this sequence converges to the class of graphs without cycles, i.e. to forests. Therefore, the class of forests is a limit class for the MAXIMUM INDEPENDENT SET problem. However, this is not a minimal limit class for the problem, which can be explained as follows.

The proof of the NP-hardness of the problem for graphs with large girth is based on a simple fact that a double subdivision of an edge in a graph G increases the size of a maximum independent set in G by exactly 1. This operation applied sufficiently many (but still polynomially many) times allows to destroy all small cycles in G, i.e. reduces the problem from an arbitrary graph G to a graph G' of girth at least k. Obviously, if G is a graph of vertex degree at most 3, then so is G', and since the problem is NP-hard for graphs of degree at most 3, we conclude that it is also NP-hard for for $(C_3, C_4, \ldots, C_k)$-free graphs of degree at most 3. This shows that the class of forests of vertex degree at most 3 is a limit class for the the MAXIMUM INDEPENDENT SET problem. However, it is still not a minimal limit class, because by the same operation (double subdivisions of edges) one can destroy small induced copies of the graph H_n shown on the left of Fig. 1. Therefore, the MAXIMUM INDEPENDENT SET problem is NP-hard in the following class for each fixed value of k:

Z_k is the class of $(C_3, \ldots, C_k, H_1, \ldots, H_k)$-free graphs of degree at most 3.

It is not difficult to see that the sequence $Z_3 \supset Z_4 \supset \ldots$ converges to the class of forests every connected component of which has the form $S_{i,j,\ell}$ represented on the right of Fig. 1. Throughout the paper we denote this class by $\mathcal{S}$, i.e. $\mathcal{S}$ is the intersection of the sequence $Z_3 \supset Z_4 \supset \ldots$.

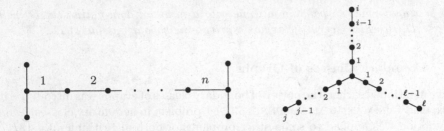

Fig. 1. Graphs H_n (left) and $S_{i,j,\ell}$ (right).

The above discussion shows that S is a limit class for the MAXIMUM INDE-
PENDENT SET problem. Moreover, in [2] it was proved that S is a *minimal* limit
class for this problem.

Definition 2. *A minimal (with respect to set inclusion) limit class for a problem
Π is called a boundary class for Π.*

The importance of the notion of boundary classes for NP-hard algorithmic
graph problems is due to the following theorem proved originally for the MAX-
IMUM INDEPENDENT SET problem in [2] (can also be found in [3] in a more
general context).

Theorem 1. *If* $P \neq NP$, *then a finitely defined class X is Π-tough if and only
if X contains a boundary class for Π.*

In what follows, we identify the first boundary class for the UPPER DOMI-
NATING SET problem. To this end, we need a number of auxiliary results. The
first of them is the following lemma dealing with limit classes, which was derived
in [2,3] as a step towards the proof of Theorem 1.

Lemma 2. *If X is a finitely defined class containing a limit class for an NP-
hard problem Π, then X is Π-tough.*

The next two results were proved in [12] and [3], respectively.

Lemma 3. *The MINIMUM DOMINATING SET problem is NP-hard in the class Z_k
for each fixed value of k.*

Theorem 2. *The class S is a boundary class for MINIMUM DOMINATING SET
problem.*

3   A Boundary Class for UPPER DOMINATION

To describe a boundary class for the UPPER DOMINATING SET problem, let us
introduce the following graph transformations. Given a graph $G = (V, E)$, we
denote by

$S(G)$ the incidence graph of G, i.e. the graph with vertex set $V \cup E$, where V and E are independent sets and a vertex $v \in V$ is adjacent to a vertex $e \in E$ in $S(G)$ if and only if v is incident to e in G. Alternatively, $S(G)$ is obtained from G by subdividing each edge e by a new vertex v_e. According to this interpretation, we call E the set of *new* vertices and V the set of *old* vertices. Any graph of the form $S(G)$ for some G will be called a *subdivision graph*.

$Q(G)$ the graph obtained from $S(G)$ by creating a clique on the set of old vertices and a clique on the set of new vertices. We call any graph of the form $Q(G)$ for some G a *Q-graph*.

The importance of Q-graphs for the UPPER DOMINATING SET problem is due to the following lemma, where we denote by $\Gamma(G)$ the size of an upper dominating set in G and by $\gamma(G)$ the size of a dominating set of minimum cardinality in G.

Lemma 4. *Let G be a graph with n vertices such that $\Gamma(Q(G)) \geq 3$. Then $\Gamma(Q(G)) = n - \gamma(G)$.*

Proof. Let D be a minimum dominating set in G, i.e. a dominating set of size $\gamma(G)$. Without loss of generality, we will assume that D satisfies Lemma 1, i.e. every vertex of D has a private neighbour outside of D. For every vertex u outside of D, consider exactly one edge, chosen arbitrarily, connecting u to a vertex in D and denote this edge by e_u. We claim that the set $D' = \{v_{e_u} : u \notin D\}$ is a minimal dominating set in $Q(G)$. By construction, D' dominates $E \cup (V - D)$ in $Q(G)$. To show that it also dominates D, assume by contradiction that a vertex $w \in D$ is not dominated by D' in $Q(G)$. By Lemma 1 we know that w has a private neighbour u outside of D. But then the edge $e = uw$ is the only edge connecting u to a vertex in D. Therefore, v_e necessarily belongs to D' and hence it dominates w, contradicting our assumption. In order to show that D' is a minimal dominating set, we observe that if we remove from D' a vertex v_{e_u} with $e_u = uw$, $u \notin D$, $w \in D$, then u becomes undominated in $Q(G)$. Finally, since $|D'| = n - |D|$, we conclude that $\Gamma(Q(G)) \geq n - |D| = n - \gamma(G)$.

Conversely, let D' be an upper dominating set in $Q(G)$, i.e. a minimal dominating set of size $\Gamma(Q(G)) \geq 3$. Then D' cannot intersect both V and E, since otherwise it contains exactly one vertex in each of these sets (else it is not minimal, because each of these sets is a clique), in which case $|D'| = 2$.

Assume first that $D' \subseteq V$. Then $V - D'$ is an independent set in G. Indeed, if G contains an edge e connecting two vertices in $V - D'$, then vertex v_e is not dominated by D' in $Q(G)$, a contradiction. Moreover, $V - D'$ is a maximal (with respect to set-inclusion) independent set in G, because D' is a minimal dominating set in $Q(G)$. Therefore, $V - D'$ is a dominating set in G of size $n - \Gamma(Q(G))$ and hence $\gamma(G) \leq n - \Gamma(Q(G))$.

Now assume $D' \subseteq E$. Let us denote by G' the subgraph of G formed by the edges (and all their endpoints) e such that $v_e \in D'$. Then:

– G' *is a spanning forest of G*, because D' covers V (else D' is not dominating in $Q(G)$) and G' is acyclic (else D' is not a minimal dominating set in $Q(G)$).

– G' *is P_4-free*, i.e. each connected component of G' is a star, since otherwise D' is not a minimal dominating set in $Q(G)$, because any vertex of D' corresponding to the middle edge of a P_4 in G' can be removed from D'.

Let D be the set of the centers of the stars of G'. Then D is dominating in G (since D' covers V) and $|D| = n - |D'|$, i.e. $\gamma(G) \leq n - \Gamma(Q(G))$, as required. □

Lemma 4 together with Theorem 2 suggest the following natural idea about a boundary class for the UPPER DOMINATING SET problem: it is the class of graphs $Q(G)$ obtained from graphs G in $\mathcal{S}$. In order to transform this idea into a formal proof, we need more notations and more auxiliary results.

For an arbitrary class X of graphs, we denote $S(X) := \{S(G) \; : \; G \in X\}$ and $Q(X) := \{Q(G) \; : \; G \in X\}$. In particular, $Q(\mathcal{G})$ is the set of all Q-graphs, where $\mathcal{G}$ is the class of all simple graphs. We observe that an induced subgraph of a Q-graph is not necessarily a Q-graph. Indeed, in a Q-graph every new vertex is adjacent to *exactly* two old vertices. However, by deleting some old vertices in a Q-graph we may obtain a graph in which a new vertex is adjacent to at most one old vertex. Therefore, $Q(X)$ is not necessarily hereditary even if X is a hereditary class. We denote by $Q^*(X)$ the hereditary closure of $Q(X)$, i.e. the class obtained from $Q(X)$ by adding to it all induced subgraphs of the graphs in $Q(X)$. Similarly, we denote by $S^*(X)$ the hereditary closure of $S(X)$.

With the above notation, our goal is proving that $Q^*(\mathcal{S})$ is a boundary class for the UPPER DOMINATING SET problem. To achieve this goal we need the following lemmas.

Lemma 5. *Let X be a monotone class of graphs such that $\mathcal{S} \not\subseteq X$, then the clique-width of the graphs in $Q^*(X)$ is bounded by a constant.*

Proof. In [16], it was proved that if $\mathcal{S} \not\subseteq X$, then the clique-width is bounded for graphs in X. It is known (see e.g. [8]) that for monotone classes, the clique-width is bounded if and only if the tree-width is bounded. By subdividing the edges of all graphs in X exactly once, we transform X into the class $S(X)$, where the tree-width is still bounded, since the subdivision of an edge of a graph does not change its tree-width. Since bounded tree-width implies bounded clique-width (see e.g. [8]), we conclude that $S(X)$ is a class of graphs of bounded clique-width. Now, for each graph G in $S(X)$ we create two cliques by complementing the edges within the sets of new and old vertices. This transforms $S(X)$ into $Q(X)$. It is known (see e.g. [11]) that local complementations applied finitely many times do not change the clique-width "too much", i.e. they transform a class of graphs of bounded clique-width into another class of graphs of bounded clique-width. Therefore, the clique-width of graphs in $Q(X)$ is bounded. Finally, the clique-width of a graph is never smaller than the clique-width of any of its induced subgraphs (see e.g. [8]). Therefore, the clique-width of graphs in $Q^*(X)$ is also bounded. □

Lemma 6. *Let $X \subseteq Q^*(\mathcal{G})$ be a hereditary class. The clique-width of graphs in X is bounded by a constant if and only if it is bounded for Q-graphs in X.*

Proof. The lemma is definitely true if $X = Q^*(Y)$ for some class Y. In this case, by definition, every non-Q-graph in X is an induced subgraph of a Q-graph from X. However, in general, X may contain a non-Q-graph H such that no Q-graph containing H as an induced subgraph belongs to X. In this case, we prove the result as follows.

First, we transform each graph H in X into a bipartite graph H' by replacing the two cliques of H (i.e. the sets of old and new vertices) with independent sets. In this way, X transforms into a class X' which is a subclass of $S^*(\mathcal{G})$. As we mentioned in the proof of Lemma 5, this transformation does not change the clique-width "too much", i.e. the clique-width of graphs in X is bounded if and only if it is bounded for graphs in X'.

By definition, $H \in X$ is a Q-graph if and only if H' is a subdivision graph, i.e. $H' = S(G)$ for some graph G. Therefore, we need to show that the clique-width of graphs in X' is bounded if and only if it is bounded for subdivision graphs in X'. In one direction, the statement is trivial. To prove it in the other direction, assume the clique-width of subdivision graphs in X' is bounded. If H' is not a subdivision, it contains new vertices of degree 0 or 1. If H' contains a vertex of degree 0, then it is disconnected, and if H' contains a vertex x of degree 1, then it has a cut-point (the neighbour of x). It is well-known that the clique-width of graphs in a hereditary class is bounded if and only if it is bounded for connected graphs in the class. Moreover, it was shown in [18] that the clique-width of graphs in a hereditary class is bounded if and only if it is bounded for 2-connected graphs (i.e. connected graphs without cut-points) in the class. Since connected graphs without cut-points in X' are subdivision graphs, we conclude that the clique-width is bounded for all graphs in X'.   □

Finally, to prove the main result of this paper, we need to show that $Q^*(\mathcal{G})$ is a finitely defined class. To show this, we first characterize graphs in $Q^*(\mathcal{G})$ as follows: a graph G belongs to $Q^*(\mathcal{G})$ if and only if the vertices of G can be partitioned into two (possibly empty) cliques U and W such that

(a) every vertex in W has at most two neighbours in U,
(b) if x and y are two vertices of W each of which has *exactly* two neighbours in U, then $N(x) \cap U \neq N(y) \cap U$.

In the proof of the following lemma, we call any partition satisfying (a) and (b) *nice*. Therefore, $Q^*(\mathcal{G})$ is precisely the class of graphs admitting a nice partition. Now we characterize $Q^*(\mathcal{G})$ in terms of minimal forbidden induced subgraphs.

Lemma 7. $Q^*(\mathcal{G}) = \text{Free}(N)$, *where N is the set of eleven graphs consisting of* $\overline{C}_3$, $\overline{C}_5$, $\overline{C}_7$ *and the eight graphs shown in Fig. 2.*

Proof. To show the inclusion $Q^*(\mathcal{G}) \subseteq \text{Free}(N)$, we first observe that $\overline{C}_3$, $\overline{C}_5$ and $\overline{C}_7$ are forbidden in $Q^*(\mathcal{G})$, since every graph in this class is co-bipartite, while $\overline{C}_3$, $\overline{C}_5$, $\overline{C}_7$ are not co-bipartite. Each of the remaining eight graphs of the set N is co-bipartite, but none of them admits a nice partition, which is a routine matter to check.

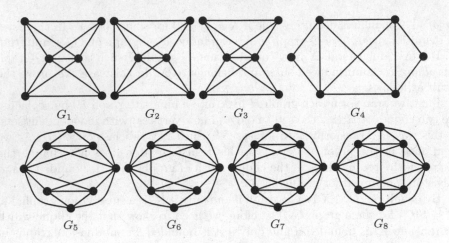

Fig. 2. Forbidden graphs for $Q^*(\mathcal{G})$

To prove the inverse inclusion $\text{Free}(N) \subseteq Q^*(\mathcal{G})$, let us consider a graph G in $\text{Free}(N)$. By definition, G contains no $\overline{C}_3$, $\overline{C}_5$, $\overline{C}_7$. Also, since G_1 is an induced subgraph of $\overline{C}_i$ with $i \geq 9$, we conclude that G contains no complements of odd cycles of length 9 or more. Therefore, G is co-bipartite. Let $V_1 \cup V_2$ be an arbitrary bipartition of $V(G)$ into two cliques. In order to show that G belongs to $Q^*(\mathcal{G})$, we split our analysis into several cases.

Case 1: G contains a K_4 induced by vertices $x_1, y_1 \in V_1$ and $x_2, y_2 \in V_2$. To analyze this case, we partition the vertices of V_1 into four subsets with respect to x_2, y_2 as follows:

A_1 is the set of vertices of V_1 adjacent to x_2 and non-adjacent to y_2,
B_1 is the set of vertices of V_1 adjacent to x_2 and to y_2,
C_1 is the set of vertices of V_1 adjacent to y_2 and non-adjacent to x_2,
D_1 is the set of vertices of V_1 adjacent neither to x_2 nor to y_2.

We partition the vertices of V_2 with respect to x_1, y_1 into four subsets A_2, B_2, C_2, D_2 analogously. We now observe the following.

(1) For $i \in \{1,2\}$, either $A_i = \emptyset$ or $C_i = \emptyset$, since otherwise a vertex in A_i and a vertex in C_i together with x_1, y_1, x_2, y_2 induce G_2.

According to this observation, in what follows, we may assume, without loss of generality, that

– $C_1 = \emptyset$ and $C_2 = \emptyset$.

We next observe that

(2) Either $A_1 = \emptyset$ or $A_2 = \emptyset$, since otherwise a vertex $a_1 \in A_1$ and a vertex $a_2 \in A_2$ together with x_1, y_1, x_2, y_2 induce either G_1 (if a_1 is not adjacent to a_2) or G_2 (if a_1 is adjacent to a_2).

Observation (2) allows us to assume, without loss of generality, that

$- A_2 = \emptyset.$

We further make the following conclusions:

(3) For $i \in \{1,2\}$, $|D_i| \leq 1$, since otherwise any two vertices of D_i together with x_1, x_2, y_1, y_2 induce G_3.

(4) If $D_1 = \{d_1\}$ and $D_2 = \{d_2\}$, then d_1 is adjacent to d_2, since otherwise $d_1, d_2, x_1, x_2, y_1, y_2$ induce G_4.

(5) If $A_1 \cup D_1 \cup D_2 \neq \emptyset$, then every vertex of B_1 is adjacent to every vertex of B_2. Indeed, assume, without loss of generality, that $z \in A_1 \cup D_1$ and a vertex $b_1 \in B_1$ is not adjacent to a vertex $b_2 \in B_2$. Then the vertices $z, b_1, b_2, x_1, x_2, y_1$ induce either G_1 (if z is not adjacent to b_2) or G_2 (if z is adjacent to b_2).

(6) Either $A_1 = \emptyset$ or $D_1 = \emptyset$, since otherwise a vertex in A_1 and a vertex in D_1 together with x_1, y_1, x_2, y_2 induce G_1.

According to (6), we split our analysis into three subcases as follows.

Case 1.1: $D_1 = \{d_1\}$. Then $A_1 = \emptyset$ (by (6)) and every vertex of B_1 is adjacent to every vertex of B_2 (by (5)). If $D_2 = \emptyset$, then $U = D_1$ and $W = B_1 \cup B_2$ is a nice partition of G (remember that $x_1, y_1 \in B_1$ and $x_2, y_2 \in B_2$).

Now assume $D_2 = \{d_2\}$ and denote by B_1^0 the vertices of B_1 nonadjacent to d_2 and by B_1^1 the vertices of B_1 adjacent to d_2. Similarly, we denote by B_2^0 the vertices of B_2 nonadjacent to d_1 and by B_2^1 the vertices of B_2 adjacent to d_1. Then $|B_1^1 \cup B_2^1| \leq 1$, since otherwise any two vertices of $B_1^1 \cup B_2^1$ together with x_1, x_2, d_1, d_2 induce G_2. But then $U = D_1 \cup D_2$ and $W = B_1 \cup B_2$ is a nice partition of G.

Case 1.2: $A_1 \neq \emptyset$. Then $D_1 = \emptyset$ (by (6)) and every vertex of B_1 is adjacent to every vertex of B_2 (by (5)). In this case, we claim that

(7) every vertex of B_2 is either adjacent to every vertex of A_1 or to none of them. Indeed, assume a vertex $b_2 \in B_2$ has a neighbour $a' \in A_1$ and a non-neighbour $a'' \in A_1$. Then $b_2, a', a'', x_1, y_1, y_2$ induce G_1.

We denote by B_2^0 the subset of vertices of B_2 that have no neighbours in A_1 and by B_2^1 the subset of vertices of B_2 adjacent to every vertex of A_1. Then

- either $|A_1| = 1$ or $|B_2^0| = 1$, since otherwise any two vertices of A_1 together with any two vertices of B_2^0 and any two vertices of B_1 induce G_3.
- if $D_2 = \{d_2\}$, then $|B_2^1| = 1$, since otherwise any two vertices of B_2^1 together with d_2, x_1, y_2 and any vertex a in A_1 induce either G_1 (if a is not adjacent to d_2)) or G_2 (if a is adjacent to d_2)).
- if $D_2 = \{d_2\}$, then d_2 has no neighbours in B_1. Indeed, if d_2 has a neighbour $b_1 \in B_1$, then vertices b_1, d_2, x_1, x_2, y_2 together with any vertex $a_1 \in A_1$ induce either G_1 (if d_2 is not adjacent to a_1) or G_2 (if d_2 is adjacent to a_1).

Therefore, either $U = A_1 \cup D_2$, $W = B_1 \cup B_2$ (if $|A_1| = 1$) or $U = B_2^0 \cup D_2$, $W = A_1 \cup B_1 \cup B_2^1$ (if $|B_2^0| = 1$) is a nice partition of G.

Case 1.3: $A_1 = \emptyset$ and $D_1 = \emptyset$. In this case, if $D_2 \neq \emptyset$, then $U = D_2$, $W = B_1 \cup B_2$ is a nice partition of G, since $B_1 \cup B_2$ is a clique (by (5)). Assume now that $D_2 = \emptyset$. If $B_1 \cup B_2$ is a clique, then G has a trivial nice partition. Suppose next that $B_1 \cup B_2$ is not a clique. If all non-edges of G are incident to a same vertex, say b (i.e. b is incident to all the edges of $\overline{G}$), then $U = \{b\}$, $W = (B_1 \cup B_2) - \{b\}$ is a nice partition of G. Otherwise, G contains a pair of non-edges $b_1' b_2' \notin E(G)$ and $b_1'' b_2'' \notin E(G)$ with all four vertices $b_1', b_1'' \in B_1$, $b_2', b_2'' \in B_2$ being distinct (i.e. $b_1' b_2'$ and $b_1'' b_2''$ form a matching in $\overline{G}$). We observe that $\{b_1', b_1'', b_2', b_2''\} \cap \{x_1, y_1, x_2, y_2\} = \emptyset$, because by definition vertices x_1, y_1, x_2, y_2 dominate the set $B_1 \cup B_2$. But then $b_1', b_1'', b_2', b_2'', x_1, y_1$ induce either G_2 (if both $b_1' b_2''$ and $b_2' b_1''$ are edges in G) or G_1 (if exactly one of $b_1' b_2''$ and $b_2' b_1''$ is an edge in G) or G_3 (if neither $b_1' b_2''$ nor $b_2' b_1''$ is an edge in G). This completes the proof of Case 1.

Case 2: G contains no K_4 with two vertices in V_1 and two vertices in V_2. We claim that in this case $V_1 \cup V_2$ is a nice partition of G. First, the assumption of case 2 implies that that no two vertices in the same part of the bipartition $V_1 \cup V_2$ have two common neighbours in the opposite part, verifying condition (b) of the definition of nice partition. To verify condition (a), it remains to prove that one of the parts V_1 and V_2 has no vertices with more than two neighbours in the opposite part. Assume the contrary and let $a_1 \in V_1$ have three neighbours in V_2 and let $a_2 \in V_2$ have three neighbours in V_1.

First, suppose a_1 is adjacent to a_2. Denote by b_2, c_2 two other neighbours of a_1 in V_2 and by b_1, c_1 two other neighbours of a_2 in V_1. Then there are no edges between b_1, c_1 and b_2, c_2, since otherwise we are in conditions of Case 1. But now $a_1, b_1, c_1, a_2, b_2, c_2$ induce a G_3.

Suppose now that a_1 is not adjacent to a_2. We denote by b_2, c_2, d_2 three neighbours of a_1 in V_2 and by b_1, c_1, d_1 three neighbours of a_2 in V_1. No two edges between b_1, c_1, d_1 and b_2, c_2, d_2 (if any) share a vertex, since otherwise we are in conditions of Case 1. But then $a_1, b_1, c_1, d_1, a_2, b_2, c_2, d_2$ induce either G_5 or G_6 or G_7 or G_8. This contradiction completes the proof of the lemma. $\quad\square$

Now we are ready to prove the main result of the paper.

Theorem 3. *If* $\mathrm{P} \neq \mathrm{NP}$, *then* $Q^*(\mathcal{S})$ *is a boundary class for the* UPPER DOMINATING SET *problem.*

Proof. From Lemmas 3 and 4 we know that UPPER DOMINATION is NP-hard in the class $Q^*(Z_k)$ for all values of $k \geq 3$. Also, it is not hard to verify that the sequence of classes $Q^*(Z_1), Q^*(Z_2)\ldots$ converges to $Q^*(\mathcal{S})$. Therefore, $Q^*(\mathcal{S})$ is a limit class for the UPPER DOMINATING SET problem. To prove its minimality, assume there is a limit class X which is properly contained in $Q^*(\mathcal{S})$. We consider a graph $F \in Q^*(\mathcal{S}) - X$, a graph $G \in Q(\mathcal{S})$ containing F as an induced subgraph (possibly $G = F$ if $F \in Q(\mathcal{S})$) and a graph $H \in \mathcal{S}$ such that $G = Q(H)$.

From the choice of G and Lemma 7, we know that $X \subseteq \text{Free}(N \cup \{G\})$, where N is the set of minimal forbidden induced subgraphs for the class $Q^*(\mathcal{G})$. Since the set N is finite (by Lemma 7), we conclude with the help of Lemma 2 that the UPPER DOMINATING SET problem is NP-hard in the class $\text{Free}(N \cup \{G\})$. To obtain a contradiction, we will show that graphs in $\text{Free}(N \cup \{G\})$ have bounded clique-width.

Denote by M the set of all graphs containing H as a spanning subgraph. Clearly $\text{Free}(M)$ is a monotone class. More precisely, it is the class of graphs containing no H as a subgraph (not necessarily induced). Since $\text{Free}(M)$ is monotone and $\mathcal{S} \not\subseteq \text{Free}(\mathcal{M})$ (as $H \in \mathcal{S}$), we know from Lemma 5 that the clique-width is bounded in $Q^*(\text{Free}(M))$.

To prove that graphs in $\text{Free}(N \cup \{G\})$ have bounded clique-width, we will show that Q-graphs in this class belong to $Q^*(\text{Free}(M))$. Let $Q(H')$ be a Q-graph in $\text{Free}(N \cup \{G\})$. Since the vertices of $Q(H')$ represent the vertices and the edges of H' and $Q(H')$ does not contain G as an induced subgraph, we conclude that H' does not contain H as a subgraph. Therefore, $H' \in \text{Free}(M)$, and hence $Q(H') \in Q(\text{Free}(M))$. By Lemma 6, this implies that all graphs in $\text{Free}(N \cup \{G\})$ have bounded clique-width. This contradicts the fact that the UPPER DOMINATING SET problem is NP-hard in this class and completes the proof of the theorem.   □

4   Conclusion

In this paper, we identified the first boundary class for the UPPER DOMINATING SET problem. Since the problem is NP-hard in the class of triangle-free graphs [1], we known (by Theorem 1) that there must exist at least one more boundary class for the problem. Revealing this class is a challenging open question.

References

1. AbouEisha, H., Hussain, S., Lozin, V., Monnot, J., Ries, B.: A dichotomy for upper domination in monogenic classes. In: Zhang, Z., Wu, L., Xu, W., Du, D.-Z. (eds.) COCOA 2014. LNCS, vol. 8881, pp. 258–267. Springer, Heidelberg (2014)
2. Alekseev, V.E.: On easy and hard hereditary classes of graphs with respect to the independent set problem. Discrete Appl. Math. **132**, 17–26 (2003)
3. Alekseev, V.E., Korobitsyn, D.V., Lozin, V.V.: Boundary classes of graphs for the dominating set problem. Discrete Math. **285**, 1–6 (2004)
4. Alekseev, V.E., Boliac, R., Korobitsyn, D.V., Lozin, V.V.: NP-hard graph problems and boundary classes of graphs. Theor. Comput. Sci. **389**, 219–236 (2007)
5. Cheston, G.A., Fricke, G., Hedetniemi, S.T., Jacobs, D.P.: On the computational complexity of upper fractional domination. Discrete Appl. Math. **27**(3), 195–207 (1990)
6. Cockayne, E.J., Favaron, O., Payan, C., Thomason, A.G.: Contributions to the theory of domination, independence and irredundance in graphs. Discrete Math. **33**(3), 249–258 (1981)

7. Courcelle, B., Makowsky, J.A., Rotics, U.: Linear time solvable optimization problems on graphs of bounded clique-width. Theor. Comput. Syst. **33**(2), 125–150 (2000)
8. Courcelle, B., Olariu, S.: Upper bounds to the clique-width of a graph. Discrete Appl. Math. **101**, 77–114 (2000)
9. Hare, E.O., Hedetniemi, S.T., Laskar, R.C., Peters, K., Wimer, T.: Linear-time computability of combinatorial problems on generalized-series-parallel graphs. In: Johnson, D.S., et al. (eds.) Discrete Algorithms and Complexity, pp. 437–457. Academic Press, New York (1987)
10. Jacobson, M.S., Peters, K.: Chordal graphs and upper irredundance, upper domination and independence. Discrete Math. **86**(1–3), 59–69 (1990)
11. Kamiński, M., Lozin, V., Milanič, M.: Recent developments on graphs of bounded clique-width. Discrete Appl. Math. **157**, 2747–2761 (2009)
12. Korobitsyn, D.V.: On the complexity of determining the domination number in monogenic classes of graphs. Diskretnaya Matematika **2**(3), 90–96 (1990). (in Russian, translation in Discrete Math. Appl. **2**(2), 191–199 (1992))
13. Korpelainen, N., Lozin, V.V., Malyshev, D.S., Tiskin, A.: Boundary properties of graphs for algorithmic graph problems. Theor. Comput. Sci. **412**, 3545–3554 (2011)
14. Korpelainen, N., Lozin, V., Razgon, I.: Boundary properties of well-quasi-ordered sets of graphs. Order **30**, 723–735 (2013)
15. Lozin, V.V.: Boundary classes of planar graphs. Comb. Probab. Comput. **17**, 287–295 (2008)
16. Lozin, V., Milanič, M.: Critical properties of graphs of bounded clique-width. Discrete Math. **313**, 1035–1044 (2013)
17. Lozin, V., Purcell, C.: Boundary properties of the satisfiability problems. Inf. Process. Lett. **113**, 313–317 (2013)
18. Lozin, V., Rautenbach, D.: On the band-, tree- and clique-width of graphs with bounded vertex degree. SIAM J. Discrete Math. **18**, 195–206 (2004)
19. Lozin, V., Zamaraev, V.: Boundary properties of factorial classes of graphs. J. Graph Theor. **78**, 207–218 (2015)
20. Murphy, O.J.: Computing independent sets in graphs with large girth. Discrete Appl. Math. **35**, 167–170 (1992)

Upper Domination: Complexity and Approximation

Cristina Bazgan[1]([⊠]), Ljiljana Brankovic[2,3], Katrin Casel[3], Henning Fernau[3], Klaus Jansen[4], Kim-Manuel Klein[4], Michael Lampis[1], Mathieu Liedloff[5], Jérôme Monnot[1], and Vangelis Th. Paschos[1]

[1] Institut Universitaire de France, Université Paris-Dauphine, PSL Research University, CNRS, UMR 7243, LAMSADE, 75016 Paris, France
{bazgan,michail.lampis,jerome.monnot,paschos}@lamsade.dauphine.fr
[2] The University of Newcastle, Callaghan, NSW 2308, Australia
ljiljana.brankovic@newcastle.edu.au
[3] Universität Trier, Fachbereich 4, Informatikwissenschaften, 54286 Trier, Germany
{casel,fernau}@uni-trier.de
[4] Universität Kiel, Institut für Informatik, 24098 Kiel, Germany
{kj,kmk}@informatik.uni-kiel.de
[5] Univ. Orléans, INSA Centre Val de Loire, LIFO EA 4022, 45067 Orléans, France
mathieu.liedloff@univ-orleans.fr

Abstract. We consider UPPER DOMINATION, the problem of finding a maximum cardinality *minimal* dominating set in a graph. We show that this problem does not admit an $n^{1-\epsilon}$ approximation for any $\epsilon > 0$, making it significantly harder than DOMINATING SET, while it remains hard even on severely restricted special cases, such as cubic graphs (APX-hard), and planar subcubic graphs (NP-hard). We complement our negative results by showing that the problem admits an $O(\Delta)$ approximation on graphs of maximum degree Δ, as well as an EPTAS on planar graphs. Along the way, we also derive essentially tight $n^{1-\frac{1}{d}}$ upper and lower bounds on the approximability of the related problem MAXIMUM MINIMAL HITTING SET on d-uniform hypergraphs, generalising known results for MAXIMUM MINIMAL VERTEX COVER.

1 Introduction

A dominating set of an undirected graph $G = (V, E)$ is a set of vertices $S \subseteq V$ such that all vertices outside of S have a neighbour in S. The problem of finding the smallest dominating set of a given graph is one of the most widely studied problems in computational complexity. In this paper, we focus on a related problem that "flips" the optimisation objective. In UPPER DOMINATION we are given a graph and we are asked to find a *maximum cardinality* dominating set that is still minimal. A dominating set is minimal if any proper subset of it is no longer dominating, that is, if it does not contain obviously redundant vertices.

Considering a MaxMin or MinMax version of a problem by "flipping" the objective is not a new idea; in fact, such questions have been posed before for

© Springer International Publishing Switzerland 2016
V. Mäkinen et al. (Eds.): IWOCA 2016, LNCS 9843, pp. 241–252, 2016.
DOI: 10.1007/978-3-319-44543-4_19

many classical optimisation problems. Some of the most well-known examples include the MINIMUM MAXIMAL INDEPENDENT SET problem [9,10,14,19] (also known as MINIMUM INDEPENDENT DOMINATING SET), the MAXIMUM MINIMAL VERTEX COVER problem [7,26] and the LAZY BUREAUCRAT problem [2,4], which is a MinMax version of KNAPSACK. The initial motivation for this type of question was rather straightforward: most classical optimisation problems admit an easy, naive heuristic algorithm which starts with a trivial solution and then gradually tries to improve it in an obvious way until it gets stuck. For example, one can produce a (maximal) independent set of a graph by starting with a single vertex and then adding vertices to the current solution while maintaining an independent set. What can we say about the worst-case performance of such a basic algorithm? Motivated by this initial question the study of MaxMin and MinMax versions of standard optimisation problems has gradually grown into a sub-field with its own interest, often revealing new insights on the structure of the original problems. UPPER DOMINATION is a natural example of this family of problems, on which somewhat fewer results are currently known. A typical pattern that often shows up in this line of research is that MaxMin versions of classical problems turn out to be much harder than the originals, especially when one considers approximation. For example, MAXIMUM MINIMAL VERTEX COVER does not admit any $n^{\frac{1}{2}-\epsilon}$ approximation, while VERTEX COVER admits a 2-approximation [7]; LAZY BUREAUCRAT is APX-hard while KNAPSACK admits a PTAS [2]; and though MINIMUM MAXIMAL INDEPENDENT SET and INDEPENDENT SET share the same (inapproximable) status, the proof of inapproximability of the MinMax version is considerably simpler, and was known long before the corresponding hardness results for INDEPENDENT SET [14].

Our first contribution is to show that this pattern also holds for UPPER DOMINATION: while DOMINATING SET admits a greedy $\ln n$ approximation, UPPER DOMINATION does not admit an $n^{1-\epsilon}$ approximation for any $\epsilon > 0$, unless P=NP. We establish this by considering the related MAXIMUM MINIMAL HITTING SET problem: given a d-uniform hypergraph, find the largest minimal set of vertices that intersects all hyperedges. Observe that the previously studied MAXIMUM MINIMAL VERTEX COVER problem is a special case of this problem for $d = 2$. We show, for any d, an approximation algorithm with ratio $n^{1-\frac{1}{d}}$, for MAXIMUM MINIMAL HITTING SET on d-uniform hypergraphs, as well as a tight $n^{1-\frac{1}{d}-\epsilon}$ inapproximability bound, exactly matching, and subsuming, the corresponding tight $\sqrt{n}$ approximation results for MAXIMUM MINIMAL VERTEX COVER given in [7]. We then obtain the inapproximability of UPPER DOMINATION by performing a reduction from an instance with sufficiently large d. We also show that UPPER DOMINATION remains hard on two restricted cases: the problem is still APX-hard on cubic graphs, and NP-hard on planar subcubic graphs. Since the problem is easy on graphs of maximum degree 2, our results completely characterise the complexity of the problem in terms of maximum degree (the best previously known result was NP-hardness for planar graphs of maximum degree 6 [1]). Given the general behavior of this type of problem, and the above results on UPPER DOMINATION in particular, the questions remains *why* are such problems typically so much harder than their original versions. Consider the following *extension* problem: Given a graph $G = (V, E)$ and a set $S \subseteq V$, does there exist a minimal dominating set *of any size* that contains S? Even though questions of this type are typically trivial for problems such as INDEPENDENT SET and LAZY BUREAUCRAT, it can be shown by a more or less easy modification of the proof of analogous results in [8,22] that in the case of UPPER DOMINATION, deciding the existence of such a minimal dominating set is NP-hard in general graphs. This helps explain the added difficulty of this problem, and more generally of problems of this type, since any natural algorithm that gradually builds a solution would have to contend with (some version of) this extension problem. In this paper we show that the extension problem for UPPER DOMINATION remains hard even for planar cubic graphs.

We complement the above negative results by giving some approximation algorithms for the problem in restricted cases. Specifically, we show that the problem admits an $O(\Delta)$-approximation on graphs with maximum degree Δ, as well as an EPTAS on planar graphs.

Previous results. It has long been known that UPPER DOMINATION is NP-complete in general [11], and even for graphs of maximum degree 6 [1]. Some polynomial-time solvable graph classes are also known. This is mainly due to the fact that on certain graph classes (like bipartite graphs) the independence number and upper domination number coincide and for those graph classes, the independence number can be computed in polynomial-time. We refer to the textbook on domination [16] for further details. We mention that the problem is polynomial for bipartite graphs [12], chordal graphs [20], generalised series-parallel graphs [15] and graphs with bounded clique-width [13]. Recently, the complexity of UPPER DOMINATION in monogenic classes of graphs defined by a single forbidden induced subgraph has led to a complexity dichotomy: if the unique forbidden induced subgraph is a P_4 or a $2K_2$ (or an induced subgraph of these), then UPPER DOMINATION is polynomial; otherwise, it is NP-complete [1].

2   Preliminaries and Combinatorial Bounds on $\Gamma(G)$

We only deal with undirected simple connected graphs $G = (V, E)$. The number of vertices $n = |V|$ is known as the order of G. As usual, $N(v)$ denotes the open neighbourhood of v, and $N[v]$ is the closed neighbourhood of v, i.e., $N[v] = N(v) \cup \{v\}$, which easily extendeds to vertex sets X, i.e., $N(X) = \bigcup_{x \in X} N(x)$ and $N[X] = N(X) \cup X$. The cardinality of $N(v)$ is known as the degree of v, denoted as $deg(v)$. The maximum degree in a graph is written as Δ. A graph of maximum degree three is called subcubic, and if all degrees equal three, it is called cubic.

Given a graph $G = (V, E)$, a subset S of V is a *dominating set* if every vertex $v \in V \setminus S$ has at least one neighbour in S, i.e., if $N[S] = V$. A dominating set is minimal if no proper subset of it is a dominating set. Likewise, a vertex set I is *independent* if $N(I) \cap I = \emptyset$. An independent set is maximal if no proper superset is independent. In the following we use classical notations: $\gamma(G)$ and

$\Gamma(G)$ are the minimum and maximum cardinalities over all minimal dominating sets in G, $\alpha(G)$ and $i(G)$ are the maximum and minimum cardinalities over all maximal independent sets, and $\tau(G)$ is the size of a minimum vertex cover, which equals $|V| - \alpha(G)$ by Gallai's identity. A minimal dominating set D of G with $|D| = \Gamma(G)$ is also known as an *upper dominating set* of G.

For any subset $S \subseteq V$ and $v \in S$ we define the private neighbourhood of v with respect to S as $pn(v, S) := N[v] \setminus N[S \setminus \{v\}]$. Any $w \in pn(v, S)$ is called a *private neighbour of v with respect to S*. S is called *irredundant* if every vertex in S has at least one private neighbour, i.e., if $|pn(v, S)| > 0$ for every $v \in S$. $\mathrm{IR}(G)$ denotes the cardinality of the largest irredundant set in G, while $\mathrm{ir}(G)$ is the cardinality of the smallest maximal irredundant set in G. We can now observe the validity of the well-known domination chain.

$$\mathrm{ir}(G) \le \gamma(G) \le i(G) \le \alpha(G) \le \Gamma(G) \le \mathrm{IR}(G)$$

The domination chain is largely due to the following two combinatorial properties: (1) Every maximal independent set is a minimal dominating set. (2) A dominating set $S \subseteq V$ is minimal if and only if $|pn(v, S)| > 0$ for every $v \in S$. Observe that v can be a private neighbour of itself, i.e., a dominating set is minimal if and only if it is also an irredundant set. Actually, every minimal dominating set is also a maximal irredundant set.

Any minimal dominating set D for a graph $G = (V, E)$ can be associated with a partition of V into four sets F, I, P, O given by: $I := \{v \in D : v \in pn(v, D)\}$, $F := D \setminus I$, $P \in \{B \subseteq N(F) \setminus D : |pn(v, D) \cap B| = 1 \text{ for all } v \in F\}$ with $|F| = |P|$, $O = V \setminus (D \cup P)$. This representation is not necessarily unique since there might be different choices for P and O, but for every partition of this kind, the following properties hold: (1) Every vertex $v \in F$ has at least one neighbour in F, called a friend. (2) The set I is an independent set in G. (3) The subgraph induced by $F \cup P$ has an edge cut set separating F and P that is also a perfect matching; hence, P is a set of private neighbours for F. (4) The neighbourhood of a vertex in I is always a subset of O, which are otherwise the outsiders. This partition is also related to a different characterisation of $\Gamma(G)$ in terms of so-called upper perfect neighbourhoods [16].

Lemma 1. *For any connected graph G with $n > 0$ vertices we have:*

$$\alpha(G) \le \Gamma(G) \le \max\left\{\alpha(G), \frac{n}{2} + \frac{\alpha(G)}{2} - 1\right\} \tag{1}$$

Lemma 2. *For any connected graph G with $n > 0$ vertices, minimum degree δ and maximum degree Δ, we have:*

$$\alpha(G) \le \Gamma(G) \le \max\left\{\alpha(G), \frac{n}{2} + \frac{\alpha(G)(\Delta - \delta)}{2\Delta} - \frac{\Delta - \delta}{\Delta}\right\} \tag{2}$$

Note that Lemma 2 generalises the earlier result of Henning and Slater on upper bounds on $\mathrm{IR}(G)$ (and hence on $\Gamma(G)$) for Δ-regular graphs G [17].

3 Hardness Results for Upper Domination

In this section we demonstrate several results that indicate that UPPER DOMINA-
TION is a rather hard problem: it does not admit any non-trivial approximation
in polynomial time, and it remains hard even in quite restricted cases.

3.1 Hardness of Approximation on General Graphs

We show that UPPER DOMINATION is hard to approximate in two steps: first, we
show that a related natural problem, MAXIMUM MINIMAL HITTING SET, is hard
to approximate, and then we show that this problem is essentially equivalent to
UPPER DOMINATION.

The MAXIMUM MINIMAL HITTING SET problem is the following: we are given
a hypergraph, that is, a base set V and a collection F of subsets of V. We wish to
find a set $H \subseteq V$ such that: (1) For all $e \in F$ we have $e \cap H \neq \emptyset$ (i.e., H is a hitting
set) (2) For all $v \in H$ there exists $e \in F$ such that $e \cap H = \{v\}$ (i.e., H is minimal)
(3) H is as large as possible. This problem generalises UPPER DOMINATION:
given a graph $G = (V, E)$, we can produce a hypergraph by keeping the same set
of vertices and creating a hyperedge for each closed neighbourhood $N[v]$ of G.
An upper dominating set of the original graph is now exactly a minimal hitting
set of the constructed hypergraph. We will also show that MAXIMUM MINIMAL
HITTING SET can be reduced to UPPER DOMINATION.

Let us note that MAXIMUM MINIMAL HITTING SET, as defined here, also generalises MAXIMUM MINIMAL VERTEX COVER, which corresponds to instances
where the input hypergraph is actually a graph. We recall that for this problem
there exists a $n^{1/2}$-approximation algorithm, while it is known to be $n^{1/2-\varepsilon}$-
inapproximable [7]. Here, we generalise this result to arbitrary hypergraphs,
taking into account the sizes of the hyperedges allowed.

Theorem 1. *For all $\varepsilon > 0, d \geq 2$, if there exists a polynomial-time approximation algorithm for* MAXIMUM MINIMAL HITTING SET *which on hypergraphs
$G = (V, F)$ where hyperedges have size exactly d has approximation ratio $n^{\frac{d-1}{d}-\varepsilon}$,
where $|V| = n$, then P=NP. This is still true for hypergraphs where $|F| \in O(|V|)$.*

Proof. Fix some constant hyperedge size d. We will present a reduction from
MAXIMUM INDEPENDENT SET, which is known to be inapproximable [18].
Specifically, for all $\varepsilon > 0$, it is known to be NP-hard to distinguish for an n-
vertex graph G if $\alpha(G) > n^{1-\varepsilon}$ or $\alpha(G) < n^\varepsilon$.

Take an instance $G = (V, E)$ of MAXIMUM INDEPENDENT SET. If $d > 2$
we begin by turning G into a d-uniform hypergraph $G' = (V, H)$ such that any
(non-trivial) hitting set of G' is a vertex cover of G and vice-versa (for $d = 2$
we simply set $G' = G$). We proceed as follows: for every edge $e \in E$ and every
$S \subseteq V \setminus e$ with $|S| = d - 2$ we construct in H the hyperedge $e \cup S$ (with size
exactly d). Thus, $|H| = O(n^d)$. Any vertex cover of G is also a hitting set of
G'. For the converse, we only want to prove that any hitting set of G' of size at
most $n - d$ is also a vertex cover of G (this is without loss of generality, since d

is a constant, so we will assume $\alpha(G) > d$). Take a hitting set C of G' with at most $n - d$ vertices; take any edge $e \in E$ and a set S with $S \subseteq V \setminus (C \cup e)$ and $|S| = d - 2$ (such a set S exists since $|V \setminus C| \geq d$). Now, $(e \cup S) \in H$, therefore C must contain a vertex of e. We thus conclude that the maximum size of $V \setminus C$, where C is a hitting set of G' is either at least $n^{1-\varepsilon}$ or at most n^ε, that is, the maximum size of $V \setminus C$ is $\alpha(G)$.

We now add some vertices and hyperedges to G' to obtain a hypergraph G''. For every set $S \subseteq V$ such that $|S| = d - 1$ and $V \setminus S$ is a hitting set of G', we add to G'' n new vertices, call them $u_{S,i}$, $1 \leq i \leq n$. Also, for each such vertex $u_{S,i}$ we add to G'' the hyperedge $S \cup \{u_{S,i}\}$, $1 \leq i \leq n$. This completes the construction. It is not hard to see that G'' has hyperedges of size exactly d, and its vertex and hyperedge set are both of size $O(n^d)$.

Let us analyse the approximability gap of this reduction. First, suppose that there is a minimal hitting set C of G' with $|V \setminus C| > n^{1-\varepsilon}$. Then, there exists a minimal hitting set of G'' with size at least $n^{d-O(d\varepsilon)}$. To see this, consider the set $C \cup \{u_{S,i} \mid S \subseteq V \setminus C, 1 \leq i \leq n\}$. This set is a hitting set, since C hits all the hyperedges of G', and for every new hyperedge of G'' that is not covered by C we select $u_{S,i}$. It is also minimal, because C is a minimal hitting set of G', and each $u_{S,i}$ selected has a private hyperedge. To calculate its size, observe that for each $S \subseteq V \setminus C$ with $|S| = d - 1$ we have n vertices. There are at least $\binom{n^{1-\varepsilon}}{d-1}$ such sets.

For the converse direction, we want to show that if $|V \setminus C| < n^\varepsilon$ for all hitting sets C of G', then any minimal hitting set of G'' has size at most $n^{1+O(d\varepsilon)}$. Consider a hitting set C' of G''. Then, $C' \cap V$ is a hitting set of G'. Let $S \subset V$ be a set of vertices such that $S \cap C' \neq \emptyset$. Then $u_{S,i} \notin C'$ for all i, because the (unique) hyperedge that contains $u_{S,i}$ also contains some other vertex of C', contradicting minimality. Now, because $V \cap C'$ is a hitting set of G' we have $|V \setminus C'| \leq n^\varepsilon$. Thus, the maximum number of different sets $S \subseteq V$ such that some $u_{S,i} \in C'$ is $\binom{n^\varepsilon}{d-1}$ and the total size of C' is at most $|C' \cap V| + n^{\varepsilon(d-1)+1} \leq n^{1+O(d\varepsilon)}$. $\square$

Corollary 1. *For any $\varepsilon > 0$* MAXIMUM MINIMAL HITTING SET *is not $n^{1-\varepsilon}$-approximable, where n is the number of vertices of the input hypergraph, unless P=NP. This is still true for hypergraphs $G = (V, F)$ where $|F| \in O(|V|)$.*

A graph is called co-bipartite if its complement is bipartite. Using Corollary 1 and the reduction of [21] from MINIMUM HITTING SET to MINIMIMUM DOMINATING SET, the following holds.

Theorem 2. *For any $\varepsilon > 0$* UPPER DOMINATION, *even restricted to co-bipartite graphs, is not $n^{1-\varepsilon}$-approximable, where n is the number of vertices of the input graph, unless P=NP.*

Note that, in fact, the inapproximability bound given in Theorem 1 is tight, for every fixed d, a fact that we believe may be of independent interest. This is shown in the following theorem, which also generalises results on MAXIMUM MINIMAL VERTEX COVER [7].

Theorem 3. *For all $d \geq 1$, there exists a polynomial-time algorithm which, given a hypergraph $G = (V, F)$ such that all hyperedges have size at most d, produces a minimal hitting set H of G with size $\Omega(n^{1/d})$. This shows an $O(n^{\frac{d-1}{d}})$-approximation for MAXIMUM MINIMAL HITTING SET on such hypergraphs.*

3.2   Hardness on Cubic and Subcubic Planar Graphs

UPPER DOMINATION is known to be NP-hard on planar graphs of maximum degree six [1]. We strengthen this result in two ways: first, we show that even for cubic graphs the problem is APX-hard; second, the problem remains NP-hard for planar subcubic graphs. We complement this hardness with an EPTAS on planar graphs.

Theorem 4. UPPER DOMINATION *is APX-hard on cubic graphs.*

Proof. (Sketch) We present a reduction from MAXIMUM INDEPENDENT SET on cubic graphs, which is APX-hard [25]. Let $G = (V, E)$ be the cubic input graph. Build G' from G by replacing every $(u, v) \in E$ by a structure of six new vertices, as shown on the right. Any $IS \subset V$ is an independent set for G if and only if G' contains an upper dominating set of cardinality $|IS| + 3|E|$.   □

Theorem 5. UPPER DOMINATION *is NP-hard on planar subcubic graphs.*

3.3   On MINIMAL DOMINATING SET EXTENSION

Algorithms working on combinatorial graph problems often try to look at local parts of the graph and then extend some part of the (final) solution that was found and fixed so far. For many maximisation problems, like UPPER IRREDUNDANCE or MAXIMUM INDEPENDENT SET, it is trivial to obtain a feasible solution that extends a given vertex set by some greedy strategy, or to know that no such extension exists. This is not true for UPPER DOMINATION, as we show next. Let us first define the problem formally.

MINIMAL DOMINATING SET EXTENSION
Input: A graph $G = (V, E)$, a set $S \subseteq V$.
Question: Does G have a minimal dominating set S' with $S' \supseteq S$?

Notice that this problem is trivial on some input with $S = \emptyset$ by using a greedy approach. If S is an independent set in G, it is also always possible to extend S to a minimal dominating set, simply by greedily extending it to a maximal independent set. If S however contains two adjacent vertices, we arrive at the problem of fixing at least one private neighbour for these vertices. This problem of preserving irredundance of the vertices in S while extending S to dominate the whole graph turns out to be a quite difficult task.

In [8] it is shown that this kind of extension of partial solutions is NP-hard for the problem of computing prime implicants of the dual of a Boolean function; a problem which can also be seen as the problem of finding a minimal hitting set for the set of prime implicants of the input function. Interpreted in this way, the proof from [8] yields NP-hardness for the minimal extension problem for 3-HITTING SET. The standard reduction from HITTING SET to DOMINATING SET however does not transfer this result to MINIMAL DOMINATING SET EXTENSION; observe that if we represent the hitting-set input-hypergraph $H = (V, F)$ with partial solution $S \subset V$ (w.l.o.g. irredundant) by $G = (V \cup F, E)$ with $E = \{(v, f): v \in V, f \in F, v \in f\} \cup (V \times V)$, the set S can always be extended to a minimal dominating set by simply adding all edge-vertices which are not dominated by S. One can repair this by adjusting this construction to forbid the edge-vertices in minimal solutions in the following way: for each edge-vertex f, add three new a_f, b_f, c_f with edges $(f, a_f), (a_f, b_f), (b_f, c_f)$ and include a_f and b_f in S. This way, f is the only choice for a private neighbour for a_f.

We will show that MINIMAL DOMINATING SET EXTENSION remains hard even for very restricted cases. Our proof is based on a reduction from the NP-complete 4-BOUNDED PLANAR 3-CONNECTED SAT problem (4P3C3SAT for short) [23], the restriction of 3-satisfiability to clauses in C over variables in V, where each variable occurs in at most four clauses and the associated bipartite graph $(C \cup X, \{(c, x) \in C \times X : (x \in c) \vee (\neg x \in c)\})$ is planar.

Theorem 6. MINIMAL DOMINATING SET EXTENSION *is NP-complete, even when restricted to planar cubic graphs.*

Proof. (Sketch) Consider an instance of 4P3C3SAT with clauses $c_1, \ldots, c_m$ and variables $v_1, \ldots, v_n$. By definition, the graph $G = (V, E)$ with $V = \{c_1, \ldots, c_m\} \cup \{v_1, \ldots, v_n\}$ and $E = \{(c_j, v_i): v_i \text{ or } \bar{v}_i \text{ is literal of } c_j\}$ is planar. Replace every vertex v_i by six new vertices $f_i^1, x_i^1, t_i^1, t_i^2, x_i^2, f_i^2$ with edges $(f_i^j, x_i^j), (t_i^j, x_i^j)$ for $j = 1, 2$.

Depending on whether v_i appears negated or non-negated in these clauses, we differentiate between the three cases depicted in Fig. 1. Observe that all other cases are rotations of these three cases and/or invert the roles of v_i and $\bar{v}_i$ and that the maximum degree of the vertices which replace v_i is three. Next, replace each clause-vertex c_j by the subgraph on the right. The vertices c_j^1, c_j^2 somehow take the role of the old vertex c_j regarding its neighbours: c_j^1 is adjacent to two of the literals of c_j and c_j^2 is adjacent to the remaining literal. This way, all vertices have degree at most three and the choices of literals to connect to c_j^1 and c_j^2 can be made such that planarity is preserved. $\square$

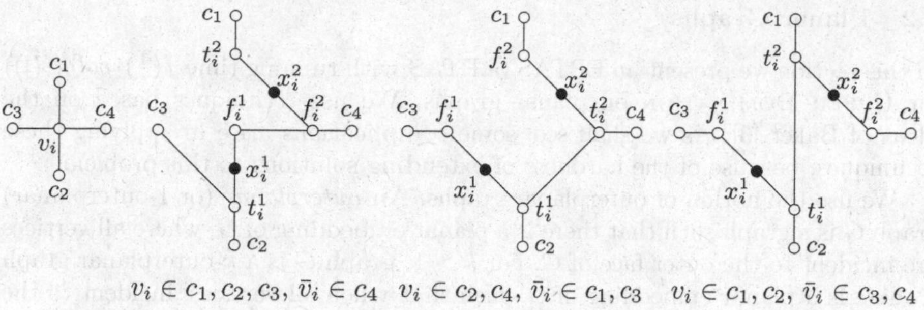
c_1
t_i^2
c_1
x_i^2
c_3
c_4
c_3
f_i^1
f_i^2
c_4
v_i
x_i^1
c_2
t_i^1
c_2
f_i^2
t_i^2
x_i^1
$v_i \in c_1, c_2, c_3, \bar{v}_i \in c_4$ $v_i \in c_2, c_4, \bar{v}_i \in c_1, c_3$ $v_i \in c_1, c_2, \bar{v}_i \in c_3, c_4$

Fig. 1. Construction of Theorem 6: A variable v_i appearing in four clauses $c_1, \ldots, c_4$, of the original instance is transformed to one of the subgraphs on the right, depending on which clauses it appears positive in. Black vertices denote elements of S.

4 Approximation Algorithms

4.1 Bounded-Degree Graphs

Unlike the general case, UPPER DOMINATION admits a simple constant factor approximation when restricted to graphs of maximum degree Δ. This follows by the fact that any dominating set in such a graph has size at least $\frac{n}{\Delta+1}$. We show that this can be improved.

Theorem 7. *Consider some graph-class $\mathcal{G}(p, \rho)$ with the following properties:*

- *One can properly colour every $G \in \mathcal{G}(p, \rho)$ with p colours in polynomial time.*
- *For any $G \in \mathcal{G}(p, \rho)$, MAXIMUM INDEPENDENT SET is ρ-approximable in polynomial time.*

Then, for every $G \in \mathcal{G}(p, \rho)$, UPPER DOMINATION is approximable in polynomial time within ratio at most $\max\left\{\rho, \frac{\Delta\rho p+\Delta-1}{2\rho\Delta}\right\}$.

The proof idea uses Eq. (2) and the fact that any maximal independent set is a minimal dominating set. We distinguish two cases, and run a different MAXIMUM INDEPENDENT SET algorithm for each case. We output the best among the computed solutions.

Any connected graph of maximum degree Δ, except a complete graph or an odd cycle, can be coloured with at most Δ colours [24]; also, MAXIMUM INDEPENDENT SET is approximable within ratio $(\Delta+3)/5$ in graphs of maximum degree Δ [5]. So, the class $\mathcal{G}(\Delta, (\Delta+3)/5)$ contains all graphs of maximum degree Δ.

Corollary 2. UPPER DOMINATION *is approximable in polynomial time within a ratio of $(6\Delta^2 + 2\Delta - 3)/10\Delta$ in general graphs.*

Theorem 7 can be improved for regular graphs where $\Gamma(G) \leqslant \frac{n}{2}$ [17].

Corollary 3. UPPER DOMINATION *in regular graphs is approximable in polynomial time within ratio $\Delta/2$.*

4.2 Planar Graphs

In this section we present an EPTAS (a PTAS with running time $f(\frac{1}{\epsilon}) \cdot poly(|I|)$) for UPPER DOMINATION on planar graphs. We use techniques based on the ideas of Baker [3]. As we shall see, some complications arise in applying these techniques, because of the hardness of extending solutions to this problem.

We use the notion of outerplanar graphs. An *outerplanar* (or 1-outerplanar) graph G is a graph such that there is a planar embedding of G, where all vertices are incident to the outer face of G. For $k > 1$, graph G is a k-outerplanar graph if there is a planar embedding of G, such that when all vertices, incident to the outer face are removed, G is a $(k-1)$-outerplanar graph. Removing stepwise the vertices that are incident to the outer face, the vertices of G can be partitioned into levels $L_1, \ldots, L_k$. We write $|L_i|$ for the number of vertices in level L_i (if $i < 1$ or $i > k$ we write $|L_i| = 0$). Bodlaender [6] proved that every k-outerplanar graph has treewidth of at most $3k - 1$. This implies the following corollary:

Corollary 4. *The maximum minimal dominating set $\Gamma(G)$ of a k-outerplanar graph G can be computed in time $f(k)n$.*

To obtain the EPTAS, we use the fact that every planar graph is k-outerplaner for some k. By removing some of the levels L_i we split the graph G into several ℓ-outerplanar subgraphs G_i of some small $\ell < k$. The maximum minimal dominating set $\Gamma(G_i)$ can be computed using the above corollary. Finally the partial solutions of G_i are merged to obtain a minimal dominating set for G. In the following theorem we analyse how the maximum of the subgraphs $\Gamma(G_i)$ correlates to the maximum $\Gamma(G)$ of the graph G.

Theorem 8. *Let $G = (V, E)$ be a k-outerplanar graph with levels $L_1, \ldots, L_k \subseteq V$. For some $i \le k$, let G_1 be the subgraph which is induced by levels $L_1, \ldots, L_{i-1}$ and let G_2 be the subgraph induced by levels $L_{i+1}, \ldots, L_k$. Then, $\Gamma(G_1) + \Gamma(G_2) \ge \Gamma(G) - \sum_{j=i-3}^{i+3} |L_j|$.*

Using the above theorem iteratively for several levels $L_{i_1}, \ldots, L_{i_s-1}$ yields the following

Corollary 5. *Let $G = (V, E)$ be a k-outerplanar graph with levels $L_1, \ldots, L_k \subseteq V$. For indices $0 = i_0 < i_1 < \ldots \le i_s = k$, let G_j be the subgraph which is induced by levels $L_{i_j}, \ldots, L_{i_{j+1}}$. Then, $\sum_{j=0}^{s-1} \Gamma(G_j) \ge \Gamma(G) - \sum_{k=0}^{s} \sum_{j=i_k-3}^{i_k+3} |L_j|$.*

The following algorithm shows how partial solutions of subgraphs can be used to obtain a minimal dominating set for the whole graph G.

Algorithm 1. *Input: A minimal dominating set of subgraphs $G_1 = (V_1, E_1)$ and $G_2 = (V_2, E_2)$ of $G = (V, E)$, which are separated by level L_i such that $V_1 \cup L_i \cup V_2 = V$.*

1. *Repeat the following steps until all vertices are covered by the dominating set.*
2. *Add vertex $v \in L_i$ which is not covered by the dominating set.*
3. *Remove vertices in $N[N[v]]$ from the dominating set until the dominating set is minimal.*

Theorem 9. *Let $G = (V, E)$ be a k-outerplanar graph with levels $L_1, \ldots, L_k \subseteq V$. For some $i \leq k$, let G_1 be the subgraph which is induced by levels $L_1, \ldots, L_{i-1}$ and let G_2 be the subgraph induced by levels $L_{i+1}, \ldots, L_k$. Let S_1 and S_2 be a minimal dominating set of G_1 and G_2, respectively. Then Algorithm 1 returns a minimal dominating set S with $|S| \geq |S_1| + |S_2| - |L_{i-1}| - |L_{i+1}|$.*

We now state our final algorithm: An EPTAS for planar UPPER DOMINATION.

Algorithm 2. *Input: A k-outerplanar graph $G = (V, E)$ for some $k \in \mathbb{N}$ and parameter ϵ.*

1. *Let $\mu = \lceil \frac{36}{\epsilon} \rceil$.*
2. *Choose x such that $0 \leq x < \mu$ and such that the following term is minimised*

$$\sum_{j \in \mathbb{N}} ((\sum_{i=-3}^{3} |L_{j\mu+x+i}|) + |L_{j\mu+x-1}| + |L_{j\mu+x+1}|)$$

3. *Let G_i be the graph induced by levels $L_{(i-1)\mu+x+1}, \ldots, L_{i\mu+x-1}$ (note that L_i with $i < 1$ or $i > k$ are empty sets) and let H_i be the graph induced by levels $L_1, \ldots, L_{i\mu+x-1}$.*
4. *Use Corollary 4 to compute the maximum minimal dominating set and its value $\Gamma(G_i)$ for each graph G_i with $0 \leq i \leq \lceil \frac{k}{\mu} \rceil$.*
5. *Apply Algorithm 1 iteratively to graph H_i and G_{i+1} with separating level $L_{i\mu+x}$ for all $0 \leq i \leq \lceil \frac{k}{\mu} \rceil$ (starting from $H_0 = G_0$) to obtain a minimal dominating set for H_{i+1}.*
6. *Return the minimal dominating set for $(H_{\lceil \frac{k}{\mu} \rceil}) = G$.*

Theorem 10. *Algorithm 2 returns a minimal dominating set S of size $|S| \geq (1 - \epsilon)\Gamma(G)$ in time bounded by $f(\frac{1}{\epsilon})n + O(n^2)$.*

Acknowledgements. We thank anonymous referees for helpful comments. We also thank our colleagues David Manlove and Daniel Meister for discussions on upper domination. Part of this research was supported by the DFG, grant FE 560/6-1.

References

1. AbouEisha, H., Hussain, S., Lozin, V., Monnot, J., Ries, B.: A dichotomy for upper domination in monogenic classes. In: Zhang, Z., Wu, L., Xu, W., Du, D.-Z. (eds.) COCOA 2014. LNCS, vol. 8881, pp. 258–267. Springer, Heidelberg (2014)
2. Arkin, E.M., Bender, M.A., Mitchell, J.S.B., Skiena, S.: The lazy bureaucrat scheduling problem. Inf. Comput. **184**(1), 129–146 (2003)
3. Baker, B.: Approximation algorithms for NP-complete problems on planar graphs. J. ACM **41**, 153–180 (1994)
4. Bender, M.A., Clifford, R., Tsichlas, K.: Scheduling algorithms for procrastinators. J. Sched. **11**(2), 95–104 (2008)
5. Berman, P., Fujito, T.: On the approximation properties of independent set problem in degree 3 graphs. In: Sack, J.-R., Akl, S.G., Dehne, F., Santoro, N. (eds.) WADS 1995. LNCS, vol. 955, pp. 449–460. Springer, Heidelberg (1995)

6. Bodlaender, H.L.: A partial k-arboretum of graphs with bounded treewidth. Theor. Comput. Sci. **209**, 1–45 (1998)
7. Boria, N., Della Croce, F.D., Paschos, V.T.: On the MAX MIN VERTEX COVER Problem. In: Kaklamanis, C., Pruhs, K. (eds.) WAOA 2013. LNCS, vol. 8447, pp. 37–48. Springer, Heidelberg (2014)
8. Boros, E., Gurvich, V., Hammer, P.L.: Dual subimplicants of positive boolean functions. Optim. Methods Softw. **10**(2), 147–156 (1998)
9. Bourgeois, N., Escoffier, B., Paschos, V.T.: Fast algorithms for MIN INDEPENDENT DOMINATING SET. In: Patt-Shamir, B., Ekim, T. (eds.) SIROCCO 2010. LNCS, vol. 6058, pp. 247–261. Springer, Heidelberg (2010)
10. Bourgeois, N., Croce, F.D., Escoffier, B., Paschos, V.T.: Fast algorithms for min independent dominating set. Discrete Appl. Math. **161**(4–5), 558–572 (2013)
11. Cheston, G., Fricke, G., Hedetniemi, S., Jacobs, D.: On the computational complexity of upper fractional domination. Discrete Appl. Math. **27**(3), 195–207 (1990)
12. Cockayne, E.J., Favaron, O., Payan, C., Thomason, A.G.: Contributions to the theory of domination, independence and irredundance in graphs. Discrete Math. **33**(3), 249–258 (1981)
13. Courcelle, B., Makowsky, A., Rotics, U.: Linear time solvable optimization problems on graphs of bounded clique-width. Theo. Comp. Syst. **33**(2), 125–150 (2000)
14. Halldórsson, M.M.: Approximating the minimum maximal independence number. Inf. Process. Lett. **46**(4), 169–172 (1993)
15. Hare, E.O., Hedetniemi, S.T., Laskar, R.C., Peters, K., Wimer, T.: Linear-time computability of combinatorial problems on generalized-series-parallel graphs. In: Johnson, D.S., et al. (eds.) Discrete Algorithms and Complexity (AP NY) (1987)
16. Haynes, T.W., Hedetniemi, S.T., Slater, P.J.: Fundamentals of Domination in Graphs. Monographs and Textbooks in Pure and Applied Mathematics, vol. 208. Marcel Dekker, New York (1998)
17. Henning, M.A., Slater, P.J.: Inequalities relating domination parameters in cubic graphs. Discrete Math. **158**(1–3), 87–98 (1996)
18. Håstad, J.: Clique is hard to approximate within $n^{1-\epsilon}$. Acta Math. **182**, 105–142 (1999)
19. Hurink, J., Nieberg, T.: Approximating minimum independent dominating sets in wireless networks. Inf. Process. Lett. **109**(2), 155–160 (2008)
20. Jacobson, M.S., Peters, K.: Chordal graphs and upper irredundance, upper domination and independence. Discrete Math. **86**(1–3), 59–69 (1990)
21. Kanté, M., Limouzy, V., Mary, A., Nourine, L.: On the enumeration of minimal dominating sets and related notions. SIAM J. Disc. Math. **28**(4), 1916–1929 (2014)
22. Kanté, M.M., Limouzy, V., Mary, A., Nourine, L., Uno, T.: Polynomial delay algorithm for listing minimal edge dominating sets in graphs. In: Dehne, F., Sack, J.-R., Stege, U. (eds.) WADS 2015. LNCS, vol. 9214, pp. 446–457. Springer, Heidelberg (2015)
23. Kratochvíl, J.: A special planar satisfiability problem and a consequence of its NP-completeness. Discrete Appl. Math. **52**, 233–252 (1994)
24. Lovász, L.: Three short proofs in graph theory. J. Combin. Theory Ser. B **19**, 269–271 (1975)
25. Papadimitriou, C.H., Yannakakis, M.: Optimization, approximation, and complexity classes. J. Comput. Syst. Sci. **43**(3), 425–440 (1991)
26. Zehavi, M.: Maximum minimal vertex cover parameterized by vertex cover. In: Italiano, G.F., Pighizzini, G., Sannella, D.T. (eds.) MFCS 2015. LNCS, vol. 9235, pp. 589–600. Springer, Heidelberg (2015)

Well-Quasi-Ordering versus Clique-Width: New Results on Bigenic Classes

Konrad K. Dabrowski[1(✉)], Vadim V. Lozin[2], and Daniël Paulusma[1]

[1] School of Engineering and Computing Sciences, Durham University, Science
Laboratories, South Road, Durham DH1 3LE, UK
{konrad.dabrowski,daniel.paulusma}@durham.ac.uk
[2] Mathematics Institute, University of Warwick, Coventry CV4 7AL, UK
v.lozin@warwick.ac.uk

Abstract. Daligault, Rao and Thomassé conjectured that if a hereditary class of graphs is well-quasi-ordered by the induced subgraph relation then it has bounded clique-width. Lozin, Razgon and Zamaraev
recently showed that this conjecture is not true for infinitely defined
classes. For finitely defined classes the conjecture is still open. It is
known to hold for classes of graphs defined by a single forbidden induced
subgraph H, as such graphs are well-quasi-ordered and are of bounded
clique-width if and only if H is an induced subgraph of P_4. For bigenic
classes of graphs i.e. ones defined by two forbidden induced subgraphs
there are several open cases in both classifications. We reduce the number of open cases for well-quasi-orderability of such classes from 12 to 9.
Our results agree with the conjecture and imply that there are only two
remaining cases to verify for bigenic classes.

1 Introduction

Well-quasi-ordering is a highly desirable property and frequently discovered concept in mathematics and theoretical computer science [16,20]. One of the most
remarkable recent results in this area is Robertson and Seymour's proof of Wagner's conjecture, which states that the set of all finite graphs is well-quasi-ordered
by the minor relation [25]. One of the first steps towards this result was the proof
of the fact that graph classes of bounded treewidth are well-quasi-ordered by the
minor relation [24] (a graph parameter π is said to be bounded for some graph
class $\mathcal{G}$ if there exists a constant c such that $\pi(G) \leq c$ for each $G \in \mathcal{G}$).

The notion of clique-width generalizes that of treewidth in the sense that
graph classes of bounded treewidth have bounded clique-width, but not necessarily vice versa. The importance of both notions is due to the fact that many algorithmic problems that are NP-hard on general graphs become polynomial-time
solvable when restricted to graph classes of bounded treewidth or clique-width.
For treewidth this follows from the meta-theorem of Courcelle [6], combined with
a result of Bodlaender [2]. For clique-width this follows from combining results
from several papers [8,15,18,23] with a result of Oum and Seymour [22].

Research supported by EPSRC (EP/K025090/1 and EP/L020408/1).

© Springer International Publishing Switzerland 2016
V. Mäkinen et al. (Eds.): IWOCA 2016, LNCS 9843, pp. 253–265, 2016.
DOI: 10.1007/978-3-319-44543-4_20

In the study of graph classes of bounded treewidth, we can restrict ourselves to minor-closed graph classes, because from the definition of treewidth it immediately follows that the treewidth of a graph is never smaller than the treewidth of its minor. This restriction, however, is not justified when we study graph classes of bounded clique-width, as the clique-width of a graph can be much smaller than the clique-width of its minor. In particular, Courcelle [7] showed that if $\mathcal{G}$ is the class of graphs of clique-width 3 and $\mathcal{G}'$ is the class of graphs obtainable from graphs in $\mathcal{G}$ by applying one or more edge contraction operations, then $\mathcal{G}'$ has unbounded clique-width. On the other hand, the clique-width of a graph is never smaller than the clique-width of any of its induced subgraphs (see, for example, [9]). This allows us to restrict ourselves to classes of graphs closed under taking induced subgraphs. Such graph classes are also known as *hereditary* classes.

It is well-known (and not difficult to see) that a class of graphs is hereditary if and only if it can be characterized by a set of minimal forbidden induced subgraphs. Due to the minimality, the set $\mathcal{F}$ of forbidden induced subgraphs is always an antichain, that is, no graph in $\mathcal{F}$ is an induced subgraph of another graph in $\mathcal{F}$. For some hereditary classes this set is finite, in which case we say that the class is *finitely defined*, whereas for other hereditary classes (such as, for instance, bipartite graphs) the set of minimal forbidden induced subgraphs forms an infinite antichain. The presence of these infinite antichains immediately shows that the induced subgraph relation is not a well-quasi-order. In fact there even exist graph classes of bounded clique-width that are not well-quasi-ordered by the induced subgraph relation: take, for example, the class of cycles, which all have clique-width at most 4. What about the inverse implication: does well-quasi-ordering imply bounded clique-width? This was stated as an open problem by Daligault, Rao and Thomassé [13] and a negative answer to this question was recently given by Lozin, Razgon and Zamaraev [21]. However, the latter authors disproved the conjecture by giving a hereditary class of graphs whose set of minimal forbidden induced subgraphs is infinite. Hence, for finitely defined classes the question remains open.

Conjecture 1. If a finitely defined class of graphs $\mathcal{G}$ is well-quasi-ordered by the induced subgraph relation, then $\mathcal{G}$ has bounded clique-width.

We emphasize that our motivation for verifying Conjecture 1 is not only mathematical but also algorithmic. Should Conjecture 1 be true, then for finitely defined classes of graphs the aforementioned algorithmic consequences of having bounded clique-width also hold for the property of being well-quasi-ordered by the induced subgraph relation.

A class of graphs is *monogenic* or H-*free* if it is characterized by a single forbidden induced subgraph H. For monogenic classes, the conjecture is true. In this case, the two notions even coincide: a class of graphs defined by a single forbidden induced subgraph H is well-quasi-ordered if and only if it has bounded clique-width if and only if H is an induced subgraph of P_4 (see, for instance, [12, 14, 19]). A class of graph is *bigenic* or (H_1, H_2)-*free* if it is characterized by two incomparable forbidden induced subgraphs H_1 and H_2. The family of bigenic classes is more diverse than the family of monogenic classes. The questions of well-quasi-orderability and having bounded clique-width still need to be resolved. Recently, considerable progress has been made towards answering the latter question for bigenic classes; see [10] for the most recent survey, which shows that there are currently eight (non-equivalent) open cases. With respect to well-quasi-orderability of bigenic classes, Korpelainen and Lozin [19] left all but 14 cases open. Since then, Atminas and Lozin [1] proved that the class of (K_3, P_6)-free graphs is well-quasi-ordered by the induced subgraph relation and that the class of $(\overline{2P_1 + P_2}, P_6)$-free graphs is not, reducing the number of remaining open cases to 12. All available results for bigenic classes verify Conjecture 1. Moreover, eight of the 12 open cases have bounded clique-width (and thus verify Conjecture 1) leaving four remaining open cases of bigenic classes for which we still need to verify Conjecture 1.

Our Results. Our first goal is to obtain more (bigenic) classes that are well-quasi-ordered by the induced subgraph relation and to support Conjecture 1 with further evidence. Our second goal is to increase our general knowledge on well-quasi-ordered graph classes and the relation to the possible boundedness of their clique-width.

Towards our first goal we prove in Sect. 4 that the class of $(\overline{2P_1 + P_2}, P_2 + P_3)$-free graphs (which has bounded clique-width [11]) is well-quasi-ordered by the induced subgraph relation. We also determine, by giving infinite antichains, two bigenic classes that are not, namely the class of $(\overline{2P_1 + P_2}, P_2 + P_4)$-free graphs, which has unbounded clique-width [11], and the class of $(\overline{P_1 + P_4}, P_1 + 2P_2)$-free graphs, for which boundedness of the clique-width is unknown. Consequently, there are nine classes of (H_1, H_2)-free graphs for which we do not know whether they are well-quasi-ordered by the induced subgraph relation, and there are two open cases left for the verification of Conjecture 1 for bigenic classes; see Open Problems 1 and 2 below. See Fig. 1 for drawings of the forbidden induced subgraphs.

Towards our second goal, we aim to develop general techniques as opposed to tackling specific cases in an ad hoc fashion. Our starting point is a very fruitful technique used for determining (un)boundedness of the clique-width of a graph class $\mathcal{G}$. We transform a given graph from $\mathcal{G}$ via a number of elementary graph operations that do not modify the clique-width by "too much" into a graph from a class for which we do know whether or not its clique-width is bounded.

It is a natural question to research how the above modification technique can be used for well-quasi-orders. The permitted elementary graph operations are vertex deletion, subgraph complementation and bipartite complementation. As we will explain in Sect. 3, these three graph operations do not preserve well-quasi-ordering. We circumvent this by checking whether these three operations preserve boundedness of a graph parameter called uniformicity, which was introduced by Korpelainen and Lozin [19]. In their paper they proved that boundedness of uniformicity is preserved by vertex deletion. Here we prove this for the remaining two graph operations. Korpelainen and Lozin [19] also showed that every graph class $\mathcal{G}$ of bounded uniformicity is well-quasi-ordered by the so-called labelled induced subgraph relation (which in turn implies that $\mathcal{G}$ is well-quasi-ordered by the induced subgraph relation). As the reverse implication does not hold, we sometimes need to rely only on the labelled induced subgraph relation. Hence, in Sect. 3 we also show that the three permitted graph operations preserve well-quasi-orderability by the labelled induced subgraph relation. We believe that the graph modification technique will also be useful for proving well-quasi-orderability of other graph classes. As such, we view the results in Sect. 3 as our second main contribution.

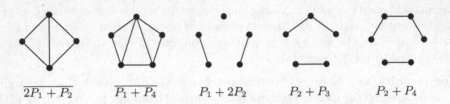

$$\overline{2P_1 + P_2} \qquad \overline{P_1 + P_4} \qquad P_1 + 2P_2 \qquad P_2 + P_3 \qquad P_2 + P_4$$

Fig. 1. The forbidden induced subgraphs considered in this paper.

Future Work. We identify several potential directions for future work starting with the two remaining bigenic classes for which Conjecture 1 must still be verified.

Open Problem 1. *Is Conjecture 1 true for the class of* (H_1, H_2)-*free graphs when:* $H_1 = K_3$ *and* $H_2 = P_2 + P_4$ *or when* $H_1 = \overline{P_1 + P_4}$ *and* $H_2 = P_2 + P_3$?

For both classes we know neither whether they are well-quasi-ordered by the induced subgraph relation nor whether their clique-width is bounded. Below we list all seven classes of (H_1, H_2)-free graphs for which we do not know whether they are well-quasi-ordered by the induced subgraph relation.

Open Problem 2. *Is the class of* (H_1, H_2)-*free graphs well-quasi-ordered by the induced subgraph relation when:*

(i) $H_1 = \overline{3P_1}$ *and* $H_2 \in \{P_1 + 2P_2, P_1 + P_5, P_2 + P_4\}$;
(ii) $H_1 = \overline{2P_1 + P_2}$ *and* $H_2 \in \{P_1 + 2P_2, P_1 + P_4\}$;
(iii) $H_1 = \overline{P_1 + P_4}$ *and* $H_2 \in \{P_1 + P_4, 2P_2, P_2 + P_3, P_5\}$.

In relation to this, we mention that the infinite antichain for $(\overline{P_1 + P_4}, P_1 + 2P_2)$-free graphs was initially found by a computer search. This computer search also showed that similar antichains do not exist for any of the remaining nine open cases. As such, constructing antichains for these cases is likely to be a challenging problem and this suggests that many of these cases may in fact be well-quasi-ordered. Some of these remaining classes have been shown to have bounded clique-width [3–5,10]. We believe that some of the structural characterizations for proving these results may be useful for showing well-quasi-orderability. Indeed, we are currently trying to prove that the class of $(K_3, P_1 + P_5)$-free graphs is well-quasi-ordered via the technique of bounding the so-called lettericity for graphs in these classes. Again, applying complementations and vertex deletions does not change the lettericity of a graph by "too much".

Another potential direction for future research is investigating linear clique-width for classes defined by two forbidden induced subgraphs. Indeed, it is not hard to show that k-uniform graphs have bounded linear clique-width. Again, we can use complementations and vertex deletions when dealing with this parameter.

2 Preliminaries

The *disjoint union* $(V(G) \cup V(H), E(G) \cup E(H))$ of two vertex-disjoint graphs G and H is denoted by $G + H$ and the disjoint union of r copies of a graph G is denoted by rG. The *complement* of a graph G, denoted by $\overline{G}$, has vertex set $V(\overline{G}) = V(G)$ and an edge between two distinct vertices if and only if these vertices are not adjacent in G. For a subset $S \subseteq V(G)$, we let $G[S]$ denote the subgraph of G *induced* by S, which has vertex set S and edge set $\{uv \mid u, v \in S, uv \in E(G)\}$. If $S = \{s_1, \ldots, s_r\}$ then, to simplify notation, we may also write $G[s_1, \ldots, s_r]$ instead of $G[\{s_1, \ldots, s_r\}]$. We use $G \setminus S$ to denote the graph obtained from G by deleting every vertex in S, i.e. $G \setminus S = G[V(G) \setminus S]$.

The graphs $C_r, K_r, K_{1,r-1}$ and P_r denote the cycle, complete graph, star and path on r vertices, respectively. For a set of graphs $\{H_1, \ldots, H_p\}$, a graph G is $(H_1, \ldots, H_p)$-*free* if it has no induced subgraph isomorphic to a graph in $\{H_1, \ldots, H_p\}$; if $p = 1$, we may write H_1-free instead of (H_1)-free.

For a graph $G = (V, E)$, the set $N(u) = \{v \in V \mid uv \in E\}$ denotes the neighbourhood of $u \in V$. A graph is *bipartite* if its vertex set can be partitioned into (at most) two independent sets. The *biclique* $K_{r,s}$ is the bipartite graph with sets in the partition of size r and s respectively, such that every vertex in one set is adjacent to every vertex in the other set. Let X be a set of vertices of a graph $G = (V, E)$. A vertex $y \in V \setminus X$ is *complete* to X if it is adjacent to every vertex of X and *anti-complete* to X if it is non-adjacent to every vertex of X. Similarly, a set of vertices $Y \subseteq V \setminus X$ is *complete* (resp. *anti-complete*) to X if every vertex in Y is complete (resp. anti-complete) to X. A vertex $y \in V \setminus X$ *distinguishes* X if y has both a neighbour and a non-neighbour in X. The set X is a *module* of G if no vertex in $V \setminus X$ distinguishes X. A module U is *non-trivial* if $1 < |U| < |V|$, otherwise it is *trivial*. A graph is *prime* if it has only trivial modules.

A *quasi order* $\leq$ on a set X is a reflexive, transitive binary relation. Two elements $x, y \in X$ in this quasi-order are *comparable* if $x \leq y$ or $y \leq x$, otherwise they are *incomparable*. A set of elements in a quasi-order is a *chain* if every pair of elements is comparable and it is an *antichain* if every pair of elements is incomparable. The quasi-order $\leq$ is a *well-quasi-order* if any infinite sequence of elements $x_1, x_2, x_3, \ldots$ in X contains a pair (x_i, x_j) with $x_i \leq x_j$ and $i < j$. Equivalently, a quasi-order is a well-quasi-order if and only if it has no infinite strictly decreasing sequence $x_1 \gneq x_2 \gneq x_3 \gneq \cdots$ and no infinite antichain.

For an arbitrary set M, let M^* denote the set of finite sequences of elements of M. Any quasi-order $\leq$ on M defines a quasi-order $\leq^*$ on M^* as follows: $(a_1, \ldots, a_m) \leq^* (b_1, \ldots, b_n)$ if and only if there is a sequence of integers $i_1, \ldots, i_m$ with $1 \leq i_1 < \cdots < i_m \leq n$ such that $a_j \leq b_{i_j}$ for $j \in \{1, \ldots, m\}$. We call $\leq^*$ the *subsequence relation*.

Lemma 1 (Higman's Lemma [17]). *If $(M, \leq)$ is a well-quasi-order then $(M^*, \leq^*)$ is a well-quasi-order.*

Labelled Induced Subgraphs and Uniformicity. To define the notion of labelled induced subgraphs, let us consider an arbitrary quasi-order $(W, \leq)$. We say that G is a *labelled* graph if each vertex v of G is equipped with an element $l_G(v) \in W$ (the *label* of v). Given two labelled graphs G and H, we say that G is a *labelled induced subgraph* of H if G is isomorphic to an induced subgraph of H and there is an isomorphism that maps each vertex v of G to a vertex w of H with $l_G(v) \leq l_H(w)$. Clearly, if $(W, \leq)$ is a well-quasi-order then a class of graphs X cannot contain an infinite sequence of labelled graphs that is strictly-decreasing with respect to the labelled induced subgraph relation. We therefore say that a class of graphs X is well-quasi-ordered by the *labelled* induced subgraph relation if it contains no infinite antichains of labelled graphs whenever $(W, \leq)$ is a *well-quasi-order*. Such a class is readily seen to be well-quasi-ordered by the induced subgraph relation as well. We will use the following three results.

Lemma 2 ([1]). *The class of P_6-free bipartite graphs is well-quasi-ordered by the labelled induced subgraph relation.*

Lemma 3 ([1]). *Let k, ℓ, m be positive integers. Then the class of $(P_k, K_\ell, K_{m,m})$-free graphs is well-quasi-ordered by the labelled induced subgraph relation.*

Lemma 4 ([1]). *Let X be a hereditary class of graphs. Then X is well-quasi-ordered by the labelled induced subgraph relation if and only if the set of prime graphs in X is. In particular, X is well-quasi-ordered by the labelled induced subgraph relation if and only if the set of connected graphs in X is.*

Let k be a natural number, let K be a symmetric square $0, 1$ matrix of order k, and let F_k be a graph on the vertex set $\{1, 2, \ldots, k\}$. Let H be the disjoint union of infinitely many copies of F_k, and for $i = 1, \ldots, k$, let V_i be the subset of $V(H)$ containing vertex i from each copy of F_k. Now we construct from H an infinite graph $H(K)$ on the same vertex set by applying a subgraph complementation to V_i if and only if $K(i, i) = 1$ and by applying bipartite complementation to a pair V_i, V_j if and only if $K(i, j) = 1$. In other words, two vertices $u \in V_i$ and $v \in V_j$ are adjacent in $H(K)$ if and only if $uv \in E(H)$ and $K(i, j) = 0$ or $uv \notin E(H)$ and $K(i, j) = 1$. Finally, let $\mathcal{P}(K, F_k)$ be the hereditary class consisting of all the finite induced subgraphs of $H(K)$.

Let k be a natural number. A graph G is *k-uniform* if there is a matrix K and a graph F_k such that $G \in \mathcal{P}(K, F_k)$. The minimum k such that G is k-uniform is the *uniformicity* of G.

The following result was proved by Korpelainen and Lozin. The class of disjoint unions of cliques is a counterexample for the reverse implication.

Theorem 1 ([19]). *Any class of graphs of bounded uniformicity is well-quasi-ordered by the labelled induced subgraph relation.*

3   Permitted Graph Operations

It is not difficult to see that if G is an induced subgraph of H, then $\overline{G}$ is an induced subgraph of $\overline{H}$. Therefore, a graph class X is well-quasi-ordered by the induced subgraph relation if and only if the set of complements of graphs in X is. In this section, we strengthen this observation in several ways. *Subgraph complementation* in a graph G is the operation of complementing a subgraph of G induced by a subset of its vertices. Applied to the entire vertex set of G, this operation coincides with the usual complementation of G. However, applied to a pair of vertices, it changes the adjacency of these vertices only. Clearly, repeated applications of this operation can transform G into any other graph on the same vertex set. Therefore, unrestricted applications of subgraph complementation may transform a well-quasi-ordered class X into a class containing infinite antichains. However, if we bound the number of applications of this operation by a constant, we preserve many nice properties of X, including well-quasi-orderability with respect to the labelled induced subgraph relation. Next, we introduce the following operations. *Bipartite complementation* in a graph G is the operation of complementing the edges between two disjoint subsets $X, Y \subseteq V(G)$. Note that applying a bipartite complementation between X and Y has the same effect as applying a sequence of three complementations: with respect to X, Y and $X \cup Y$. Finally, we define the following operation: *Vertex deletion* in a graph G is the operation of removing a single vertex v from a graph, together with any edges incident to v.

Let $k \geq 0$ be a constant and let γ be a graph operation. A graph class $\mathcal{G}'$ is (k, γ)-*obtained* from a graph class $\mathcal{G}$ if (i) every graph in $\mathcal{G}'$ is obtained from a graph in $\mathcal{G}$ by performing γ at most k times, and (ii) for every $G \in \mathcal{G}$ there exists at least one graph in $\mathcal{G}'$ obtained from G by performing γ at most k times. We say that γ *preserves* well-quasi-orderability by the labelled induced subgraph relation if for any finite constant k and any graph class $\mathcal{G}$, any graph class $\mathcal{G}'$ that is (k, γ)-obtained from $\mathcal{G}$ is well-quasi-ordered by this relation if and only if $\mathcal{G}$ is.

Lemma 5. *The following operations preserve well-quasi-orderability by the labelled induced subgraph relation:*

(i) Subgraph complementation,
(ii) Bipartite complementation and
(iii) Vertex deletion.

Proof. We start by proving the lemma for subgraph complementations. Let X be a class of graphs and Y be a set of graphs obtained from X by applying a subgraph complementation to each graph in X. More precisely, for each graph $G \in X$ we choose a set Z_G of vertices in G; we let G' be the graph obtained from G by applying a complementation with respect to the subgraph induced by Z_G and we let Y be the set of graphs G' obtained in this way. Clearly it is sufficient to show that X is well-quasi-ordered by the labelled induced subgraph relation if and only if Y is.

Suppose that X is not well-quasi-ordered under the labelled induced subgraph relation. Then there must be a well-quasi-order $(L, \leq)$ and an infinite sequence of graphs $G_1, G_2, \ldots$ in $\mathcal{X}$ with vertices labelled with elements of L, such that these graphs form an infinite antichain under the labelled induced subgraph relation. Let $(L', \leq')$ be the quasi-order with $L' = \{(k, l) : k \in \{0, 1\}, l \in L\}$ and $(k, l) \leq' (k', l')$ if and only if $k = k'$ and $l \leq l'$ (so L' is the disjoint union of two copies of L, where elements of one copy are incomparable with elements in the other copy). Note that $(L', \leq')$ is a well-quasi-order since $(L, \leq)$ is a well-quasi-order.

For each graph G_i in this sequence, with labelling l_i, we construct the graph G'_i (recall that G'_i is obtained from G_i by applying a complementation on the vertex set Z_{G_i}). We label the vertices of $V(G'_i)$ with a labelling l'_i as follows: set $l'_i(v) = (1, l_i(v))$ if $v \in Z_{G_i}$ and set $l'_i(v) = (0, l_i(v))$ otherwise.

We claim that when $G'_1, G'_2, \ldots$ are labelled in this way they form an infinite antichain with respect to the labelled induced subgraph relation. Indeed, suppose for contradiction that G'_i is a labelled induced subgraph of G'_j for some $i \neq j$. This means that there is a injective map $f : V(G'_i) \to V(G'_j)$ such that $l'_i(v) \leq' l'_j(f(v))$ for all $v \in V(G'_i)$ and $v, w \in V(G'_i)$ are adjacent in G'_i if and only if $f(v)$ and $f(w)$ are adjacent in G'_j. Now since $l'_i(v) \leq' l'_j(f(v))$ for all $v \in V(G'_i)$, by the definition of $\leq'$ we conclude the following: $l_i(v) \leq l_j(f(v))$ for all $v \in V(G'_i)$ and $v \in Z_{G_i}$ if and only if $f(v) \in Z_{G_j}$.

Suppose $v, w \in V(G_i)$ with $w \notin Z_{G_i}$ (v may or may not belong to Z_{G_i}) and note that this implies $f(w) \notin Z_{G_j}$. Then v and w are adjacent in G_i if and only if v and w are adjacent in G'_i if and only if $f(v)$ and $f(w)$ are adjacent in G'_j if and only if $f(v)$ and $f(w)$ are adjacent in G_j.

Next suppose $v, w \in Z_{G_i}$, in which case $f(v), f(w) \in Z_{G_j}$. Then v and w are adjacent in G_i if and only if v and w are non-adjacent in G'_i if and only if $f(v)$ and $f(w)$ are non-adjacent in G'_j if and only if $f(v)$ and $f(w)$ are adjacent in G_j.

It follows that f is an injective map $f : V(G_i) \to V(G_j)$ such that $l_i(v) \leq l_j(f(v))$ for all $v \in V(G_i)$ and $v, w \in V(G_i)$ are adjacent in G_i if and only if $f(v)$ and $f(w)$ are adjacent in G_j. In other words G_i is a labelled induced subgraph of G_j. This contradiction means that if $G_1, G_2, \ldots$ is an infinite antichain then $G'_1, G'_2, \ldots$ must also be an infinite antichain.

Therefore, if the class X is not well-quasi-ordered by the labelled induced subgraph relation then neither is Y. Repeating the argument with the roles of $G_1, G_2, \ldots$ and $G'_1, G'_2, \ldots$ reversed shows that if Y is not well-quasi-ordered under the labelled induced subgraph relation then neither is X. This completes the proof for subgraph complementations.

Since a bipartite complementation is equivalent to doing three subgraph complementations one after another, the result for bipartite complementations follows. Hence it remains to prove the result for vertex deletions. Let X be a class of graphs and let Y be a set of graphs obtained from X by deleting exactly one vertex z_G from each graph G in X. We denote the obtained graph by $G - z_G$. Clearly it is sufficient to show that X is well-quasi-ordered by the labelled induced subgraph relation if and only if Y is.

Suppose that Y is well-quasi-ordered by the labelled induced subgraph relation. We will show that X is also a well-quasi-order by this relation. For each graph $G \in X$, let G' be the graph obtained from G by applying a bipartite complementation between $\{z_G\}$ and $N(z_G)$, so z_G is an isolated vertex in G'. Let Z be the set of graphs obtained in this way. By Lemma 5.(ii), Z is a well-quasi-order by the labelled induced subgraph relation if and only if X is. Suppose G_1, G_2 are graphs in Z with vertices labelled from some well-quasi-order $(L, \leq)$. Then for $i \in \{1, 2\}$ the vertex z_{G_i} has a label from L and the graph $G_i - z_{G_i}$ belongs to Y. Furthermore if $G_1 - z_{G_1}$ is a labelled induced subgraph of $G_2 - z_{G_2}$ and $l_{G_1}(z_{G_1}) \leq l_{G_2}(z_{G_2})$ then G_1 is a labelled induced subgraph of G_2. Now by Lemma 1 it follows that Z is well-quasi-ordered by the labelled induced subgraph relation. Therefore X is also well-quasi-ordered by this relation.

Now suppose that Y is not well-quasi-ordered by the labelled induced subgraph relation. Then Y contains an infinite antichain $G_1, G_2, \ldots$ with the vertices of G_i labelled by functions l_i which takes values in some well-quasi-order $(L, \leq)$. For each G_i, let G'_i be a corresponding graph in X, so $G_i = G'_i - z_{G'_i}$. Then in G'_i we label $z_{G'_i}$ with a new label $*$ and label all other vertices $v \in V(G'_i)$ with the same label as that used in G_i. We make this new label $*$ incomparable to all the other labels in L and note that the obtained quasi order $(L \cup \{*\}, \leq)$ is also a well-quasi-order. It follows that $G'_1, G'_2, \ldots$ is an antichain in X when labelled in this way. Therefore, if Y is not well-quasi-ordered by the labelled induced subgraph relation then X is not either. This completes the proof. □

The above lemmas only apply to well-quasi-ordering with respect to the *labelled* induced subgraph relation. Indeed, if we take a cycle and delete a vertex, complement the subgraph induced by an edge or apply a bipartite complementation to two adjacent vertices, we obtain a path. However, while the set of cycles is an infinite antichain with respect to the induced subgraph relation, the set of paths is not.

We now show that our graph operations do not change uniformicity by "too much" either. The result for vertex deletion this was proved by Korpelainen and Lozin. We omit the proof of the remaining two operations.

Lemma 6. *Let G be a graph of uniformicity k. Let G', G'' and G''' be graphs obtained from G by applying one vertex deletion, subgraph complementation or bipartite complementation, respectively. Let ℓ', ℓ'' and ℓ''' be the uniformicities of G, G' and G'', respectively. Then the following three statements hold:*

(i) $\ell' < k < 2\ell' + 1$ [19];

(ii) $\frac{k}{2} \leq \ell'' \leq 2k$;

(iii) $\frac{k}{3} \leq \ell''' \leq 3k$.

4   One New WQO Class and Two New Non-WQO Classes

In this section we show that $(\overline{2P_1 + P_2}, P_2 + P_3)$-free graphs are well-quasi-ordered by the labelled induced subgraph relation. We divide the proof into several sections, depending on whether or not the graphs under consideration contain certain induced subgraphs or not. We follow the general scheme that Dabrowski, Huang and Paulusma [11] used to prove that this class has bounded clique-width, but we will also need a number of new arguments. We first consider graphs containing a K_5 and state the following lemma (proof omitted).

Lemma 7. *The class of $(\overline{2P_1 + P_2}, P_2 + P_3)$-free graphs that contain a K_5 is well-quasi-ordered by the labelled induced subgraph relation.*

By Lemma 7, we may restrict ourselves to looking at K_5-free graphs in our class. We now consider the case where these graphs have an induced C_5 (proof omitted).

Lemma 8. *The class of $(\overline{2P_1 + P_2}, P_2 + P_3, K_5)$-free graphs that contain an induced C_5 has bounded uniformicity.*

By Lemmas 7 and 8, we may restrict ourselves to looking at (K_5, C_5)-free graphs in our class. We need the following structural result (proof omitted).

Lemma 9. *Let G be a $(\overline{2P_1 + P_2}, P_2 + P_3, K_5, C_5)$-free graph containing an induced C_4. Then by deleting at most 17 vertices and applying at most two bipartite complementations, we can modify G into the disjoint union of a $P_2 + P_3$-free bipartite graph and a 3-uniform graph.*

Since $P_2 + P_3$ is an induced subgraph of P_6, it follows that every $P_2 + P_3$-free graph is P_6-free. Combining Lemma 9 with Theorem 1 and Lemmas 2, 4, 5.(ii) and 5.(iii) we therefore obtain the following corollary.

Corollary 1. *The class of connected $(\overline{2P_1 + P_2}, P_2 + P_3, K_5, C_5)$-free graphs with an induced C_4 is well-quasi-ordered by the labelled induced subgraph relation.*

Theorem 2. *The class of $(\overline{2P_1 + P_2}, P_2 + P_3)$-free graphs is well-quasi-ordered by the labelled induced subgraph relation.*

Proof. Graphs in the class under consideration containing an induced subgraph isomorphic to K_5, C_5 or C_4 are well-quasi-ordered by the labelled induced subgraph relation by Lemmas 7 and 8 and Corollary 1, respectively. The remaining graphs form a subclass of $(P_6, K_5, K_{2,2})$-free graphs, since $C_4 = K_{2,2}$ and $P_2 + P_3$ is an induced subgraph of P_6. By Lemma 3, this class of graphs is well-quasi-ordered by the labelled induced subgraph relation. Therefore, the class of $(\overline{2P_1 + P_2}, P_2 + P_3)$-free graphs is well-quasi-ordered by the labelled induced subgraph relation.   □

Our final two results show that the classes of $(\overline{2P_1 + P_2}, P_2 + P_4)$-free graphs and $(\overline{P_1 + P_4}, P_1 + 2P_2)$-free graphs are not well-quasi-ordered by the induced subgraph relation. The antichain used to prove the first of these cases was previously used by Atminas and Lozin to show that the class of $(\overline{2P_1 + P_2}, P_6)$-free graphs is not well-quasi-ordered with respect to the induced subgraph relation. Because of this, we can show show a stronger result for the first case (proof omitted).

Theorem 3. *The class of $(\overline{2P_1 + P_2}, P_2 + P_4, P_6)$-free graphs is not well-quasi-ordered by the induced subgraph relation.*

Theorem 4. *The class of $(\overline{P_1 + P_4}, P_1 + 2P_2)$-free graphs is not well-quasi-ordered by the induced subgraph relation.*

Proof. Let $n \geq 3$ be an integer. Consider a cycle C_{4n}, say $x_1 - x_2 - \cdots - x_{4n} - x_1$. We partition the vertices of C_{4n} into the set $X = \{x_i \mid i \equiv 0 \text{ or } 1 \mod 4\}$ and $Y = \{x_i \mid i \equiv 2 \text{ or } 3 \mod 4\}$. Next, we apply a complementation to each of X and Y, so that in the resulting graph X and Y each induce a clique on $2n$ vertices with a perfect matching removed. Let G_{4n} be the resulting graph.

Suppose, for contradiction that G_{4n} contains an induced $P_1 + 2P_2$. Without loss of generality, the set X must contain three of the vertices v_1, v_2, v_3 of the $P_1 + 2P_2$. Since every component of $P_1 + 2P_2$ contains at most two vertices, without loss of generality we may assume v_1 is non-adjacent to both v_2 and v_3. However, every vertex of $G_{4n}[X]$ has exactly one non-neighbour in X. This contradiction shows that G_{4n} is indeed $(P_1 + 2P_2)$-free.

Every vertex in X has exactly one neighbour in Y and vice versa. This means that any K_3 in G_{4n} must lie entirely in $G_{4n}[X]$ or $G_{4n}[Y]$. Since $G_{4n}[X]$ or $G_{4n}[Y]$ are both complements of perfect matchings and every vertex of $\overline{P_1 + P_4}$ lies in one of three induced K_3's, which are pairwise non-disjoint, it follows that G_{4n} is $\overline{P_1 + P_4}$-free.

It remains to show that the graphs G_{4n} form an infinite antichain with respect to the induced subgraph relation. Since $n \geq 3$, every vertex in X (resp. Y) has at least two neighbours in X (resp. Y) that are pairwise adjacent. Therefore, given x_1, we can determine which vertices lie in X and which lie in Y. Every vertex in X (resp. Y) has a unique neighbour in Y (resp. X) and a unique non-neighbour in X (resp. Y). Therefore, by specifying which vertex in G_{4n} is x_1, we uniquely determine $x_2, \ldots, x_{4n}$. Suppose G_{4n} is an induced subgraph of G_{4m} for some $m \geq 3$. Then $n \leq m$ due to the number of vertices. By symmetry, we may assume that the induced copy of G_{4n} in G_{4m} has vertex x_1 of G_{4n} in the position of vertex x_1 in G_{4m}. Then the induced copy of G_{4n} must have vertices $x_2, \ldots, x_{4n}$ in the same position as $x_2, \ldots, x_{4n}$ in G_{4m}, respectively. Now x_1 and x_{4n} are non-adjacent in G_{4n}. If $n < m$ then x_1 and x_{4n} are adjacent in G_{4m}, a contradiction. We conclude that if G_{4n} is an induced subgraph of G_{4m} then $n = m$. In other words $\{G_{4n} \mid n \geq 3\}$ is an infinite antichain with respect to the induced subgraph relation. $\square$

References

1. Atminas, A., Lozin, V.V.: Labelled induced subgraphs and well-quasi-ordering. Order **32**(3), 313–328 (2015)
2. Bodlaender, H.L.: A linear-time algorithm for finding tree-decompositions of small treewidth. SIAM J. Comput. **25**(6), 1305–1317 (1996)
3. Brandstädt, A., Kratsch, D.: On the structure of $(P_5,$ gem)-free graphs. Discrete Appl. Math. **145**(2), 155–166 (2005)
4. Brandstädt, A., Le, H.-O., Mosca, R.: Gem- and co-gem-free graphs have bounded clique-width. Int. J. Found. Comput. Sci. **15**(1), 163–185 (2004)
5. Brandstädt, A., Le, H.-O., Mosca, R.: Chordal co-gem-free and $(P_5,$ gem)-free graphs have bounded clique-width. Discrete Appl. Math. **145**(2), 232–241 (2005)
6. Courcelle, B.: The monadic second-order logic of graphs III: tree-decompositions, minor and complexity issues. Informatique Thorique et Appl. **26**, 257–286 (1992)
7. Courcelle, B.: Clique-width and edge contraction. Inf. Process. Lett. **114**(1–2), 42–44 (2014)
8. Courcelle, B., Makowsky, J.A., Rotics, U.: Linear time solvable optimization problems on graphs of bounded clique-width. Theory Comput. Syst. **33**(2), 125–150 (2000)
9. Courcelle, B., Olariu, S.: Upper bounds to the clique width of graphs. Discrete Appl. Math. **101**(1–3), 77–114 (2000)
10. Dabrowski, K.K., Dross, F., Paulusma, D.: Colouring diamond-free graphs. In: Pagh, R. (ed.) SWAT 2016. LIPIcs, vol 53, pp. 16:1–16:14. Schloss Dagstuhl–Leibniz-Zentrum für Informatik, Dagstuhl (2016)
11. Dabrowski, K.K., Huang, S., Paulusma, D.: Bounding clique-width via perfect graphs. J. Comput. Syst. Sci. (2016, in press). doi:10.1016/j.jcss.2016.06.007
12. Dabrowski, K.K., Paulusma, D.: Clique-width of graph classes defined by two forbidden induced subgraphs. Comput. J. **59**(5), 650–666 (2016)
13. Daligault, J., Rao, M., Thomassé, S.: Well-quasi-order of relabel functions. Order **27**(3), 301–315 (2010)
14. Damaschke, P.: Induced subgraphs and well-quasi-ordering. J. Graph Theory **14**(4), 427–435 (1990)
15. Espelage, W., Gurski, F., Wanke, E.: How to solve NP-hard graph problems on clique-width bounded graphs in polynomial time. In: Brandstädt, A., Le, V.B. (eds.) WG 2001. LNCS, vol. 2204, pp. 117–128. Springer, Heidelberg (2001)
16. Finkel, A., Schnoebelen, P.: Well-structured transition systems everywhere!. Theoret. Comput. Sci. **256**(1–2), 63–92 (2001)
17. Higman, G.: Ordering by divisibility in abstract algebras. Proc. Lond. Math. Soc. **s3–2**(1), 326–336 (1952)
18. Kobler, D., Rotics, U.: Edge dominating set and colorings on graphs with fixed clique-width. Discrete Appl. Math. **126**(2–3), 197–221 (2003)
19. Korpelainen, N., Lozin, V.V.: Two forbidden induced subgraphs and well-quasi-ordering. Discrete Math. **311**(16), 1813–1822 (2011)
20. Kruskal, J.B.: The theory of well-quasi-ordering: a frequently discovered concept. J. Comb. Theory Ser. A **13**(3), 297–305 (1972)
21. Lozin, V.V., Razgon, I., Zamaraev, V.: Well-quasi-ordering does not imply bounded clique-width. In: Proceedings of WG 2015. LNCS, vol. 9224 (2015, to appear)
22. Oum, S.-I., Seymour, P.D.: Approximating clique-width and branch-width. J. Comb. Theory Ser. B **96**(4), 514–528 (2006)

23. Rao, M.: MSOL partitioning problems on graphs of bounded treewidth and clique-width. Theoret. Comput. Sci. **377**(1–3), 260–267 (2007)
24. Robertson, N., Seymour, P.: Graph minors. IV. tree-width and well-quasi-ordering. J. Comb. Theory Ser. B **48**(2), 227–254 (1990)
25. Robertson, N., Seymour, P.: Graph minors. XX. Wagner's conjecture. J. Comb. Theory Ser. B **92**(2), 325–357 (2004)

Sufficient Conditions for Tuza's Conjecture on Packing and Covering Triangles

Xujin Chen[✉], Zhuo Diao, Xiaodong Hu, and Zhongzheng Tang

Institute of Applied Mathematics, AMSS, Chinese Academy of Sciences,
Beijing 100190, China
{xchen,diaozhuo,xdhu,tangzhongzheng}@amss.ac.cn

Abstract. Given a simple graph $G = (V, E)$, a subset of E is called a triangle cover if it intersects each triangle of G. Let $\nu_t(G)$ and $\tau_t(G)$ denote the maximum number of pairwise edge-disjoint triangles in G and the minimum cardinality of a triangle cover of G, respectively. Tuza conjectured in 1981 that $\tau_t(G)/\nu_t(G) \leq 2$ holds for every graph G. In this paper, using a hypergraph approach, we design polynomial-time combinatorial algorithms for finding small triangle covers. These algorithms imply new sufficient conditions for Tuza's conjecture on covering and packing triangles. More precisely, suppose that the set $\mathscr{T}_G$ of triangles covers all edges in G. We show that a triangle cover of G with cardinality at most $2\nu_t(G)$ can be found in polynomial time if one of the following conditions is satisfied: (i) $\nu_t(G)/|\mathscr{T}_G| \geq \frac{1}{3}$, (ii) $\nu_t(G)/|E| \geq \frac{1}{4}$, (iii) $|E|/|\mathscr{T}_G| \geq 2$.

Keywords: Triangle cover · Triangle packing · Linear 3-uniform hypergraphs · Combinatorial algorithms

1 Introduction

Graphs considered in this paper are undirected, simple and finite (unless otherwise noted). Given a graph $G = (V, E)$ with vertex set $V(G) = V$ and edge set $E(G) = E$, for convenience, we often identify a triangle in G with its edge set. A subset of E is called a *triangle cover* if it intersects each triangle of G. Let $\tau_t(G)$ denote the minimum cardinality of a triangle cover of G, referred to as the *triangle covering number* of G. A set of pairwise edge-disjoint triangles in G is called a *triangle packing* of G. Let $\nu_t(G)$ denote the maximum cardinality of a triangle packing of G, referred to as the *triangle packing number* of G. It is clear that $1 \leq \tau_t(G)/\nu_t(G) \leq 3$ holds for every graph G. Our research is motivated by the following conjecture raised by Tuza [11] in 1981.

Conjecture 1 (Tuza's Conjecture [11]). $\tau_t(G)/\nu_t(G) \leq 2$ holds for every graph G.

The conjecture is still unsolved in general. If it is true, then the upper bound 2 is sharp as shown by K_4 and K_5 – the complete graphs of orders 4 and 5. Throughout, by *extremal graphs* we mean graphs G with $\tau_t(G)/\nu_t(G) = 2$.

Research supported in part by NNSF of China under Grant No. 11531014 and 11222109.

© Springer International Publishing Switzerland 2016
V. Mäkinen et al. (Eds.): IWOCA 2016, LNCS 9843, pp. 266–277, 2016.
DOI: 10.1007/978-3-319-44543-4_21

Related Work. The only known universal upper bound smaller than 3 was given by Haxell [7], who showed that $\tau_t(G)/\nu_t(G) \leq 66/23 = 2.8695...$ for all graphs G. Haxell's proof [7] implies a polynomial-time algorithm for finding a triangle cover of cardinality at most $66/23$ times that of some maximal triangle packing.

Other partial results on Conjecture 1 concern special classes of graphs. Tuza [12] confirmed the conjecture for planar graphs, K_5-free chordal graphs and graphs with n vertices and at least $7n^2/16$ edges. The proof for planar graphs [12] gives an elegant polynomial-time algorithm for finding a triangle cover in planar graphs with cardinality at most twice that of some maximal triangle packing. The validity of Conjecture 1 on the class of planar graphs was later generalized by Krivelevich [9] to the class of graphs without $K_{3,3}$-subdivision. Haxell and Kohayakawa [8] showed that $\tau_t(G)/\nu_t(G) \leq 2 - \epsilon$ for tripartite graphs G, where $\epsilon > 0.044$. Haxell et al. [6] proved that every K_4-free planar graph G satisfies $\tau_t(G)/\nu_t(G) \leq 1.5$.

Regarding the tightness of the conjectured upper bound 2, Tuza [12] noticed that infinitely many extremal graphs exist. Cui et al. [5] characterized planar extremal graphs – they are edge-disjoint unions of K_4's plus possibly some vertices and edges that are not in any triangles. Baron and Kahn [1] proved that Conjecture 1 is asymptotically tight for dense graphs.

Fractional and weighted variants of Conjecture 1 were also studied. Krivelevich [9] confirmed two fractional versions of the conjecture: $\tau_t(G) \leq 2\nu_t^*(G)$ and $\tau_t^*(G) \leq 2\nu_t(G)$ hold for all graphs G, where $\tau_t^*(G)$ and $\nu_t^*(G)$ are the values of a minimum fractional triangle cover and a maximum fractional triangle packing of G, respectively. The result was generalized by Chapuy et al. [3] to the weighted case, which amounts to packing and covering triangles in multigraphs G_w (obtained from G by adding multiple edges). The authors [3] showed that $\tau_t(G_w) \leq 2\nu_t^*(G_w) - \nu_t^*(G_w)/6 + 1$ and $\tau_t^*(G_w) \leq 2\nu_t(G_w)$; the arguments imply an LP-based 2-approximation algorithm for finding a minimum weighted triangle cover in graph G.

Our Contributions. Along a different line, we establish new sufficient conditions for validity of Conjecture 1 by comparing the triangle packing number, the number of triangles and the number of edges. Given a graph G, we use

$$\mathscr{T}_G = \{E(T) : T \text{ is a triangle in } G\}$$

to denote the set consisting of the (edge sets of) triangles in G. Without loss of generality, we focus on the graphs where every edge is contained in some triangle. These graphs are called *irreducible*.

Theorem 1. *Let $G = (V, E)$ be an irreducible graph. Then a triangle cover of G with cardinality at most $2\nu_t(G)$ can be found in polynomial time, which implies $\tau_t(G) \leq 2\nu_t(G)$, if one of the following conditions is satisfied: (i) $\nu_t(G)/|\mathscr{T}_G| \geq \frac{1}{3}$, (ii) $\nu_t(G)/|E| \geq \frac{1}{4}$, and (iii) $|E|/|\mathscr{T}_G| \geq 2$.*

The primary idea behind the theorem is simple: any one of conditions (i) – (iii) allows us to remove at most $\nu_t(G)$ edges from G to make the resulting graph G' satisfy $\tau_t(G') = \nu_t(G')$; the removed edges and the edges in a minimum triangle cover of G' form a triangle cover of G with size at most $\nu_t(G) + \nu_t(G') \leq 2\nu_t(G)$. The idea is realized by establishing new results on linear 3-uniform hypergraphs (see Sect. 2); the most important one states that such a hypergraph could be made acyclic by removing a number of vertices that is no more than a third of the number of its edges. A key observation here is that hypergraph $(E, \mathscr{T}_G)$ is linear and 3-uniform.

To show the qualities of conditions (i) – (iii) in Theorem 1, we obtain the following result which complements to the constants $\frac{1}{3}$, $\frac{1}{4}$ and 2 in these conditions with $\frac{1}{4}$, $\frac{1}{5}$ and $\frac{3}{2}$, respectively.

Theorem 2. *Conjecture 1 holds for every graph if there exists some real $\delta > 0$ such that Conjecture 1 holds for every irreducible graph G satisfying one of the following inequalities: $\nu_t(G)/|\mathscr{T}_G| \geq \frac{1}{4} - \delta$, $\nu_t(G)/|E| \geq \frac{1}{5} - \delta$, and $|E|/|\mathscr{T}_G| \geq \frac{3}{2} - \delta$.*

It is worthwhile pointing out that strengthening Theorem 1, our arguments actually establish stronger results for linear 3-uniform hypergraphs.

Theorem 3. *Let $\mathcal{H} = (\mathcal{V}, \mathcal{E})$ be a linear 3-uniform hypergraph without isolated vertices. If $\nu(\mathcal{H})/|\mathcal{E}| \geq \frac{1}{3}$ or $|\mathcal{V}|/|\mathcal{E}| \geq 2$, then a transversal of $\mathcal{H}$ with cardinality at most $2\nu(\mathcal{H})$ can be found in polynomial time, which implies $\tau(\mathcal{H}) \leq 2\nu(\mathcal{H})$.*

The rest of paper is organized as follows. Section 2 proves theoretical and algorithmic results on linear 3-uniform hypergraphs concerning feedback sets, which are main technical tools for establishing new sufficient conditions for Tuza's conjecture in Sect. 3. Section 4 concludes the paper with extensions and future research directions. Omitted deals and proofs can be found in the full version of the paper [4].

2   Hypergraphs

This section develops hypergraph tools for studying Conjecture 1. The theoretical and algorithmic results are of interest in their own right.

Let $\mathcal{H} = (\mathcal{V}, \mathcal{E})$ be a hypergraph with vertex set $\mathcal{V}$ and edge set $\mathcal{E}$. For convenience, we use $\|\mathcal{H}\|$ to denote the number $|\mathcal{E}|$ of edges in $\mathcal{H}$. If hypergraph $\mathcal{H}' = (\mathcal{V}', \mathcal{E}')$ satisfies $\mathcal{V}' \subseteq \mathcal{V}$ and $\mathcal{E}' \subseteq \mathcal{E}$, we call $\mathcal{H}'$ a *sub-hypergraph* of $\mathcal{H}$, and write $\mathcal{H}' \subseteq \mathcal{H}$. For each $v \in \mathcal{V}$, the *degree* $d_{\mathcal{H}}(v)$ is the number of edges in $\mathcal{E}$ that contain v. We say v is an *isolated vertex* of $\mathcal{H}$ if $d_{\mathcal{H}}(v) = 0$. Let $k \in \mathbb{N}$ be a positive integer. Hypergraph $\mathcal{H}$ is called *k-regular* if $d_{\mathcal{H}}(u) = k$ for each $u \in \mathcal{V}$, and *k-uniform* if $|e| = k$ for each $e \in \mathcal{E}$. Hypergraph $\mathcal{H}$ is *linear* if $|e \cap f| \leq 1$ for any pair of distinct edges $e, f \in \mathcal{E}$.

A vertex-edge alternating sequence $v_1 e_1 v_2 ... v_k e_k v_{k+1}$ of $\mathcal{H}$ is called a *path* (of *length k*) between v_1 and v_{k+1} if $v_1, v_2, ..., v_{k+1} \in \mathcal{V}$ are distinct, $e_1, e_2, ..., e_k \in \mathcal{E}$ are distinct, and $\{v_i, v_{i+1}\} \subseteq e_i$ for each $i \in [k] = \{1, ..., k\}$. Hypergraph $\mathcal{H}$ is said to be *connected* if there is a path between any pair of distinct vertices in $\mathcal{H}$. A maximal connected sub-hypergraph of $\mathcal{H}$ is called a *component* of $\mathcal{H}$.

A vertex-edge alternating sequence $\mathcal{C} = v_1 e_1 v_2 e_2 ... v_k e_k v_1$, where $k \geq 2$, is called a *cycle* (of length k) if $v_1, v_2, ..., v_k \in \mathcal{V}$ are distinct, $e_1, e_2, ..., e_k \in \mathcal{E}$ are distinct, and $\{v_i, v_{i+1}\} \subseteq e_i$ for each $i \in [k]$, where $v_{k+1} = v_1$. We consider the cycle $\mathcal{C}$ as a sub-hypergraph of $\mathcal{H}$ with vertex set $\cup_{i \in [k]} e_i$ and edge set $\{e_i : i \in [k]\}$. For any $\mathcal{S} \subset \mathcal{V}$ (resp. $\mathcal{S} \subset \mathcal{E}$), we write $\mathcal{H} \backslash \mathcal{S}$ for the sub-hypergraph of $\mathcal{H}$ obtained from $\mathcal{H}$ by deleting all vertices in $\mathcal{S}$ and all edges incident with some vertices in $\mathcal{S}$ (resp. deleting all edges in $\mathcal{E}$ and keeping vertices). If $\mathcal{S}$ is a singleton set $\{s\}$, we write $\mathcal{H} \backslash s$ instead of $\mathcal{H} \backslash \{s\}$. For any $\mathcal{S} \subseteq 2^{\mathcal{V}}$, the hypergraph $(\mathcal{V}, \mathcal{E} \cup \mathcal{S})$ is often written as $\mathcal{H} \oplus \mathcal{S}$ if $\mathcal{S} \cap \mathcal{E} = \emptyset$.

A vertex (resp. edge) subset of $\mathcal{H}$ is called a *feedback vertex set* or FVS (resp. *feedback edge set* or FES) of $\mathcal{H}$ if it intersects the vertex (resp. edge) set of every cycle of $\mathcal{H}$. A vertex subset of $\mathcal{H}$ is called a *transversal* of $\mathcal{H}$ if it intersects every edge of $\mathcal{H}$. Let $\tau_c^{\mathcal{V}}(\mathcal{H})$, $\tau_c^{\mathcal{E}}(\mathcal{H})$ and $\tau(\mathcal{H})$ denote, respectively, the minimum cardinalities of a FVS, a FES, and a transversal of $\mathcal{H}$. A *matching* of $\mathcal{H}$ is an nonempty set of pairwise disjoint edges of $\mathcal{H}$. Let $\nu(\mathcal{H})$ denote the maximum cardinality of a matching of $\mathcal{H}$. It is easy to see that $\tau_c^{\mathcal{V}}(\mathcal{H}) \leq \tau_c^{\mathcal{E}}(\mathcal{H})$, $\tau_c^{\mathcal{V}}(\mathcal{H}) \leq \tau(\mathcal{H})$ and $\nu(\mathcal{H}) \leq \tau(\mathcal{H})$. Our discussion will frequently use the trivial observation that if no cycle of $\mathcal{H}$ contains any element of some subset $\mathcal{S}$ of $\mathcal{V} \cup \mathcal{E}$, then $\mathcal{H}$ and $\mathcal{H} \backslash \mathcal{S}$ have the same set of FVS's, and $\tau_c^{\mathcal{V}}(\mathcal{H}) = \tau_c^{\mathcal{V}}(\mathcal{H} \backslash \mathcal{S})$. The following theorem is one of our main contributions.

Theorem 4. *Let $\mathcal{H}$ be a linear 3-uniform hypergraph. Then $\tau_c^{\mathcal{V}}(\mathcal{H}) \leq \|\mathcal{H}\|/3$.*

Proof. Suppose that the theorem failed. We take a counterexample $\mathcal{H} = (\mathcal{V}, \mathcal{E})$ with $\tau_c^{\mathcal{V}}(\mathcal{H}) > |\mathcal{E}|/3$ such that $\|\mathcal{H}\| = |\mathcal{E}|$ is as small as possible. Obviously $|\mathcal{E}| \geq 3$. Without loss of generality, we can assume that $\mathcal{H}$ has no isolated vertices. Since $\mathcal{H}$ is linear, any cycle in $\mathcal{H}$ is of length at least 3.

If there exists some $e \in \mathcal{E}$ which does not belong to any cycle of $\mathcal{H}$, then $\tau_c^{\mathcal{V}}(\mathcal{H}) = \tau_c^{\mathcal{V}}(\mathcal{H} \backslash e)$. The minimality of $\mathcal{H} = (\mathcal{V}, \mathcal{E})$ implies $\tau_c^{\mathcal{V}}(\mathcal{H} \backslash e) \leq (|\mathcal{E}| - 1)/3$, giving $\tau_c^{\mathcal{V}}(\mathcal{H}) < |\mathcal{E}|/3$, a contradiction. So we have

(1) Every edge in $\mathcal{E}$ is contained in some cycle of $\mathcal{H}$.

If there exists some $v \in \mathcal{V}$ with $d_{\mathcal{H}}(v) \geq 3$, then $\tau_c^{\mathcal{V}}(\mathcal{H} \backslash v) \leq (|\mathcal{E}| - d_{\mathcal{H}}(v))/3 \leq (|\mathcal{E}| - 3)/3$, where the first inequality is due to the minimality of $\mathcal{H}$. Given a minimum FVS $\mathcal{S}$ of $\mathcal{H} \backslash v$, it is clear that $\mathcal{S} \cup \{v\}$ is a FVS of $\mathcal{H}$ with size $|\mathcal{S}| + 1 = \tau_c^{\mathcal{V}}(\mathcal{H} \backslash v) + 1 \leq |\mathcal{E}|/3$, a contradiction to $\tau_c^{\mathcal{V}}(\mathcal{H}) > |\mathcal{E}|/3$. So we have

(2) $d_{\mathcal{H}}(v) \leq 2$ for all $v \in \mathcal{V}$.

Suppose that there exists some $v \in \mathcal{V}$ with $d_{\mathcal{H}}(v) = 1$. Let $e_1 \in \mathcal{E}$ be the unique edge that contains v. Recall from (1) that e_1 is contained in a cycle $\mathcal{C} = v_1 e_1 v_2 e_2 v_3 \cdots e_k v_1$, where $k \geq 3$. By (2), we have $d_{\mathcal{H}}(v_i) = 2$ for all $i \in [k]$. In particular $d_{\mathcal{H}}(v_1) = d_{\mathcal{H}}(v_2) = 2 > d_{\mathcal{H}}(v)$ implies $v \notin \{v_1, v_2\}$, and in turn $v_1, v_2, v \in e_1$ enforces $e_1 = \{v_1, v, v_2\}$. Let $\mathcal{S}$ be a minimum FVS of $\mathcal{H}' = \mathcal{H} \backslash \{e_1, e_2, e_3\}$. It follows from (2) that

$$\mathcal{H} \backslash v_3 \subseteq \mathcal{H} \backslash \{e_2, e_3\} = \mathcal{H}' \oplus e_1,$$

and in $\mathcal{H}' \oplus e_1$, edge e_1 intersects at most one other edge, and therefore is not contained in any cycle. Thus $\mathcal{S}$ is a FVS of $\mathcal{H}' \oplus e_1$, and hence a FVS of $\mathcal{H} \setminus v_3$, implying that $\{v_3\} \cup \mathcal{S}$ is a FVS of $\mathcal{H}$. We deduce that $|\mathcal{E}|/3 < \tau_c^\nu(\mathcal{H}) \leq |\{v_3\} \cup \mathcal{S}| \leq 1 + |\mathcal{S}|$. Therefore $\tau_c^\nu(\mathcal{H}') = |\mathcal{S}| > (|\mathcal{E}| - 3)/3 = \|\mathcal{H}'\|/3$ shows a contradiction to the minimality of $\mathcal{H}$. Hence the vertices of $\mathcal{H}$ all have degree at least 2, which together with (2) gives

(3) $\mathcal{H}$ is 2-regular.

Let $\mathcal{C} = (\mathcal{V}_c, \mathcal{E}_c) = v_1 e_1 v_2 e_2 \ldots v_k e_k v_1$ be a shortest cycle in $\mathcal{H}$, where $k \geq 3$. For each $i \in [k]$, suppose that $e_i = \{v_i, u_i, v_{i+1}\}$, where $v_{k+1} = v_1$.

Because $\mathcal{C}$ is a shortest cycle, for each pair of distinct indices $i, j \in [k]$, we have $e_i \cap e_j = \emptyset$ if and only if e_i and e_j are not adjacent in $\mathcal{C}$, i.e., $|i - j| \notin \{1, k - 1\}$. This fact along with the linearity of $\mathcal{H}$ says that $v_1, v_2, \ldots, v_k, u_1, u_2, \ldots, u_k$ are distinct. By (3), each u_i is contained in a unique edge $f_i \in \mathcal{E} \setminus \mathcal{E}_c$, $i \in [k]$. We distinguish among three cases depending on the values of $k \pmod 3$. In each case, we construct a proper sub-hypergraph $\mathcal{H}'$ of $\mathcal{H}$ with $\|\mathcal{H}'\| < \|\mathcal{H}\|$ and $\tau_c^\nu(\mathcal{H}') > \|\mathcal{H}'\|/3$ which shows a contradiction to the minimality of $\mathcal{H}$.

CASE 1. $k \equiv 0 \pmod 3$: Let $\mathcal{S}$ be a minimum FVS of $\mathcal{H}' = \mathcal{H} \setminus \mathcal{E}_c$. Setting $\mathcal{V}_* = \{v_i : i \equiv 0 \pmod 3, i \in [k]\}$ and $\mathcal{E}_* = \{e_i : i \equiv 1 \pmod 3, i \in [k]\}$, it follows from (3) that

$$\mathcal{H} \setminus \mathcal{V}_* \subseteq (\mathcal{H} \setminus \mathcal{E}_c) \oplus \mathcal{E}_* = \mathcal{H}' \oplus \mathcal{E}_*,$$

and in $\mathcal{H}' \oplus \mathcal{E}_*$, each edge in $\mathcal{E}_*$ intersects exactly one other edge, and therefore is not contained in any cycle. Thus $(\mathcal{H}' \oplus \mathcal{E}_*) \setminus \mathcal{S}$ is also acyclic, so is $(\mathcal{H} \setminus \mathcal{V}_*) \setminus \mathcal{S}$, saying that $\mathcal{V}_* \cup \mathcal{S}$ is a FVS of $\mathcal{H}$. We deduce that $|\mathcal{E}|/3 < \tau_c^\nu(\mathcal{H}) \leq |\mathcal{V}_* \cup \mathcal{S}| \leq k/3 + |\mathcal{S}|$. Therefore $\tau_c^\nu(\mathcal{H}') = |\mathcal{S}| > (|\mathcal{E}| - k)/3 = \|\mathcal{H}'\|/3$ shows a contradiction.

CASE 2. $k \equiv 1 \pmod 3$: Consider the case where $f_1 \neq f_3$ or $f_2 \neq f_4$. Relabeling the vertices and edges if necessary, we may assume without loss of generality that $f_1 \neq f_3$. Let $\mathcal{S}$ be a minimum FVS of $\mathcal{H}' = \mathcal{H} \setminus (\mathcal{E}_c \cup \{f_1, f_3\})$. Set $\mathcal{V}_* = \emptyset$, $\mathcal{E}_* = \emptyset$ if $k = 4$ and $\mathcal{V}_* = \{v_i : i \equiv 0 \pmod 3, i \in [k] - [3]\}$, $\mathcal{E}_* = \{e_i : i \equiv 1 \pmod 3, i \in [k] - [6]\}$ otherwise. In any case we have $|\mathcal{V}_*| = (k - 4)/3$ and

$$\mathcal{H} \setminus (\{u_1, u_3\} \cup \mathcal{V}_*) \subseteq (\mathcal{H} \setminus (\mathcal{E}_c \cup \{f_1, f_3\})) \oplus (\{e_2, e_4\} \cup \mathcal{E}_*) = \mathcal{H}' \oplus (\{e_2, e_4\} \cup \mathcal{E}_*).$$

Note from (3) that in $\mathcal{H}' \oplus (\{e_2, e_4\} \cup \mathcal{E}_*)$, each edge in $\{e_2, e_4\} \cup \mathcal{E}_*$ can intersect at most one other edge, and therefore is not contained in any cycle. Thus $(\mathcal{H}' \oplus (\{e_2, e_4\} \cup \mathcal{E}_*)) \setminus \mathcal{S}$ is also acyclic, so is $(\mathcal{H} \setminus (\{u_1, u_3\} \cup \mathcal{V}_*)) \setminus \mathcal{S}$. Thus $\{u_1, u_3\} \cup \mathcal{V}_* \cup \mathcal{S}$ is a FVS of $\mathcal{H}$, and $|\mathcal{E}|/3 < \tau_c^\nu(\mathcal{H}) \leq |\{u_1, u_3\} \cup \mathcal{V}_* \cup \mathcal{S}| \leq 2 + |\mathcal{V}_*| + |\mathcal{S}| = (k + 2)/3 + |\mathcal{S}|$. This gives $\tau_c^\nu(\mathcal{H}') = |\mathcal{S}| > (|\mathcal{E}| - k - 2)/3 = \|\mathcal{H}'\|/3$, a contradiction.

Consider the case where $f_1 = f_3$ and $f_2 = f_4$. As u_1, u_2, u_3, u_4 are distinct and $|f_1| = |f_2| = 3$, we have $f_1 \neq f_2$. Observe that $u_1 e_1 v_2 e_2 v_3 e_3 u_3 f_3 u_1$ is a cycle in $\mathcal{H}$ of length 4. The minimality of k enforces $k = 4$. Therefore $\mathcal{E}_c \cup \{f_1, f_2\}$

consist of 6 distinct edges. Let $\mathcal{S}$ be a minimum FVS of $\mathcal{H}' = \mathcal{H} \setminus (\mathcal{E}_c \cup \{f_1, f_2\})$. It follows from (3) that

$$\mathcal{H} \setminus \{u_2, u_4\} \subseteq (\mathcal{H} \setminus (\mathcal{E}_c \cup \{f_1, f_2\})) \oplus \{e_1, e_3, f_1\} = \mathcal{H}' \oplus \{e_1, e_3, f_1\}.$$

In $\mathcal{H}' \oplus \{e_1, e_3, f_1\}$, both e_1 and e_3 intersect only one other edge, which is f_1, and any cycle through f_1 must contain e_1 or e_3. It follows that none of e_1, e_3, f_1 is contained by a cycle of $\mathcal{H}' \oplus \{e_1, e_3, f_1\}$. Thus $(\mathcal{H}' \oplus \{e_1, e_3, f_1\}) \setminus \mathcal{S}$ is acyclic, so is $(\mathcal{H} \setminus \{u_2, u_4\}) \setminus \mathcal{S}$, saying that $\{u_2, u_4\} \cup \mathcal{S}$ is a FVS of $\mathcal{H}$. Hence $|\mathcal{E}|/3 < \tau_c^\nu(\mathcal{H}) \le |\{u_2, u_4\} \cup \mathcal{S}| \le 2 + |\mathcal{S}|$. In turn $\tau_c^\nu(\mathcal{H}') = |\mathcal{S}| > (|\mathcal{E}| - 6)/3 = \|\mathcal{H}'\|/3$ shows a contradiction.

CASE 3. $k \equiv 2 \pmod 3$: Let $\mathcal{S}$ be a minimum FVS of $\mathcal{H}' = \mathcal{H} \setminus (\mathcal{E}_c \cup \{f_1\})$. Setting $\mathcal{V}_* = \{v_i : i \equiv 1 \pmod 3, i \in [k] - [3]\}$ and $\mathcal{E}_* = \{e_i : i \equiv 2 \pmod 3, i \in [k]\}$, we have $|\mathcal{V}_*| = (k-2)/3$ and

$$\mathcal{H} \setminus (\{u_1\} \cup \mathcal{V}_*) \subseteq (\mathcal{H} \setminus (\mathcal{E}_c \cup \{f_1\})) \oplus \mathcal{E}_* = \mathcal{H}' \oplus \mathcal{E}_*$$

In $\mathcal{H}' \oplus \mathcal{E}_*$, each edge in $\mathcal{E}_*$ intersects at most one other edge, and therefore is not contained in any cycle. Thus $(\mathcal{H}' \oplus \mathcal{E}_*) \setminus \mathcal{S}$ is acyclic, so is $(\mathcal{H} \setminus (\{u_1\} \cup \mathcal{V}_*)) \setminus \mathcal{S}$. Hence $\{u_1\} \cup \mathcal{V}_* \cup \mathcal{S}$ is a FVS of $\mathcal{H}$, yielding $|\mathcal{E}|/3 < \tau_c^\nu(\mathcal{H}) \le |\{u_1\} \cup \mathcal{V}_* \cup \mathcal{S}| \le 1 + (k-2)/3 + |\mathcal{S}|$ and a contradiction $\tau_c^\nu(\mathcal{H}') = |\mathcal{S}| > (|\mathcal{E}| - k - 1)/3 = \|\mathcal{H}'\|/3$.

The combination of the above three cases complete the proof. □

The upper bound $\|\mathcal{H}\|/3$ in Theorem 4 is best possible. See Fig. 1 for illustrations of five linear 3-uniform hypergraphs attaining the upper bound. It is easy to prove that the maximum degree of every extremal hypergraph (those $\mathcal{H}$ with $\tau_c^\nu(\mathcal{H}) = \|\mathcal{H}\|/3$) is at most three. Despite a number of attempts, we did not find any extremal hypergraph other than those in Fig. 1. It would be interesting to characterize all extremal hypergraphs for Theorem 4.

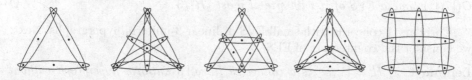

Fig. 1. Some linear 3-uniform hypergraphs $\mathcal{H}$ with $\tau_c^\nu(\mathcal{H}) = \|\mathcal{H}\|/3$.

The proof of Theorem 4 actually gives a recursive combinatorial algorithm (Algorithm 1) for finding in polynomial time a FVS of size at most $\|\mathcal{H}\|/3$ on a linear 3-uniform hypergraph $\mathcal{H}$.

Note that Algorithm 1 never visits isolated vertices (it only scans along the edges of the current hypergraph). The number of iterations performed by the algorithm is upper bounded by $|\mathcal{E}|$. Since $\mathcal{H}$ is 3-uniform, the condition in any step is checkable in $O(|\mathcal{E}|^2)$ time. One can use the breadth first search algorithm to find a cycle in stated in Step 7 or Step 9 in $O(|\mathcal{E}|^2)$ time. Thus Algorithm 1 runs in $O(|\mathcal{E}|^3)$ time.

ALGORITHM 1. Fvs($\cdot$) for finding FVS's of linear 3-uniform hypergraphs

---

Input: A linear 3-uniform hypergraph $\mathcal{H} = (\mathcal{V}, \mathcal{E})$.

Output: Fvs($\mathcal{H}$), which is a FVS of $\mathcal{H}$ with cardinality at most $\|\mathcal{H}\|/3$.

1. **If** $|\mathcal{E}| \le 2$ **Then** Return $\emptyset$
2. **Else If** $\exists\, s \in \mathcal{V} \cup \mathcal{E}$ s.t. s is not contained in any cycle of $\mathcal{H}$
3.   **Then** Return Fvs($\mathcal{H} \setminus s$)
4.   **If** $\exists\, s \in \mathcal{V}$ s.t. $d_{\mathcal{H}}(s) \ge 3$
5.   **Then** Return $\{s\} \cup$ Fvs($\mathcal{H} \setminus s$)
6.   **If** $\exists\, v \in \mathcal{V}$ s.t. $d_{\mathcal{H}}(v) = 1$
7.   **Then** Let $v_1 e_1 v_2 e_2 v_3 \cdots e_k v_1$ be a cycle of $\mathcal{H}$ s.t. $e_1 = \{v_1, v_2, v\}$
8.   Return $\{v_3\} \cup$ Fvs($\mathcal{H} \setminus \{e_1, e_2, e_3\}$)
9.   Let $(\mathcal{V}_c, \mathcal{E}_c) = v_1 e_1 v_2 e_2 \ldots v_k e_k v_1$ be a shortest cycle in $\mathcal{H}$
10.   For each $i \in [k]$, let $u_i \in \mathcal{V}_c$, $f_i \in \mathcal{E} \setminus \mathcal{E}_c$ be s.t. $\{u_i, v_i, v_{i+1}\} = e_i$, $u_i \in f_i$
11.   **If** $k \equiv 0 \pmod 3$
12.   **Then** Return $\{v_i : i \equiv 0 \pmod 3, i \in [k]\} \cup$ Fvs($\mathcal{H} \setminus \mathcal{E}_c$)
13.   **If** $k \equiv 1 \pmod 3$
14.   **Then If** $f_1 \ne f_3$ or $f_2 \ne f_4$
15.   **Then** Relabel vertices & edges if necessary to make $f_1 \ne f_3$
16. $\mathcal{V}_* \leftarrow \{v_i : i \equiv 0 \pmod 3, i \in [k] - [3]\}$
17.   Return $\{u_1, u_3\} \cup \mathcal{V}_* \cup$ Fvs($\mathcal{H} \setminus \mathcal{E}_c \setminus \{f_1, f_3\}$)
18.   **Else** Return $\{u_2, u_4\} \cup$ Fvs($\mathcal{H} \setminus \mathcal{E}_c \setminus \{f_1, f_2\}$)
19.   **If** $k \equiv 2 \pmod 3$
20.   **Then** $\mathcal{V}_* \leftarrow \{v_i : i \equiv 1 \pmod 3, i \in [k] - [3]\}$
21.   Return $\{u_1\} \cup \mathcal{V}_* \cup$ Fvs($\mathcal{H} \setminus (\mathcal{E}_c \cup \{f_1\})$)

Corollary 1. *Given any linear 3-uniform hypergraph $\mathcal{H}$, Algorithm 1 finds in $O(\|\mathcal{H}\|^3)$ time a FVS of $\mathcal{H}$ with size at most $\|\mathcal{H}\|/3$.* $\square$

Corollary 1 concerns with small FVS of linear 3-uniform hypergraphs. Next, we consider the counterpart of FES.

Lemma 1. *If $\mathcal{H} = (\mathcal{V}, \mathcal{E})$ is a connected linear 3-uniform hypergraph without cycles, then $|\mathcal{V}| = 2|\mathcal{E}| + 1$.*

Proof. We prove by induction on $|\mathcal{E}|$. The base case where $|\mathcal{E}| = 0$ is trivial. Inductively, we assume that $|\mathcal{E}| \ge 1$ and the lemma holds for all connected acyclic linear 3-uniform hypergraph of edges fewer than $\mathcal{H}$. Take arbitrary $e \in \mathcal{E}$. Since $\mathcal{H}$ is connected, acyclic and 3-uniform, $\mathcal{H} \setminus e$ contains exactly three components $\mathcal{H}_i = (\mathcal{V}_i, \mathcal{E}_i)$, $i = 1, 2, 3$. Note that for each $i \in [3]$, hypergraph $\mathcal{H}_i$ with $|\mathcal{E}_i| < |\mathcal{E}|$ is connected, linear, 3-uniform and acyclic. By the induction hypothesis, we have $|\mathcal{V}_i| = 2|\mathcal{E}_i| + 1$ for $i = 1, 2, 3$. It follows that $|\mathcal{V}| = \sum_{i=1}^{3} |\mathcal{V}_i| = 2 \sum_{i=1}^{3} |\mathcal{E}_i| + 3 = 2|\mathcal{E}| + 1$. $\square$

Given any hypergraph $\mathcal{H} = (\mathcal{V}, \mathcal{E})$, we can easily find a minimal (not necessarily minimum) FES in $O(|\mathcal{E}|^2)$ time: Go through the edges of the trivial FES

$\mathcal{E}$ in any order, and remove the edge from the FES immediately if the edge is redundant. The redundancy test can be implemented using Depth First Search.

Lemma 2. *Let* $\mathcal{H} = (\mathcal{V}, \mathcal{E})$ *be a linear 3-uniform hypergraph with p components. If $\mathcal{F}$ is a minimal FES of $\mathcal{H}$, then $|\mathcal{F}| \leq 2|\mathcal{E}| - |\mathcal{V}| + p$. In particular, $\tau_c^{\varepsilon}(\mathcal{H}) \leq 2|\mathcal{E}| - |\mathcal{V}| + p$.*

Proof. Suppose that $\mathcal{H} \setminus \mathcal{F}$ contains exactly k components $\mathcal{H}_i = (\mathcal{V}_i, \mathcal{E}_i)$, $i = 1, \ldots, k$. It follows from Lemma 1 that $|\mathcal{V}_i| = 2|\mathcal{E}_i| + 1$ for each $i \in [k]$. Thus $|\mathcal{V}| = {}_{i \in [k]} |\mathcal{V}_i| = 2 {}_{i \in [k]} |\mathcal{E}_i| + k = 2(|\mathcal{E}| - |\mathcal{F}|) + k$, which means $2|\mathcal{F}| = 2|\mathcal{E}| - |\mathcal{V}| + k$. To establish the lemma, it suffices to prove $k \leq |\mathcal{F}| + p$.

In case of $|\mathcal{F}| = 0$, we have $\mathcal{F} = \emptyset$ and $k = p = |\mathcal{F}| + p$. In case of $|\mathcal{F}| \geq 1$, suppose that $\mathcal{F} = \{e_1, \ldots, e_{|\mathcal{F}|}\}$. Because $\mathcal{F}$ is a minimal FES of $\mathcal{H}$, for each $i \in [|\mathcal{F}|]$, there is a cycle $\mathcal{C}_i$ in $\mathcal{H} \setminus (\mathcal{F} \setminus \{e_i\})$ such that $e_i \in \mathcal{C}_i$, and $\mathcal{C}_i \setminus e_i$ is a path in $\mathcal{H} \setminus \mathcal{F}$ connecting two of the three vertices in e_i. Considering $\mathcal{H} \setminus \mathcal{F}$ being obtained from $\mathcal{H}$ be removing $e_1, e_2, \ldots, e_{|\mathcal{F}|}$ sequentially, for $i = 1, \ldots, |\mathcal{F}|$, since $|e_i| = 3$, the presence of path $\mathcal{C}_i \setminus e_i$ implies that the removal of e_i can create at most one more component. Therefore we have $k \leq p + |\mathcal{F}|$ as desired. $\square$

Given a hypergraph $\mathcal{H}$, let $M_{\mathcal{H}}$ be the $\mathcal{V} \times \mathcal{E}$ incidence matrix. If $\mathcal{H}$ is acyclic, then $M_{\mathcal{H}}$ falls within the class of *restricted totally unimodular* matrices, and a minimum transversal and a maximum matching of $\mathcal{H}$ can be found using Yanakakis's combinatorial algorithm [13] based on the current best combinatorial algorithms for the b-matching problem and the maximum weighted independent set problem on bipartite multigraphs [10].

Theorem 5 ([2,13]). *Let $\mathcal{H}$ be a hypergraph with n non-isolated vertices and m edges. If $\mathcal{H}$ has no cycle, then $\tau(\mathcal{H}) = \nu(\mathcal{H})$, and a minimum transversal and a maximum matching of $\mathcal{H}$ can be found in $O(n(m + n \log n) \log n)$ time.* $\square$

3 Triangle Packing and Covering

This section establishes several new sufficient conditions for Conjecture 1, and provides their algorithmic implications on finding small triangle covers. Section 3.1 deals with graphs of high triangle packing numbers. Section 3.2 investigates irreducible graphs with many edges.

To each graph $G = (V, E)$, we associate a hypergraph $\mathcal{H}_G = (E, \mathscr{T}_G)$, referred to as *triangle hypergraph* of G. Since G is simple, it is easy to see that $\mathcal{H}_G$ is 3-uniform and linear, $\nu(\mathcal{H}_G) = \nu_t(G)$ and $\tau(\mathcal{H}_G) = \tau_t(G)$. Note that $\|\mathcal{H}_G\| = |\mathscr{T}_G| < \min\{|V|^3, |E|^3\}$, and $|E| \leq 3|\mathscr{T}_G|$ if G is irreducible, i.e., $\cup_{T \in \mathscr{T}_G} E(T) = E$. Note that the number of non-isolated vertices of $\mathcal{H}_G$ is upper bounded by $3\|\mathcal{H}_G\| = 3|\mathscr{T}_G|$.

3.1 Graphs with Many Edge-Disjoint Triangles

We investigate Conjecture 1 for graphs with large triangle packing numbers, which are firstly compared with the number of triangles, and then with the number of edges.

Theorem 6. *If a graph G and a real number $c \in (0,1]$ satisfy $\nu_t(G)/|\mathscr{T}_G| \geq c$, then a triangle cover of G with size at most $\frac{3c+1}{3c}\nu_t(G)$ can be found in $O(|\mathscr{T}_G|^3)$ time, which implies $\tau_t(G)/\nu_t(G) \leq 1 + \frac{1}{3c}$.*

Proof. We consider the triangle hypergraph $\mathcal{H}_G = (E, \mathscr{T}_G)$ of G which is 3-uniform and linear. By Corollary 1, we can find in $O(|\mathscr{T}_G|^3)$ time a FVS $\mathcal{S}$ of $\mathcal{H}_G$ with $|\mathcal{S}| \leq |\mathscr{T}_G|/3$. Since $\nu(\mathcal{H}_G) = \nu_t(G) \geq c|\mathscr{T}_G|$, it follows that $|\mathcal{S}| \leq \nu(\mathcal{H}_G)/(3c)$. As $\mathcal{H}_G \setminus \mathcal{S}$ is acyclic, Theorem 5 enables us to find in $O(|\mathscr{T}_G|^2 \log^2 |\mathscr{T}_G|)$ time a minimum transversal $\mathcal{R}$ of $\mathcal{H}_G \setminus \mathcal{S}$ such that $|\mathcal{R}| = \tau(\mathcal{H}_G \setminus \mathcal{S}) = \nu(\mathcal{H}_G \setminus \mathcal{S})$. We observe that $\mathcal{S} \cup \mathcal{R} \subseteq E$ and $G \setminus (\mathcal{S} \cup \mathcal{R})$ is triangle-free. Hence $\mathcal{S} \cup \mathcal{R}$ is a triangle cover of G with size

$$|\mathcal{S} \cup \mathcal{R}| \leq \frac{\nu(\mathcal{H}_G)}{3c} + \nu(\mathcal{H}_G \setminus \mathcal{S}) \leq \frac{3c+1}{3c}\nu(\mathcal{H}_G) = \frac{3c+1}{3c}\nu_t(G),$$

which proves the theorem.   □

The special case of $c = 1/3$ in the above theorem gives the following result providing a new sufficient condition for Conjecture 1.

Corollary 2. *If graph G satisfies $\nu_t(G)/|\mathscr{T}_G| \geq 1/3$, then $\tau_t(G)/\nu_t(G) \leq 2$.*   □

The mapping from the lower bound c in the condition $\nu_t(G)/|\mathscr{T}_G| \geq c$ to the upper bound $1 + \frac{1}{3c}$ in the conclusion $\tau_t(G)/\nu_t(G) \leq 1 + \frac{1}{3c}$ of Theorem 6 shows a kind of trade-off. In Corollary 2, $c = \frac{1}{3}$ maps to $1 + \frac{1}{3c} = 2$ hitting the boundary of Conjecture 1. It remains to study graphs G with $\nu_t(G)/|\mathscr{T}_G| < \frac{1}{3}$. The next theorem (Theorem 7) says that we only need to take care of graphs G with $\nu_t(G)/|\mathscr{T}_G| \in (\frac{1}{4} - \epsilon, \frac{1}{3})$, where ϵ can be any arbitrarily small positive number. So, in some sense, to settle Conjecture 1, we only have a gap of $\frac{1}{3} - \frac{1}{4} = \frac{1}{12}$ to be bridged. Interestingly, for $c = \frac{1}{4}$, we have $1 + \frac{1}{3c} = \frac{7}{3} = 2.333...$, which is much better than the best known general bound 2.87 due to Haxell [7].

Theorem 7. *If there exists some real $\delta > 0$ such that Conjecture 1 holds for every graph G with $\nu_t(G)/|\mathscr{T}_G| \geq 1/4 - \delta$, then Conjecture 1 holds for every graph.*

Proof. If $\delta \geq \frac{1}{4}$, the theorem is trivial. We consider $0 < \delta < \frac{1}{4}$. As the set of rational numbers is dense, we may assume $\delta \in \mathbb{Q}$ and $1/4 - \delta = i/j$ for some $i, j \in \mathbb{N}$. Therefore $i/j < 1/4$ gives $4i + 1 \leq j$, i.e., $4 + 1/i \leq j/i$. It remains to prove that for any graph G with $\nu_t(G) < (i/j)|\mathscr{T}_G|$ there holds $\tau_t(G) \leq 2\nu_t(G)$.

Write k for the positive integer $i|\mathscr{T}_G| - j \cdot \nu_t(G)$. Let G' be the disjoint union of G and k copies of K_4. Clearly, $|\mathscr{T}_{G'}| = |\mathscr{T}_G| + k|\mathscr{T}_{K_4}| = |\mathscr{T}_G| + 4k$, $\tau_t(G') =$

$\tau_t(G) + k \cdot \tau_t(K_4) = \tau_t(G) + 2k$ and $\nu_t(G') = \nu_t(G) + k \cdot \nu_t(K_4) = \nu_t(G) + k$. It follows that

$$
\begin{aligned}
(i/j)|\mathscr{T}_{G'}| &= (i/j)(|\mathscr{T}_G| + 4k) \\
&= (i/j)((k + j \cdot \nu_t(G))/i + 4k) \\
&= (i/j)(j \cdot \nu_t(G)/i + (4 + 1/i)k) \\
&\leq \nu_t(G) + k \\
&= \nu_t(G')
\end{aligned}
$$

where the inequality is guaranteed by $4 + 1/i \leq j/i$. So $\nu_t(G') \geq (1/4 - \delta)|\mathscr{T}_{G'}|$ together with the hypothesis of the theorem implies $\tau_t(G') \leq 2\nu_t(G')$, i.e., $\tau_t(G) + 2k \leq 2(\nu_t(G) + k)$, giving $\tau_t(G) \leq 2\nu_t(G)$ as desired. $\square$

Next, we discuss the sufficient condition that compares the triangle packing number with the number of edges. It is based on the fact that every graph G can be made bipartite (and thus triangle-free) in polynomial time by removing at most half of its edges. Therefore $\tau_t(G) \leq |E(G)|/2$, which implies the following result.

Corollary 3. *If $G = (V, E)$ is a graph such that $\nu_t(G)/|E| \geq c$ for some $c > 0$, then $\tau_t(G)/\nu_t(G) \leq 1/(2c)$. In particular, if $\nu_t(G)/|E| \geq 1/4$, then $\tau_t(G)/\nu_t(G) \leq 2$.* $\square$

Thus if $\nu_t(G)/|E| \geq c$ for some $c > 0$, then a triangle cover of G with size at most $\nu_t(G)/(2c)$ can be found in polynomial time. Complementary to Corollary 2 whose condition $\nu_t(G)/|\mathscr{T}_G| \geq 1/3$ mainly takes care of sparse graphs, the second statement of Corollary 3 applies to many dense graphs, including complete graphs on 25 or more vertices.

Similar to Corollary 2 and Theorem 7, by which our future investigation space on Conjecture 1 shrinks to interval $(\frac{1}{4} - \epsilon, \frac{1}{3})$ w.r.t. $\nu_t(G)/|\mathscr{T}_G|$, Corollary 3 and the following Theorem 8 narrow the interval w.r.t. $\nu_t(G)/|E|$ to $(\frac{1}{5} - \epsilon, \frac{1}{4})$. Moreover, when taking $c = \frac{1}{5}$ in Corollary 3. we obtain $\frac{1}{2c} = 2.5$, still better than the general bound 2.87 of Haxell [7].

Theorem 8. *If there exists some real $\delta > 0$ such that Conjecture 1 holds for every graph G with $\nu_t(G)/|E| \geq 1/5 - \delta$, then Conjecture 1 holds for every graph.* $\square$

Proof. We use the similar trick to that in proving Theorem 7; we add a number of complete graphs on five (instead of four) vertices. We may assume $\delta \in (0, \frac{1}{5}) \cap \mathbb{Q}$ and $1/5 - \delta = i/j$ for some $i, j \in \mathbb{N}$. Therefore $i/j < 1/5$ and the integrality of i, j imply $5 + 1/i \leq j/i$. To prove Conjecture 1 for each graph G with $\nu_t(G) < (i/j)|E|$, we write $k = i|E| - j \cdot \nu_t(G) \in \mathbb{N}$. Let $G' = (V', E')$ be the disjoint union of G and k copies of K_5's. Then $|E'| = |E| + 10k$, $\tau_t(G') = \tau_t(G) + k \cdot \tau_t(K_5) = \tau_t(G) + 4k$, $\nu_t(G') = \nu_t(G) + k \cdot \nu_t(K_5) = \nu_t(G) + 2k$, and

$$
(i/j)|E'| = (i/j)(|E| + 10k) = (i/j)(j \cdot \nu_t(G)/i + (10 + 1/i)k) \leq \nu_t(G) + 2k = \nu_t(G')
$$

where the inequality is guaranteed by $10 + 1/i \leq 2j/i$. So $\nu_t(G') \geq (1/5 - \delta)|E'|$ together with the hypothesis the theorem implies $\tau_t(G') \leq 2\nu_t(G')$, i.e., $\tau_t(G) + 4k \leq 2(\nu_t(G) + 2k)$, giving $\tau_t(G) \leq 2\nu_t(G)$ as desired. $\square$

3.2   Graphs with Many Edges on Triangles

Each graph has a unique maximum irreducible subgraph. Conjecture 1 is valid for a graph if and only the conjecture is valid for its maximum irreducible subgraph. In this section, we study sufficient conditions for Conjecture 1 on irreducible graphs that bound the number of edges from below in terms of the number of triangles.

Theorem 9. *If $G = (V, E)$ is an irreducible graph such that $|E|/|\mathcal{T}_G| \geq 2$, then a triangle cover of G with cardinality at most $2\nu_t(G)$ can be found in $O(|\mathcal{T}_G|^2 \log^2 |\mathcal{T}_G|)$ time, which implies $\tau_t(G)/\nu_t(G) \leq 2$.*

Proof. Let p be the number of components of the linear 3-uniform hypergraph $\mathcal{H} = (E, \mathcal{T}_G)$ associated to G. By Lemma 2, we can find in $O(|\mathcal{T}_G|^2)$ time a minimal FES $\mathcal{F}$ of $\mathcal{H}$ such that $|\mathcal{F}| \leq 2|\mathcal{T}_G| - |E| + p \leq p$. Since G is irreducible, we see that $\mathcal{H}$ has no isolated vertices, i.e., every component of $\mathcal{H}$ has at least one edge. Thus $\nu(\mathcal{H}) \geq p \geq |\mathcal{F}|$. For the acyclic hypergraph $\mathcal{H} \setminus \mathcal{F}$, By Lemma 5 we may found in $O(|\mathcal{T}_G|^2 \log^2 |\mathcal{T}_G|)$ time a minimum transversal $\mathcal{R}$ of $\mathcal{H} \setminus \mathcal{F}$ such that

$$|\mathcal{R}| = \tau(\mathcal{H} \setminus \mathcal{F}) = \nu(\mathcal{H} \setminus \mathcal{F}).$$

Observe that $\mathcal{R} \subseteq E$ and $\mathcal{F} \subseteq \mathcal{T}_G$. If $\mathcal{F} = \emptyset$, set $\mathcal{S} = \emptyset$, else for each $F \in \mathcal{F}$, take $e_F \in E$ with $e_F \in F$, and set $\mathcal{S} = \{e_F : F \in \mathcal{F}\}$. It is clear that $\mathcal{R} \cup \mathcal{S}$ is a transversal of $\mathcal{H}$ (i.e., a triangle cover of G) with cardinality $|\mathcal{R} \cup \mathcal{S}| \leq \nu(\mathcal{H} \setminus \mathcal{F}) + |\mathcal{F}| \leq 2\nu(\mathcal{H}) = 2\nu_t(G)$, establishing the theorem.   □

We observe that the graphs G that consist of a number of triangles sharing a common edge satisfy $|E(G)| \geq 2|\mathcal{T}_G|$, and $\nu_t(G) < |\mathcal{T}_G|/3$ when $|\mathcal{T}_G| \geq 4$. So in some sense, Theorem 9 works as a supplement of Corollary 2 for sparse graphs.

Along the same line as in the previous subsection, Theorem 9 and the following Theorem 10 jointly narrow the interval w.r.t. $|E(G)|/|\mathcal{T}_G|$ to $(1.5 - \epsilon, 2)$ for future study of Conjecture 1 on graph G.

Theorem 10. *If there exists some real $\delta > 0$ such that Conjecture 1 holds for every irreducible graph $G = (V, E)$ with $|E|/|\mathcal{T}_G| \geq 3/2 - \delta$, then Conjecture 1 holds for every graph.*   □

4   Conclusion

Using tools from hypergraphs, we design polynomial-time combinatorial algorithms for finding a small triangle covers in graphs, which particularly imply several sufficient conditions for Conjecture 1. The high level idea of these algorithms is to remove *some edges* from G so that the triangle hypergraph of the remaining graph is *acyclic* (see the proofs of Theorems 4 and 9), which guarantees that the remaining graph has equal triangle covering number and triangle packing number, and a minimum triangle cover of the remaining graph is computable in polynomial time (see Theorem 5). It is well-known that the acyclic condition in Theorem 5 could be weakened to odd-cycle-freeness [13]. So our sufficient conditions could be (significantly) improved if we can remove (much) *fewer edges* from G such that the triangle hypergraph of the remaining graph is *odd-cycle free*.

Acknowledgements. The authors are indebted to Dr. Gregory J. Puleo and Dr. Zbigniew Lonc for their invaluable comments and suggestions.

References

1. Baron, J.D., Kahn, J.: Tuza's conjecture is asymptotically tight for dense graphs. arXiv preprint (2014). arXiv:1408.4870
2. Berge, C.: Hypergraphs: Combinatorics of Finite Sets. Elsevier, New York (1989)
3. Chapuy, G., DeVos, M., McDonald, J., Mohar, B., Scheide, D.: Packing triangles in weighted graphs. SIAM J. Discrete Math. **28**(1), 226–239 (2014)
4. Chen, X., Diao, Z., Hu, X., Tang, Z.: Sufficient conditions for tuza's conjecture on packing and covering triangles. arXiv preprint (2016). arXiv:1605.01816
5. Cui, Q., Haxell, P., Ma, W.: Packing and covering triangles in planar graphs. Graphs Comb. **25**(6), 817–824 (2009)
6. Haxell, P., Kostochka, A., Thomassé, S.: Packing and covering triangles in K_4-free planar graphs. Graphs Comb. **28**(5), 653–662 (2012)
7. Haxell, P.E.: Packing and covering triangles in graphs. Discrete Math. **195**(1), 251–254 (1999)
8. Haxell, P.E., Kohayakawa, Y.: Packing and covering triangles in tripartite graphs. Graphs Comb. **14**(1), 1–10 (1998)
9. Krivelevich, M.: On a conjecture of Tuza about packing and covering of triangles. Discrete Math. **142**(1), 281–286 (1995)
10. Schrijver, A.: Combinatorial Optimization: Polyhedra and Efficiency. Springer Science & Business Media, Heidelberg (2003)
11. Tuza, Z.: Conjecture in: finite and infinite sets. In: Proceedings of Colloque Mathematical Society Jnos Bolyai, Eger, Hungary, North-Holland, p. 888 (1981)
12. Tuza, Z.: A conjecture on triangles of graphs. Graphs. Comb. **6**(4), 373–380 (1990)
13. Yannakakis, M.: On a class of totally unimodular matrices. Math. Oper. Res. **10**(2), 280–304 (1985)

Dynamic Programming

Linear Time Algorithms for Happy Vertex Coloring Problems for Trees

N. R. Aravind, Subrahmanyam Kalyanasundaram,
and Anjeneya Swami Kare$^{(\boxtimes)}$

Department of Computer Science and Engineering Indian Institute of Technology,
Hyderabad, India
{aravind,subruk,cs14resch01002}@iith.ac.in

Abstract. Given an undirected graph $G = (V, E)$ with $|V| = n$ and a vertex coloring, a vertex v is *happy* if v and all its neighbors have the same color. An edge is *happy* if its end vertices have the same color. Given a partial coloring of the vertices of the graph using k colors, the *Maximum Happy Vertices* (also called k-MHV) problem asks to color the remaining vertices such that the number of happy vertices is maximized. The *Maximum Happy Edges* (also called k-MHE) problem asks to color the remaining vertices such that the number of happy edges is maximized. For arbitrary graphs, k-MHV and k-MHE are NP-Hard for $k \geq 3$. In this paper we study these problems for trees. For a fixed k we present linear time algorithms for both the problems. In general, for any k the proposed algorithms take $O(nk \log k)$ and $O(nk)$ time respectively.

Keywords: Happy vertex · Happy edge · Graph coloring · Coloring trees

1 Introduction

Graph coloring problems are well studied in literature. The traditional vertex coloring problem asks to color the vertices of the graph using minimum number of colors such that the adjacent vertices get different colors. There are many variants of coloring problems. Recently, Zhang and Li [10] studied a coloring problem in which adjacent vertices are allowed to get same color. The proposed problems have applications related to homophyly in networks (see Chapter 4 of [4]).

Given an undirected graph $G = (V, E)$ and a vertex coloring, a vertex is *happy* if the vertex and all its adjacent vertices have the same color and *unhappy* otherwise. An edge is *happy* if its end vertices have the same color and *unhappy* otherwise.

For $S \subseteq V$, let $c_p : S \rightarrow \{1, 2, \ldots, k\}$ be a partial vertex coloring. A coloring $c_f : V \rightarrow \{1, 2, \ldots, k\}$ is an *extended full coloring* for c_p, if $c_f(v) = c_p(v), \forall v \in S$.

A.S. Kare — Faculty member of University of Hyderabad. This work is carried out as part of his PhD program at IIT Hyderabad.

© Springer International Publishing Switzerland 2016
V. Mäkinen et al. (Eds.): IWOCA 2016, LNCS 9843, pp. 281–292, 2016.
DOI: 10.1007/978-3-319-44543-4_22

Given an $S \subseteq V$ and a partial coloring c_p, *Maximum Happy Vertices* (MHV) (respectively, *Maximum Happy Edges* (MHE)) problem asks to find an extended full coloring c such that the number of happy vertices (respectively, edges) is maximized. As k is also an input parameter, the problem is also referred to as k-MHV (respectively, k-MHE).

Definition 1. *Multiway-Cut*

(Instance) We are given an undirected graph $G = (V, E)$ and a terminal set $S = \{s_1, s_2, \ldots, s_k\} \subseteq V$.

(Goal) Find a set of edges $C \subseteq E$ with minimum cardinality whose removal disconnects all the terminals from each other.

Definition 2. *Multiway-Uncut*

(Instance) We are given an undirected graph $G = (V, E)$ and a terminal set $S = \{s_1, s_2, \ldots, s_k\} \subseteq V$.

(Goal) Find a partition $\{V_1, V_2, \ldots, V_k\}$ of V such that each partition contains exactly one terminal and the number of edges not cut by the partition is maximized.

The k-MHE problem is a generalization of the Multiway Uncut problem [7] which is the complement of Multiway Cut problem [1,2]. The Multiway Uncut problem is a special case of k-MHE problem in which there is exactly one pre-colored vertex (terminal) for each color.

Both k-MHV and k-MHE problems are NP-Hard [10] for $k \geq 3$ for arbitrary graphs. In [10], $O(mn^7 \log n)$ and $O(\min\{n^{\frac{2}{3}}m, m^{\frac{3}{2}}\})$ time algorithms are presented for 2-MHV and 2-MHE respectively. Towards this end, the authors of [10] used techniques such as minimizing sub modular functions (2-MHV) [6] and max-flow algorithms (2-MHE) [5]. Zhang and Li [10] presented approximation algorithms with approximation ratios $\max\{\frac{1}{k}, \Omega(\Delta^{-3})\}$ and $\frac{1}{2}$ for k-MHV and k-MHE respectively. Here, Δ is the maximum degree of the graph. Later, Zhang et al. [9] presented approximation algorithms with approximation ratios $\frac{1}{\Delta+1}$ and $(\frac{1}{2} + \frac{\sqrt{2}}{4} f(k)) \geq 0.8535$ for k-MHV and k-MHE respectively.

1.1 Our Results

Apart from the results in [9,10], the MHV and MHE problems does not seem to be addressed for any class of graphs. In this paper, we study these problems for trees. We propose dynamic programming based algorithms for both k-MHV and k-MHE. For an arbitrary k, the proposed algorithms take $O(nk \log k)$ and $O(nk)$ time respectively. When k is fixed, the algorithms run in linear time. We also extend our algorithms to generate all the optimal colorings of the tree. Generating each optimal coloring takes polynomial time.

Using the result from [2] we observe that, for an arbitrary k, the k-MHE problem is NP-Hard for planar graphs. Using the result from [3] we infer that, when the number of pre-colored vertices is bounded, the k-MHE problem can be solved in linear time for graphs with bounded branch width.

The rest of the paper is organized as follows: In Sect. 2 we discuss the algorithm for the k-MHV problem, in Sect. 3 we discuss the algorithm for the k-MHE problem and the related observations. We conclude with Sect. 4. Throughout the paper we assume that the input graph is a tree (T). We use integers from 1 to k to denote the colors.

2 Algorithm for k-MHV Problem

We root the tree at an arbitrary vertex. Let T_v denotes the subtree rooted at a vertex v. Before presenting the algorithm we give a simple reduction rule, which can be executed in linear time.

Rule 1: If a leaf vertex is uncolored, remove it and count the leaf vertex as happy.

We can give the color of its parent to the uncolored leaf to make it happy. Hence, without loss of generality we can assume that all the leaves are colored.

We process the vertices of the rooted tree according to post order traversal. At each vertex v, we maintain a list of $2k$ integer values. The maximum value of these $2k$ values gives the maximum number of happy vertices in T_v, the sub tree rooted at v. The maximum value of the $2k$ values associated with the root gives us the maximum number of happy vertices of the tree. The corresponding optimal coloring can also be traced back in reverse direction. The list of $2k$ values defined as follows, for $1 \leq i \leq k$:

- $T_v[i, H]$: The maximum number of happy vertices in the subtree T_v, when v is colored i and is happy in T_v. That is, when v and all its children are colored i. Note that, here we focus on v being happy in the subtree T_v. The vertex v can become unhappy in the tree T because its parent gets another color.
- $T_v[i, U]$: The maximum number of happy vertices in T_v, when v is colored i and is unhappy in T_v. That is, when one or more children of v are colored with a color other than i.

Note that, if a vertex or some of its children are already colored, then some of the $2k$ values are invalid. We use -1 to denote an invalid value. We keep these $2k$ values in an array to access any specific item in constant time. The values are indexed in the order, $T_v[1, H], T_v[1, U], T_v[2, H], T_v[2, U], \ldots, T_v[k, H], T_v[k, U]$.

The following expressions are defined to simplify some of the equations:

- $T_v[i, *]$: The maximum number of happy vertices in the subtree T_v, when v is colored i. v may be happy or unhappy. That is:

$$T_v[i, *] = \max\{T_v[i, H], T_v[i, U]\}. \tag{1}$$

- $T_v[i, -]$: The maximum number of happy vertices in T_v excluding v, when v is colored i.

$$T_v[i, -] = \max\{T_v[i, H] - 1, T_v[i, U]\}. \tag{2}$$

– $T_v[\bar{\imath}, *]$: The maximum number of happy vertices in the subtree T_v, when v is colored with color other than i.

$$T_v[\bar{\imath}, *] = \max_{r \neq i}\{T_v[r, *]\}. \tag{3}$$

– $T_v[\bar{\imath}, -]$: The maximum number of happy vertices in the subtree T_v excluding v, when v is colored with color other than i.

$$T_v[\bar{\imath}, -] = \max_{r \neq i}\{T_v[r, -]\}. \tag{4}$$

– $T_v[*, *]$: The maximum number of happy vertices in T_v. That is:

$$T_v[*, *] = \max\{T_v[1, *], T_v[2, *], \ldots, T_v[k, *]\}. \tag{5}$$

Now we explain the process to compute these $2k$ values at each vertex. As a leaf vertex is pre-colored, it is always happy alone as a subtree with a single vertex. Only one out of $2k$ values is valid. Suppose the color of the leaf is i, then the only valid value is $T_v[i, H] = 1$.

The following subsections consider the case when v is a non leaf vertex. Let $v_1, v_2, \ldots, v_d$ be the children of v. The values $T_v[i, H]$ and $T_v[i, U]$ are invalid, if v is pre-colored with a color $r \neq i$. Otherwise, we compute $T_v[i, H]$ and $T_v[i, U]$ as follows:

2.1   Computing $T_v[i, H]$

Computing $T_v[i, H]$ has two cases:

Algorithm 1. Computing $T_v[i, H]$

1: **procedure** COMPUTETVH(v, i)
2:     **if** $\forall v_j, T_{v_j}[i, *] \neq -1$ **then**
3:         return $(1 + \sum_{v_j} T_{v_j}[i, *])$ ▷ Case 2
4:     **else**
5:         return -1 ▷ Case 1
6:     **end if**
7: **end procedure**

Case 1: For some child v_j, $T_{v_j}[i, *] = -1$.
This means that the child v_j is pre colored with a color other than i. In this case, v becomes unhappy when it gets color i. So $T_v[i, H]$ is invalid.
Case 2: For every child v_j, $T_{v_j}[i, *] > -1$.
In this case, we use the following equation to compute $T_v[i, H]$.

$$T_v[i, H] = 1 + \sum_{v_j} T_{v_j}[i, *]. \tag{6}$$

Algorithm 2. Computing $T_v[i, U]$

1: **procedure** COMPUTETVU(v, i)
2:      **if** every child v_j is pre-colored with color i **then**
3:          return -1 ▷ Case 1
4:      **else if** $\exists v_{j'}$ child of v such that $T_{v_{j'}}[*, *] \neq T_{v_{j'}}[i, *]$ **then**
5:          return $(\sum_{v_j} \max\{T_{v_j}[1, -], \ldots, T_{v_j}[i, *], \ldots, T_{v_j}[k, -]\})$ ▷ Case 2
6:      **else**   ▷ Case 3
7:          **for each** child v_j **do**
8:              diff$(v_j, i) \leftarrow T_{v_j}[i, *] - T_{v_j}[\bar{\imath}, -]$
9:          **end for**
10:        $v_\ell \leftarrow \operatorname{argmin}_{v_j} \operatorname{diff}(v_j, i)$
11:        $q \leftarrow \operatorname{argmax}_{r \neq i} T_{v_\ell}[r, -]$
12:        return $(T_{v_\ell}[q, -] + \sum_{v_j \neq v_\ell} T_{v_j}[i, *])$
13:      **end if**
14: **end procedure**

2.2 Computing $T_v[i, U]$

Computing $T_v[i, U]$ has three cases:

Case 1: Every child v_j is pre colored with color i.
In this case, we cannot make v unhappy by giving color i to v. Hence $T_v[i, U]$ is invalid.

Case 2: For some child $v_{j'}$, $T_{v_{j'}}[*, *] \neq T_{v_{j'}}[i, *]$.
That is, the child $v_{j'}$ has color $r \neq i$ in the optimal coloring of $T_{v_{j'}}$. When v is colored i and $v_{j'}$ is colored r, irrespective of the colors of the other children, v will certainly be unhappy. In this case, we use the following expression to compute $T_v[i, U]$.

$$T_v[i, U] = T_{v_{j'}}[r, -] + \sum_{\substack{v_j \text{ child of } v, \\ v_j \neq v_{j'}}} \max\{T_{v_j}[1, -], \ldots, T_{v_j}[i, *], \ldots, T_{v_j}[k, -]\}$$

$$\tag{7}$$

$$= \sum_{v_j \text{ child of } v} \max\{T_{v_j}[1, -], \ldots, T_{v_j}[i, *], \ldots, T_{v_j}[k, -]\}. \tag{8}$$

Case 3: For every child v_j, $T_{v_j}[*, *] = T_{v_j}[i, *]$.
For each v_j, if we pick $T_{v_j}[i, *]$, v will become happy, but we need v to be unhappy. To avoid this situation, for some child we pick a value with color other than i as follows:

For each v_j, we define diff(v_j, i) as follows:

$$\operatorname{diff}(v_j, i) = T_{v_j}[i, *] - T_{v_j}[\bar{\imath}, -]. \tag{9}$$

We pick the child (say v_ℓ) with minimum $\text{diff}(v_j, i)$ value. Suppose, $T_{v_\ell}[\bar{\imath}, -] = T_{v_\ell}[q, -]$, we replace $T_{v_\ell}[i, *]$ with $T_{v_\ell}[q, -]$. The new expression is:

$$T_v[i, U] = T_{v_\ell}[q, -] + \sum_{v_j \neq v_\ell} T_{v_j}[i, *]. \tag{10}$$

Algorithm 3. Algorithm for MHV problem

1: **for each** $v \in V$ in post order **do**
2:     **for** $i = 1$ to k **do**
3:         **if** v is a leaf **then**
4:             **if** $\text{color}(v) = i$ **then**
5:                 $T_v[i, H] \leftarrow 1$
6:                 $T_v[i, U] \leftarrow -1$
7:             **else**
8:                 $T_v[i, H] \leftarrow -1$
9:                 $T_v[i, U] \leftarrow -1$
10:            **end if**
11:        **else**
12:            **if** v is pre-colored and $\text{color}(v) \neq i$ **then**
13:                $T_v[i, H] \leftarrow -1$
14:                $T_v[i, U] \leftarrow -1$
15:            **else**
16:                $T_v[i, H] \leftarrow \textsc{ComputeTvH}(v, i)$
17:                $T_v[i, U] \leftarrow \textsc{ComputeTvU}(v, i)$
18:            **end if**
19:        **end if**
20:    **end for**
21: **end for**

Theorem 1. *There is an $O(nk \log k)$ time algorithm for the k-MHV problem for trees.*

Proof. We evaluate the time spent at a particular vertex v to compute $T_v[i, H]$ and $T_v[i, U]$, for $1 \leq i \leq k$. Let $v_1, v_2, \ldots, v_d$ be the children of v.

Computing $T_v[i, H]$: The $T_{v_j}[i, H]$ and $T_{v_j}[i, U]$ values are accessible in constant time for each child v_j. Time to compute $T_v[i, H]$, $\forall 1 \leq i \leq k$ is:

$$\sum_{1 \leq i \leq k} O(d) = O(kd). \tag{11}$$

Computing $T_v[i, U]$: We sort the $2k$ values in descending order. For any child v_j, $T_{v_j}[i, *]$ is available in constant time from the original array. From the sorted array $T_{v_j}[*, *]$ and $T_{v_j}[\bar{\imath}, *]$ are available in constant time. Hence $T_v[i, U]$, $\forall 1 \leq i \leq k$ can be computed in:

$$O(dk \log k) + \sum_{1 \leq i \leq k} O(d) = O(dk \log k). \tag{12}$$

Hence the total time is:

$$\sum_v dk + dk \log k \le \sum_v 2dk \log k = 2k \log k \sum_v d = O(nk \log k). \qquad (13)$$

$\square$

The correctness of the value $T_v[*, *]$ for every vertex v implies the correctness of the algorithm. The correctness of the value $T_v[*, *]$ follows from the correctness of the $2k$ values $T_v[1, H], T_v[1, U], T_v[2, H], T_v[2, U], \ldots, T_v[k, H], T_v[k, U]$ associated with v.

Theorem 2. *Algorithm 3 correctly computes the values $T_v[i, H]$ and $T_v[i, U]$ for every v and $1 \le i \le k$.*

Proof. We prove the theorem by using induction on the size of the subtrees. For a leaf vertex v, the algorithm correctly computes the values $T_v[i, H]$ and $T_v[i, U]$ for $1 \le i \le k$. Since the leaf vertices are pre-colored, each leaf vertex has only one valid value (this value being 1).

For a non-leaf vertex v, let $v_1, v_2, \ldots, v_d$ be the children of v. By induction on the size of the sub-trees, all the $2k$ values associated with each child v_j of v are correctly computed. Let x be the value computed by the algorithm for $T_v[i, H]$ (or $T_v[i, U]$) for any color i. If x is not the optimal value, it will contradict the optimality of at least one value of a child of v. Hence the algorithm correctly computes the values $T_v[i, H]$ and $T_v[i, U]$ for every v and $1 \le i \le k$. $\square$

2.3 Generating All Optimal Happy Vertex Colorings

Our algorithm can also be extended to generate all the optimal happy vertex colorings of the tree. Among the $2k$ values associated with a vertex v, there may be multiple values equal to the optimal value. So, while generating optimal happy vertex coloring, we can chose any of these values to generate a different optimal coloring. For example, let $T_v[i, H]$ be an optimal value for the vertex v. Let v_j be a child of v with both $T_{v_j}[i, H]$ and $T_{v_j}[i, U]$ are optimal. So, we can generate one optimal coloring by picking $T_{v_j}[i, H]$ and another optimal coloring by picking $T_{v_j}[i, U]$. There may be exponentially many optimal colorings, but, generating each optimal coloring takes polynomial time (linear time for fixed k).

3 Algorithm for k-MHE Problem

Before presenting the algorithm we give simple reduction rules, which can be executed in linear time.

Rule 2: Let v be a pre-colored vertex with degree more than 1. Let $v_1, v_2, \ldots, v_d$ be the neighbours of v in T. We can divide T into d edge disjoint subtrees $T_1, T_2, \ldots, T_d$ and all these trees share only the vertex v.

$$k\text{-MHE}(T) = k\text{-MHE}(T_1) + k\text{-MHE}(T_2) + \cdots + k\text{-MHE}(T_d). \qquad (14)$$

With the application of Rule 2, without loss of generality we can assume that T does not have a pre-colored vertex with degree more than 1.

Now, we root the tree at an arbitrary vertex with degree more than 1.

Rule 3: (Similar to Rule 1 in Sect. 2) If a leaf vertex is uncolored, remove it and count the edge connecting the leaf vertex as happy.

With Rule 2 and Rule 3, without loss of generality, all the leaves of the rooted tree T are pre-colored and no non-leaf vertex is pre-colored.

Our algorithm for k-MHE problem has two phases. In the first phase, we visit the vertices according to post order traversal and populate a list of tentative colors for each vertex. In the second phase we visit the vertices according to pre-order traversal and assign a color for each vertex.

Algorithm 4. Phase 1 of the algorithm

1: **procedure** POPULATETENTATIVECOLORS(T)
2:     **for each** $v \in V$ in post order **do**
3:         **if** v is a leaf **then**
4:             $L(v) \leftarrow color(v)$
5:         **else** $\triangleright$ Let $v_1, v_2, \ldots, v_d$ be the children of v
6:             $frequency[1..k] \leftarrow \{0\}$
7:             **for each** child v_j of v **do**
8:                 **for each** color $c \in L(v)$ **do**
9:                     $frequency[c] \leftarrow frequency[c] + 1$
10:                **end for**
11:            **end for**
12:            $max \leftarrow 0$
13:            **for** $i = 1$ to k **do**
14:                **if** $frequency[i] > max$ **then**
15:                    $max \leftarrow frequency[i]$
16:                **end if**
17:            **end for**
18:            **for** $i = 1$ to k **do**
19:                **if** $frequency[i] = max$ **then**
20:                    $L(v) \leftarrow L(v) \cup \{i\}$
21:                **end if**
22:            **end for**
23:        **end if**
24:    **end for**
25: **end procedure**

Phase 1: We visit the vertices according to post order traversal. At each vertex v, we keep a list of tentative colors to assign to the vertex v in the optimal solution. The size of this list is at most k. Let $L(v)$ denote the list of tentative colors associated with the vertex v.

If the vertex v is a leaf, as the leaf vertex is pre-colored, we add that pre-color to $L(v)$. Otherwise, let $v_1, v_2, \ldots, v_d$ be the children of v. The list of tentative colors $L(v_j)$ for each vertex v_j are already computed. For each child v_j, we traverse the list $L(v_j)$ and compute the frequency of occurrences of each color in the multiset that is union of the lists. Let frequency(i) denote the frequency of color i. We add all the colors with maximum frequency to $L(v)$. The process is captured in Algorithm 4.

Algorithm 5. Phase 2 of the algorithm

1: **procedure** ATTACHCOLORS(T, L)   ▷ Fixing color to vertices
2:     **for each** $v \in V$ in pre order **do**
3:         **if** $|L(v)| = 1$ **then**
4:             color(v) ← Only element of $L(v)$
5:         **else if** color(parent(v)) $\in L(v)$ **then**
6:             color(v) ← color(parent(v))
7:         **else**
8:             color(v) ← Any element of $L(v)$
9:         **end if**
10:     **end for**
11: **end procedure**

Phase 2: We visit the vertices according to pre-order traversal to assign a color to each vertex. Let v be the vertex in pre-order. If $|L(v)| = 1$, then we fix the color of v to the only color in $L(v)$. Otherwise, we check if the color of the parent of v is present in $L(v)$, and assign it to v if present. Otherwise, we pick any arbitrary color from $L(v)$ and assign it to v. The process is captured in Algorithm 5.

Theorem 3. *There is an $O(nk)$ time algorithm for the k-MHE problem for trees.*

Proof. At each vertex with degree d, we perform $O(kd)$ time in the Phase 1 and $O(k)$ time in the Phase 2. The time complexity is:

$$\sum_v O(kd) = O(nk). \tag{15}$$

□

The correctness of the algorithm can be proved using induction on the size of the sub-tree similar to Theorem 2.

3.1 Generating All Optimal Happy Edge Colorings

Our algorithm can be extended to generate all the optimal happy edge colorings. We keep a list of tentative colors at each vertex. At a vertex v, if the color(parent(v)) is present in $L(v)$, then, we assign the color(parent(v)) to v in the optimal coloring. Otherwise, we can generate a different optimal coloring for each color in $L(v)$. Here we point out that, this scheme may miss out some optimal colorings when color(parent(v)) is not present in $L(v)$ but present in the set of colors with frequency one less than the maximum frequency. In this case, we can assign the color(parent(v)) to v even though the color(parent(v)) is not present in $L(v)$. A special case of this scenario is when there is a vertex v where all its children have distinct colors (the maximum frequency being 1). Even though the color(parent(v)) not present in $L(v)$, we can assign the color(parent(v)) to v as it has zero frequency at v.

There may be exponentially many optimal happy edge colorings. Generating each optimal coloring takes polynomial time (linear time for fixed k).

3.2 k-MHE for Planar Graphs and Graphs with Bounded Branch Width

The Multiway-Cut problem is NP-Hard for planar graphs [2] when k, the number of terminals, is not fixed. This implies the following theorem on hardness of k-MHE for planar graphs for an arbitrary k.

Theorem 4. *For an arbitrary k, the k-MHE problem is NP-Hard for planar graphs.*

In [8], Robertson and Seymour introduced the notions of tree width and branch width. They showed that these two quantities are always within a constant factor of each other. Many graph problems that are NP-Hard for general graphs have been shown to be solvable in polynomial time for graphs with bounded tree width or equivalently bounded branch width. For more formal definitions of branch width and tree width we refer the readers to [8].

Definition 3. *Multi-Multiway Cut*
(Instance) We are given an undirected graph $G = (V, E)$ and c sets of vertices $S_1, S_2, \ldots, S_c$.
(Goal) Find a set of edges $C \subseteq E$ with minimum cardinality whose removal disconnects every pair of vertices in each set S_i.

When $c = 1$, the Multi-Multiway Cut problem is equivalent to Multiway Cut problem. The k-MHE problem can also be formulated as a Multi-Multiway Cut problem, by creating vertex sets with every pair of pre-colored vertices with different colors. In [3], Deng et al. studied the Multi-Multiway Cut problem for graphs with bounded branch width and presented an $O(b^{2b+2}.2^{2bc}.|G|)$ time algorithm, where b is the branch width of the graph and c is the number of vertex sets. The algorithm runs in linear time when the branch width and the number of vertex sets are fixed.

Theorem 5. *When the branch width of the graph and the number of pre-colored vertices are bounded, there is a linear time algorithm for the k-MHE problem.*

Proof. Let the number of pre-colored vertices be p and the branch width be b. For this instance of k-MHE, we can formulate a Multi-Multiway Cut problem with at most p^2 vertex sets. Hence, the k-MHE problem can be solved in time $O(b^{2b+2}.2^{2bp^2}.|G|)$. Hence, when both the number of pre-colored vertices and the branch width are constants, the k-MHE problem can be solved in linear time. □

4   Conclusions

In this paper, we study the Maximum Happy Vertices (k-MHV) and Maximum Happy Edges (k-MHE) problems for trees. We have presented $O(nk \log k)$ and $O(nk)$ time algorithms for k-MHV and k-MHE problems respectively. Our algorithms run in linear time when k is fixed. Our algorithms can be extended to generate all the optimal colorings of the tree.

As a future direction, it is interesting to study the hardness of the k-MHV problem for planar graphs. For fixed k, the Multiway Cut problem has a polynomial time algorithm for planar graphs [2]. So, for planar graphs and when k is fixed, polynomial time algorithms might be possible for k-MHV and k-MHE. Finding a linear time algorithm for graphs with bounded tree width (branch width) without the constraint on the number of pre-colored vertices is another direction.

Acknowledgement. We thank the anonymous reviewers for their detailed reviews and suggestions.

References

1. Chopra, S., Rao, M.R.: On the multiway cut polyhedron. Networks **21**(1), 51–89 (1991)
2. Dahlhaus, E., Johnson, D.S., Papadimitriou, C.H., Seymour, P.D., Yannakakis, M.: The complexity of multiway cuts (extended abstract). In: Proceedings of the Twenty-fourth Annual ACM Symposium on Theory of Computing, STOC 1992, pp. 241–251 (1992)
3. Deng, X., Lin, B., Zhang, C.: Multi-multiway cut problem on graphs of bounded branch width. In: Fellows, M., Tan, X., Zhu, B. (eds.) FAW-AAIM 2013. LNCS, vol. 7924, pp. 315–324. Springer, Heidelberg (2013)
4. Easley, D., Kleinberg, J.: Networks, Crowds, and Markets: Reasoning About a Highly Connected World. Cambridge University Press, New York (2010)
5. Even, S., Tarjan, R.E.: Network flow and testing graph connectivity. SIAM J. Comput. **4**(4), 507–518 (1975)
6. Iwata, S., Fleischer, L., Fujishige, S.: A combinatorial strongly polynomial algorithm for minimizing submodular functions. J. ACM **48**(4), 761–777 (2001)
7. Langberg, M., Rabani, Y., Swamy, C.: Approximation algorithms for graph homomorphism problems. In: Díaz, J., Jansen, K., Rolim, J.D.P., Zwick, U. (eds.) APPROX 2006 and RANDOM 2006. LNCS, vol. 4110, pp. 176–187. Springer, Heidelberg (2006)

8. Robertson, N., Seymour, P.: Graph minors. X. Obstructions to tree-decomposition. J. Comb. Theory, Ser. B **52**(2), 153–190 (1991)

9. Zhang, P., Jiang, T., Li, A.: Improved approximation algorithms for the maximum happy vertices and edges problems. In: Xu, D., Du, D., Du, D. (eds.) COCOON 2015. LNCS, vol. 9198, pp. 159–170. Springer, Heidelberg (2015)

10. Zhang, P., Li, A.: Algorithmic aspects of homophyly of networks. Theor. Comput. Sci. **593**, 117–131 (2015)

Speeding up Dynamic Programming in the Line-Constrained k-median

Paweł Gawrychowski[1] and Łukasz Zatorski[2(✉)]

[1] University of Haifa, Haifa, Israel
[2] Institute of Computer Science, University of Wrocław, Wroclaw, Poland
lzatorski@gmail.com

Abstract. In the planar k-median problem we are given a set of demand points and want to open up to k facilities as to minimize the sum of the transportation costs from each demand point to its nearest facility. In the line-constrained version the medians are required to lie on a given line. We present a new dynamic programming formulation for this problem, based on constructing a weighted DAG over a set of median candidates. We prove that, for any convex distance metric and any line, this DAG satisfies the concave Monge property. This allows us to construct efficient algorithms in L_∞ and L_1 and any line, while the previously known solution (Wang and Zhang, ISAAC 2014) works only for vertical lines. We also provide an asymptotically optimal $\mathcal{O}(n)$ solution for the case of $k = 1$.

Keywords: k-median · Dynamic programming · Monge property

1 Introduction

The planar k-median problem is a variation of the well-known facility location problem. For a given set P of *demand points*, we want to find a set Q of k *facilities*, such that the sum of all transportation costs from a demand point to its closest facility is minimized. Each $p \in P$ is associated with its own (positive) cost per unit of distance to assigned facility, denoted $w(p)$. Formally, we want to minimize:

$$S(P) = \sum_{p \in P} \min_{q \in Q} w(p) \cdot d(p, q)$$

Because the problem is NP-hard for many metrics [7], we further restrict it by introducing a *line-constraint* on the set Q. We require that all facilities should belong to a specified *facility line* χ defined by an equation $ax + by = c$, where $a, b, c \in \mathbb{R}$ and $a \cdot b \neq 0$. Such a constraint is natural when all facilities are by design placed along a path that can be locally treated as linear, e.g., pipeline, railroad, highway, country border, river, longitude or latitude.

For $k = 1$ we obtain the line-constrained 1-median problem. Despite the additional restriction, the complexity of this simplest variant strongly depends

© Springer International Publishing Switzerland 2016
V. Mäkinen et al. (Eds.): IWOCA 2016, LNCS 9843, pp. 293–305, 2016.
DOI: 10.1007/978-3-319-44543-4_23

on the metric. For a point $p \in \mathbb{R}^2$, let $x(p)$ and $y(p)$ denote its x- and y-coordinate. The most natural metric is the Euclidean distance, where $L_2(p, q) = \sqrt{(x(p) - x(q))^2 + (y(p) - y(q))^2}$. It is known that even for 5 points, it is not possible to construct the 1-median with a ruler and compass. It can also be proven that the general, line-constrained and 3-dimension versions of the k-median problem are not solvable over the field of rationals [2]. Hence it is natural to consider also other distance functions, for example:

Chebyshev distance $L_\infty(p, q) = \max\{|x(p) - x(q)|, |y(p) - y(q)|\}$,
Manhattan distance $L_1(p, q) = |x(p) - x(q)| + |y(p) - y(q)|$,
squared Euclidean distance $L_2^2(p, q) = (x(p) - x(q))^2 + (y(p) - y(q))^2$.[1]

All these distances functions have been recently considered by Wang and Zhang [10] in the context of line-constrained k-median problem. They designed efficient algorithms based on a reduction to the minimum weight k-link path problem. However, their L_1 and L_∞ solutions work only in the special case of a horizontal facility line.

We provide a different dynamic programming formulation of the problem that works for any facility line χ in L_1 and L_∞. The new formulation can also be seen as a minimum weight k-link path in a DAG, where the weights are Monge. However, looking up the weight of an edge in this DAG is more expensive. We show how to implement edge lookups in $\mathcal{O}(\log n)$ after $\mathcal{O}(n \log n)$ time and space preprocessing which then allows us to apply the SMAWK algorithm [1] or, if $k = \Omega(\log n)$, the algorithm of Schieber [9] to obtain the following complexities.

Metric	Facility line	Time complexity
Wang and Zhang [10]		
L_1	horizontal	$\min\{\mathcal{O}(nk), n2^{\mathcal{O}(\sqrt{\log k \log \log n})} \log n\}$
L_∞	horizontal	$\min\{\mathcal{O}(nk \log n), n2^{\mathcal{O}(\sqrt{\log k \log \log n})} \log^2 n\}$
Our results		
L_1	general	$\min\{\mathcal{O}(nk \log n), n2^{\mathcal{O}(\sqrt{\log k \log \log n})} \log n\}$
L_∞	general	$\min\{\mathcal{O}(nk \log n), n2^{\mathcal{O}(\sqrt{\log k \log \log n})} \log n\}$

In L_∞, our general solution is faster than the one given by Wang and Zhang for the special case of horizontal facility line. We also provide a specialized procedure solving the problem for $k = 1$ in linear time.

2 Preliminaries

A basic tool for speeding up dynamic programming is the so-called Monge property. It can often be used to improve the time complexity by an order of magnitude, especially in geometric problems.

---

[1] This is not a metric.

Definition 1. *A weight function w is concave Monge if, for all $a < b$ and $c < d$, $w(a, c) + w(b, d) \leq w(b, c) + w(a, d)$.*

Dynamic programming can often be visualized as finding the row minima in a $n \times n$ matrix. Naively, this takes $\mathcal{O}(n^2)$ time. However, Aggarwal et al. [1] showed how to decrease the time complexity to $\mathcal{O}(n)$ if the matrix has the so-called total monotonicity property, which is often established through the Monge property. Their method is usually referred to as the SMAWK algorithm. There is a deep connection between SMAWK and other methods for speeding up dynamic programming, such as the Knuth-Yao inequality used for building optimal binary search trees, as observed by Bein et al. [3].

Let D be a DAG on n nodes $0, 1, \ldots, n - 1$ with a concave Monge weight function $w(i, j)$ defined for $0 \leq i < j < n$ that corresponds to the weight of the edge $\langle i, j \rangle$. A minimum diameter path in D is a path from 0 to $n - 1$ with the minimum weight. Galil and Park showed how to find such path in optimal $\mathcal{O}(n)$ time using the SMAWK algorithm [5]. A minimum weight k-link path is a minimum weight path from 0 to $n - 1$ consisting of exactly k edges (links).

Lemma 2. *Minimum weight k-link path can be found in $\mathcal{O}(nk)$ and, for $k = \Omega(\log n)$, $n2^{\mathcal{O}(\sqrt{\log k \log \log n})}$ time.*

Proof. To obtain $\mathcal{O}(nk)$ time complexity, we iteratively compute minimum weight 1-link, 2-link, $\ldots$, $(k-1)$-link and finally k-link paths from 0 to every other node. This can be seen as k layers of dynamic programming, each requiring only $\mathcal{O}(n)$ time thanks to the SMAWK algorithm. Alternatively, $n2^{\mathcal{O}(\sqrt{\log k \log \log n})}$ time algorithm for $k = \Omega(\log n)$ was given by Schieber [9]. $\qquad\square$

The weights of the edges in our DAG will be computed on-the-fly with orthogonal queries. We will use the following tool: preprocess a given set of n weighted points in a plane for computing the sum of the weights of all points in a given query range $[x, +\infty] \times [y, +\infty]$. We call this problem orthogonal range sum. The following is well-known.

Lemma 3. *There exists a data structure for the orthogonal range sum problem that can be built in $\mathcal{O}(n \log n)$ time and answers any query in $\mathcal{O}(\log n)$ time.*

Proof. We convert the points into a sequence by sorting them according to their x-coordinates (without losing the generality, these coordinates are all distinct) and writing down the corresponding y-coordinates. The y-coordinates are further normalized by replacing with the ranks on a sorted list of all y-coordinates (again, we assume that they are all distinct). Hence we obtain a sequence of length n over an alphabet $[n]$, where each character has its associated weight. We build a wavelet tree [6] of this sequence in $\mathcal{O}(n \log n)$ time. Each node of the wavelet tree is augmented with an array of partial sums of the prefixes of its subsequence. Given an orthogonal query, we first normalize it by looking at the sorted list of all x- and y-coordinates. Then we traverse the wavelet tree starting from the root and accumulate appropriate partial sums. The details can be found in [8]. $\square$

3   Normalizing Problem Instances

L_1 and L_∞ metrics are equivalent, which can be seen by rotating the plane by 45°. Hence from now on we will work in L_1 metric. This simplification was not possible in the previous approach [10], since it required the facility line to be horizontal, which is no longer true after rotation.

We further modify the problem instance so that the line χ is expressed in a slope intercept form $y = ax$, where $a \in [0, 1]$, and all coordinates of points in P are non-negative. This is always possible by reflecting along the horizontal axis, then along the line $y = x$, and finally translating. Such transformations do not modify the distances in L_1, so computing the k-median solution Q for the transformed instance gives us the answer for the original instance. Because any solution Q can be transformed so that the x-coordinates of all facilities are distinct without increasing the cost, we will consider only such solutions and identify each facility with its x-coordinate.

4   Computing 1-median

Let $D(p, x)$ be the weighted distance between $p \in P$ and $(x, a \cdot x) \in \chi$:

$$D(p, x) = w(p) \cdot d(p, (x, a \cdot x))$$

Whenever we say that $p \in P$ is closer to coordinate x_i than x_j, we mean that $D(p, x_i) < D(p, x_j)$. For a set of points $A \subseteq P$, $D(A, x)$ is the sum of weighted distances:

$$D(A, x) = \sum_{p \in A} w(p) \cdot d(p, (x, a \cdot x))$$

The 1-median is simply $\min_{x \in \mathbb{R}} D(P, x)$.

A function $f : \mathbb{R} \to \mathbb{R}$ is convex if the line segment between any two points on its graph lies above or on the graph. Such functions have the following properties:

1. $f(x) = |x - y|$ is convex for any y.
2. if $f(x)$ is convex, then $g(x) = c \cdot f(x)$ is convex for any positive c.
3. if $f(x)$ and $g(x)$ are convex, then $h(x) = f(x) + g(x)$ is also convex.

Lemma 4. *For any point p, $D(p, x)$ is convex. For any set of points P, $D(P, x)$ is also convex.*

Proof. Consider any point $p \in P$. From the definition:

$$D(p, x) = w(p) \cdot L_1(p, (x, a \cdot x)) = w(p) \cdot (|x(p) - x| + |y(p) - a \cdot x|).$$

This is a sum of absolute values functions multiplied by the (positive) weight of p. Hence by the properties of convex functions $D(p, x)$ is convex. Then $D(P, x)$ is also convex since it is a sum of convex functions over $p \in P$.   □

Since $D(p, x)$ is convex, any of its local minima is a global minimum. Similarly to the function $f(x) = |x|$, it is only semi-differentiable. Its derivative $D'(p, x)$ is a staircase nondecreasing function, undefined for at most two values $x = x_1$ and $x = x_2$. We call x_1 and x_2 the *median candidates* and for convenience assume that $D'(p, x)$ is equal to its right derivative there. When $a = 0$ or $p \in \chi$, $D'(p, x)$ has exactly one median candidate $x_1 = x(p)$, that is the minimum. Otherwise, there are two median candidates $x_1 = x(p)$ and $x_2 = \frac{y(p)}{a}$. For $a \in (0, 1)$, x_1 is the only minimum, whereas for $a = 1$ every value in range $[x_1, x_2]$ is a minimum. Because the derivative of a sum of functions is the sum of their derivatives, $D'(P, x)$ can only change at a median candidate of some $p \in P$. This means that a minimum of $D(p, x)$ corresponds to one of at most $2n$ median candidates of P. In other words, there exists a solution $(x, y) \in \chi$, such that $x = x(p)$ or $y = y(p)$ for some $p \in P$. From now on, we use $\mathcal{M}(P)$ to denote the set of median candidates of P. $\mathcal{M}(P)$ can be computed in $\mathcal{O}(n)$ time by simply iterating over $p \in P$ and adding $x = x(p)$ and $x = \frac{y(p)}{a}$ to the result (note that this might give us a multiset, i.e., some median candidates might be included multiple times).

Theorem 5. *We can solve line-constrained* 1-*median problem in* $\mathcal{O}(n)$ *time.*

Proof. Because $D'(p, x)$ is nondecreasing, we can binary search for the largest x such that $D'(p, x) \le 0$. Then we return x as the solution. In every step of the binary search we use the median selection algorithm [4] to narrow down the current search range $X = (x_{\text{left}}, x_{\text{right}})$. At the beginning of every step:

1. M is a multiset of all median candidates of P that are in X.
2. S contains all points from P with at least one median candidate in M.
3. $r = D'(P \setminus S, x)$ for some $x \in X$.

We select the median x_m of M and compute $D'(P, x_m)$. If $D'(p, x_m) > 0$, we continue the search in (x_{left}, x_m), and otherwise in (x_m, x_{right}), updating S and M accordingly. Eventually $x_{\text{left}} = x_{\text{right}}$ and we return x_{left}.

The key observation is that when a point p is removed from S, it does no longer have a median candidate within X and its $D'(p, x)$ remains constant in all further computations. This means that $D'(P \setminus S, x)$ is constant for all $x \in X$ and r can be updated after removing every point p from S in $\mathcal{O}(1)$ time. x_m can be found in $\mathcal{O}(|M|)$ time. Calculating $D'(P, x_m) = r + D'(S, x_m)$ then takes $\mathcal{O}(1 + |S|)$ time. For a point p to be in S, one of its median candidates must belong to M, so $|S| \le |M|$. Hence the complexity of a single iteration is $\mathcal{O}(|M|)$. After each iteration the size of M decreases by a factor of two, so the running time is described by $T(n) = \mathcal{O}(n) + T(n/2)$, which solves to $\mathcal{O}(n)$. $\square$

Theorem 6. *We can calculate* $D(P, x)$ *for every* $x \in \mathcal{M}(P)$ *in* $\mathcal{O}(n \log n)$ *time.*

Proof. The elements of $\mathcal{M}(P)$ can be sorted in $\mathcal{O}(n \log n)$ time, and we can assume that every point generates exactly two median candidates. Let $\mathcal{M}(P) = \{x_1, x_2, \dots, x_{2n}\}$, where $x_i \le x_{i+1}$ for all $i = 1, 2, \dots, 2n - 1$. Recall that $D'(P, x) = D'(P, x_i)$ for any $x \in (x_i, x_{i+1})$. We compute $D(p, x_1)$ together with

$D'(P, x_1)$ in $\mathcal{O}(n)$ time. Then all other $D(p, x_i)$ are computed sequentially for $i = 2, 3, \ldots, 2n$ in $\mathcal{O}(1)$ time each using the formula:

$$D(P, x_i) = D(P, x_{i-1}) + D'(P, x_{i-1}) \cdot (x_i - x_{i-1})$$
$$D'(P, x_i) = D'(P, x_{i-1}) + 2 \cdot w(p) \cdot \sigma$$

where x_i is generated by the point p, $\sigma = 1$ if $x_i = x(p)$ and $\sigma = a$ otherwise. □

5 Computing k-median

Consider now any optimal solution Q of the k-median problem for the given set of weighted points P. For any facility $q \in Q$, let P_q be the set of points of P assigned to q. By interchanging the order of the summation, Q should minimize

$$\sum_{q \in Q} \sum_{p \in P_q} w(p) \cdot d(p, q).$$

Hence q must be an optimal solution of the 1-median problem for P_q. Since replacing q will not increase the sum of distances of points in $P \setminus P_q$, q can be chosen to be a median candidate of P_q. We deduce that there exists an optimal solution Q' such that

$$\bigvee_{q \in Q'} q \in \mathcal{M}(P_q) \subseteq \mathcal{M}(P).$$

For $k \geq \min(n, |\mathcal{M}(P)|)$, every $p \in P$ can be assigned to its closest possible facility. Such an assignment can be easily computed in $\mathcal{O}(n)$ time. If we are required to return exactly k medians, then we add enough additional points to $\mathcal{M}(P)$. From now on, we assume that $k < \min(n, |\mathcal{M}(P)|)$. Thus there exists an optimal k-median solution, where all facilities are 1-median candidates of P.

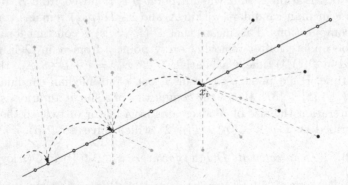

Fig. 1. Path in the DAG ending at the candidate x_i. Dashed lines represent current assignment of points from P to the closest chosen facility.

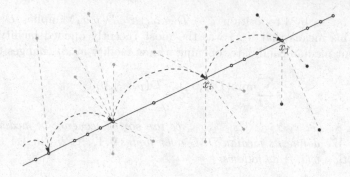

Fig. 2. We follow the edge $\langle i, j \rangle$. All (black) points now assigned to x_j were previously assigned to x_i, see Fig. 1.

By arranging all median candidates in a sequence according to their x-coordinates, we can view choosing k facilities as selecting a $(k + 1)$-link path in a DAG between two artificial elements infinitely to the left and to the right of the sequence, called source and sink, respectively.

Imagine that we traverse the sequence from left to right while deciding if we should open a new facility at the current median candidate, see Fig. 1. Initially, all points are assigned to the artificial facility source and the cost of the current solution S is set to $+\infty$. If we decide to open a new facility at the current median candidate x_j, for every $p \in P$ we check if x_j is closer to p than the facility p is currently assigned to. If so, we reassign p to x_j, see Fig. 2.

We claim that $p \in P$ can be closer to x_j than the facility p is currently assigned to only if the currently assigned facility is the most recently chosen facility x_i, that is, the current solution does not contain any facilities between x_i and x_j. Assuming that the claim holds, we define the weight of an edge $\langle \text{source}, i \rangle$ to be $D(P, x_i)$, and the weight of an internal edge $\langle i, j \rangle$ to be total decrease of the cost after giving each point $p \in P$ the possibility to switch from x_i to x_j. Finally, the weight of an edge $\langle j, \text{sink} \rangle$ is 0. Then selecting k medians corresponds to selecting an $(k + 1)$-link from source to sink in the DAG. However, we need to show the claim. To this end we consider the following properties of convex functions:

Proposition 7. *For any convex function f and $a < b < c$:*

(a) If $f(c) < f(b)$ then $f(b) < f(a)$.
(b) If $f(c) < f(a)$ then $f(b) < f(a)$.

Proof. Assume otherwise for any of the two implications. This means that $f(a) \leq f(b) > f(c)$ and the segment AC where $A = (a, f(a))$ and $C = (c, f(c))$ lies below $f(b)$, contradicting the assumption that the function f is convex.   □

Now we can prove the claim. Consider a point $p \in P$ such that its currently assigned facility is x_i and, for some $k > i$, facility x_k was not selected as a better option. Then, for any $j > k$, facility x_j cannot be a better option either, because

$x_i < x_k < x_j$ so by Proposition 7(a) $D(p, x_i) \le D(p, x_k)$ implies $D(p, x_j) \ge D(p, x_k)$. This means that if x_i was the most recently opened facility and x_j is the current median candidate, opening a new facility at x_j changes the total cost by

$$\sum_{p \in P} \min(D(p, x_j) - D(p, x_i), 0).$$

Definition 8. *Let* $x_1, x_2, \ldots, x_{n-1}, x_{2n}$ *be the sorted sequence of median candidates of* P. *We define its median DAG over nodes* $0, 1, \ldots, 2n, 2n + 1$ *with edge weight function* $w(i, j)$ *as follows:*

$$w(i, j) = \begin{cases} \infty & \textit{if } i = 0 \textit{ and } j = 2n + 1, \\ 0 & \textit{if } i > 0 \textit{ and } j = 2n + 1, \\ D(P, x_j) & \textit{if } i = 0 \textit{ and } j \in \{1, 2, \ldots, 2n\}, \\ \sum_{p \in P} \min(D(p, x_j) - D(p, x_i)), 0) & \textit{otherwise.} \end{cases}$$

The total cost of any k-median solution is equal to the sum of weights on its corresponding path of length $k + 1$ between 0 and $2n + 1$, so finding k-median reduces to finding the minimum weight $(k + 1)$-link path in the median DAG.

Because a sum of Monge functions is also Monge, to prove that $w(i, j)$ is Monge we argue that $w_p(i, j)$ is Monge, where $w(i, j) = \sum_{p \in P} w_p(i, j)$ and:

$$w_p(i, j) = \begin{cases} \infty & \text{if } i = 0 \text{ and } j = 2n + 1, \\ 0 & \text{if } i > 0 \text{ and } j = 2n + 1, \\ D(p, x_j) & \text{if } i = 0 \text{ and } j \in \{1, 2, \ldots, 2n\}, \\ \min(D(p, x_j) - D(p, x_i)), 0) & \text{otherwise.} \end{cases}$$

Proposition 9. *For any convex function* f, *if* $a < b < c$ *then:*

$$\min(f(c) - f(a), 0) \le \min(f(c) - f(b), 0).$$

Proof. If $f(c) \ge f(b)$ then the right side of the equation is equal to 0 and left side is non-positive. If $f(c) < f(b)$ then by Proposition 7(a) also $f(b) < f(a)$, so

$$\min(f(c) - f(a), 0) \le f(c) - f(a) < f(c) - f(b) = \min(f(c) - f(b), 0)$$

so the claim holds.   □

Proposition 10. *For any convex function* f, *if* $a < b < c$ *then*

$$f(b) + \min(f(c) - f(a), 0) \le f(c) + \min(f(b) - f(a), 0).$$

Proof. If $f(b) \ge f(a)$ then by Proposition 7(a) also $f(c) \ge f(b)$. Hence also $f(c) \ge f(a)$ and

$$f(b) + \min(f(c) - f(a), 0) = f(b) \le f(c) = f(c) + \min(f(b) - f(a), 0)$$

so the property holds. Otherwise, $f(b) < f(a)$ and the property becomes

$$f(b) + \min(f(c) - f(a), 0) \leq f(c) + f(b) - f(a)$$

which is always true due to $\min(f(c) - f(a), 0) \leq f(c) - f(a)$.   □

Proposition 11. *For any convex function f, if $a < b < c < d$ then*

$$\min(f(c) - f(a), 0) + \min(f(d) - f(b), 0) \leq \min(f(d) - f(a), 0) + \min(f(c) - f(b), 0).$$

Proof. If $f(d) \geq f(a)$, then

$$\min(f(d) - f(b), 0) \leq 0 = \min(f(d) - f(a), 0).$$

Combined with Proposition 9 applied to $a < b < c$ we obtain the claim. Otherwise, $f(d) < f(a)$ and by Proposition 7(b) applied to $a < c < d$ also $f(c) < f(a)$, so the property becomes

$$f(c) + \min(f(d) - f(b), 0) \leq f(d) + \min(f(c) - f(b), 0)$$

which holds by Proposition 10 applied to $b < c < d$.   □

Theorem 12. *For any point p, $w_p(i, j)$ is concave Monge.*

Proof. Consider any $s, t, u, v \in [0, 2n + 1]$ such that $s < t < u < v$. We need to prove that for any $p \in P$:

$$w_p(s, u) + w_p(t, v) \leq w_p(s, v) + w_p(t, u).$$

Case 1. $s = 0$ and $v = 2n + 1$
Straightforward, since $w_p(s, v) = \infty$ and all other edges have finite weights.
Case 2. $s > 0$ and $v = 2n + 1$

$$
\begin{aligned}
w_p(s, u) + w_p(t, v) &= w_p(s, u) + 0 \\
&= \min(D(p, u) - D(p, s), 0) \\
&\overset{9}{\leq} \min(D(p, u) - D(p, t), 0) \\
&= 0 + w_p(t, u) \\
&= w_p(s, v) + w_p(t, u)
\end{aligned}
$$

Case 3. $s = 0$ and $v < 2n + 1$

$$
\begin{aligned}
w_p(s, u) + w_p(t, v) &= D(p, u) + \min(D(p, v) - D(p, t), 0) \\
&\overset{10}{\leq} D(p, v) + \min(D(p, u) - D(p, t), 0) \\
&= w_p(s, v) + w_p(t, u)
\end{aligned}
$$

Case 4. $s > 0$ and $v < 2n + 1$

$$w_p(s, u) + w_p(t, v) = \min(D(p, u) - D(p, s), 0) + \min(D(p, v) - D(p, t), 0)$$
$$\overset{11}{\leq} \min(D(p, v) - D(p, s), 0) + \min(D(p, u) - D(p, t), 0)$$
$$= w_p(s, v) + w_p(t, u)$$

So in all cases $w_p(s, u) + w_p(t, v) \leq w_p(s, v) + w_p(t, u)$ and hence $w_p(i, j)$ is concave Monge.   □

In order to apply the known algorithms for finding minimum weight k-link path in the k-median problem, we need to answer queries for $w(i, j)$.

Lemma 13. *After $\mathcal{O}(n \log n)$ time and space preprocessing, we can answer queries for $w(i, j)$ in $\mathcal{O}(\log n)$ time per query.*

Proof. All edges from the source can be computed in $\mathcal{O}(n \log n)$ time via Theorem 6. All edges to sink have zero weight. It remains to show how to calculate the weight of an internal edge $\langle i, j \rangle$. Consider the set of points $p \in P$ that are closer to x_j than to x_i:

$$V(i, j) = \{(x, y) \in P : |x - x_i| + |y - a \cdot x_i| > |x - x_j| + |y - a \cdot x_j|\}$$

By definition, $w(i, j) = D(V(i, j), x_j) - D(V(i, j), x_i)$. We describe how to compute $D(V(i, j), x_i)$. $D(V(i, j), x_j)$ can be computed using the formula:

$$D(V(i, j), x_j) = D(P, x_j) - D(P \setminus V(i, j), x_j)$$

where $D(P, x_j)$ is the already preprocessed weight of the edge $\langle \text{source}, j \rangle$, and $D(P \setminus V(i, j), x_j)$ can be calculated by rotating the plane by $180°$ and using the same method as the one described below.

First we argue that if $(x, y) \in V(i, j)$ then $x > x_i$. Otherwise

$$|y - a \cdot x_i| - |y - a \cdot x_j| > x_j - x_i \geq a \cdot x_j - a \cdot x_i \geq 0$$

and we obtain a contradiction in each of the three cases:

1. $y < a \cdot x_i$ then the inequality becomes $a \cdot x_i - a \cdot x_j > 0$ but $x_i < x_j$.
2. $y \in [a \cdot x_i, a \cdot x_j)$ then the inequality becomes $2y > 2a \cdot x_j$ but $y < a \cdot x_j$.
3. $y > a \cdot x_j$ then the inequality becomes $a \cdot x_j - a \cdot x_i > a \cdot x_j - a \cdot x_i$.

We partition $V(i, j)$ into $V_1(i, j)$ and $V_2(i, j)$ with a horizontal line $y = a \cdot x_i$:

$$V_1(i, j) = V(i, j) \cap \{(x, y) : y \geq a \cdot x_i\}$$
$$V_2(i, j) = V(i, j) \cap \{(x, y) : y < a \cdot x_i\}.$$

The median candidate $(x_i, a \cdot x_i)$ is on the left and bottom of all points in $V_1(i, j)$ and on the left and top of all points in $V_2(i, j)$. Consider the minimum area rectangle enclosing P with sides parallel to the coordinate axes, and enumerate

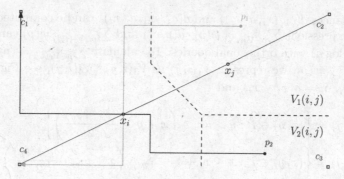

Fig. 3. Shortest route in L_1 from p_2 to c_1 and from p_1 to c_4 passing through the median candidate x_i.

its corners clockwise starting from the top left as c_1, c_2, c_3, c_4. In L_1 metric, one of the shortest routes from any point in $V_1(i,j)$ to the bottom left corner point c_4 goes via x_i, see Fig. 3. Therefore our desired sum of distances to x_i can be described in respect to c_4 as:

$$D(V_1(i,j), x_i) = \sum_{p \in V_1(i,j)} w(p) \cdot d(p, (x_i, a \cdot x_i))$$

$$= \left(\sum_{p \in V_1(i,j)} w(p) \cdot d(p, c_4) \right) - \left(d(c_4, (x_i, a \cdot x_i)) \cdot \sum_{p \in V_1(i,j)} w(p) \right).$$

Similarly, one of the shortest routes from any point in $V_2(i,j)$ to c_1 goes via x_i:

$$D(V_2(i,j), x_i) = \left(\sum_{p \in V_2(i,j)} w(p) \cdot d(p, c_1) \right) - \left(d(c_1, (x_i, a \cdot x_i)) \cdot \sum_{p \in V_2(i,j)} w(p) \right).$$

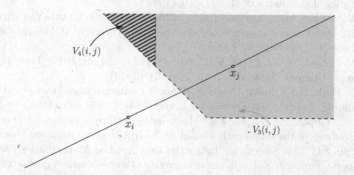

Fig. 4. V_1 represented as the gray V_3 minus the striped V_4.

The distances $d(c_1, (x_i, a \cdot x_i))$ and $d(c_4, (x_i, a \cdot x_i))$ can be computed in $\mathcal{O}(1)$ time. The expressions $\sum_{p \in V_2(i,j)} w(p) \cdot d(p, c_1)$ and $\sum_{p \in V_2(i,j)} w(p)$ can be evaluated in $\mathcal{O}(\log n)$ with orthogonal queries. To calculate $\sum_{p \in V_1(i,j)} w(p) \cdot d(p, c_4)$ and $\sum_{p \in V_1(i,j)} w(p)$, we represent $V_1(i,j)$ as $V_3(i,j) \setminus V_4(i,j)$, see Fig. 4 where $\delta_x = x_j - x_i$, $\delta_y = a(x_j - x_i)$ and

$$V_3(i,j) = \left\{ (x,y) \in P : y > ax_i \wedge \left(x + y > \frac{(\delta_x + \delta_y)}{2} \right) \right\}$$

$$V_4(i,j) = \left\{ (x,y) \in P : x \le x_i + \delta_x - \delta_y \wedge \left(x + y > \frac{(\delta_x + \delta_y)}{2} \right) \right\}.$$

Now each of $V_2(i,j)$, $V_3(i,j)$ and $V_4(i,j)$ is defined by an intersection of two half-planes. By transforming every point $p \in P$ into $(x(p)+y(p), y(p))$ for $V_3(i,j)$ and into $(x(p)+y(p), x(p))$ for $V_4(i,j)$, we can assume that the lines defining the half-planes are parallel to the coordinate axes. Hence each sum can be calculated with orthogonal queries in $\mathcal{O}(\log n)$ time and $\mathcal{O}(n \log n)$ time and space preprocessing by Lemma 3.   □

We reduced the line-constrained k-median problem in L_1 to the minimum k-link path problem. The weight of any edge can be retrieved in $\mathcal{O}(\log n)$ time by decomposing it into a constant number of orthogonal queries. By plugging in an appropriate algorithm for the minimum k-link path problem, we obtain the final theorem.

Theorem 14. *We can solve the line-constrained k-median problem in L_1 and L_∞ using $\mathcal{O}(kn \log n)$ time or, if $k = \Omega(\log n)$, $n2^{\mathcal{O}(\sqrt{\log k \log \log n})} \log n$ time.*

References

1. Aggarwal, A., Klawe, M.M., Moran, S., Shor, P., Wilber, R.: Geometric applications of a matrix-searching algorithm. Algorithmica **2**(1–4), 195–208 (1987)
2. Bajaj, C.: The algebraic degree of geometric optimization problems. Discrete & Computational Geometry **3**(1), 177–191 (1988)
3. Bein, W., Golin, M.J., Larmore, L.L., Zhang, Y.: The Knuth-Yao quadrangle-inequality speedup is a consequence of total monotonicity. ACM Transactions on Algorithms (TALG) **6**(1), 17 (2009)
4. Blum, M., Floyd, R.W., Pratt, V.R., Rivest, R.L., Tarjan, R.E.: Time bounds for selection. J. Comput. Syst. Sci. **7**(4), 448–461 (1973)
5. Galil, Z., Park, K.: A linear-time algorithm for concave one-dimensional dynamic programming. Information Processing Letters **33**(6), 309–311 (1990)
6. Grossi, R., Gupta, A., Vitter, J.S.: High-order entropy-compressed text indexes. In: Proceedings of the fourteenth annual ACM-SIAM symposium on Discrete algorithms. pp. 841–850. Society for Industrial and Applied Mathematics (2003)
7. Megiddo, N., Supowit, K.J.: On the complexity of some common geometric location problems. SIAM journal on computing **13**(1), 182–196 (1984)

8. Navarro, G., Russo, L.M.S.: Space-Efficient Data-Analysis Queries on Grids. In: Asano, T., Nakano, S., Okamoto, Y., Watanabe, O. (eds.) ISAAC 2011. LNCS, vol. 7074, pp. 323–332. Springer, Heidelberg (2011)
9. Schieber, B.: Computing a minimum weight k-link path in graphs with the concave Monge property. Journal of Algorithms **29**(2), 204–222 (1998)
10. Wang, H., Zhang, J.: Line-Constrained k-Median, k-Means, and k-Center Problems in the Plane. In: Ahn, H.-K., Shin, C.-S. (eds.) ISAAC 2014. LNCS, vol. 8889, pp. 3–14. Springer, Heidelberg (2014)

Combinatorial Algorithms

SOBRA - Shielding Optimization for BRAchytherapy

Guillaume Blin[1], Marie Gasparoux[1,2], Sebastian Ordyniak[3(✉)], and Alexandru Popa[4]

[1] Univ. Bordeaux, LaBRI, CNRS UMR 5800, 33400 Talence, France
[2] DIRO, Univ. Montréal, C.P. 6128 Succ. Centre-Ville,
Montreal, QC H3C 3J7, Canada
[3] Institute of Information Systems, TU Wien, Vienna, Austria
`sordyniak@gmail.com`
[4] Department of Computer Science, University of Bucharest, Bucharest, Romania

Abstract. In this paper, we study a combinatorial problem arising in the development of innovative treatment strategies and equipment using tunable shields in internal radiotherapy. From an algorithmic point of view, this problem is related to circular integer word decomposition into circular binary words under constraints. We consider several variants of the problem, depending on constraints and parameters and present exact algorithms, polynomial time approximation algorithms and NP-hardness results.

1 Introduction

In France, every year, almost 200,000 patients are treated by radiotherapy as part of their cancer treatment. This kind of therapy uses ionizing radiation aiming at controlling or killing malignant cells as a curative procedure or as part of adjuvant therapy and is widely used (in 2/3 of the cancer treatments). While internal radiotherapy treatments are currently widespread and considered as routine, there is still room for related innovative developments. The aim is to concentrate the radiation beams as precisely as possible towards the tumor site while sparing as much as possible the nearby healthy tissues, such as skin or vital organs (the so-called organs at risk from radiation).

Brachytherapy – also sometimes named Curietherapy – refers to a short distance (*brachys* in Greek) treatment of cancer with radiation from small, encapsulated radionuclide sources (also called *seeds*). These radioactive seeds are used to deliver a high dose to the tissues close to the source. It is characterized by strong dose gradients, *i.e.*, the dose becomes negligible in a very short distance from the source (about 10 % decay per mm) [7]. Such a treatment is given by placing sources directly into or near the volume to be treated. The dose is then

This research was partially supported by PoPRA project funded by Conseil Régional d'Aquitaine and European FEDER and IdEx Bordeaux. We thank Maxence Ronzié for initial assistance in the very first step of the project.

© Springer International Publishing Switzerland 2016
V. Mäkinen et al. (Eds.): IWOCA 2016, LNCS 9843, pp. 309–320, 2016.
DOI: 10.1007/978-3-319-44543-4_24

delivered continuously, either over a short period of time (temporary implants) or over the lifetime of the source to a complete decay (permanent implants). There are many different techniques and sources available.

In this contribution, we focus on *High Dose Rate* (HDR) implants. HDR brachytherapy is a form of internal radiation which temporarily exposes abnormal tissue to a high amount of radiation. Under Computed Tomography and Fluoroscopy guidance, a bronchoscope or a needle is used to deliver a catheter into a position at the tumor site. The other end of this catheter is connected to a computerized machine. This machine passes a small radioactive metal seed through the catheter. The catheter guides the seed to the tumor site. The seed moves step by step through the catheter in order to cover the whole tumor site. The time spent at each position – also known as *dwell time* – is used to control the radiation dose distribution accross the tumor. The overall effect of HDR brachytherapy is to deliver short and precise amounts of high-dose radiation to a tumor while minimizing healthy tissue exposure. After a series of treatment sessions, the catheter is removed leaving no radioactive seeds in the body.

One of the main drawbacks of this technique comes from the lack of precise modulation of the irradiation field and thus of conformation to the shape of the tumor site. In this paper, we aim at studying the benefit of an innovative modulation technique in brachytherapy using tunable shields (as done in external radiotherapy). This approach will allow accumulating both the temporal modulation currently used and the shielding modulation. The aim is to provide treatment of better accuracy by adapting more precisely to the tumor shape. Indeed, currently, the modulation of the radiation source is done by controlling the time spent at each position by the source along the catheter. The main problem is that, at any position, the irradiation is uniform and can be represented as a cylinder surrounding the catheter. This shape does not always conform to the relative placement of the tumor and organs at risk (i.e., in the radiation field). In this contribution, we consider modulating a unique radioactive source using a gear inspired by external radiotherapy.

The use of the shield will allow to preserve, for a given position along the catheter, some part of the surrounding area. The so-called rotating shield brachytherapy (RSBT) was conceptually proposed by Ebert in 2002 [3]. In RSBT, the dose is delivered through a partially shielded radiation source in an optimized step-shot fashion (as done in classical brachytherapy treatment) to improve tumor dose conformity. The intensity of radiation is modulated by the amount of time the shield is pointed in a given direction. RSBT [5,6,13] and other intensity-modulated brachytherapy techniques such as dynamic modulated brachytherapy (DMBT) [10–12] were further studied with the aim of improving intracavitary brachytherapy dose distributions for rectal and cervical cancer. We will first focus on a peculiar type of shield which have been briefly described in the patent [9] and studied in [4]. It corresponds to a set of shield segments forming a cylinder that can be individually retracted to produce circumferentially limited radiation output, directed radially. According to the way the sources are introduced in the patient body, and the physical constraints of the material, it is not possible to build sector of size as small, and thus as high resolution, as wanted. Therefore, using the possible rotation of the equipment, the aim is to find a sequence of sectors configurations that allows delivering a dose distribution as near as possible to the prescribed one. The corresponding algorithmic aspects are unexplored and the goal of this paper is to conduct an algorithmic study which will guide the final development of the equipment. From an algorithmic point of view, the problem is related to circular integer word decomposition into circular binary words under constraints. In Sect. 2 we formally introduce the considered problem and we present an overview of the results.

2   A Formal Model for HDR Brachytherapy with Shields

Considering each dwell position of the irradiation source (denoted I), our main objective is to deliver to each part of the surrounding volume its proper irradiation dose. For this purpose, we will use a paddle-based shielding equipment P of K paddles (also referred as sectors for ease) that can stop the radiation going through when they are not retracted. We will consider the surrounding volume to be treated as a circular volume of interest divided in N subvolumes. In the following, a treatment plan for a given dwell position will be defined as a sequence of T shield configurations $((P^1, \tau^1), (P^2, \tau^2), \ldots, (P^T, \tau^T))$ where P^t, $1 \le t \le T$, is a paddle configuration and τ^t is its dwell time. Each paddle configuration is represented as a binary string $P^t = p_0^t p_1^t \ldots p_{K-1}^t$ where p_k^t represents the state (open or closed) of the sector k of P^t. An open sector of the shield (paddle retracted allowing radiation going through) is represented by a 1, while a closed one (paddle is out and radiation is stopped) is represented by a 0.

For each given step (P^t, τ^t) in the treatment plan, a corresponding *received dose* D^t by the surrounding volume is defined as a string of integers $D^t = d_0^t d_1^t \ldots d_{N-1}^t$ where d_n^t corresponds to the total irradiation time the subvolume n was exposed to during this step. Roughly, it corresponds to the contribution of the corresponding treatment step to the whole treatment plan. For ease, when parameters P^t and τ^t are not needed for comprehension we may omit them and only write D. Regarding the entire treatment plan, we will denote *the prescribed doses* as a string of nonnegative integers $\hat{D} = \hat{d}_0 \hat{d}_1 \ldots \hat{d}_{N-1}$ where $\hat{d}_n$ corresponds to the total irradiation time needed to achieve the right dose for the subvolume n. We will moreover denote *the total received doses* as a string of integers $D = d_0 d_1 \ldots d_{N-1}$, such that for all $d_n \in D$, $d_n = \sum_{1 \le t \le T} d_n^t$.

For ease and without loss of generality, we assume that each shield sector is associated to $w = N/K$ consecutive subvolumes, and, for simplicity, that K divides N (so w is an integer). By default, each shield sector p_k will be associated to $D_k = D[k \cdot w, \ (k+1) \cdot w - 1] = d_{k \cdot w} d_{k \cdot w+1} \ldots d_{k \cdot w+w-1}$ of length w (see example Fig. 1a). We can remark that $D = D_0 D_1 \ldots D_K$. Informally, one may see P and D as circular strings, P placed inside D and representing a mask that can stop the radiation from going through (see Fig. 1b, with a counterclockwise indexation).

Let us consider the practical case where one is applying a given shielded configuration (represented by P) on a patient (represented by D) for a given

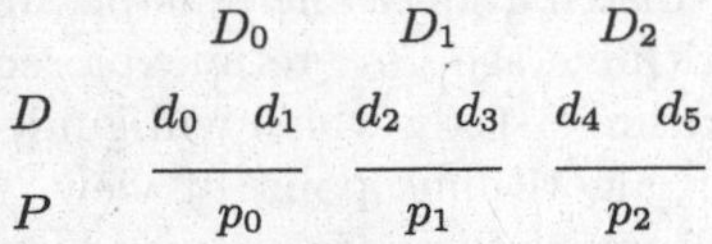

	D_0		D_1		D_2	
D	d_0	d_1	d_2	d_3	d_4	d_5
P		p_0		p_1		p_2

(a) Linear representation

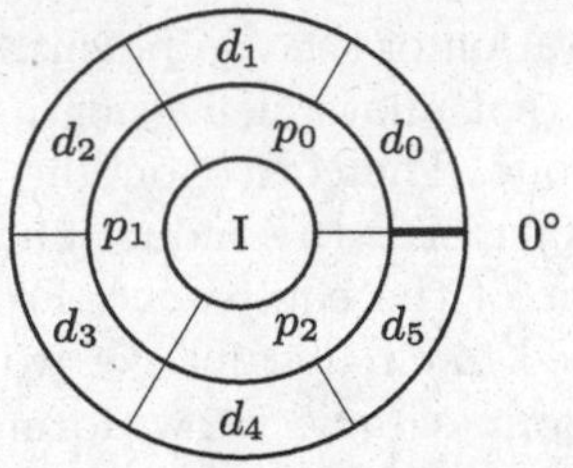

(b) Circular representation

Fig. 1. Relation between P and D $(K = 3, N = 6)$

amount of time τ (expressed in a given unit of time). Let us denote $D(P, \tau) = d_0 d_1 \ldots d_{N-1}$ the string of integers obtained by applying radiation for a time τ to D through the mask P. We consider that $p_k = 1$ (resp. $p_k = 0$) denotes applying radiation (resp. no radiation applied) to the area D_k. Moreover, $d_n = \tau$ (resp. $d_n = 0$) if radiation is applied to the volume n for a time τ (resp. if no radiation is applied there). In other words, each subvolume associated to an open sector (represented by a 1) is irradiated τ units of time, while volume associated to a closed sector (represented by a 0) is left in its previous state.

One may consider several variants of the problem, depending on constraints and parameters. First of all, the shield configuration can be considered as fixed or dynamic (one fixed mask or a minimal number of chosen masks) and provided with or without rotation capabilities (this last property is not considered here). These properties are related to manufacturing purposes and constraints. We moreover consider allowing or not irradiation overdoses ($d_n > \hat{d}_n$). Indeed, in practice, it is convenient to overdose a tumor region while one should try to not overdose regions of organs at risk. From a combinatorial point of view, there are two parameters that alter the overall treatment time; namely, the sum of the irradiation times and the number of configurations (as a transition between two configurations will require some time). In the following, we will consider variants of the problem based on the previous observations. In the first two variants, the input consists of only one shield configuration that is given and fixed. The goal is to decide what is the optimum amount of radiation that can be applied when allowing or disallowing overdoses. As proven in Sect. 3, these variants of the problem are polynomial time solvable.

*Problem 1 (*FixMask*).* Given a prescribed dose represented as a string of non-negative integers $\hat{D} = \hat{d}_0 \hat{d}_1 \ldots \hat{d}_{N-1}$ and a fixed shield configuration represented as a binary string $P = p_0 p_1 \ldots p_{K-1}$, find the dwell time τ minimizing $\sum_{n=0}^{N-1} |\hat{d}_n - d_n|$ with $D = D(P, \tau)$.

While in the FixMask variant of the problem, $\hat{d}_n - d_n$ can be negative – that is overdoses are allowed – in the FixMask$^+$ variant, we moreover impose that $\forall n < N, d_n \leq \hat{d}_n$ – thus, forbidding overdoses. We now consider variants of the problem where multiple shield configurations are allowed. As mentioned previously, two different criteria can be optimized in such a treatment plan. One would like to either achieve the optimal difference between the prescribed dose and the actual total delivered dose using a minimal number of shield configurations or given an upper bound on the number of shield configurations, achieving the minimum reachable difference. Formally, the problems are defined as follows.

*Problem 2 (*MinFixMasks$_{\text{OPT}}$*).* Given two nonnegative integers K and *diff* and a string of integers $\hat{D} = \hat{d}_0 \hat{d}_1 \ldots \hat{d}_{N-1}$ (with N being a multiple of K), find a treatment plan $((P^1, \tau^1), (P^2, \tau^2), \ldots, (P^T, \tau^T))$ minimizing T such that $\sum_{n=0}^{N-1} \left| \hat{d}_n - d_n \right| \leq \textit{diff}$, where $\forall d_n \in D, d_n = \sum_{1 \leq t \leq T} d_n^t$.

*Problem 3 (*MinFixMasks$_{\text{BOUND}}$*).* Given two nonnegative integers K and $T_{\max}$ and a string of integers $\hat{D} = \hat{d}_0 \hat{d}_1 \ldots \hat{d}_{N-1}$ (with N being a multiple of K), find a treatment plan $((P^1, \tau^1), (P^2, \tau^2), \ldots, (P^T, \tau^T))$ where $T < T_{\max}$ minimizing $\sum_{n=0}^{N-1} \left| \hat{d}_n - d_n \right|$, where $\forall d_n \in D, d_n = \sum_{1 \leq t \leq T} d_n^t$.

Similarly to FixMask$^+$, in MinFixMasks$_{\text{OPT}}^+$ and MinFixMasks$_{\text{BOUND}}^+$ variants of the problem, we moreover impose that $\forall n < N, d_n \leq \hat{d}_n$ – thus, forbidding overdoses.

Our results can be summarized as follows. We will show in Sect. 3 that the problems FixMask and FixMask$^+$ can be solved in polynomial time. We will then show in Sect. 4 that all of {MinFixMasks$_{\text{OPT}}$, MinFixMasks$_{\text{OPT}}^+$, MinFixMasks$_{\text{BOUND}}$, MinFixMasks$_{\text{BOUND}}^+$} can be solved in quasi-polynomial time if $\hat{d}_{max}$ is bounded by a polynomial in the number of prescribed doses, where $\hat{d}_{max}$ is the maximum prescribed dose to a subvolume of the patient. In the same section we will also show that the problems MinFixMasks$_{\text{OPT}}$ and MinFixMasks$_{\text{OPT}}^+$ can be approximated in polynomial time within a factor of $\log \hat{d}_{max}$ of the optimum. Finally, we will show in Sect. 5 that the problems MinFixMasks$_{\text{OPT}}$ and MinFixMasks$_{\text{BOUND}}$ are NP-complete.

3   Polynomial Results

In this section, we show that the variants of the problem where the shield configuration is given and fixed are solvable in polynomial time. Clearly, for a fixed masked, the doses associated to closed paddles cannot be brought closer to the corresponding prescribed doses and will thus not be considered.

Theorem 1. FixMask$^+$ *can be solved in $\mathcal{O}(N)$ time.*

Proof. Because we are not allowed to apply overdoses, we obtain that the maximum and also the optimum irradiation time is equal to the minimum of all prescribed doses $\hat{d}_j$ of $\hat{D}$ for which the corresponding paddle is open. Since the minimum of these doses can be obtained in linear time, the result follows.   □

The main observation required to show that FixMask can also be solved in polynomial time is given in the following lemma, which can be considered folklore and is stated here only for the convenience of the reader.

Lemma 1. *For a sequence S of natural numbers and a natural number x, consider the function $f(x)$ such that $f(x) = \sum_{s \in S} |s - x|$. Then $f(x)$ has a unique minimum, which is only reached by any number x in between the at most two medians of S. Moreover, for any x not between the at most two medians of S, the function $f(x)$ decreases with the distance of x to a median of S.*

The above lemma implies that an optimum dwell time for an instance of FixMask is a median of the subsequence of $\hat{D}$ containing all prescribed doses for which the paddles are open.

Theorem 2. FixMask *can be solved in $\mathcal{O}(N)$.*

Proof. Because of Lemma 1 the best possible value that we can achieve for $\sum_{n=0}^{N-1} |\hat{d}_n - d_n|$ is obtained by setting the dwell time τ_0 to any median of the subsequence of $\hat{D}$ containing only the prescribed doses for which the paddles are open (in the given mask). It is known [1] that a median of n numbers can be found in linear time.   □

4   Quasi-polynomial Algorithms for MinFixMasks

In this section, we present exact algorithms for all variants of the MinFixMasks problem. The presented algorithms run in quasi-polynomial time if the values of the prescribed patient doses are bounded by a polynomial in the number of prescribed doses. As a by-product we show that the problems MinFixMasks$_{\text{OPT}}$ and MinFixMasks$_{\text{OPT}}^+$ can be approximated in polynomial-time within a factor of $\log \hat{d}_{\max}$ of the optimum where $\hat{d}_{\max}$ is the maximum prescribed dose to a subvolume of the patient, *i.e.*, $\hat{d}_{\max} := \max_{\hat{d}_n \in \hat{D}} \hat{d}_n$. We first show that it is sufficient to consider treatment plans where the applied dwell times are pairwise distinct.

Lemma 2. *For any instance of* MinFixMasks$_{\text{OPT}}$, MinFixMasks$_{\text{OPT}}^+$, MinFixMasks$_{\text{BOUND}}$, *and* MinFixMasks$_{\text{BOUND}}^+$ *there is an optimal solution $((P^1, \tau^1), (P^2, \tau^2), \ldots, (P^T, \tau^T))$ satisfying $\tau^i \neq \tau^j$ for every i and j with $1 \leq i \neq j \leq T$.*

Proof. Let $\mathcal{P} = ((P^1, \tau^1), \ldots, (P^T, \tau^T))$ be an optimal solution of an instance $\mathcal{I}$ of any of the mentioned variants of the MinFixMasks problem. We will show that we can transform $\mathcal{P}$ into an equivalent treatment plan that does not use any dwell time more than once. Let i and j with $1 \leq i \neq j \leq T$ be such that $\tau^i = \tau^j$. Let (i) $\tau_*^j = 2\tau^j$, (ii) the binary string P_*^i be obtained by the *XOR* of binary strings P^i and P^j, and (iii) the binary string P_*^j be obtained by the *AND* of binary strings P^i and P^j. Then the treatment plan obtained from $\mathcal{P}$ by replacing (P^i, τ^i) with (P_*^i, τ^i) and (P^j, τ^j) with (P_*^j, τ_*^j) is also an optimal solution of $\mathcal{I}$. Moreover, by applying this procedure iteratively we eventually obtain an optimal solution of $\mathcal{I}$ such that all dwell times are pairwise distinct.   □

Let S be a set of dwell times. We say that S is *complete* if it contains a subset S' for every number $1 \leq i \leq \hat{d}_{max}$ such that $i = \sum_{s \in S'} s$. We say that a treatment plan is *S-restricted* if it uses only dwell times from S and each of them at most once.

Lemma 3. *Let S be a set of dwell times. Then an S-restricted treatment plan minimizing $\sum_{n=0}^{N-1} \left| \hat{d}_n - d_n \right|$ can be found in time $O((\hat{d}_{max})^2 |S| + Kw + K\hat{d}_{max})$. Moreover, the same applies to an S-restricted treatment satisfying the additional constraint that $\hat{d}_n - d_n \geq 0$ for every $0 \leq n \leq N-1$. Finally, if S is complete then the S-restricted treatment plans returned by the above algorithms are optimal among all (not necessarily S-restricted) treatment plans.*

Lemma 4. *There is a treatment plan minimizing $\sum_{n=0}^{N-1} \left| \hat{d}_n - d_n \right|$ using at most $\lfloor \log \hat{d}_{max} \rfloor + 1$ steps. Moreover, such a treatment plan can be found in polynomial time. The same holds for a treatment plan minimizing $\sum_{n=0}^{N-1} \left| \hat{d}_n - d_n \right|$ under the additional constraint that $\hat{d}_n - d_n \geq 0$ for every n with $0 \leq n \leq N-1$.*

Proof. Because the set $S = \{ 2^i : 0 \leq i \leq \lfloor \log \hat{d}_{max} \rfloor \}$ is complete and has size $\lfloor \log \hat{d}_{max} \rfloor + 1$, this follows immediately from Lemma 3. $\qquad \square$

Because any non-trivial instance of $\text{MinFixMasks}_{\text{OPT}}$ and $\text{MinFixMasks}_{\text{OPT}}^+$ require at least one step, we obtain the following corollary from the above lemma.

Corollary 1. $\text{MinFixMasks}_{\text{OPT}}$ *and* $\text{MinFixMasks}_{\text{OPT}}^+$ *can be approximated in polynomial time within a factor of $\log \hat{d}_{max}$ of the optimum.*

We are now ready to show our main theorem of this section.

Theorem 3. $\text{MinFixMasks}_{\text{OPT}}$, $\text{MinFixMasks}_{\text{OPT}}^+$, $\text{MinFixMasks}_{\text{BOUND}}$, *and* $\text{MinFixMasks}_{\text{BOUND}}^+$ *can be solved in time $O(\hat{d}_{max}^{\lfloor \log \hat{d}_{max} \rfloor + 1} ((\hat{d}_{max})^2 (\lfloor \log \hat{d}_{max} \rfloor + 1) + Kw + K\hat{d}_{max}))$.*

Proof. The algorithm goes over all sets S containing at most $\lfloor \log \hat{d}_{max} \rfloor + 1$ (respectively at most $\min\{ \lfloor \log \hat{d}_{max} \rfloor + 1, T_{\max} \}$ in the case of $\text{MinFixMasks}_{\text{BOUND}}$ and $\text{MinFixMasks}_{\text{BOUND}}^+$) dwell times between 1 and $\hat{d}_{max}$. For every such set S, the algorithm then uses Lemma 3 to compute the optimal (the meaning of optimal here depends on the considered problem) S-restricted treatment plan. Finally, in the case of $\text{MinFixMasks}_{\text{OPT}}$ and $\text{MinFixMasks}_{\text{OPT}}^+$ the algorithm returns a shortest treatment plan satisfying $\sum_{n=0}^{N-1} \left| \hat{d}_n - d_n \right| \leq diff$ and in the case of $\text{MinFixMasks}_{\text{BOUND}}$ and $\text{MinFixMasks}_{\text{BOUND}}^+$ returns a treatment plan minimizing $\sum_{n=0}^{N-1} \left| \hat{d}_n - d_n \right|$ found for any of the considered sets S. The stated running time of the algorithm follows because there are at most $\hat{d}_{max}^{\lfloor \log \hat{d}_{max} \rfloor + 1}$ such sets S and because of Lemma 3 for each set S, we require time at most $O((\hat{d}_{max})^2 |S| + Kw + K\hat{d}_{max})$. The correctness of the algorithm follows from Lemmas 2, 3 and 4. $\qquad \square$

Corollary 2. $\textsc{MinFixMasks}_{\text{OPT}}$, $\textsc{MinFixMasks}_{\text{OPT}}^{+}$, $\textsc{MinFixMasks}_{\text{BOUND}}$, *and* $\textsc{MinFixMasks}_{\text{BOUND}}^{+}$ *can be solved in quasi-polynomial time if* $\hat{d}_{max}$ *is bounded by a polynomial in the number of prescribed doses.*

5 Hardness of MinFixMasks$_{\text{opt}}$ and MinFixMasks$_{\text{bound}}$

In this section, we show that $\textsc{MinFixMasks}_{\text{OPT}}$ and $\textsc{MinFixMasks}_{\text{BOUND}}$ are NP-complete already for $w = 2$ (recall that $w = N/K$). Observe that the decision version of the problems $\textsc{MinFixMasks}_{\text{OPT}}$ and $\textsc{MinFixMasks}_{\text{BOUND}}$ are the same, i.e., given a target sequence $\hat{D} = \hat{d}_0 \ldots \hat{d}_{N-1}$ and integers K, *diff*, and $T_{\max}$, determine whether there is a treatment plan $((P^1, \tau^1), (P^2, \tau^2), \ldots, (P^T, \tau^T))$ such that $T \leq T_{\max}$ and $\sum_{n=0}^{N-1} \left| \hat{d}_n - d_n \right| \leq \textit{diff}$.

Our proof uses a reduction from the $\textsc{Monotone}$ 1-3 SAT problem (proven to be NP-complete in [8]): given a boolean formula $\phi = \{c_1, c_2, \ldots\}$ in 3-CNF of $|\phi|$ clauses built on a set $V = \{v_1, v_2, \ldots\}$ of $|V|$ variables, such that its clauses contain only unnegated literals, does there exist a truth assignment on V satisfying ϕ such that each clause is satisfied by exactly one of its three literals?

Given any instance (ϕ, V) of $\textsc{Monotone}$ 1-3 SAT problem, we build an instance of the decision version of $\textsc{MinFixMasks}_{\text{OPT}}$ and $\textsc{MinFixMasks}_{\text{BOUND}}$ as follows. For all $i \in [1, |V|]$, let q_i be an integer value computed using the following recurrence formula: $q_i = 1 + 2 \times \sum_{j=1}^{i-1}(1 + q_j)$ with $q_1 = |V|$. For each variable $v_i \in V$, we build the sequence $V_i = (q_i, 1 + q_i)$. For each clause $c_m = (v_a, v_b, v_c) \in \phi$, we build the sequence C_m composed of two copies of $(q_a + q_b + q_c + 2)$. For each pair (v_i, v_j), $i < j \leq |V|$, we build the sequence $V_{i,j} = (q_i + q_j, q_i + q_j + 2)$. Let $V_{*,j}$ be the concatenation of $V_{1,j}, V_{2,j}, \ldots V_{j-1,j}$. The sequence $\hat{D}$ is obtained by concatenating in order $V_1 \ V_2 \ldots V_{|V|} \ C_1 \ C_2 \ldots C_{|\phi|} \ V_{*,2} \ V_{*,3} \ldots V_{*,|V|}$. We finally set $K = |V| + |\phi| + \frac{|V| \cdot (|V|-1)}{2}$, $N = 2 \cdot K$ (*i.e.*, $w = 2$), $\textit{diff} = |V|^2$, and $T_{\max} = |V|$. An illustration is given in Fig. 2.

$$\underbrace{5 \ 6}_{V_1} \ \underbrace{13 \ 14}_{V_2} \ \underbrace{41 \ 42}_{V_3} \ \underbrace{125 \ 126}_{V_4} \ \underbrace{377 \ 378}_{V_5} \ \underbrace{61^2}_{C_1} \ \underbrace{545^2}_{C_2} \ \underbrace{517^2}_{C_3} \ \underbrace{145^2}_{C_4} \ \underbrace{18 \ 20}_{V_{1,2}} \ \underbrace{46 \ 48}_{V_{1,3}} \ldots \underbrace{502 \ 504}_{V_{4,5}}$$

Fig. 2. Example of an instance of $\textsc{MinFixMasks}$ considering the boolean formula $\phi = (v_1, v_2, v_3) \wedge (v_3, v_4, v_5) \wedge (v_2, v_4, v_5) \wedge (v_1, v_2, v_4)$ which only admits one optimal solution ($v_1 = v_5 = \textbf{true}$ and $v_2 = v_3 = v_4 = \textbf{false}$). For ease of notation, v^x will denote x occurrences of the element v (thus 61^2 corresponds to 61 61) and most elements $V_{i,j}$ have been omitted.

Let us start by showing some important properties for any treatment plan of the constructed instance. Let $((P^1, \tau^1), (P^2, \tau^2), \ldots, (P^T, \tau^T))$ be a treatment plan that is a solution for an instance constructed from the given formula. We say that a step t *contributes* to a sequence V_i if the block V_i is irradiated at step

t (the ith bit of its mask P^t is set to 1) and t *minimizes* a sequence V_i if the step t is the last one of the treatment plan contributing to V_i: at step $t - 1$ and before, V_i did not reach its minimum yet, at step $t + 1$ and after, V_i cannot be lowered. Note that a sequence is minimized at exactly one step.

Lemma 5. *For every treatment plan it holds that $\sum_{n=0}^{N-1} \left| \hat{d}_n - d_n \right| \geq$ diff $= |V|^2$. Moreover, any treatment plan for which $\sum_{n=0}^{N-1} \left| \hat{d}_n - d_n \right| =$ diff, uses at least $|V|$ steps. Finally, for every treatment plan that uses at most $|V|$ steps, it holds that every step minimizes at most one sequence V_i.*

Using the reduction defined above, we are now ready to show the main theorem of this section.

Theorem 4. *The MINFIXMASKS$_{\text{OPT}}$ and MINFIXMASKS$_{\text{BOUND}}$ problems are NP-complete when $w = 2$.*

Proof. Clearly, both problems are contained in NP, since there is always an optimal solution of length at most $\lfloor \log \hat{d}_{max} \rfloor + 1$ (see also Lemma 4).

We will show the correctness of the reduction from MONOTONE 1-3 SAT to the decision versions of MINFIXMASKS$_{\text{OPT}}$ and MINFIXMASKS$_{\text{BOUND}}$ given above the theorem.

($\Rightarrow$) Let τ be an assignment satisfying ϕ such that each clause is satisfied by exactly one of its literals. We will construct a treatment plan $\mathcal{P} = ((P^1, \tau^1), (P^2, \tau^2), \dots, (P^{|V|}, \tau^{|V|}))$ satisfying $\sum_{n=0}^{N-1} \left| \hat{d}_n - d_n \right| =$ diff $= |V|^2$ as follows.

For all $1 \leq n \leq |V|$, τ^n is defined by setting $\tau^n = q_n$ if $\tau(v_n) =$ **true** and setting $\tau^n = 1 + q_n$ otherwise. Each P^n is obtained by concatenating three substrings corresponding to the V_i's, C_m's and $V_{i,j}$'s as follows: $P^n = P_V^n P_C^n P_{V_*}^n$ where $P_V^n = 0^{n-1} 10^{|V|-n}$, $P_C^n = \text{In}(n,1) \text{In}(n,2) \dots \text{In}(n,|\phi|)$ ($\text{In}(n,m)$ is 1 if $v_n \in c_m$, 0 otherwise) and $P_{V_*}^n = P_{V_{*,2}}^n \dots P_{V_{*,|V|}}^n$. Each $P_{V_{*,i}}^n$ is defined accordingly to i and n as $P_{V_{*,i}}^n = 0^{i-1}$ if $i < n$; $P_{V_{*,i}}^n = 1^{n-1}$ if $i = n$ and $P_{V_{*,i}}^n = 0^{n-1} 10^{i-1-n}$ otherwise.

By construction, $\mathcal{P}$ applies total dwell time of either q_n or $1 + q_n$ each V_n. Moreover, any C_m corresponding to a clause (v_a, v_b, v_c) receives a total dwell time of $q_a + q_b + q_c + 2$, since by hypothesis exactly one of $\{v_a, v_b, v_c\}$ is true in our assignment: that is either $q_a + (1 + q_b) + (1 + q_c)$ or $(1 + q_a) + q_b + (1 + q_c)$ or $(1 + q_a) + (1 + q_b) + q_c$. Finally, a total dwell time $\tau^i + \tau^j$ such that $q_i + q_j \leq \tau^i + \tau^j \leq q_i + q_j + 2$ has been applied to each $V_{i,j}$, lowering its cost to 2. Thus, any MONOTONE 1-3 SAT solution over ϕ gives us an optimal solution for our instance of MINFIXMASKS using $|V|$ shield configurations.

($\Leftarrow$) Let $\mathcal{P} = ((P^1, \tau^1), (P^2, \tau^2), \dots, (P^T, \tau^T))$ be a solution, i.e., it holds that $\sum_{n=0}^{N-1} |\hat{d}_n - d_n| \leq$ diff $= |V|^2$ and $T \leq T_{\max}$. It follows from Lemma 5 that $\sum_{n=0}^{N-1} |\hat{d}_n - d_n| =$ diff $= |V|^2$ and moreover $T = T_{\max} = |V|$.

Because we have $|V|$ sequences V_i that need to be minimized at some step of $\mathcal{P}$ and $\mathcal{P}$ has $|V|$ steps, we obtain from Lemma 5 that exactly one sequence

V_i is minimized at any step of $\mathcal{P}$. W.l.o.g., we can assume that $\mathcal{P}$ is ordered in such a way that each sequence V_i is minimized at step i. In the next proposition, we prove that the dwell time at step i can take one of two possible values, thus corresponding to a true/false assignment of variable i.

Proposition 1. *For any i, $\tau^i \in \{q_i, 1 + q_i\}$.*

Proof. We prove the result by induction. Because of our assumption on the ordering of $\mathcal{P}$, we obtain that $q_i - \sum_{j=1}^{i-1} \tau^j \leq \tau^i \leq 1 + q_i$ for each step i. Thus, for the first induction step, it holds that $q_1 \leq \tau^1 \leq 1 + q_1$, so $\tau^1 \in \{q_1, 1 + q_1\}$.

Considering the step j, and the sequence $V_{j-1,j} = (q_{j-1} + q_j, q_{j-1} + q_j + 1)$, no step after j can contribute to $V_{j-1,j}$ since

$$\tau^{j+1} \geq q_{j+1} - \sum_{i=1}^{j}(1 + q_i) \geq 1 + 2 \times \sum_{i=1}^{j}(1 + q_i) - \sum_{i=1}^{j}(1 + q_i)$$
$$> q_{j-1} + q_j + 2$$

Moreover, $\sum_{i=1}^{j-1} \tau^i \leq \sum_{i=1}^{j-1}(1 + q_i) < q_j$, so the contribution of step j is mandatory to minimize $V_{j-1,j}$ which induces that

$$\tau^j \geq q_{j-1} + q_j - \sum_{i=1}^{j-1} \tau^i \geq q_{j-1} + q_j - \sum_{i=1}^{j-1}(1 + q_i)$$
$$\geq q_j - 1 - \sum_{i=1}^{j-2}(1 + q_i)$$

Suppose now that there exists $k \geq 2$ such that $\tau^j \geq q_{j-1} - 1 - \sum_{i=1}^{j-k}(1 + q_i)$. Consider then $V_{j-k,j} = (q_{j-k} + q_j, q_{j-k} + q_j + 2)$. Applying a similar reasoning as before, we conclude that the contribution of step j is mandatory, and, with our last lower bound over τ^j that

$$q_{j-k} + q_j - \tau^j \leq q_{j-k} + q_j - \left(q_j - 1 - \sum_{i=1}^{j-k}(1 + q_i) \right)$$
$$\leq q_{j-k} + 1 + \sum_{i=1}^{j-k}(1 + q_i) < q_{j-k+1}$$

Thus, steps strictly between $j - k$ and j cannot contribute. Therefore, the only steps able to contribute are j and 1 to $j - k$:

$$\tau^j \geq q_{j-k} + q_j - \sum_{i=1}^{j-k} \tau^i \geq q_{j-k} + q_j - \sum_{i=1}^{j-k}(1 + q_i)$$
$$\geq q_j - 1 - \sum_{i=1}^{j-(k+1)}(1 + q_i)$$

We obtain a greater lower bound for τ^j. This reasoning can be applied as long as $V_{j-k,j}$ exists, that is as long as $j - k \geq 1$. The last application ($k = j - 1$, $k + 1 = j$) leads to

$$\tau^j \;\geq\; q_j - 1 - \sum_{i=1}^{j-j}(1 + q_i) \;\geq\; q_j - 1$$

On the whole, we obtain $\tau^j \in \{q_j - 1, q_j, 1 + q_j\}$. Moreover, if $\tau^j = q_j - 1$, then step j is not enough to minimize $V_j = (q_j, 1 + q_j)$ (an amount of 1 or 2 is missing). But we can only use the dwell times of the treatment plan, and the lowest one is $\tau^1 \in \{|V|, |V| + 1\}$, where $|V|$ is the number of variables in ϕ (so $|V| \geq 3$). Thus, $\tau^j = q_j - 1$ is impossible. This leads to $\tau^j \in \{q_j, 1 + q_j\}$ for any step j. ∎

To complete our proof, it remains to show that a sequence $C_m = (q_a + q_b + q_c + 2, q_a + q_b + q_c + 2)$ cannot be minimized by other steps, except those corresponding to an assignment to the variables a, b and c.

Proposition 2. *Minimizing a sequence $C_m = (q_a + q_b + q_c + 2, q_a + q_b + q_c + 2)$ implies the contribution of exactly the steps a, b, and c.*

Proof. W.l.o.g. let $a < b < c$. To minimize C_m, we need to apply a total amount of exactly $\tau = q_a + q_b + q_c + 2$. Since $\tau^{c+1} \geq q_{c+1} \geq 1 + 2 \times \sum_{i=1}^{c}(1 + q_i) > \tau$, step $c + 1$ or higher cannot contribute to C_m. Thus the contribution of step c is mandatory since $\sum_{i=1}^{c-1}\tau^i \leq \sum_{i=1}^{c-1}(1 + q_i) < 1 + 2 \times \sum_{i=1}^{c-1}(1 + q_i) < q_c$. Similarly, $\tau^{b+1} \geq q_{b+1} > \tau - \tau^c$, so steps strictly between b and c cannot contribute, and $\sum_{i=1}^{b-1}\tau^i < q_b \leq \tau - \tau^c$ inducing that the contribution of step b is mandatory. Finally, $\tau^{a+1} \geq q_{a+1} > \tau - \tau^c - \tau^b$, so steps strictly between a and b cannot contribute, implying that the contribution of step a is mandatory since $\sum_{i=1}^{a-1}\tau^i < q_a \leq \tau - \tau^c - \tau^b$. ∎

Gathering the previous results, we have an optimal solution to our MIN-FIXMASKS instance if and only if each sequence C_m corresponding to a clause (v_a, v_b, v_c) receives exactly the dwell times received by the sequences V_a, V_b and V_c. Moreover, each of theses V_i receives either q_i or $1 + q_i$ as a (total) dwell time. Finally, minimizing C_m implies that exactly one of the three V_i receives the lowest of its two possible values. This corresponds to a truth assignment over ϕ such that each of its clauses contains exactly one true variable. □

6 Conclusions and Future Work

We gave the first rigorous algorithmic study of the recently introduced rotating shield brachytherapy. Our analysis led to efficient algorithms as well matching hardness results. For future work we plan to explore further variants of the problem, e.g., variants resulting from a rotation of the shield.

References

1. Blum, M., Floyd, R.W., Pratt, V.R., Rivest, R.L., Tarjan, R.E.: Time bounds for selection. J. Comput. Syst. Sci. **7**(4), 448–461 (1973)
2. Cormen, T., Leiserson, C., Rivest, R., Stein, C.: Introduction to Algorithms. MIT Press, Cambridge (2014)
3. Ebert, M.A.: Possibilities for intensity-modulated brachytherapy: technical limitations on the use of non-isotropic sources. Phys. Med. Biol. **47**(14), 2495 (2002)
4. Liu, Y., Flynn, R.T., Kim, Y., Dadkhah, H., Bhatia, S.K., Buatti, J.M., Xu, W., Wu, X.: Paddle-based rotating-shield brachytherapy. Med. Phys. **42**(10), 5992–6003 (2015)
5. Liu, Y., Flynn, R.T., Kim, Y., Yang, W., Wu, X.: Dynamic rotating-shield brachytherapy. Med. Phys. **40**(12), 121703 (2013)
6. Liu, Y., Flynn, R.T., Yang, W., Kim, Y., Bhatia, S.K., Sun, W., Wu, X.: Rapid emission angle selection for rotating-shield brachytherapy. Med. Phys. **40**(5), 051720 (2013)
7. Potter, R., Haie-Meder, C., Limbergen, E.V., Barillot, I., Brabandere, M.D., Dimopoulos, J., Dumas, I., Erickson, B., Lang, S., Nulens, A., Petrow, P., Rownd, J., Kirisits, C.: Recommendations from gynaecological (GYN) GEC ESTRO working group (ii): concepts and terms in 3D image-based treatment planning in cervix cancer brachytherapy–3D dose volume parameters and aspects of 3D image-based anatomy, radiation physics, radiobiology. Radiother. Oncol. **78**(1), 67–77 (2006)
8. Schaefer, T.J.: The complexity of satisfiability problems. In: Proceedings of the Tenth Annual ACM Symposium on Theory of Computing, STOC 1978, pp. 216–226. ACM, New York (1978)
9. Smith, P., Klein, M., Hausen, H., Lovoi, P.: Radiation therapy apparatus with selective shielding capability, January 2008. US Patent App. 11/471,277
10. Webster, M., Scanderbeg, D., Watkins, T., Stenstrom, J., Lawson, J., Song, W.: SU-F-BRA-11: dynamic modulated brachytherapy (DMBT): concept, design, and application. Med. Phys. **38**(6), 3702–3702 (2011)
11. Webster, M.J., Devic, S., Vuong, T., Yup Han, D., Park, J.C., Scanderbeg, D., Lawson, J., Song, B., Tyler Watkins, W., Pawlicki, T., Song, W.Y.: Dynamic modulated brachytherapy (DMBT) for rectal cancer. Med. Phys. **40**(1) (2013)
12. Webster, M.J., Scanderbeg, D.J., Watkins, W.T., Stenstrom, J., Lawson, J.D., Song, W.Y.: Dynamic modulated brachytherapy (DMBT): concept, design, and system development. Brachytherapy **10**(Suppl. 1), S33–S34 (2011). Abstracts of the 32nd Annual Meeting of the American Brachytherapy Society April 14–16, 2011
13. Yang, W., Kim, Y., Wu, X., Song, Q., Liu, Y., Bhatia, S.K., Sun, W., Flynn, R.T.: Rotating-shield brachytherapy for cervical cancer. Phys. Med. Biol. **58**(11), 3931 (2013)

A Bit-Scaling Algorithm for Integer Feasibility in UTVPI Constraints

K. Subramani and Piotr Wojciechowski(⊠)

LCSEE, West Virginia University, Morgantown, WV, USA
k.subramani@mail.wvu.edu, pwojciec@mix.wvu.edu

Abstract. In this paper, we discuss a new model-generating algorithm for integer feasibility in a system of Unit Two Variable Per Inequality (UTVPI) constraints (**IF**). Recall that a UTVPI constraint is a linear constraint of the form: $a \cdot x + b \cdot y \leq c$, where $a, b \in \{0, 1, -1\}$ and $c \in \mathbb{Z}$. These constraints arise in a number of application domains including but not limited to program verification (array bounds checking and abstract interpretation), operations research (packing and covering) and logic programming. Over the years, several algorithms have been proposed for the **IF** problem. Most of these algorithms are based on two inference rules, viz. the *transitive rule* and the *tightening rule*. None of these algorithms are bit-scaling, i.e., the running times of these algorithms are parameterized only by the number of variables and the number of constraints in the UTVPI system. We introduce a novel algorithm for the **IF** problem, which is based on a collection of new insights. These insights areused to design a new bit-scaling algorithm for **IF** that runs in $O(\sqrt{n} \cdot m \cdot \log C)$ time, where n denotes the number of variables, m denotes the number of constraints and C denotes the largest absolute values of all the constants defining the system.

1 Introduction

In this paper, we discuss a new model-generating algorithm for integer feasibility in a system of Unit Two Variable Per Inequality (UTVPI) constraints (**IF**). Recall that a UTVPI constraint is a constraint of the form: $a \cdot x + b \cdot y \leq c$, where $a, b \in \{0, 1, -1\}$ and $c \in \mathbb{Z}$. A conjunction of such constraints constitutes a UTVPI constraint system (UCS) and can be represented in matrix form as: $\mathbf{A} \cdot \mathbf{x} \leq \mathbf{c}$, where $\mathbf{A}$ has m rows and n columns. These constraints arise in a number of application domains including but not limited to program verification (array bounds checking and abstract interpretation) [10], operations research

K. Subramani—This work was supported by the Air Force Research Laboratory under US Air Force contract FA8750-16-3-6003. The views expressed are those of the authors and do not reflect the official policy or position of the Department of Defense or the U.S. Government.

P. Wojciechowski—This research was supported in part by the National Science Foundation through Award CCF-1305054.

© Springer International Publishing Switzerland 2016
V. Mäkinen et al. (Eds.): IWOCA 2016, LNCS 9843, pp. 321–333, 2016.
DOI: 10.1007/978-3-319-44543-4_25

(packing and covering) and logic programming [9]. Over the years, several algorithms have been proposed for the **IF** problem [8–11]. Most of these algorithms are based on two inference rules, viz. the *transitive rule* and the *tightening rule*.

1.

$$\frac{a \cdot x_i + b \cdot x_j \leq c_{ij} \qquad -b \cdot x_j + c' \cdot x_k \leq c_{jk}}{a \cdot x_i + c' \cdot x_k \leq c_{ij} + c_{jk}}$$

This rule is called the transitive rule and it is solution preserving.

2.

$$\frac{a \cdot x_i + b \cdot x_j \leq c_{ij} \qquad a \cdot x_i - b \cdot x_j \leq c'_{ij}}{a \cdot x_i \leq \lfloor \frac{c_{ij} + c'_{ij}}{2} \rfloor} \tag{1}$$

This rule is called the tightening rule and it is lattice-point preserving.

None of these algorithms are bit-scaling, i.e., the running times of these algorithms are parameterized only by the number of variables and the number of constraints in the UTVPI system [1]. However, in many practical application the constants defining the UTVPI constraints are rather small [15]. It is therefore worthwhile to investigate whether one can design an algorithm whose running time can be parameterized in terms of the defining constants of the constraints, in addition to the cardinalities of the variable and constraint sets. This question is answered affirmatively in this paper. Our work should be contrasted with the work on bit-scaling algorithms for difference constraints [6].

We introduce a novel algorithm for the **IF** problem which is based on a collection of new insights, which permit the transformation of a linearly feasible UTVPI system into a 2CNF formula. We then use the fact that 2CNF formulas are decidable in linear time [2]. This transformation also allows us to make certain observations about the length of certificates of integer infeasibility for UTVPI constraints.

These insights are used to design a new bit-scaling algorithm for **IF** that runs in $O(\sqrt{n} \cdot m \cdot \log C)$ time, where n denotes the number of variables, m denotes the number of constraints and C denotes the largest absolute values of all the constants defining the system. Additionally, the algorithm is certifying, i.e., if the input UTVPI system is integer infeasible, then it provides an easily verifiable certificate that affirms the integer infeasibility. This is an improvement over the model generating algorithm in [10] which runs in $O(m \cdot n + n^2 \cdot \log n)$ time.

The principal contributions of this paper are as follows: 1. New insights into the connection between linear and integer feasibility in UTVPI constraints. 2. The design and analysis of a new algorithm for modeling integer feasibility in UTVPI constraints. 3. Establishing a link between the length of integer refutations in linearly feasible UTVPI systems and the length of resolution refutations in 2CNF formulas.

The rest of this paper is organized as follows: Sect. 2 defines the problem being studied. It also includes a brief discussion of related work in the literature. The new algorithm is described in Sect. 3. We conclude in Sect. 4, by summarizing our contributions and identifying avenues for future research.

2   Statement of Problem

In this section, we formally describe the problem under consideration and define the types of refutations studied in this paper.

Let $\mathbf{U} : \mathbf{A} \cdot \mathbf{x} \leq \mathbf{c}$ denote a system of UTVPI constraints. Observe that each row of $\mathbf{A}$ has at most two non-zero entries and all entries belong to the set $\{0, 1, -1\}$.

We are interested in checking if $\mathbf{U}$ is integer feasible, i.e., whether there exists a solution to the problem:
$\mathbf{x} \in \mathbb{Z}^n : \mathbf{A} \cdot \mathbf{x} \leq \mathbf{c}$.

We assume that $\mathbf{U}$ is linear feasible; i.e., the program $\mathbf{x} \in \mathbb{R}^n : \mathbf{A} \cdot \mathbf{x} \leq \mathbf{c}$ is feasible. Clearly, if this is not the case then the $\mathbf{U}$ is integer infeasible.

Definition 1. *A read-once refutation is a refutation in which we can use each constraint only once. However, we can re-derive constraints as long as we never reuse constraints originally in the system.*

Not all constraint classes have a read-one refutation.

Definition 2. *A dag-like refutation is a refutation in which we can use any constraint multiple times. A dag-like refutation has length k if k unique constraints are used by the refutation.*

Definition 3. *A tree-like refutation is a refutation in which we can use derived constraints only once. However, we can use the constraints in the original system multiple times and re-derive constraints. A tree-like refutation has length k if k constraints are used by the refutation (counting duplicates).*

It is important to note that tree-like refutations are **complete** in that every unsatisfiable CNF formula and linear system has a tree-like refutation. This follows from the fact that every refutation (tree-like or not) can be arranged so that each use of a resolvent can be accomplished by making copies of the input constraints. This transformation from an unconstrained refutation to a tree-like refutation can cause an exponential blow up in the size of the refutation [13].

The first known decision procedure for checking the integer feasibility of a system of UTVPI constraints is detailed in [9]. This algorithm processes a set of UTVPI constraints with the goal of finding its transitive and tightening closure. Such a closure is essentially a finite representation of all possible UTVPI constraints that can be inferred from the input set of constraints (also see [3]). In other words, it finds all possible deductions from the initial set of constraints, including rounded constraints which can be forced into integral solutions. It then checks to see if the system of constraints thus generated, is feasible by virtue of having no contradictions. The algorithm runs in $O(m \cdot n^2)$ time and uses $O(n^2)$ space. Furthermore, it is not certifying. [8] improves on the approach in [9] from an ease-of-implementation standpoint, by combining the transitive and tightening closures into a single step. However, the additional wrinkle does not improve the asymptotic complexity of the algorithm in [9]; nor does it provide certificates.

The algorithm in [10] (henceforth, the Lahiri algorithm) is the fastest known algorithm to date, for deciding integer feasibility in UTVPI systems. We will elaborate on their method, in order to provide the proper background to contrast our procedures.

The Lahiri algorithm begins by converting each constraint into a pair of difference constraints with positive and negative versions of each involved variable. For instance, a sum constraint, say, $x_i + x_j \leq c_{ij}$ is converted into the following difference constraint pair: $x_i^+ - x_j^- \leq c_{ij}$ and $x_j^+ - x_i^- \leq c_{ij}$. Once all constraints are thus converted, the converted constraint system is represented by a constraint network as detailed in [4]. For instance, the constraint $x_j^- - x_i^- \leq c_{ij}$ results in an edge $x_j^- \xleftarrow{c_{ij}} x_i^-$. The resulting edges are then tightened by converting edges of the form $x_i \xleftarrow{c_{ii}} x_i$, where c_{ii} is odd to $x_i \xleftarrow{c_{ii}-1} x_i$, in order to ensure integral solutions. A negative cycle detection subroutine (such as the Bellman-Ford algorithm) then determines whether the system is satisfiable.

We note that in order for the Lahiri algorithm to produce a model, it must compute the transitive and tightening closure of the original constraint system, *even* when such a set of constraints is known to be satisfiable. Indeed, it uses a procedure similar to the one in [8,9] to find bounds for all variables and assign values to them. A naive implementation of this algorithm runs in $O(n^3)$ time and uses $O(n^2)$ space. Utilizing Johnson's algorithm for implementing the transitive closure [4], the time complexity can be improved to $O(m \cdot n + n^2 \cdot \log n)$, while maintaining $O(n^2)$ space complexity. However, even the improved algorithm is more expensive (asymptotically) to the ideal $O(m \cdot n)$ time and $O(m + n)$ space complexity of the non-model generating decision algorithm.

Recently, there has been some work on *incremental* satisfiability of UTVPI constraints. For instance, [14] describes an algorithm for incremental (integer) satisfiability checking in UTVPI constraints. Their algorithm adds a single constraint to a set of UTVPI constraints in $O(m + n \cdot \log n)$ time. Incremental algorithms are extremely important from the perspective of SAT Modulo Theories [12].

In this paper, we focus on designing a bit-scaling algorithm for UTVPI constraints, along the lines of Goldberg's bit-scaling algorithms for difference constraints [6].

We now provide a brief summary of how Goldberg's Algorithm operates. First we introduce the concept of ϵ-feasibility.

Definition 4. *A price vector* **f** *is* ϵ**-feasible** *for a constraint network G, if for every edge* $e = (x_i, x_j)$ *in* **G**, $weight(e) + f_i - f_j \geq -\epsilon$.

Note that an ϵ-feasible price vector is allowed to violate the constraints corresponding to **G**. However, these constraints cannot be violated by more than ϵ. If **G** has no ϵ-feasible price function then it has no feasible price function.

The algorithm in [6] starts with ϵ such that:

1. ϵ is a power of 2.
2. For all edges $e \in$ **G** weight$(e) \geq -\epsilon$.

The algorithm then refines ϵ-feasible solutions into $\frac{\epsilon}{2}$-feasible solutions until either:

1. It reaches a value of ϵ for which no ϵ-feasible solution is found. In this case, **G** has a negative cycle and so the original system has no linear (or integer) solutions.
2. It finds a 0-feasible price vector **f**. In this case, **f** can be used to find a half integral solution to the original system.

Each stage of the refinement process takes $O(\sqrt{n} \cdot m)$ time and **f** is refined at most $\log_2 C$ times. Thus, the algorithm has an overall running time of $O(\sqrt{n} \cdot m \cdot \log_2 C)$.

3 The New Algorithm

In this section, we provide a bit-scaling algorithm for UTVPI constraints.

Our algorithm requires the transformation of the input UCS into a constraint network as described in [10].

This transformation is handled by Algorithm 3.1.

MAKE-GRAPH (System of UTVPI constraints S)

1: Let **G** be a constraint network.
2: **for** $i = 1 \ldots n$ **do**
3:   Add the vertices x_i^+ and x_i^- to **G**.
4: **end for**
5: **for** Every constraint e in S **do**
6:   **if** e is of the form $x_i + x_j \leq c_k$ **then**
7:     Add the edge $x_i^- \rightarrow x_j^+$ to **G** with weight c_k.
8:     Add the edge $x_j^- \rightarrow x_i^+$ to **G** with weight c_k.
9:   **end if**
10:   **if** e is of the form $x_i - x_j \leq c_k$ **then**
11:     Add the edge $x_i^- \rightarrow x_j^-$ to **G** with weight c_k.
12:     Add the edge $x_j^+ \rightarrow x_i^+$ to **G** with weight c_k.
13:   **end if**
14:   **if** e is of the form $-x_i + x_j \leq c_k$ **then**
15:     Add the edge $x_i^+ \rightarrow x_j^+$ to **G** with weight c_k.
16:     Add the edge $x_j^- \rightarrow x_i^-$ to **G** with weight c_k.
17:   **end if**
18:   **if** e is of the form $-x_i - x_j \leq c_k$ **then**
19:     Add the edge $x_i^+ \rightarrow x_j^-$ to **G** with weight c_k.
20:     Add the edge $x_j^+ \rightarrow x_i^-$ to **G** with weight c_k.
21:   **end if**
22:   **if** e is of the form $x_i \leq c_k$ **then**
23:     Add the edge $x_i^- \rightarrow x_i$ to **G** with weight $2 \cdot c_k$.
24:   **end if**
25:   **if** e is of the form $-x_i \leq c_k$ **then**
26:     Add the edge $x_i^+ \rightarrow x_i^-$ to **G** with weight $2 \cdot c_k$.
27:   **end if**
28: **end for**
29: **return** **G** as a constraint network.

Algorithm 3.1: MAKE-GRAPH

[10] transforms the input UTVPI system into a constraint network as follows: Consider the following constraint system.

$$x_1 + x_3 \leq 0 \qquad x_2 - x_3 \leq -7 \qquad x_4 - x_2 \leq 3$$
$$-x_1 - x_4 \leq 5 \qquad x_1 \leq 6 \tag{2}$$

For each variable, two vertices (a positive version and a negative version) are added to the constraint network. For instance, corresponding to the variable x_i, we create the vertices x_i^+ and x_i^-. Each constraint is replaced by a pair of equivalent constraints. For instance, a difference constraint $x_i - x_j \leq c$ is replaced by the two constraints $x_i^+ - x_j^+ \leq c$ and $x_j^- - x_i^- \leq c$. The exception is for absolute constraints, each of which is simply converted to a single equivalent constraint. For instance, $x_i \leq c$ yields $x_i^+ - x_i^- \leq 2 \cdot c$. Once all the equivalent constraints have been determined, they are represented in a directed graph, as discussed in [4]. It is thus seen that the constraint network constructed as per [10] has $2 \cdot n$ vertices (assuming n variables in the constraint system) and up to $2 \cdot m$ edges (assuming m constraints in the original constraint system). The resultant constraint network is called the potential graph. Figure 1 shows the potential graph, corresponding to System (2).

We are now ready to present our bit-scaling algorithm.

Algorithm 3.2 divides the process of obtaining an integer solution to a system of UTVPI constraints into several steps.

First the system of UTVPI constraints, $\mathbf{A} \cdot \mathbf{x} \leq \mathbf{c}$, is converted into a constraint network. This is the same process used in [10] and is described in greater detail in Algorithm 3.1. Note that $\mathbf{G}$ has two vertices, x_i^+ and x_i^-, corresponding to each variable. Thus, $\mathbf{f}$ will have $2 \cdot n$ values with $f_{2 \cdot i - 1}$ as the price of vertex x_i^+ and $f_{2 \cdot i}$ as the price of vertex x_i^-.

Linear feasibility is then determined using Goldberg's Bit-Scaling Algorithm. This is the same process used in [6]. We refer to this algorithm as GOLDBERG($\mathbf{G}$). If $\mathbf{A} \cdot \mathbf{x} \leq \mathbf{c}$ is not linearly feasible, then it is not integer feasible and it is returned as such. However, if it is linearly feasible, then we can construct a linear solution $\mathbf{d}$. Note that every element of $\mathbf{d}$ is an integer multiple of $\frac{1}{2}$, thus $\mathbf{d}$ is a *half-integral* solution.

$\mathbf{A} \cdot \mathbf{x} \leq \mathbf{c}$ is then transformed into a system of 2CNF clauses, $\Phi(\mathbf{v})$. This process is done by Algorithm 3.3.

From the original system and half-integral solution $\mathbf{d}$, we can construct a new system of UTVPI constraints, $\mathbf{A} \cdot \mathbf{x} \leq \mathbf{c}'$, as follows:

1. Replace each constraint $x_i + x_j \leq c_k$ with $x_i + x_j \leq c_k - (d_i + d_j)$.
2. Replace each constraint $x_i - x_j \leq c_k$ with $x_i - x_j \leq c_k - (d_i - d_j)$.
3. Replace each constraint $-x_i + x_j \leq c_k$ with $-x_i + x_j \leq c_k - (-d_i + d_j)$.
4. Replace each constraint $-x_i - x_j \leq c_k$ with $-x_i - x_j \leq c_k - (-d_i - d_j)$.

Note that, by construction, $\mathbf{c}' = \mathbf{c} - \mathbf{A} \cdot \mathbf{d} \geq \mathbf{0}$. This corresponds to the process of re-weighting difference constraints with a potential function.

We can now reduce the number of constraints by focusing on the constraints of the form $\pm x_i \pm x_j \leq 0$ in $\mathbf{A} \cdot \mathbf{x} \leq \mathbf{c}'$ such that $d_i, d_j \notin \mathbb{Z}$. Let $\mathbf{A}' \cdot \mathbf{x}' \leq \mathbf{0}$ be the system of these constraints.

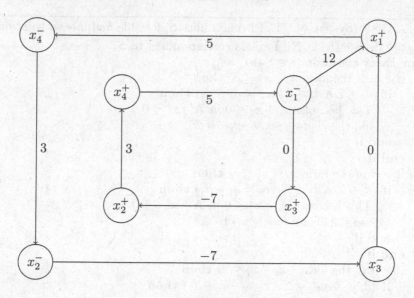

Fig. 1. Example potential graph.

UTVPI-SCALING (System of UTVPI constraints S)

1: Let $\mathbf{d}$ denote a linear solution to S.

2: Let $\mathbf{y}$ denote an integer solution to S.
3: $\mathbf{G} \leftarrow$ MAKE-GRAPH(S).
4: $\mathbf{f} \leftarrow$ GOLDBERG($\mathbf{G}$).
5: **if** $\mathbf{f}$ is a feasible price function for $\mathbf{G}$ **then**
6:    **for** $i = 1 \ldots n$ **do**
7:       $d_i \leftarrow \frac{f_{2 \cdot i - 1} - f_{2 \cdot i}}{2}$.
8:    **end for**
9: **else**
10:    **return** S is not linear feasible.
11:    {Thus, S is not integer feasible.}
12: **end if**
13: $\Phi \leftarrow$ MAKE-2CNF(S, $\mathbf{d}$).

14: **if** Φ is satisfiable **then**
15:    $\mathbf{v} \leftarrow$ satisfying assignment to Φ.
16: **else**
17:    **return** S is not integer feasible.
18: **end if**
19: **for** $i = 1 \ldots n$ **do**
20:    **if** $d_i \in \mathbb{Z}$ **then**
21:       $y_i \leftarrow d_i$
22:    **else if** v_i is **true then**
23:       $y_i \leftarrow d_i + \frac{1}{2}$
24:    **else**
25:       $y_i \leftarrow d_i - \frac{1}{2}$
26:    **end if**
27: **end for**
28: **return** $\mathbf{y}$ as an integer solution to S.

Algorithm 3.2: UTVPI-SCALING

From $\mathbf{A}' \cdot \mathbf{x}' \leq \mathbf{0}$, we can construct a system, $\Phi(\mathbf{v})$, of 2CNF clauses which is satisfiable if and only if $\mathbf{A} \cdot \mathbf{x} \leq \mathbf{c}$ is integer feasible. $\Phi(\mathbf{v})$ also has the property that any proof of unsatisfiability for $\Phi(\mathbf{v})$ can be easily converted into a proof of integer infeasibility of the same length for $\mathbf{A} \cdot \mathbf{x} \leq \mathbf{c}$ and vice-versa.

MAKE-2CNF (System of UTVPI constraints S, feasible half-integer solution $\mathbf{d}$)

1: Let Φ denote the 2CNF formula corresponding to S.
2: **for** Every constraint $e \in S$ **do**
3:   **if** e is of the form $x_i + x_j \leq c_k$ **then**
4:     **if** $(d_i \notin \mathbb{Z} \wedge d_j \notin \mathbb{Z} \wedge c_k = d_i + d_j)$ **then**
5:       {This becomes $x_i + x_j \leq 0$ in $\mathbf{A}' \cdot \mathbf{x}' \leq \mathbf{0}$.}
6:       Add the clause $(\neg v_i \vee \neg v_j)$ to Φ.
7:     **end if**
8:   **end if**
9:   **if** e is of the form $x_i - x_j \leq c_k$ **then**
10:     **if** $(d_i \notin \mathbb{Z} \wedge d_j \notin \mathbb{Z} \wedge c_k = d_i - d_j)$ **then**
11:       {This becomes $x_i - x_j \leq 0$ in $\mathbf{A}' \cdot \mathbf{x}' \leq \mathbf{0}$.}
12:       Add the clause $(\neg v_i \vee v_j)$ to Φ.
13:     **end if**
14:   **end if**
15:   **if** e is of the form $-x_i + x_j \leq c_k$ **then**
16:     **if** $(d_i \notin \mathbb{Z} \wedge d_j \notin \mathbb{Z} \wedge c_k = -d_i + d_j)$ **then**
17:       {This becomes $-x_i + x_j \leq 0$ in $\mathbf{A}' \cdot \mathbf{x}' \leq \mathbf{0}$.}
18:       Add the clause $(v_i \vee \neg v_j)$ to Φ.
19:     **end if**
20:   **end if**
21:   **if** e is of the form $-x_i - x_j \leq c_k$ **then**
22:     **if** $(d_i \notin \mathbb{Z} \wedge d_j \notin \mathbb{Z} \wedge c_k = -d_i - d_j)$ **then**
23:       {This becomes $-x_i - x_j \leq 0$ in $\mathbf{A}' \cdot \mathbf{x}' \leq \mathbf{0}$.}
24:       Add the clause $(v_i \vee v_j)$ to Φ.
25:     **end if**
26:   **end if**
27: **end for**
28: **return** Φ as a system of 2CNF clauses.

Algorithm 3.3: MAKE-2CNF

Note that, $\mathbf{A} \cdot \mathbf{x} \leq \mathbf{c}'$ and $\mathbf{A}' \cdot \mathbf{x}' \leq \mathbf{0}$ are not actually constructed by Algorithm 3.3. However, these systems are used to prove the correctness of the algorithms.

In $\Phi(\mathbf{v})$, each v_i corresponds to an x_i which assumes a non-integer value in the linear solution $\mathbf{d}$. v_i being **true** corresponds to x_i being rounded up, while **false** corresponds to x_i being rounded down. This action is performed by the final step of Algorithm 3.2.

If $\Phi(\mathbf{v})$ is feasible, then the values of each v_i correspond to a rounding needed to make an integral solution to the original system $\mathbf{A} \cdot \mathbf{x} \leq \mathbf{c}$. Similarly, if $\Phi(\mathbf{v})$ infeasible, then no such rounding is possible and $\mathbf{A} \cdot \mathbf{x} \leq \mathbf{c}$ is integer infeasible.

3.1   Resource Analysis

Algorithm 3.2 can be broken up into several parts. The complexity of each part can be considered independently.

1. First, Algorithm 3.2 finds a linear solution. This is accomplished by running Goldberg's Bit-Scaling Algorithm on the constraint network construction in [10]. This takes $O(\sqrt{n} \cdot m \cdot \log C)$ time [6].
2. Then, Algorithm 3.3 converts the system into a system of 2CNF clauses. This consists of checking each constraint in the system and performing a series of constant time operations to generate the 2CNF clause. Thus, this takes $O(m)$ time.
3. Then, Algorithm 3.2 generates a feasible solution to the 2-SAT system or declares the system infeasible. This can be done in $O(n + m)$ time [2].
4. Finally, Algorithm 3.2 generates a feasible integer solution to the UTVPI system or declares the system not integer feasible. This is done by utilizing the 2-SAT solution and initial linear solution and runs in $O(n)$ time.

Thus, Algorithm 3.2 generates a feasible integer solution to the UTVPI system or declares the system not integer feasible in $O(\sqrt{n} \cdot m \cdot \log C)$ time.

3.2   Proof of Correctness

We now establish the correctness of the reduction from linearly feasible UTVPI to 2-SAT.

We first show that the limitations on the constraints used in the reduction do not eliminate any proofs of integer infeasibility. Note that all proof in this section apply only to linearly feasible systems of UTVPI constraints.

Theorem 1. *If $\mathbf{A} \cdot \mathbf{x} \leq \mathbf{c}$ has a proof of integer infeasibility, then the constraints forming that proof correspond to constraints of the form $\pm x_i \pm x_j \leq 0$ in $\mathbf{A} \cdot \mathbf{x} \leq \mathbf{c}'$.*

Proof. Let $\mathbf{d}$ be a half-integral solution to $\mathbf{A} \cdot \mathbf{x} \leq \mathbf{c}$, and let x_i be a variable such that $d_i \notin \mathbb{Z}$. Since $\mathbf{A} \cdot \mathbf{x} \leq \mathbf{c}$ is not integer feasible there exist no solutions with $x_i = \lceil d_i \rceil$ or $x_i = \lfloor d_i \rfloor$. Thus, we can derive the constraints $x_i + x_i \leq 2 \cdot d_i$ and $-x_i - x_i \leq -2 \cdot d_i$.

Consider the constraints in $\mathbf{A} \cdot \mathbf{x} \leq \mathbf{c}$ added together to obtain $x_i + x_i \leq 2 \cdot d_i$. When we add the corresponding constraints in $\mathbf{A} \cdot \mathbf{x} \leq \mathbf{c}'$ we obtain

$$x_i + x_i \leq 2 \cdot d_i - (d_i + d_i) = 0.$$

All constraints in $\mathbf{A} \cdot \mathbf{x} \leq \mathbf{c}'$ have $c'_k \geq 0$. Thus, every constraint involved in this new sum must have $c'_k = 0$. The same holds true for the constraints used to derive $-x_i - x_i \leq -2 \cdot d_i$. Thus, the constraints used to establish the integer infeasibility of $\mathbf{A} \cdot \mathbf{x} \leq \mathbf{c}$ correspond to constraints in $\mathbf{A} \cdot \mathbf{x} \leq \mathbf{c}'$ such that $c'_k = 0$.

Theorem 2. *If $\mathbf{A} \cdot \mathbf{x} \leq \mathbf{c}$ has a proof of integer infeasibility, then the constraints in that proof involve only variables x_j such that $d_j \notin \mathbb{Z}$.*

Proof. From Theorem 1, we have that every constraint involved in the proof of integer infeasibility must correspond to a constraint with $c'_k = 0$ in $\mathbf{A} \cdot \mathbf{x} \le \mathbf{c}'$.

For any constraint $x_j + x_l \le 0$ in $\mathbf{A} \cdot \mathbf{x} \le \mathbf{c}'$, d_j and d_l must both be integral or both be non-integral. Otherwise,

$$c'_k = c_k \pm d_j \pm d_l \notin \mathbb{Z}.$$

A proof of integer infeasibility for $\mathbf{A} \cdot \mathbf{x} \le \mathbf{c}$ consists of establishing bounds on a variable x_i with $d_i \notin \mathbb{Z}$. Thus, all constraints in that proof must involve only variables x_j such that $d_j \notin \mathbb{Z}$.

Together these two theorems imply that to find a proof of integer infeasibility we only need to focus on constraints in $\mathbf{A} \cdot \mathbf{x} \le \mathbf{c}'$ such that $c'_k = 0$ and involving variables x_i for which d_i is non-integral.

Theorem 3. *A 2CNF clause can be resolved from $\Phi(\mathbf{v})$ if and only if the corresponding UTVPI constraint can be derived from $\mathbf{A}' \cdot \mathbf{x}' \le \mathbf{0}$.*

Proof. The inference rule used in the resolution of 2CNF clauses is

$$\frac{(l_i \vee l_j) \qquad\qquad (\neg l_j \vee l_k)}{(l_i \vee l_k)}$$

for literals l_i, l_j and l_k.

Let us consider the case where $l_i = v_i$, $l_j = v_j$, and $l_k = v_j$. If we look at the corresponding constraints in $\mathbf{A}' \cdot \mathbf{x}' \le \mathbf{0}$, then we see that, in this case, the clauses correspond to the constraints $-x_i - x_j \le 0$ and $x_j - x_k \le 0$ yielding $-x_i - x_k \le 0$. This is exactly what would be obtained from applying the transitive inference rule. It is easy to see that the reverse also holds.

The cases corresponding to the other possible assignments to the literals l_i, l_j, and l_k are handled similarly.

We can now establish the correctness of the reduction.

Theorem 4. *$\Phi(\mathbf{v})$ is satisfiable if and only if $\mathbf{A} \cdot \mathbf{x} \le \mathbf{c}$ has an integer solution.*

Proof. Assume that $\Phi(\mathbf{v})$ is unsatisfiable. Thus, we can derive the clauses (v_i) and $(\neg v_i)$ for some variable v_i. These clauses correspond to the constraints $x_i + x_i \le 0$ and $-x_i - x_i \le 0$. Thus, from Theorem 3, these constraints are derivable from $\mathbf{A}' \cdot \mathbf{x}' \le \mathbf{0}$. Since d_i is an odd multiple of $\frac{1}{2}$, these correspond to the constraints

$$x_i + x_i \le 2 \cdot d_i = 2 \cdot \lfloor d_i \rfloor + 1 \text{ and } -x_i - x_i \le -2 \cdot d_i = -2 \cdot \lfloor d_i \rfloor - 1.$$

These constraints are derivable from $\mathbf{A} \cdot \mathbf{x} \le \mathbf{c}$.

When we tighten these constraints, we get $x_i \le \lfloor d_i \rfloor$ and $-x_i \le -\lfloor d_i \rfloor - 1$. Summing these two constraints yields $0 \le -1$. Thus, showing that $\mathbf{A} \cdot \mathbf{x} \le \mathbf{c}$ is infeasible.

Now assume that $\Phi(\mathbf{v})$ is satisfiable. Let $\mathbf{v}'$ be a boolean vector such that $\Phi(\mathbf{v}')$ is **true**. From $\mathbf{v}'$ and $\mathbf{d}$, we can construct the vector $\mathbf{r}$ as follows:

1. If $d_i \in \mathbb{Z}$, then set $r_i = 0$.
2. If $d_i \notin \mathbb{Z}$ and v_i' is **true**, then set $r_i = \frac{1}{2}$.
3. If $d_i \notin \mathbb{Z}$ and v_i' is **false**, then set $r_i = -\frac{1}{2}$.

We now show that $\mathbf{A} \cdot \mathbf{r} \leq \mathbf{c}'$. Let $\pm x_i \pm x_j \leq c_k'$ be a constraint in $\mathbf{A} \cdot \mathbf{x} \leq \mathbf{c}'$. Let us examine all possible cases:

1. $c_k' \geq 1$: We have that $\pm r_i \pm r_j \leq \frac{1}{2} + \frac{1}{2} = 1 \leq c_k'$. Thus, the constraint is satisfied by $\mathbf{r}$.
2. $c_k' = 0$ and $d_i \in \mathbb{Z}$: From the proof of Theorem 2, we know that $d_j \in \mathbb{Z}$. Thus, $r_i = r_j = 0$ and $\pm r_i \pm r_j = 0 = c_k'$. Thus, the constraint is satisfied by $\mathbf{r}$.
3. $c_k' = 0$ and $d_i \notin \mathbb{Z}$: From the proof of Theorem 2, we know that $d_j \notin \mathbb{Z}$. In this case, we look at each possible constraint individually.
   (a) $x_i + x_j \leq 0$: By construction, the clause $(\neg v_i \vee \neg v_j)$ is in $\Phi(\mathbf{v})$. Thus, v_i' or v_j' must be **false**. This means that $r_i = -\frac{1}{2}$ or $r_j = -\frac{1}{2}$. In either case, we have that
   $$r_i + r_j \leq -\frac{1}{2} + \frac{1}{2} = 0 = c_k'.$$
   Thus, the constraint is satisfied by $\mathbf{r}$.
   (b) $x_i - x_j \leq 0$: By construction, the clause $(\neg v_i \vee v_j)$ is in $\Phi(\mathbf{v})$. Thus, v_i' must be **false** or v_j' must be **true**. This means that $r_i = -\frac{1}{2}$ or $r_j = \frac{1}{2}$. In either case, we have that
   $$r_i - r_j \leq -\frac{1}{2} + \frac{1}{2} = 0 = c_k'.$$
   Thus, the constraint is satisfied by $\mathbf{r}$.
   (c) $-x_i + x_j \leq 0$: By construction, the clause $(v_i \vee \neg v_j)$ is in $\Phi(\mathbf{v})$. Thus, v_i' must be **true** or v_j' must be **false**. This means that $r_i = \frac{1}{2}$ or $r_j = -\frac{1}{2}$. In either case, we have that
   $$-r_i + r_j \leq -\frac{1}{2} + \frac{1}{2} = 0 = c_k'.$$
   Thus, the constraint is satisfied by $\mathbf{r}$.
   (d) $-x_i - x_j \leq 0$: By construction, the clause $(v_i \vee v_j)$ is in $\Phi(\mathbf{v})$. Thus, v_i' or v_j' must be **true**. This means that $r_i = \frac{1}{2}$ or $r_j = \frac{1}{2}$. In either case, we have that
   $$-r_i - r_j \leq -\frac{1}{2} + \frac{1}{2} = 0 = c_k'.$$
   Thus, the constraint is satisfied by $\mathbf{r}$.

By the construction of $\mathbf{r}$, we have $\mathbf{d} + \mathbf{r} \in \mathbb{Z}^n$. We also have that $\mathbf{c}' = \mathbf{c} - \mathbf{A} \cdot \mathbf{d}$. Thus,

$$\mathbf{A} \cdot (\mathbf{d} + \mathbf{r}) = \mathbf{A} \cdot \mathbf{d} + \mathbf{A} \cdot \mathbf{r} \leq \mathbf{A} \cdot \mathbf{d} + \mathbf{c}' = \mathbf{c}.$$

This means that $(\mathbf{d} + \mathbf{r})$ is a valid integer solution to $\mathbf{A} \cdot \mathbf{x} \leq \mathbf{c}$.

4 Conclusion

The primary contribution of this paper is the design and analysis of a new bit-scaling algorithm for the problem of checking integer feasibility in UTVPI constraints. On a UTVPI system over n variables and m constraints, the bit-scaling algorithm runs in time $O(\sqrt{n} \cdot m \cdot \log C)$, where C is the maximum absolute value of all the defining constants in the UTVPI system. Thus, there are now bit-scaling algorithms for obtaining integer solutions to systems of UTVPI constraints and difference constraints [6]. As remarked earlier, the algorithm is certifying in that in the event the given system is infeasible, it provides an easily checkable certificate that certifies the infeasibility. Of course, in the event the given system is satisfiable, the output is a lattice point, whose appropriateness can be checked in linear time. Additionally, the following results are documented in the journal version of the paper:

1. We establish a link between the lengths of integer refutations in linearly feasible systems of UTVPI constraints and the lengths of resolution refutations in 2CNF formulas.
2. We will use the UCS network construction described in [16], as opposed to [10]. This requires a significant modification to Goldberg's algorithm.

References

1. Ahuja, R.K., Magnanti, T.L., Orlin, J.B.: Network Flows: Theory, Algorithms and Applications. Prentice-Hall, New Jersey (1993)
2. Tarjan, R.E., Aspvall, B., Plass, M.F.: A linear time algorithm for testing the truth of certain quantified boolean formulas. Inform. Process. Lett. **8**(3), 121–123 (1979)
3. Bagnara, R., Hill, P.M., Zaffanella, E.: Weakly-relational shapes for numeric abstractions: improved algorithms and proofs of correctness. Formal Meth. Syst. Des. **35**(3), 279–323 (2009)
4. Cormen, T.H., Leiserson, C.E., Rivest, R.L., Stein, C.: Introduction to Algorithms. MIT Press, Cambridge (2001)
5. Dantzig, G.B., Eaves, B.C.: Fourier-Motzkin elimination and its dual. J. Comb. Theory (A) **14**, 288–297 (1973)
6. Goldberg, A.V.: Scaling algorithms for the shortest paths problem. SIAM J. Comput. **24**(3), 494–504 (1995)
7. Hochbaum, D.S., Seffi-Naor, J.: Simple, fast algorithms for linear, integer programs with two variables per inequality. SIAM J. Comput. **23**(6), 1179–1192 (1994)
8. Harvey, W., Stuckey, P.J.: A unit two variable per inequality integer constraint solver for constraint logic programming. In: Proceedings of the 20th Australasian Computer Science Conference, pp. 102–111 (1997)
9. Jaffar, J., Maher, M.J., Stuckey, P.J., Yap, H.C.: Beyond Finite Domains. In: Proceedings of the Second International Workshop on Principles and Practice of Constraint Programming (1994)
10. Lahiri, S.K., Musuvathi, M.: An efficient decision procedure for UTVPI constraints. In: Gramlich, B. (ed.) FroCos 2005. LNCS (LNAI), vol. 3717, pp. 168–183. Springer, Heidelberg (2005)

11. Miné, A.: The octagon abstract domain. Higher-Order Symbolic Comput. **19**(1), 31–100 (2006)
12. Nieuwenhuis, R., Oliveras, A., Tinelli, C.: Abstract DPLL and abstract DPLL modulo theories. In: Baader, F., Voronkov, A. (eds.) LPAR 2004. LNCS (LNAI), vol. 3452, pp. 36–50. Springer, Heidelberg (2005)
13. Pitassi, T., Urquhart, A.: The complexity of the hajós calculus. In: 33rd Annual Symposium on Foundations of Computer Science, Pittsburgh, Pennsylvania, USA, 24–27, October 1992, pp. 187–196 (1992)
14. Schutt, A., Stuckey, P.J.: Incremental satisfiability and implication for UTVPI constraints. INFORMS J. Comput. **22**(4), 514–527 (2010)
15. Seshia, S.A., Subramani, K., Bryant, R.E.: On solving boolean combinations of UTVPI constraints. J. Satisfiability Boolean Model. Comput. **3**(12), 67–90 (2007)
16. Subramani, K., Wojciechowski, P.: An optimal certifying algorithm for lattice point feasibility in a system of UTVPI constraints. Algorithmica (Accepted, 2016 in press)

Limits of Greedy Approximation Algorithms for the Maximum Planar Subgraph Problem

Markus Chimani, Ivo Hedtke[✉], and Tilo Wiedera

Theoretical Computer Science, Osnabrück University, Osnabrück, Germany
{markus.chimani,ivo.hedtke,twiedera}@uni-osnabrueck.de

Abstract. The Maximum Planar Subgraph (MPS) problem asks for a planar subgraph with maximum edge cardinality of a given undirected graph. It is known to be MaxSNP-hard and the currently best known approximation algorithm achieves a ratio of 4/9.

We analyze the general limits of approximation algorithms for MPS, based either on planarity tests or on greedy inclusion of certain subgraphs. On the one hand, we cover upper bounds for the approximation ratios. On the other hand, we show NP-hardness for thereby arising subproblems, which hence must be approximated themselves. We also provide simpler proofs for two already known facts.

1   Introduction

The *Maximum Planar Subgraph* (MPS) problem is to determine a planar subgraph of a graph G, such that it has maximum edge cardinality. The related, yet easier, *maximal* planar subgraph problem asks for a planar subgraph to which no further edge of G can be added without destroying planarity. MPS is MaxSNP-hard and the best possible approximation ratio is unknown. The strongest known approximation algorithm has a tight (w.r.t. its analysis) ratio of 4/9 [2].

In the following, we always consider simple undirected connected graphs. Let n be the number of nodes. By Euler's formula, planar graphs have at most $3n-6$ edges and $2n-4$ (triangular) faces. We call any planar graph that has this exact number of edges a *triangulation*. It is trivial to achieve an approximation ratio of 1/3 by picking any spanning tree (with thus $n-1$ edges).

The approximation of MPS has received significant attention despite the fact that recent advances are scarce. Both Cimikowski [3] and Zelikovsky [13] presented algorithmic ideas that were never completed. Poranen [11] conjectured that two algorithms based on iteratively selecting triangles (building on top of [2]) would achieve the ratio of 4/9 which turned out to be false [7, Sect. 56.6].

Cimikowski [4] also showed that several sophisticated specific algorithms achieve an approximation ratio of only 1/3 for MPS. This includes an algorithm based on the planarity test by Hopcroft and Tarjan [8]. Later, Hsu [9] extended

---

M. Chimani and T. Wiedera—Supported by the German Research Foundation (DFG) project CH 897/2-1.

© Springer International Publishing Switzerland 2016
V. Mäkinen et al. (Eds.): IWOCA 2016, LNCS 9843, pp. 334–346, 2016.
DOI: 10.1007/978-3-319-44543-4_26

another DFS-based planarity test [1,12] to compute a maximal planar subgraph in linear time without considering any approximation properties.

We are interested in bounding approximation ratios of general classes of algorithms that are based on common underlying ideas like those sketched above. Thereby, we hope to point out promising directions for new algorithmic ideas.

Outline. In Sect. 2 we introduce the basic notation and give an alternative, simpler proof that a maximal planar subgraph approximates MPS by 1/3. In Sect. 3 we show that it remains NP-hard to find a maximum planar subgraph that contains a given DFS/BFS/spanning tree. Note that these problems differ from MPS and it is not immediately clear that they are NP-hard. We also show that the approximation ratio of algorithms based on such an idea is at most 2/3. This very general result includes all known MPS heuristics based on planarity tests, in particular also Hsu's algorithm [9].[1] Our argumentation also allows an alternative, much shorter, proof for the NP-hardness of MPS itself, which is presented in Sect. 4. Finally, we consider several variants to generalize the best known approximation algorithms [2]. Again, we prove corresponding hardness results and bounds for the approximation ratio. In particular, this rules out several ideas along the lines of [11] to achieve improved approximation ratios.

2 Preliminaries and Maximality

An edge between nodes u and v is denoted by uv. For a subset X of nodes or edges, $G[X]$ denotes the induced subgraph. A k-path is a path with k edges. For two nodes u and v we define a *u-v-bundle* $\mathcal{B}^t_{u,v}$ *of thickness* t as a set of t parallel 2-paths between u and v; the new inner nodes $\mathcal{I}(\mathcal{B}^t_{u,v})$ have degree 2. For convenience, we write $[k] := \mathbb{Z}_k$; addition and subtraction are modulo k.

While the following statement is known to be true, we provide a simpler instance than the original source [6]. They use a 3-colorable planar triangulated graph extended by $\Theta(n)$ edges that form three cycles on the node partitions induced by the coloring. Our argument is based on a K_5.

Observation 1. *A maximal planar subgraph of a given graph G yields an approximation ratio of at most 1/3 for the MPS problem on G.*

Proof. Consider the complete graph K_5 on 5 nodes. We construct G by replacing a single edge vu by $\mathcal{B}^\ell_{v,u}$, and adding a Hamiltonian path $P = p_1, \ldots, p_\ell$ for the nodes $\mathcal{I}(\mathcal{B}^\ell_{v,u})$. Let $S := E(K_5) \setminus \{vu\} \cup \{vp_k \mid k \text{ odd}\} \cup \{p_k u \mid k \text{ even}\}$. S is a maximal planar subgraph of G since adding any edge yields a K_5 subdivision (cf. Fig. 1). An MPS H can be obtained by removing any one edge outside of $\mathcal{B}^\ell_{v,u}$ from G. The approximation ratio is thus at most $\lim_{\ell \to \infty} |S|/|E(H)| = 1/3$. □

---

[1] The proof by Cimikowski [4] for the Hopcroft-Tarjan based heuristic exploits the specific embedding of backedges and cannot be generalized to arbitrary algorithms based on DFS trees.

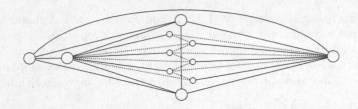

Fig. 1. (cf. Observation 1) Maximal planar subgraph for $\ell = 6$.

3   Algorithms Inspired by Planarity Tests

First, we focus on DFS- and BFS-based algorithms providing hardness results and bounds for families of approximation algorithms. We denote the problem of finding a maximum planar subgraph that contains a given DFS (or BFS) tree by *MPS-DFS* (or *MPS-BFS*, respectively). In particular, any known algorithm based on a planarity test in fact solves MPS-DFS heuristically.

A *k-book* is a collection of k half-planes (*pages*) that share a common boundary (*spine*). A *k-book embedding* is an embedding of a graph into a k-book such that the vertices are placed on the spine, every edge is drawn on a single page, and no two edges cross each other. Consider a circle with straight-line chords $\mathcal{C}$. A *circle graph* is the intersection graph of the latter: $\mathcal{C}$ are its nodes, two nodes are adjacent iff their chords cross. The *overlap graph* is the graph where each chord is an edge, and the chords' end nodes are connected by a Hamiltonian cycle according to the original drawing. For a given circle graph $G = (V, E)$ and $c, k \in \mathbb{N}$, the problem of finding a subset $V' \subseteq V$, such that $|V'| \geq k$ and $G[V']$ is c-colorable is the *c-Colorable Induced Subgraph problem for Circle Graphs* (*c-CIG*). It is NP-hard for $c \geq 2$ [5]. Clearly, any solution for c-CIG corresponds to a c-book embedding of the respective overlap graph; the circle corresponds to the spine and each color class is embedded in its own page.

Theorem 1. *MPS-DFS is NP-hard. Furthermore, there are (infinitely many) graphs G that allow a DFS tree T_v for each possible start node v such that MPS-DFS on each (G, T_v) is NP-hard.*

Proof. We perform a reduction from 2-CIG to MPS-DFS. Let (G, k) be an instance for 2-CIG and $C = (W, F)$ be the corresponding overlap graph. Let $n := |W|$, $m := |F|$, and $\pi \colon [n] \to W$ denote the cyclic order of W induced by C. Let $B_i := \mathcal{B}^m_{\pi_i, \pi_{i+1}}$ denote a π_i-π_{i+1}-bundle of thickness m and $B'_i := B_i \cup \{\pi_i \pi_{i+1}\}$. We construct the input graph $D := (\bigcup_{i \in [n]} V(B_i), F \cup E_B)$ for MPS-DFS, with $E_B := \bigcup_{i \in [n]} E(B'_i)$; see Fig. 2. The set $T := \{\pi_i \pi_{i+1}, u \pi_{i+1} \mid 0 \leq i < n - 1, u \in \mathcal{I}_i\} \cup \{u \pi_{n-1} \mid u \in \mathcal{I}_{n-1}\}$, where $\mathcal{I}_i := \mathcal{I}(B_i)$, is a DFS tree of D.[2] We show that the

---

[2] We start at π_0 with $\pi_0 \pi_1$. Next, we pick all edges of B_0 that are incident to π_1 since the $\mathcal{I}(B_0)$-vertices lead only to π_0 (visited). We iterate this until we arrive at $\pi_{n-2} \pi_{n-1}$. Finally, we pick all edges connecting π_{n-1} with $\mathcal{I}(B_{n-2})$ and $\mathcal{I}(B_{n-1})$.

2-CIG instance (G, k) has a solution if and only if D has a planar subgraph of size $\xi := k + n(2m + 1)$ that contains T.

(If) Assume there is a planar embedded subgraph S of D that contains T and has ξ edges. D contains $m + n(2m + 1)$ edges. Removing more than m edges from D yields a graph with less than ξ edges. Thus, there are at least $m + 1$ edges from each B_i' in S. Consequently, for each pair of nodes π_i, π_{i+1} there is a path within B_i' connecting them. Hence we have a cycle through $\pi_0, \pi_1, \ldots, \pi_{n-1}, \pi_0$, splitting $E(S) \setminus E_B$, the edge set of S corresponding to chords, into an inside and an outside partition. Since S is planar, this induces a 2-book embedding of those edges. Thus, we have a solution of 2-CIG on (G, k) as $|E(S)| - |E_B| = k$.

(Only If) Assume there is a solution for 2-CIG on (G, k). This corresponds to a 2-book embedding of a subgraph $C' := (W, E')$ of C, where the vertices W are placed on the spine according to π, and $|E'| \geq k$. Adding E_B to C' yields a planar graph. Note that $T \subseteq E_B$ and C' contains $|E_B| + k \geq \xi$ edges.

Note that the proof works independently of the DFS start node since π is cyclic and π_0 can be chosen arbitrarily. $\qquad\square$

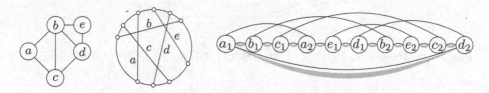

Fig. 2. (cf. Theorem 1) The circle graph G on the left with the respective overlap graph in the middle and a schematic depiction of the input graph D for MPS-DFS with ordering $\pi_0 = a_1, \pi_1 = b_1, \ldots$ on the right (bundles sketched in gray).

We will see that any algorithm adding edges to an arbitrary DFS tree has an approximation ratio of at most $2/3$. However, the second part of the theorem above shows that we cannot simply iterate over all possible start nodes to find a tractable MPS-DFS instance and use this to approximate MPS.

Theorem 2. *An optimal solution to MPS-DFS yields an approximation ratio of at most 2/3 for the corresponding MPS problem.*

Proof. Given a number $p \geq 4$, consider the following graph $G := (V, S \cup \{\tilde{e}\} \cup T)$ with $V := \{u_1, \ldots, u_p, v_1, \ldots, v_p\}$, and $S := \bigcup_{i=1}^{p-1} \{u_i u_{i+1}, v_i v_{i+1}\}$. The edges in S form two disjoint paths, both of length $p - 1$. Let T be an edge set that triangulates $G[S]$. Note that this is possible (cf. Fig. 3) in a way such that

$$\forall e \in T: e = u_i v_j \wedge |i - j| \leq 2. \tag{1}$$

Finally, we define $\tilde{e} := u_p v_1$. Observe that $|T| = 4p - 4$ and $P := S \cup \{\tilde{e}\}$ forms a Hamiltonian path. Assume that the DFS on G returns P. We prove that any planar subgraph H of G that contains P can have at most half of the edges of T.

Any such graph can be constructed by successively inserting edges of T into $G[P]$. After each step the there are at least two faces f_1, f_2 that have exactly one edge of T on their boundary: Initially, adding the first edge to $G[P]$ yields two such faces. If the next edge is embedded neither in f_1 nor in f_2 the invariant holds. Otherwise, the edge is embedded in, say, f_1. Then f_1 is split into two faces, one of which becomes the new f_1. For each edge $t \in T$, $P \cup \{t\}$ has a cycle of length at least $p - 2$, which follows from Eq. (1) by construction of P. We conclude that H has two faces of degree at least $p - 2$, and at least $2p - 10$ edges are missing for H to be a triangulation.

The edges $E(G) \setminus \{\tilde{e}\}$ form an MPS. We conclude that MPS-DFS approximates MPS by a ratio of at most $\lim_{p \to \infty}(|P| + |T| - (2p - 10))/|E(G) \setminus \{\tilde{e}\}| = 2/3$.   □

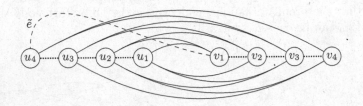

Fig. 3. (cf. Theorem 2) Drawing of the graph G with $p = 4$. Edges of S are dotted.

The result above shows that the approximation ratio of DFS-based algorithms is bounded from above by 2/3. We wonder if this is caused by the special structure of DFS trees or if this can be extended, for example to BFS-based algorithms:

Theorem 3. *MPS-BFS is NP-hard. Furthermore, there are (infinitely many) graphs G that allow a BFS tree T_v for each possible start node v such that MPS-BFS on each (G, T_v) is NP-hard.*

Proof. We give a reduction from Hamiltonian cycle (HC) to MPS-BFS. Let $G = (V, E)$ be an instance for HC, $n := |V|$, $m := |E|$, s a new node, and $B_v := \mathcal{B}_{v,s}^{m+1}$ for each $v \in V$. We construct an input graph $G' := (V', E')$ for MPS-BFS, where $V' := \{r\} \cup V \cup V_B$, $V_B := \bigcup_{v \in V} V(B_v)$, $E' := E_r \cup E \cup E_B$, $E_r := \{rv \mid v \in V\}$, and $E_B := \bigcup_{v \in V} E(B_v)$, cf. Fig. 4(a). G' contains $2 + n(m + 2)$ nodes and $m + n(2m + 3)$ edges. Choose $u \in V$ and $\tilde{p} \in \mathcal{I}(B_u)$ arbitrarily. We define $T := \{\tilde{p}s\} \cup E(G'[V' \setminus \{s\}]) \setminus E$, a BFS tree of G'.[3] We show that G has a HC if and only if G' has a planar subgraph of size $\xi := n(2m + 4)$ that contains T.

(If) Given a planar subgraph H of G' with ξ edges that contains T. There are at most $m - 1$ edges of G' not in H since $|E'| - m < \xi$. Thus, for each

---

[3] Starting at r (level 0) includes all edges of E_r. E cannot be taken since all of V lies on level 1. Each node $v \in V$ is connected to all of $\mathcal{I}(B_v)$, which lie on level 2. Only s remains, which is connected to $\tilde{p}$—the first investigated node on level 2.

bundle at least one 2-path is part of H. It follows that there can be at most n edges of E in H since H is planar. Consequently, $|E(H)| \leq k - m + |E'|$ where $k := |E \cap E(H)| \leq n$. Assuming $k < n$ leads to $|E(H)| < \xi$, a contradiction. By planarity of H we observe that $H[V]$ forms a Hamiltonian cycle in G.

(Only if) Given a Hamiltonian cycle C on G. We construct a planar subgraph $H := G'[T \cup C \cup E_B]$ that contains T (by construction) and has ξ edges. Note that adding C to T yields a planar graph since $H[\{r\} \cup V]$ forms a wheel graph. Likewise, adding E_B preserves planarity since $G'[E_B]$ is planar and contains a face with all nodes of V that allows an arbitrary ordering of those nodes.

Finally, we show the independence of the start node. From the above input graph G', we construct G'' by replacing each edge $rv \in E_r$ with $\mathcal{B}_{r,v}^{m+1}$ (cf. Fig. 4(b)), and replacing each edge of E with a path containing 5 new edges where each of the 4 new nodes is also connected to r with a new edge. Note that we can reach all nodes of V in at most 4 BFS levels, independent of the start node. Consequently, none of the 5-paths that correspond to edges in E can be fully contained in the resulting BFS tree. We conclude that any BFS tree constructed in the above way allows a reduction from HC to MPS-BFS. □

As for DFS trees, we have that any algorithm adding edges to an arbitrary BFS tree has an approximation ratio of at most 2/3.

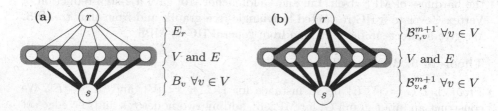

Fig. 4. (a) (cf. Theorem 3) Schematic drawing of G' for $|V| = 6$. Thick edges depict bundles of $m + 1$ parallel 2-paths. **(b)** (cf. Theorems 3 and 6) The analogously constructed graph for the MPS hardness proof.

Theorem 4. *An optimal solution to MPS-BFS yields an approximation ratio of at most 2/3 for the corresponding MPS problem.*

Proof. Let $G = (V, E)$ denote a triangulated graph that allows a 3-coloring $\phi \colon V \to [3]$ of the nodes, for example an even cycle C with two new nodes adjacent to all of C. We define the input graph $G' := (\{s, s_0, s_1, s_2\} \cup V, E \cup T)$ for MPS-BFS with $T := \{ss_i \mid i \in [3]\} \cup \{s_{\phi(v)}v \mid v \in V\}$. T is a BFS tree rooted at s. By construction, every triangle in G' requires 3 nodes of V of different color. We can add at most one triangle to T, as a $K_{3,3}$-subdivision would arise otherwise, see Fig. 5. Hence, the number of triangular faces in any planar subgraph H of G' that contains T is bounded by a constant, independent of $|V|$. Thus, the upper bound on the approximation ratio converges from above to 2/3 for large $|V|$. □

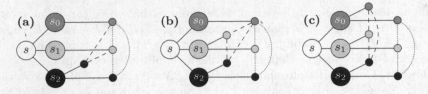

Fig. 5. (cf. Theorem 4) Arising $K_{3,3}$-subdivisions after adding two triangles to the BFS tree. One triangle is dotted, the other is dashed. From left to right: **(a)** both triangles share two nodes, **(b)** both triangles share a single node, **(c)** the triangles are disjoint.

Since any DFS or BFS tree is also a spanning tree, we have:

Corollary 5. *It is NP-hard to find a maximum planar subgraph that contains a given spanning tree. Likewise, an optimal solution to this problem approximates MPS with at most 2/3.*

4   MPS Is NP-hard: A Simple Proof

Inspired by our proof that MPS-BFS is NP-hard, we can give a shorter proof for the hardness of MPS itself. Liu and Geldmacher [10] gave a 2-step-reduction of Vertex Cover to a HC restricted to triangle-free graphs and from that to MPS. We give a direct simple reduction from general HC to MPS.

Theorem 6. *MPS is NP-hard.*

Proof. Let $G = (V, E)$ be an instance for HC, $n := |V|$, and $m := |E|$. We construct an input graph G' for MPS by adding two nodes r, s and the edge set $E_{\mathrm{B}} := \bigcup_{v \in V} (\mathcal{B}_{r,v}^{m+1} \cup \mathcal{B}_{v,s}^{m+1})$ (cf. Fig. 4(b)). Note that G' contains $2 + n(2m + 3)$ nodes and $m + 4n(m + 1)$ edges. We show that G has a Hamiltonian cycle if and only if G' has a planar subgraph of size $\xi := |E_{\mathrm{B}}| + n$.

(If) Given a planar subgraph H of G' with ξ edges. There are at most $m - 1$ edges of G' not in H since $|E(G')| - m < \xi$. Thus, for each bundle in E_{B} at least one 2-path is part of H. It follows that there can be at most n edges of E in H as H is planar. Consequently, $|E(H)| \leq k - m + |E(G')|$ where $k \leq n$ equals the number of edges of E in H. Assuming $k < n$ leads to $|E(H)| < \xi$, a contradiction. By planarity of H we observe that $H[V]$ forms a Hamiltonian cycle in G.

(Only if) Given a Hamiltonian cycle C on G. The graph $H := G'[C \cup E_{\mathrm{B}}]$ has ξ edges and is planar.   □

Consider an MPS instance $G = (V, E)$. We can construct G' by replacing every edge in E with a path of length $k := \lceil p/3 \rceil$. Now all cycles in G' contain at least p nodes, and $\mathrm{OPT}(G') = \mathrm{OPT}(G) + (k - 1)|E|$. We conclude:

Corollary 7. *MPS remains NP-hard for graphs with any given girth.*

Algorithm 1. Cactus Algorithm

Input: connected simple graph $G = (V, E)$
edge set $S_1 := \varnothing$
while $\exists$ triangle $T \subseteq E$ whose nodes are in 3 different components of (V, S_1) **do**
$\quad | \quad S_1 := S_1 \cup T$
$S_2 := S_1$
while $\exists$ edge $e \in E$ whose nodes are in different components of (V, S_2) **do**
$\quad | \quad S_2 := S_2 \cup \{e\}$
return S_2

---

5 Algorithms Inspired by Cactus Structures

The (greedy) *Cactus Algorithm*, see Algorithm 1, for MPS was developed by Călinescu et al. [2] and first constructs a cactus subgraph S_1 consisting of triangles joined at single nodes. The resulting structure S_2 achieves a tight approximation ratio of $7/18$. When the first phase of the algorithm is replaced to find a cactus structure of maximum cardinality (which requires the use of a graphic matroid parity subalgorithm), the approximation ratio can be improved to $4/9$. One may either search for an algorithm with a better approximation guarantee or for an algorithm with an approximation ratio better than $7/18$ that requires only simple operations (in contrast to the matroid-based algorithm), possibly again based on a greedy scheme. Poranen proposed two algorithms that greedily select triangles and conjectured approximation ratios of at least $4/9$ [11]. However, both conjectures were refuted by Fernandes et al. [7, Sect. 56.6]. We show that related, more general classes of algorithms are not suited to achieve the desired approximation guarantee or have an approximation ratio of at most $1/2$.

It is fairly natural to ask for a more sophisticated yet easily implementable greedy selection of the triangles to build a cactus. We first investigate algorithms that greedily select either edges or triangles in an "intuitively smart" manner. Given a graph G and a subgraph $G' \subseteq G$, we say that an edge $e \in E(G)$ is *forbidden* in G' if and only if $G' + e$ is non-planar. Similarly, we call an edge set $F \subseteq E(G)$ forbidden iff there is a forbidden edge $f \in F$.

The algorithm that iteratively picks an edge (or triangle) that minimizes the number of resulting forbidden edges (or triangles), is called *Greedy Edge Selection* (GES) (or *Greedy Triangle Selection* (GTS), respectively).

Theorem 8. *GES has a tight approximation ratio of $1/3$.*

Proof. Let $p \geq 4$. Define $H_p := (V, E_H)$ with $V := \{v_\ell^i \mid \ell \in [p], i \in [3]\}$ and $E_H := \{v_\ell^i v_\ell^{i+1} \mid 0 \leq \ell \leq p - 1, i \in [3]\} \cup \{v_{\ell-1}^i v_\ell^i \mid 1 \leq \ell \leq p - 1, i \in [3]\} \cup \{v_\ell^i v_{\ell-1}^{i+1} \mid 1 \leq \ell \leq p - 1, i \in [3]\}$, cf. Fig. 6(a). We define $\Lambda(v_\ell^i) := \ell$ as the *level* of v_ℓ^i. Note that H_p is a triangulation and 4-colorable with the coloring $\phi(v_\ell^i) := (3\ell + i) \mod 4$. For any color $c \in [4]$, let $V_c := \{v \in V \mid \phi(v) = c\}$ be the c-colored node partition induced by ϕ. We denote the increasing order

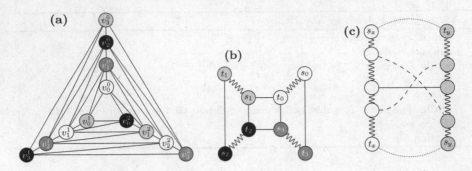

Fig. 6. (cf. Theorem 8) **(a)** The graph H_p for $p = 4$. **(b)** The outerplanar graph X_p on $V(H_p)$. **(c)** Inserting independent edges whose endpoints are non-adjacent between V_x and V_y in X_p.

of V_c according to Λ by π^c. For each of the four colors we define the (new) path $P_c := \{\pi_i^c \pi_{i+1}^c \mid 1 \le i < |V_c|\}$. The lowest and highest level node of a path P_c is denoted by s_c and t_c, respectively. We obtain the graph X_p on the nodes V by adding $\{t_0 s_1, s_1 t_2, t_2 s_3, s_3 t_0, s_0 t_3, t_1 s_2\}$ to the paths P_c, cf. Fig. 6(b).

Consider the graph $G := H_p \cup X_p$ (over the common node set V) as our input. The triangulation H_p is an MPS of G. The graph X_p is outerplanar. Thus, we can add any single edge planarly to X_p, and X_p can arise during GES since none of its edges was forbidding any other edges. By showing that we can only add a constant number of edges to X_p while preserving planarity we bound the approximation ratio by $\lim_{p\to\infty}(|E(X_p)| + \text{const})/(|E(H_p)|) = 1/3$.

We can ignore all edges incident to nodes $\{s_c, t_c \mid c \in [4]\}$: this is a constant number of edges since we have bounded degree (independent of p). Given two colors x, y, there are at most two faces in any embedding of X_p that have P_x and P_y on their boundary. Traversing any such face will visit the nodes along both paths in the same order (either $s_x \to t_x$ and $s_y \to t_y$; or $t_x \to s_x$ and $t_y \to s_y$). Let $E_{xy} \subseteq (V_x \times V_y) \cap E_H$ be an arbitrary set of independent edges whose endpoints are non-adjacent in X_p. The orderings π^x and π^y induce two orderings of E_{xy}. By construction of H_p we have $|\Lambda(v) - \Lambda(w)| \le 1$ for all $vw \in E_H$. Hence, we observe that the above two orderings of E_{xy} are in fact identical. It follows that we can insert at most one edge of E_{xy} into each of the at most two suitable faces of X_p, cf. Fig. 6(c). The number of color pairs is constant. Thus, for any color pair (x, y) and suitable face f, the insertable edges $E'_{xy} \subseteq (V_x \times V_y) \cap E_H$ need to be either adjacent, or incident to adjacent nodes. Since G has bounded degree, we can only add a constant number of edges to X_p. $\square$

Theorem 9. *Any algorithm that selects the edges picked by GTS has an approximation ratio of at most* $7/18$.

Proof. Let G be the graph of the proof of Theorem 8 for an arbitrary but fixed $p \ge 5$, and $n_p := |V(G)| = 3p$. Again, we speak of the paths P_c for the colors

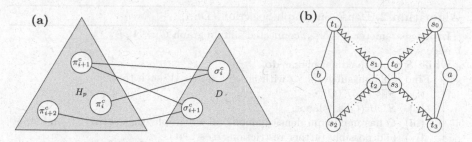

Fig. 7. (cf. Theorem 9 **(a)** Schematic structure of G' showing only some nodes of color c. **(b)** The outerplanar graph X'_p.

$c \in [4]$, and the outerplanar subgraph X_p of G. Our initial argument is based on the same principle as before. Coarsely speaking we replace the edges of the paths P_c by new triangles, preserve outerplanarity, and extend H_p by a similar structure on the newly inserted nodes:

Let D be a copy of H_{p-1} where we delete the node v_{p-2}^2. Note that D is triangulated with the exception of one face of degree 5. As in the proof above, this graph is 4-colorable which induces the node partitions $D_c := V_c(D)$ for $c \in [4]$. D can be seen as a copy of H_p where one node of each color $(v_{p-1}^0, v_{p-1}^1, v_{p-1}^2, v_{p-2}^2)$ is removed. We keep the notation of the ordering of nodes V_c in X_p by π^c and denote the analogous ordering of the nodes in the newly introduced partitions D_c by σ^c. Let $X'_p := (V(X_p) \cup V(D) \cup \{a, b\}, E(X_p) \cup E_\triangle \cup \{s_1 s_3, s_0 a, a t_3, t_1 b, b s_2\})$ with $E_\triangle := \{\pi_i^c \sigma_i^c, \sigma_i^c \pi_{i+1}^c \mid c \in [4], \pi_i^c \pi_{i+1}^c \in P_c\}$, see Fig. 7. I.e. $E_\triangle$ consists of a level-monotone Hamiltonian path for each color class.

Let $G' := (V(X'_p), E(X'_p) \cup E(H_p) \cup E(D))$. The graph $J := H_p \cup D$ is a planar subgraph of G'. Every edge in X'_p is part of a triangle and the graph remains outerplanar. Thus, we can add any single triangle planarly to X'_p, and X'_p could arise during GTS on G'. Analogous to the proof for Theorem 8, we can only add a constant number of edges to X'_p while preserving planarity.

Let F_J denote the set of triangular faces in J. We obtain the graph G'' from G' by inserting new nodes v_f of degree 3 for all $f \in F_J$, connecting v_f with the nodes on the boundary of f. Let $L := \{v_f \mid f \in F_J\}$ denote the newly inserted nodes and E_L the incident edges. Considering G'' as the input for GTS, similar to above, the number of edges that we can add to X'_p while preserving planarity is bounded by $|L| + \text{const}$: Any edge in E_L is part of a 2-path u_1-w-u_2, where $u_i \in V(G')$, $\phi(u_1) \neq \phi(u_2)$, and $w \in L$. On the other hand, $J \cup (L, E_L)$ remains planar. We conclude that the approximation ratio is at most

$$\lim_{p \to \infty} \frac{|E(X'_p)| + |L| + \text{const}}{|E(J)| + |E_L|} = \lim_{p \to \infty} \frac{(3n_p + \text{const}) + (4n_p + \text{const}) + \text{const}}{(6n_p + \text{const}) + (12n_p + \text{const})}.$$

□

Corollary 10. *Any algorithm that first selects an arbitrary (possibly maximal) set of triangles has an approximation ratio of at most 7/18.*

Algorithm 2. Dense Subgraph Selection (DSS)

Input: parameter $k \in \mathbb{N}_{\geq 3}$, connected simple graph $G = (V, E)$
edge set $S := \varnothing$
while S is not maximal planar **do**
  Find a planar subgraph Q with up to k nodes W such that
  (i) $S[W] \subsetneqq E(Q)$,
  (ii) $S \cup E(Q)$ is planar,
  (iii) Q has maximum density among all subgraphs that satisfy (i) and (ii),
  (iv) and possibly further restrictions (see text).
  $S := S \cup E(Q)$
return S

Observe that this bound matches the one of Algorithm 1 [2]. Similar to any DFS- and BFS-based algorithms, it remains NP-hard to determine a maximum set of edges that can be added planarly to a selected set of triangles. We will show a more general result in Theorem 11.

We investigate the selection of dense subgraphs, which is a natural generalization of triangle-based algorithms such as GTS. Given an edge set S and a node set W, we define $S[W]$ as the edges of S that connect nodes of W. Let the *density* of a graph (V, E) be defined as $|E|/|V|$, the edges per node.

We denote Algorithm 2 by DSS. In its most general form (DSS-U) we do not pose any further restrictions (iv) on the selection of dense subgraphs: they may overlap arbitrarily. A restricted version of this algorithm, called DSS-D, requires the subgraphs Q in the loop to be node disjoint to the current structure S. Similarly, we denote by DSS-C the algorithm with the restriction that the nodes of Q are pairwise disconnected in the current structure S.

Theorem 11. *Consider any MPS instance G. It remains NP-hard to find a maximum planar subgraph of G under either the restriction that it contains* (a) *the solution S of DSS-D, or the restriction that it contains* (b) *the solution S of DSS-C, respectively.*

Proof. (a) Given an arbitrary triangle-free graph $G = (V, E)$, we construct G' by adding $k - 1$ nodes V_v for each $v \in V$, such that $G_v := G'[\{v\} \cup V_v]$ is triangulated. Let $S := \bigcup_{v \in V} E(G_v)$. Note that each G_v is a graph on k nodes with maximal density and that any other subgraph of G' has strictly lower density. Consequently, the algorithm selects each G_v to S. Thus, any subgraph H' of G' that contains S corresponds to a subgraph H of G with $|E(H')| = |E(H)| + n(3k - 6)$. MPS is NP-hard on triangle-free graphs, see Corollary 7.

(b) Consider a graph G together with a spanning tree T. We know from Corollary 5 that it is NP-hard to find a maximum set of edges that can be added planarly to T. Replacing each edge of T with a triangulated subgraph on k nodes in G yields an instance where Algorithm 2 can select exactly the structures corresponding to T. Thus, finding a maximum set of edges that can be added to the selected structure remains NP-hard, independent of k. $\qquad \square$

Note that Theorem 11 for DSS-C and $k = 3$ is the above claimed hardness result for algorithms based on triangle selection.

Theorem 12. *For any fixed $k \geq 3$, DSS-U has an approximation ratio of at most $1/2$. For any fixed $k \geq 7$ any variant of DSS that poses arbitrary restrictions (iv) on the cut of Q with S has an approximation ratio of at most $1/2$.*

Proof. First assume that $k \geq 7$. Let $F := \{f_0, \ldots, f_3\}$ denote the set of faces of a K_4, δ_i the set of nodes incident to face f_i and $\kappa := k - 7$. We define $F' := F \setminus \{f_0\}$ and $\{b, t, u_0\} =: \delta_0$. We construct $G = (V, E)$ with $V := V(K_4) \cup \{w_i \mid f_i \in F'\} \cup \{u_{i+1} \mid i \in [\kappa]\}$, $E := E(K_4) \cup E_W \cup E_U$, $E_W := \{w_i v \mid f_i \in F', v \in \delta_i\}$, and $E_U := \bigcup_{i=1}^{\kappa} \{bu_i, u_i t, u_i u_{i-1}\}$. Note that G is triangulated, planar, and contains exactly k nodes. Furthermore, we cannot connect any nodes w_i, w_j, $i \neq j$, while preserving planarity. We define the input graph G' as $(V \cup L, E \cup E_L)$, where $L := \{s_1, \ldots, s_\ell\}$ and $E_L := \bigcup_{i \in [\ell]} \{s_i w_1, s_i w_2, s_i w_3\}$ (cf. Fig. 8), for some $\ell \geq 7$.

The algorithm may pick a graph Q that is the entire triangulated subgraph G in its first iteration, since G contains exactly k nodes. Thus, nodes in L can only be added with a single edge and we thus pick at most $1/3$ of E_L. On the other hand, a planar subgraph $H \subseteq G$ can be obtained by picking every edge in E except for the edge of K_4 incident to f_1 and f_2. Then, each node in L can be connected with w_1 and w_2 (picking $2/3$ of E_L), giving $H' \subseteq G'$. We conclude that the approximation ratio is at most $\lim_{\ell \to \infty} (|Q| + \ell)/|H'| = \lim_{\ell \to \infty} (|E| + \ell)/(|E| - 1 + 2\ell) = 1/2$.

For $k < 7$ we construct the graph G as for $k = 7$ where DSS-U may still return a subgraph containing G, independent of k and ℓ. $\qquad\square$

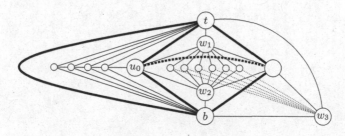

Fig. 8. (cf. Theorem 12) Schematic drawing of the input graph for $\ell = 6$ and $k = 4$. The K_4 subgraph is highlighted by thick edges. Dotted edges are not included in H_2.

References

1. Boyer, J.M., Myrvold, W.J.: On the cutting edge: Simplified $O(n)$ planarity by edge addition. J. Graph Algorithms Appl. **8**(2), 241–273 (2004)
2. Călinescu, G., Fernandes, C., Finkler, U., Karloff, H.: A better approximation algorithm for finding planar subgraphs. J. Algorithms **27**, 269–302 (1998)

3. Cimikowski, R.: Graph planarization and skewness (unpublished). http://citeseerx.ist.psu.edu/viewdoc/summary?doi=10.1.1.68.9958
4. Cimikowski, R.J.: An analysis of some heuristics for the maximum planar subgraph problem. In: Clarkson, K.L. (ed.) Proceedings of 6th SODA, pp. 322–331 (1995)
5. Cong, J., Liu, C.: On the k-layer planar subset and topological via minimization problems. IEEE Trans. CAD Integr. Circ. Syst. **10**(8), 972–981 (1991)
6. Dyer, M., Foulds, L., Frieze, A.: Analysis of heuristics for finding a maximum weight planar subgraph. Eur. J. Oper. Res. **20**(1), 102–114 (1985)
7. Fernandes, C.G., Călinescu, G.: Maximum planar subgraph. In: Gonzalez, T. (ed.) Handbook of Approximation Algorithms and Metaheur. Chapman & Hall/CRC, Boca Raton (2007)
8. Hopcroft, J.E., Tarjan, R.E.: Efficient planarity testing. J. ACM **21**(4), 549–568 (1974)
9. Hsu, W.-L.: A linear time algorithm for finding a maximal planar subgraph based on PC-trees. In: Wang, L. (ed.) COCOON 2005. LNCS, vol. 3595, pp. 787–797. Springer, Heidelberg (2005)
10. Liu, P.C., Geldmacher, R.C.: On the deletion of nonplanar edges of a graph. In: Proceedings of 10th Southeastern International Conference on Combinatorics, Graph Theory, and Computing, Congress Numbers XXIII-XXIV, pp. 727–738. Utilitas Mathematica, Winnipeg, Manitoba (1979)
11. Poranen, T.: Two new approximation algorithms for the maximum planar subgraph problem. Acta Cybern. **18**(3), 503–527 (2008)
12. Shih, W., Hsu, W.: A new planarity test. Theor. Comput. Sci. **223**(1–2), 179–191 (1999)
13. Zelikovsky, A.: Improved approximations of maximum planar subgraph (unpublished). http://citeseerx.ist.psu.edu/viewdoc/summary?doi=10.1.1.71.304

Exact Algorithms for Weighted Coloring in Special Classes of Tree and Cactus Graphs

Robert Benkoczi, Ram Dahal$^{(\boxtimes)}$, and Daya Ram Gaur

Department of Mathematics and Computer Science, University of Lethbridge,
4401 University Dr W, Lethbridge, AB T1K 6T5, Canada
{robert.benkoczi,ram.dahal,daya.gaur}@uleth.ca
http://www.uleth.ca

Abstract. We study the weighted vertex coloring problem (WVCP) in binary trees and a restricted class of cactus graphs we called cactus paths. WVCP is a generalization of the vertex coloring problem where a color class of a feasible coloring is assigned a cost equal to the largest weight of a vertex from the color class. The objective is to find a feasible coloring which minimizes the sum of the color costs assigned to each color class. We improve the exact algorithms for solving WVCP on binary trees and propose new and efficient algorithms for WVCP on cycles and cactus paths with maximum degree three. Our work extends the results of Kavitha and Mestre [8]. Our algorithms have a time complexity of $O(n \log^2 n)$ for cactus paths and $O(n^2 \log n)$ for binary trees.

Keywords: Vertex coloring · Max coloring · Weighted coloring · Scheduling · Binary trees · Cactus graph · Dynamic programming · Spine tree decomposition

1 Introduction

The vertex coloring problem in a graph (VCP) is a fundamental combinatorial optimization problem that dates back to the work of Francis Guthrie (1831–1899) who conjectured that a planar graph can be colored with four colors [4]. VCP, which seeks to assign the smallest number of colors to the vertices of a graph so that no two adjacent vertices receive the same color, has motivated the work of many famous mathematicians. The problem arises in several practical applications as well, such as scheduling, timetabling, computing derivatives, and frequency assignment in second generation cellular networks [11].

A channel allocation problem in fourth generation WiMAX cellular networks can be modeled as a generalization of the VCP [13]. In the weighted vertex coloring problem WVCP, the vertices of the graph are assigned positive weights and a color i of a feasible colouring is assigned a cost equal to the maximum weight among the vertices colored with color i. The goal is to find a feasible coloring for which the sum of the color costs is minimized. It is not difficult to find an example showing that a coloring with the smallest number of colors is not necessarily an optimal solution to the WVCP.

© Springer International Publishing Switzerland 2016
V. Mäkinen et al. (Eds.): IWOCA 2016, LNCS 9843, pp. 347–358, 2016.
DOI: 10.1007/978-3-319-44543-4_27

WVCP has some interesting properties. It is strictly more difficult than the classical VCP. For example, WVCP is NP-hard for some classes of graphs, such as bi-partite and interval graphs, for which the VCP can be solved exactly in polynomial time [6]. This has motivated a number of research articles investigating polynomial time algorithms for special classes of graphs. Paths and several restrictions of tree and bi-partite graphs have been investigated and there was an effort to improve the running time for paths [8,12,15]. The case of trees is particularly interesting. For arbitrary trees, an algorithm with complexity $O(n^{\Theta(\log n)})$ that relies on a procedure to solve the list coloring problem [10], follows from the work of Guan and Zhu [6]. Moreover, Araujo *et al.* show that this algorithm is optimal in some sense (thus WVCP on arbitrary trees is most likely not in P unless $P = NP$). WVCP for arbitrary trees is unlikely to be NP-hard as there is a quasi polytime algorithm [1]. For any class of trees whose maximum degree is bounded by a constant, the quasi polytime algorithm is polynomial. In particular, for binary trees, the algorithm runs in time $O(n^4)$ which can be improved to $O(n^3 \log n)$ by a straightforward application of binary search.

Our Contributions: In this paper, we propose a new algorithm for solving WVCP on binary trees and cactus paths graphs with maximum degree bounded by a constant. For the class of cactus paths, our method solves WVCP in time $O(n \log^2 n)$ and for binary trees in time $O(n^2 \log n)$ by the use of the *Spine Tree Decomposition* [2]. To the best of our knowledge these results are new.

1.1   Problem Definition and Notation

Let $G(V, E, w)$ be a vertex-weighted undirected graph with vertex set V, edge set E, and weight function $w : V \to \mathbb{R}$. We denote by w_v the weight of vertex v. A color class (an independent set) is a set of non-adjacent vertices in a graph. Consider a feasible coloring X of G using k colors (a k-coloring) as a partition of the vertex set V into k sets $X = \{\alpha_1, \alpha_2, ..., \alpha_k\}$ where α_i is a color class in G. We define the cost of color i by $w(\alpha_i) = \max_{v \in \alpha_i} w_v$, and the cost of coloring X by $w(X) = \sum_{i=1}^{k} w(\alpha_i)$.

Problem 1. *Weighted vertex coloring (WVCP) [6]: Given a vertex-weighted undirected graph G, find a feasible coloring X of G for which $w(X)$ is minimum.*

When $w_v = 1$ for all $v \in V$, WVCP reduces to the proper vertex coloring problem (VCP). It follows that WVCP is strongly NP-hard [5]. Interestingly, an optimal solution to WVCP may use more colors then that of unweighed vertex coloring problem. Consider the following example in Fig. 1

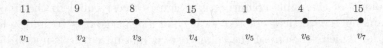

Fig. 1. Two coloring is not optimal. Greedy is not optimal.

Example: In Fig. 1, any 2-coloring has cost 30. While coloring v_2, v_4 and v_7 by color 1 and coloring v_1, v_3 and v_6 by color 2 and v_5 by color 3 gives the WVCP with the weight of $15 + 11 + 1 = 27$. A simple greedy strategy based on sorting the vertices, has cost of $15 + 9 + 8 = 32$. In the greedy strategy v_1, v_4, v_7 are colored with color 1 and v_2, v_6 are colored with color 2 and v_5, v_3 are coloured with color 3.

Applications: WVCP arises in the scheduling of data transmissions in a time division multiple access (TDMA) wireless network [12,14]. One example of such TDMA technology is the Worldwide Interoperability for Microwave Access (WiMAX) standard which is responsible for a large portion of the data mobility services today. The WiMAX standard does not specify any channel allocation algorithms. This is to allow the most flexible and efficient use of resources possible. Moreover, the duration of the time slots in the WiMAX standard need not be uniform. Given a set of clients with different bandwidth requirement, a channel allocation scheme in a mesh network using WiMAX technology seeks to group the transmissions in such a way that interference does not occur and the channel is used efficiently. Such a schedule corresponds to a WVCP problem where the graph that represents interference in the network. The vertex weights correspond to the bandwidth requirement for the devices participating in the communication. In particular, line, ring and tree topologies are commonly used in telecommunication networks [9], hence our results on trees and cactus paths may be of interest in this domain as well.

WVCP also occurs in batch scheduling, an important problem in distributed computing [7]. Here, a set of jobs is to be scheduled in parallel on a large number of processors. However, several jobs may require access to the same resource. These jobs cannot be scheduled in parallel. A parallel schedule with minimum make-span can be formed by solving a WVCP problem. The jobs are the nodes in the graph and edges represent conflicts between jobs and the weight of nodes represent the processing time of the job.

The Matrix Decomposition Problem in Time Division Multiple Access Traffic Assignment can also be modeled as WVCP. In this problem, a traffic matrix is decomposed into k mode matrices such that: no more than one non-zero element in each column and row; and each non-zero entry of the original traffic matrix should appear in one and only one matrix mode of the decomposition. The cost of a mode matrix is determined by the maximum of its non-zero elements. The objective is to minimize the sum of mode matrices. This problem can be modeled as WVCP by creating a graph where all the non-zero entries of the given matrix are vertices with their weights; and add an edge between the vertices if the vertices are in the same row or the column. A color class in the generated graph corresponds to a mode matrix and its weight is determined by the weight of the vertex. Riberio *et al.* [17] gave an exact algorithm based on column generation. Prais *et al.* [16] proposed a heuristic approach based on a Greedy Randomized Adaptive Search Procedure.

Related Work: There are several results on WVCP when restricted to paths. Guan and Zhu [6] were the first to propose an $O(n^4)$ time algorithm for WVCP on paths. Since the maximum number of colors needed for WVCP is not larger than $\Delta + 1$, where Δ is the maximum degree. So, three colors suffice for a path. The running time was subsequently improved by Escoffier *et al.* [3] to $O(n^2)$. Their approach is as follows. Consider the colors labelled in such a way that $w_1 \geq w_2 \geq w_3$, where w_i is the cost of color i. The value w_1 is fixed and equal to the maximum weight in the graph. They enumerate all possible values for w_2. All of the vertices with weight greater than w_2 must be colored 1. Now, consider two vertices u and v that must be colored 1. Two vertices between u and v are colored other than color 1. If there exists an even number of vertices between u and v, introduce third color otherwise color the vertices with two colors $(1, 2)$. Halldorsson *et al.* [7] further reduced the complexity of the algorithm to $O(n \log n)$. For each fixed value of w_2, they find the minimum value of w_3 in $O(\log n)$ time [7].

Finally, Kavitha and Mestre [8] gave linear time algorithm on paths and skinny trees. They assume the vertices can be sorted independently. Trees for which the set of vertices of degree at least 3 forms an independent set are referred to a skinny tree. Their idea is to find a set of candidate values for the cost of color 3. Then, by preprocessing the weights at odd and even positions on the path, one can determine, in amortized constant time, the costs for colors 1 and 2. The idea works for skinny trees as well.

Apart from Kavitha and Mestre's algorithm for skinny trees and the general algorithm based on list coloring originating from the work of Guan and Zhu [6], no other algorithms for trees are known. In the following section we describe a new algorithm based on computing, in a bottom up manner, the set of weight values corresponding to feasible weighted colorings of sub-trees of the given tree.

2   An Exact Algorithm for Binary Trees

The approach described by Guan and Zhu in [6] can be adapted to solve WVCP for binary trees in polytime. The idea is to enumerate all possible values for the costs of the colours of the WVCP and use a procedure for list coloring to test if a coloring with such color costs is feasible.

Our approach is different. Rather than testing feasibility with a procedure for list coloring, we compute the set of values for w_i for which list coloring is feasible. We call this set of values the *feasible weight set*. In the next section, we characterize this set and we show that its complexity is linear in the size of the binary tree.

2.1   Feasible Weight Sets

Consider a vertex weighted binary tree $T = (V, E, w)$ where $w : V \to \mathbb{R}$. We wish to represent the set of weight values $w_i : 1 \leq i \leq 4$ for binary trees $\Delta \leq 3$ for which a feasible WVCP coloring of T exists where the cost of color i is denoted w_i, and $w_1 \geq w_2 \geq w_3 \geq w_4$. Naturally, $w_1 = \max\limits_{v \in V} w(v)$ is fixed. Consider w_2 is also fixed. For this case, we are concerned with representing the set of values for w_3 and w_4. There are $O(n)$ possible choices for w_3 and w_4 and therefore, the size of the feasible weight set is $O(n^2)$. The important point here is that, although the size is $O(n^2)$ we will represent this set by a geometric construction with complexity $O(n)$.

We represent the feasible weight set by points in the two dimensional space with coordinate axes w_3 and w_4 (see Fig. 2). Because $w_2 \geq w_3 \geq w_4$, the feasible set is contained in the upper triangular region in the figure.

We discuss a few simple properties of the feasible weight set to help build our intuition. We first remark that point (w_2, w_2) is always part of the feasible weight set. This is because there is a feasible coloring with $w_3 = w_4 = w_2$. In this case, there are two types of vertices in tree T: vertices v with $w_v > w_2$ which must be colored 1, and vertices with $w_v \leq w_2$ which can be colored with any of the four colors. If the choice of w_2 is feasible, then the vertices of type 1 must form an independent set. If we remove these vertices from T, we obtain a forest which can be colored with any two colors from the set of allowed colors $\{2, 3, 4\}$.

Another interesting point is the origin O in Fig. 2. The origin is part of the feasible weight set if and only if there exists a feasible two coloring with color weights w_1 and w_2. Of course, this is not true for all values of w_2. Given the two observations above, the feasible weight set corresponds to a set of points inside $\triangle OPQ$ possibly separated from the origin by a polygonal line which we call *boundary line of the feasible weight set* (Fig. 3). We characterize this boundary line and claim that it has a complexity of $O(n)$. To do this, we first prove the following simple lemma.

Lemma 1. *Let $W(T)$ denote the set of weights of tree T. Let $A = (a, b)$ be a point in the feasible weight set. Then any point $Z = (x, y)$ with $x \in W(T)$, $y \in W(T)$, $x \geq a$, and $y \geq b$ is also in the feasible weight set.*

Proof. The proof is immediate. If the list coloring problem is feasible for point A ($w_3 = a$ and $w_4 = b$), then it must also be feasible for point Z since the list of allowed colors for the list coloring instance at point Z contains the lists of allowed colors from the instance at point A. □

Lemma 1 implies that, if the origin is not inside the feasible weight set, then the boundary line separating the feasible weight set from the origin is an x-y monotone polygonal line with axis parallel line segments, with one endpoint on line segment $[PQ]$ and the other endpoint on line segment $[OQ]$ (see Fig. 3). For this reason, we also call the boundary line *the staircase*. The staircase contains $O(n)$ points, at most two points for each possible value of w_3. If the origin is part of the feasible weight set, we consider the vertical line segment $[OP]$ to be the boundary line. In this case, the feasible weight set is the set of $O(n^2)$ points inside $\triangle OPQ$. Consequently, we claim the following that defines the representation of a feasible weight set.

Corollary 1 (Feasible weight set). *The feasible weight set for a fixed w_1 and w_2 is uniquely determined by the boundary line that starts at horizontal line $w_3 = w_2$ and ends at diagonal line $w_3 = w_4$.*

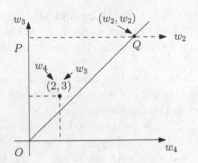

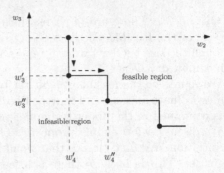

Fig. 2. The region in w_3–w_4 space enclosing the feasible weight set.

Fig. 3. Boundary line for the feasible weight set.

2.2   An Efficient Dynamic Progamming (DP) Algorithm for Binary Trees

Given the feasible weight set for fixed w_2 for the entire tree, a simple procedure can determine the optimal solution to the WVCP by traversing the boundary line of this set in $O(n)$ time and computing the values $w_3 + w_4$ for every corner point on the boundary line. A *corner point* is a point on the boundary line where the boundary line of w_3 and w_4 intersects. The feasible weight set of any arbitrary binary tree can be efficiently computed if we use the Spine Tree Decomposition (STD) [2]. In the STD, the given rooted binary tree is recursively decomposed into two or more sub-trees called components. The decomposition is determined by a path computed in such a way than no component adjacent to this path is excessively large in size. This fact allows the decomposition to be balanced. This path is called the *spine*. Each node on the spine is considered as leaf node in a balanced binary search tree built on top of the spine. A path connecting any two search trees nodes can be called a super-path. Two important properties of the STD are given below for reference.

Theorem 1. *Any two tree nodes x and y are connected, in the STD data structure, by a super-path of search tree nodes of length $O(\log n)$ where n is the number of vertices.*

Theorem 2. *An STD data structure can be constructed in $O(n)$ time and uses $O(n)$ space.*

Weight Set Computation: We compute feasible weight sets in a bottom-up fashion. We find the optimal solution to the WVCP from the feasible weight sets of the entire tree which are computed at the root of the STD, r_{SD} (see Fig. 4). We associate feasible weight sets with the internal nodes in the search trees of the STD. These sets represent the set of feasible weights assigned to colors 2 through 4 for which a feasible coloring exists in the sub-tree corresponding to the internal search tree node. If v represents an internal search tree node, the sub-tree T_v corresponding to v has a special structure. It is connected with the rest of tree T by at most two boundary vertices (vertices of a spine). We compute the feasible weight sets for T_v under the constraint that the two boundary vertices are assigned a prescribed pair of colors. This means that the DP algorithm computes a feasible vertex set that corresponds to colors assigned to at most *two* spine nodes. Hence, the number of feasible weight sets computed per STD node is $4^2 = 16$.

Our computation proceeds, in a bottom up fashion, and associates feasible weight sets to the nodes of the STD search trees, one search tree at a time. Each such calculation is performed for w_3 and w_4 for a fixed value of w_2. The calculation is then repeated for another value of w_2 and so on, starting with the smallest possible value for w_2 until w_2 equals w_1. Let $S_{ij}(v)$ denote the staircase for feasible weight set of sub-tree T_v when the two boundary nodes are colored with colors $i : 1 \le i, j \le 4$ respectively, the optimal solution to the WVCP is obtained by computing the cost $w_3 + w_4$ on each cornor point on the staircase $S_i(r_{SD})$ for all (i,j) combinations of colors. This can be done in time $O(n)$ as shown in Sect. 2.1.

The following lemma, proved in [8] provides a starting value for w_2.

Lemma 2. *In every valid coloring, we must have $w_2 \ge \max\{\min(w_u, w_v) : (u, v) \in E\}$ [8].*

Base Case: In this section we show how to compute the feasible weight for a leaf node v in the STD tree. In this case, then, v is on the spine of the STD search tree. We distinguish the following two cases.

Case 1: Node v has degree three. In this case, the feasible weight sets associated with v are obtained from the root of the search tree corresponding to the child spine that node v starts. No additional processing is required.
Case 2: Node v has degree two. We have the following situations:
  i: $w_v > w_2$. Then v must be colored 1 and $S_i(v) = \emptyset$ for $i \ge 2$ and $S_1(v)$ corresponds to line segment $[OP]$ in Fig. 5 (no restriction).
  ii: $w_v \le w_2$. Then $S_1(v)$ and $S_2(v)$ again correspond to line segment $[OP]$, no restriction. However, $S_3(v)$ corresponds to staircase in Fig. 6 and staircase $S_4(v)$ corresponds to staircase in Fig. 7.

Recursive step: We distinguish three cases.
Case 1: When the parent node v has two leaf nodes x and y. Thus v has at most twelve possible color combinations. Let $S_j(v)$ represents the staircase function for color $j : 1 \le j \le 4$. Let $k = 1, \ldots, 4$.

$$S_{jk}(v) = S_j(x) \cap S_k(y), \forall j, \forall k, j \neq k$$

Case 2: In this case, we calculate the feasible weight sets of the parent node v whose immediate neighbour x is a leaf node of STD and y is an internal node of the search tree. Let $p = 1, \ldots, 4$ and $q = 1, \ldots, 4$:

$$S_{pq}(v) = \bigcup_{j \neq p} (S_p(x) \cap S_{jq}(y))$$

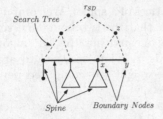

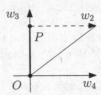

Fig. 4. Spine Tree Decomposition; x and y are leaves of the search tree

Fig. 5. The staircase line is segment $[OP]$.

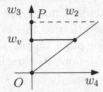

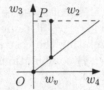

Fig. 6. Base case for $S_3(v)$

Fig. 7. Base case for $S_4(v)$

Case 3: Here, we calculate the feasible weight sets of node v whose children x and y are internal nodes of the search tree.

$$S_{pq}(v) = \bigcup_{j \neq k} (S_{pj}(x) \cap S_{kq}(y))$$

When $w_v \geq w_2$ we have to further intersect $S_j(x) : 1 \leq j \leq 4$ with $\max\{w_3, w_v\}$ if j is the third color or with $\max\{w_4, w_v\}$ if j is the fourth color (Figs. 8 and 9).

Analysis: We already know that the complexity of the feasible weight sets is linear in the size of the tree they are defined on. It is not difficult to notice that the number of steps needed to compute unions or intersections of feasible weight sets is proportional to the total complexity of the boundary lines of the feasible sets being merged. The lines are x-y monotone and one can carefully traverse the two lines in the same direction and compute their intersection in amortized constant time per point visited. We thus state the following theorem whose proof is to be provided in the full version.

Theorem 3. *The WVCP problem in arbitrary binary trees can be solved exactly in time $O(n^2 \log n)$.*

3   An Exact Algorithm for Cactus Paths

A cactus graph is a connected graph where any two cycles have at most one vertex in common. In this section we give an exact algorithm for a special class

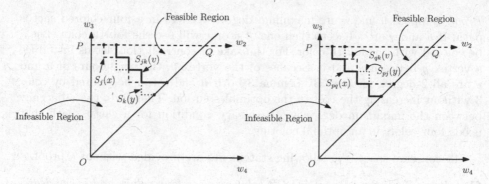

Fig. 8. Combining the feasible weight sets $S_j(x)$ and $S_k(y)$.

Fig. 9. Computing $S_{pq}(v)$ from the sixteen temporary staircase lines $S_{jk}(v)$; only two temporary staircase lines are depicted.

of cactus graphs, the *cactus paths* with maximum degree three. A cactus path is a cactus graph which generates a path graph by contracting all cycles. A contraction of a cycle means converting a cycle into a single node. In general, cactus graphs generate trees upon contraction of cycles.

Results of Guan and Zhu [6] imply that a cactus path with maximum degree three is 4-colorable. Figure 10 illustrates that four colors are needed in an optimal solution. The minimum three coloring has cost $8 + 4 + 4 = 16$, whereas the four coloring has cost $8 + 4 + 2 + 1 = 15$.

We now establish several properties that characterize the optimal solution to the WVCP in cactus graphs with maximum degree three. These properties are essential in establishing the correctness of the proposed algorithm.

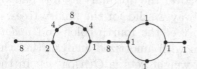

Fig. 10. A cactus path

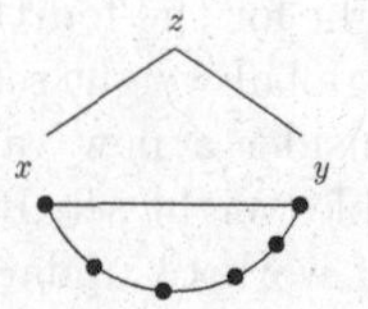

Fig. 11. A base case of a cactus path

Lemma 3. *Let G be a graph with maximum degree Δ. If all of the maximum degree vertices of G have neighbors with constant degree k where $k < \Delta$, then the optimal solution to the WVCP uses at most Δ colors [6,8].*

Lemma 4. *If the optimal solution to the WVCP on a cactus path with maximum degree Δ uses four colors, then at least two vertices of maximum degree are adjacent.*

Proof. Suppose u and v are maximum degree vertices in a four colored cactus path. If u and v are adjacent then one of u or v will use the fourth color. Let u be colored with the fourth color. For the sake of contradiction, let us introduce a vertex y between u and v. Because of the vertex y, the neighbours of u and v are all 2 degree vertices. By Lemma 3 both u and v can be colored by color 3 without increasing the cost of the optimal solution. Therefore, the adjacency between the maximum degree is a necessary condition for a cactus path that needs four colors in an optimal coloring.   □

To conserve space, the following statements are provided without a proof.

Corollary 2. *If the optimal WVCP solution uses four colors and every degree three vertex has at most one neighbor with degree three, then the color 4 candidate vertices are the degree three vertices with a neighbor with bigger neighbor.*

3.1   Algorithm

In this section we summarize the $O(n \log^2 n)$ algorithm for cactus paths.

We associate a node with every cycle of the cactus path. We also associate a node with every component left after removal of all the cycles from the cactus path. According to the structure of the cactus paths, these components must be paths or isolated vertices if they exist. We arrange these nodes so that consecutive nodes correspond to components (cycles or paths) that are adjacent. We consider a balanced binary tree whose leaves are the nodes just created. Like on the previous sections, we consider feasible solutions with at most four colors where w_i represents the cost of color i and $w_1 \geq w_2 \geq w_3 \geq w_4$. Unlike the algorithm for trees, we fix the cost of the fourth colour and we represent feasible weight sets for the cactus graph components in the space w_2 and w_3. We merge these feasible weight sets, in a bottom-up fashion, given the balanced binary tree. At the root of the tree, we obtain the feasible weight sets for the entire cactus path and a fixed weight for the fourth color. We recover the optimal (w_2, w_3) pair by processing this global weight set in the same way we did for the binary trees.

We then consider a new value for the cost of the fourth color and *update* the global feasible weight set. It can be shown with amortized analysis that the total amount of work to update the feasible weight sets is $O(n \log^2 n)$. In the following paragraph, we describe the procedure to compute the feasible weight sets for the components of the cactus paths.

Basis step: The feasible weight sets for a component are computed for the base case where the boundary vertices of the component are colored with predetermined colors. Let x and y be two boundary vertices of a component and consider any color assignment for these vertices (see Fig. 11). If the component is a cycle and we remove the boundary vertices x and y, then the cycle is split into two different paths. We use the algorithm of Kavitha and Mestre [8] to obtain a feasible weight set in w_2 and w_3 space for each one of these two paths. We then compute the feasible weight set for the cycle by intersecting the feasible weight sets obtained for the two paths.

Recursive step: Once the feasible weight sets of the components of the cactus graph are available, we proceed to merge them according to the balanced binary tree. We use the same union and intersection operations as for the WVCP algorithm on binary trees (details omitted). Since the complexity of the boundary of the feasible weight sets is linear in the size of the corresponding cactus components, it can be shown that the entire computation takes $O(n \log n)$ time.

Algorithm:

Step 1: Compute, the components of the cactus path in $O(n)$ time.

Step 2: Determine the vertices that are candidates for the fourth color using Lemma 2. Sort these vertices in non-decreasing order of their weights in list L.

Step 3: For $w_4 \in (\{0\} \cup L)$ do:

    a: Compute or update the feasible weight set by traversing the balanced binary tree bottom-up.

    b: Traverse the boundary of the feasible solution to obtain the best solution for the fixed value of w_4.

Step 4: Return the best solution obtained over all values of w_4.

We note that updating the feasible weight sets is dictated by the weight of candidate vertex to be colored with a four color (Lemma 2). The update requires union of the existing feasible sets with an additional feasible weight set. The complexity of the feasible weight set is proportional to the size of the component adjacent to the candidate color four vertex. The union can be performed in time $O(S \log n)$ where S is the size of the component, irrespective of the complexity of the feasible weight set being updated using binary search. Since a component is involved in update at most once and the height of the balanced binary tree is $O(\log n)$, it follows that the total update time for all values of w_4 takes $O(n \log^2 n)$ time.

Theorem 4. *The WVCP problem in cactus paths can be solved exactly in time $O(n \log^2 n)$.*

Theorem 5. *The WVCP problem in cycles can be solved exactly in time $O(n)$. Assume the vertices can be sorted independently.*

4 Conclusion

In this paper, we consider the weighted coloring problem in binary trees and cactus paths. It is well known that for WVCP, any tree with the maximum degree bounded by a constant, can be in polynomial time. However, the running time of this algorithm is quite high, $O(n^3 \log n)$ for binary trees and cactus paths. For certain applications such as solving robust versions of WVCP, this can be slow. We describe a more powerful method that computes the coloring for binary trees in time only $O(n^2 \log n)$ using the spine tree decomposition. We then discuss how the method can be extended to cactus paths to give an algorithm with time complexity $O(n \log^2 n)$. We conjecture that using feasible weight sets, an algorithm with sub-quadratic time complexity for the binary trees can be designed.

Acknowledgements. Robert Benkoczi's and Daya Ram Gaur's research was supported in part by individual NSERC Discovery Grants. Ram Dahal's research was supported by the PIMS and the School of Graduate Studies, University of Lethbridge. We would like to thank the anonymous reviewers for their insightful comments that help improve the quality of the paper.

References

1. Araujo, J., Nisse, N., Pérennes, S.: Weighted coloring in trees. SIAM J. Discrete Math. **28**(4), 2029–2041 (2014)
2. Benkoczi, R.R.: Cardinality constrained facility location problems in trees. Ph.D. Thesis, Simon Fraser University (2004)
3. Escoffier, B., Monnot, J., Paschos, V.T.: Weighted coloring: further complexity and approximability results. Inf. Process. Lett. **97**(3), 98–103 (2006)
4. Fritsch, R., Peschke, J., Fritsch, G.: The Four-Color Theorem: History, Topological Foundations, and Idea of Proof. Springer, New York (2012)
5. Gary, M.R., Johnson, D.S.: Computers and Intractability A Guide to the Theory of np-Completeness. WH Freman and Co, New York (1979)
6. Guan, D.J., Xuding, Z.: A coloring problem for weighted graphs. Inf. Process. Lett. **61**(2), 77–81 (1997)
7. Halldórsson, M.M., Shachnai, H.: Batch coloring flat graphs and thin. In: Gudmundsson, J. (ed.) SWAT 2008. LNCS, vol. 5124, pp. 198–209. Springer, Heidelberg (2008)
8. Kavitha, T., Mestre, J.: Max-coloring paths: tight bounds and extensions. J. Comb. Optim. **24**(1), 1–14 (2012)
9. Khan, N., Pal, A., Pal, M.: Edge colouring of cactus graphs. Adv. Model. Optim **11**(4), 407–421 (2009)
10. Kratochvil, J., Tuza, Z.: Algorithmic complexity of list colorings. Discrete Appl. Math. **50**(3), 297–302 (1994)
11. Lewis, R.: A Guide to Graph Colouring, Algorithms and Applications. Springer, Switzerland (2015)
12. Malaguti, E., Monaci, M., Toth, P.: Models and heuristic algorithms for a weighted vertex coloring problem. J. Heuristics **15**(5), 503–526 (2009)
13. McDiarmid, C., Reed, B.: Channel assignment and weighted coloring. Networks **36**(2), 114–117 (2000)
14. Mishra, A., Banerjee, S., Arbaugh, W.: Weighted coloring based channel assignment for wlans. SIGMOBILE Mob. Comput. Commun. Rev. **9**(3), 19–31 (2005)
15. Pemmaraju, S.V., Raman, R.: Approximation algorithms for the max-coloring problem. In: Caires, L., Italiano, G.F., Monteiro, L., Palamidessi, C., Yung, M. (eds.) ICALP 2005. LNCS, vol. 3580, pp. 1064–1075. Springer, Heidelberg (2005)
16. Prais, M., Ribeiro, C.C.: Reactive GRASP: an application to a matrix decomposition problem in TDMA traffic assignment. INFORMS J. Comput. **12**(3), 164–176 (2000)
17. Ribeiro, C.C., Minoux, M., Penna, M.C.: An optimal column-generation-with-ranking algorithm for very large scale set partitioning problems in traffic assignment. Eur. J. Oper. Res. **41**(2), 232–239 (1989)

Graph Algorithms

Finding Cactus Roots in Polynomial Time

Petr A. Golovach[1], Dieter Kratsch[2], Daniël Paulusma[3],
and Anthony Stewart[3(✉)]

[1] Department of Informatics, University of Bergen, P.B. 7803, 5020 Bergen, Norway
`petr.golovach@ii.uib.no`
[2] Laboratoire d'Informatique Théorique et Appliquée, Université de Lorraine,
57045 Metz Cedex 01, France
`dieter.kratsch@univ-lorraine.fr`
[3] School of Engineering and Computing Sciences, Durham University,
Durham DH1 3LE, UK
`{daniel.paulusma,a.g.stewart}@durham.ac.uk`

Abstract. A cactus is a connected graph in which each edge belongs to at most one cycle. A graph H is a cactus root of a graph G if H is a cactus and G can be obtained from H by adding an edge between any two vertices in H that are of distance 2 in H. We show that it is possible to test in $O(n^4)$ time whether an n-vertex graph G has a cactus root.

1 Introduction

Squares and square roots are well-known concepts in graph theory that have been studied first from a structural perspective [22,24] but later also from an algorithmic perspective, as we will discuss. The *square* $G = H^2$ of a graph $H = (V_H, E_H)$ is the graph with vertex set $V_G = V_H$, such that any two distinct vertices $u, v \in V_H$ are adjacent in G if and only if u and v are of distance at most 2 in H. A graph H is a *square root* of G if $G = H^2$. It is a straightforward exercise to check that there exist graphs with no square root, graphs with a unique square root as well as graphs with many square roots.

In this paper we consider square roots from an algorithmic point of view. The corresponding recognition problem, which asks whether a given graph admits a square root, is called the SQUARE ROOT problem. Our research is motivated by the result of Motwani and Sudan [21] who proved in 1994 that SQUARE ROOT is NP-complete. Afterwards, SQUARE ROOT was shown to be polynomial-time solvable for various graph classes, such as K_4-free graphs (trivial), planar graphs [18], or more general, any non-trivial minor-closed graph class [23], block graphs [16], line graphs [19], trivially perfect graphs [20], threshold graphs [20], graphs of maximum degree 6 [3], 3-degenerate graphs [11] and (K_r, P_t)-free graphs for any two integers $r, t \geq 1$ [11]. It was also shown that SQUARE ROOT is NP-complete for chordal graphs [13]. We refer to [3,4,10] for a number of parameterized complexity results on SQUARE ROOT. The computational hardness of SQUARE ROOT also led to the following natural research question:

Supported by EPSRC (EP/G043434/1), ERC (267959) and ANR project AGAPE.

© Springer International Publishing Switzerland 2016
V. Mäkinen et al. (Eds.): IWOCA 2016, LNCS 9843, pp. 361–372, 2016.
DOI: 10.1007/978-3-319-44543-4_28

Is it possible to test in polynomial time whether a given graph has a square root that belongs to some specified graph class $\mathcal{H}$?

It has been shown that such a polynomial-time algorithm exists if $\mathcal{H}$ is the class of trees [18], proper interval graphs [13], bipartite graphs [12], block graphs [16], strongly chordal split graphs [17], graphs with girth at least g for any fixed $g \geq 6$ [9], ptolemaic graphs [14], 3-sun-free split graphs [14] (see [15] for an extension of the latter result to other subclasses of split graphs). In contrast, NP-completeness of this problem has been shown if $\mathcal{H}$ is the class of split graphs [13], chordal graphs [13], graphs of girth at least 4 [9] or graphs of girth at least 5 [8].

Our Result. We consider the class of all graphs being a cactus as $\mathcal{H}$. A connected graph is a *cactus* if every edge of it is contained in at most one cycle. We give an $O(n^4)$-time algorithm that tests whether an n-vertex graph has a cactus root. Our result is motivated by the nontrivial question whether squares of planar graphs can be recognized in polynomial time. The known result that squares of trees, which form a subclass of the class of cactuses, can be recognized in polynomial time [18] can be seen as a first step in solving this problem. As every cactus is planar, our result could be seen as a second step in solving it. On a side note, cactuses are not a subclass of any of the other aforementioned classes of which the squares can be recognized in polynomial time.

We prove our result by analyzing, in Sect. 3, the structure of squares of cactuses. In this way we are able to recognize vertices of the input graph G that are cut-vertices in any cactus root (if such a square root exists) together with a set of compulsory edges and a set of forbidden edges of any cactus root of G. In this way we can reduce, in Sect. 4, the graph G to a number of smaller instances such that G has a cactus root if and only if each of these smaller instances has a cactus root. Showing that each of the smaller instances has bounded treewidth and observing that we can solve the problem in linear time on any graph class of bounded treewidth completes the proof.

We observe that in several variants of the SQUARE ROOT problem where the aim is to find some type of sparse square root [1,8,9,18], such a square root is unique or unique up to isomorphism. This uniqueness can be exploited and as such is very helpful for finding the square root. However, this is not the case for cactus roots: Fig. 1 shows a graph that has two non-isomorphic cactus roots.

In Sect. 5 we discuss some directions of future work.

2   Preliminaries

We consider only finite undirected graphs without loops and multiple edges. We refer to the textbook of Diestel [7] for any undefined graph terminology.

Basic Graph Terminology. We denote the vertex set of a graph G by V_G and the edge set by E_G. The subgraph of G induced by a subset $U \subseteq V_G$ is denoted by $G[U]$. The graph $G - U$ is the graph obtained from G after removing the vertices of U. If $U = \{u\}$, we also write $G - u$. Similarly, we denote the graph

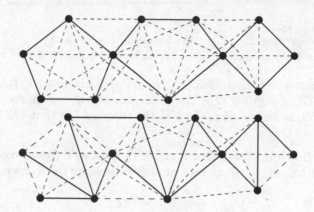

Fig. 1. A graph with non-isomorphic square cactus roots. The edges of the cactus roots are shown by solid lines, whereas the other edges are shown by dashed lines.

obtained from G after deleting a set of edges S (an edge e) by $G - S$ ($G - e$ respectively).

Let G be a graph. A *connected component* of G is a maximal connected subgraph. The *distance* $\mathrm{dist}_G(u, v)$ between a pair of vertices u and v of G is the number of edges of a shortest path between them. The diameter $\mathrm{diam}(G)$ of G is the maximum distance between two vertices of G. The *open neighborhood* of a vertex $u \in V_G$ is defined as $N_G(u) = \{v \mid uv \in E_G\}$, and its *closed neighborhood* is defined as $N_G[u] = N_G(u) \cup \{u\}$. Two (adjacent) vertices u, v are said to be *true twins* if $N_G[u] = N_G[v]$. A vertex v is *simplicial* if $N_G[v]$ is a clique, that is, if there is an edge between any two vertices of $N_G[v]$. The *degree* of a vertex $u \in V_G$ is defined as $d_G(u) = |N_G(u)|$. The maximum degree of G is $\Delta(G) = \max\{d_G(v) \mid v \in V_G\}$. A vertex of degree 1 is said to be a *pendant* vertex. If v is a pendant vertex, then we say that the unique edge incident to u is a *pendant* edge.

A vertex u is a *cut vertex* of a connected graph G with at least two vertices if $G - u$ is disconnected. An inclusion-maximal induced subgraph of G that has no cut vertex is called a *block*. Recall that a connected graph G is a cactus if each edge of G is contained in at most one cycle. This implies the following well-known property.

Observation 1. *Each block of a cactus with at least two vertices is either a K_2 (an edge) or a cycle.*

A *tree decomposition* of a graph G is a pair (T, X) where T is a tree and $X = \{X_i \mid i \in V_T\}$ is a collection of subsets (called *bags*) of V_G such that the following three conditions hold:

(i) $\bigcup_{i \in V_T} X_i = V_G$,

(ii) for each edge $xy \in E_G$, $x, y \in X_i$ for some $i \in V_T$, and

(iii) for each $x \in V_G$ the set $\{i \mid x \in X_i\}$ induces a connected subtree of T.

The *width* of a tree decomposition $(\{X_i \mid i \in V_T\}, T)$ is $\max_{i \in V_T} \{|X_i| - 1\}$. The *treewidth* $\mathbf{tw}(G)$ of a graph G is the minimum width over all tree decompositions of G. If T is restricted to be a path, then we say that (X, T) is a *path decomposition* of G.

Problem Definition. Recall that a graph H is called a cactus root of a graph G if H is a cactus and a square root of G. We consider the following problem:

CACTUS ROOT

Input: a graph G.

Question: is there a cactus H with $H^2 = G$?

We also need to define the following more general variant introduced in [3] for general square roots:

CACTUS ROOT WITH LABELS

Input: a graph G and sets of edges $R, B \subseteq E_G$.

Question: is there a cactus H with $H^2 = G$, $R \subseteq E_H$ and $B \cap E_H = \emptyset$?

By choosing $R = B = \emptyset$ we see that CACTUS ROOT is indeed a special case of CACTUS ROOT WITH LABELS.

3   A Number of Structural Observations and Lemmas

In this section we state three observations and prove seven lemmas. We will use these results, which are all structural, for the design of our $O(n^4)$ time algorithm for CACTUS ROOT presented in Sect. 4.

The first observation is known and easily follows from the definition of the treewidth.

Observation 2. *For a cactus G, $\mathbf{tw}(G) \leq 2$.*

The second observation gives an upper bound for the treewidth of the square of a graph; it follows from the well-known fact that we can transform every tree decomposition (T, X) of a graph G into a tree decomposition of G^2 by adding, to each bag X_i of T, all the neighbors of every vertex from X_i.

Observation 3. *For a graph G, $\mathbf{tw}(G^2) \leq (\mathbf{tw}(G) + 1)\Delta(G) - 1$.*

Let H be a square root of a graph G. We say that H is a *minimal* square root of G if $H^2 = G$ but any proper subgraph of H is not a square root of G. Note that the two cactus roots displayed in Fig. 1 are both minimal. Since any connected subgraph of a cactus is a cactus, we can make the following observation.

Observation 4. *If a graph G has a cactus root, then G has a minimal cactus root.*

A block of a graph G is called a *leaf block* if it contains at most one cut vertex of G. This leads to our first lemma.

Lemma 1. *If a cactus H is a minimal square root of a graph G, then H has no leaf block that is a triangle.*

Proof. Suppose that a cactus H is a minimal square root of G such that a triangle with vertices x, y, z is a leaf block of H. As a leaf block contains at most one cut vertex of H by definition, we may assume that y and z are not cut vertices of H. Let $H' = H - yz$. It is straightforward to verify that $H'^2 = G$, contradicting the minimality of H.   □

Suppose that u and v are pendant vertices of a square root H of G and that u and v are adjacent to the same vertex of $H - \{u, v\}$. Then, in G, u and v are simplicial vertices and true twins. We use this observation in the proof of the following lemma.

Lemma 2. *Let H be a minimal cactus root of a graph G. If G contains at least six simplicial vertices that are pairwise true twins, then at least one of these vertices is a pendant vertex of H.*

Proof. Let H be a minimal cactus root of a graph G that contains a set X of six simplicial vertices that are pairwise true twins. The vertices of X cannot all belong to the same block of H, because such a block would be a cycle with at least six vertices (by Observation 1) and any two vertices of this block could not be true twins of G. Hence, there is a cut vertex u of H such that there exist two vertices $x, y \in X$ that are in distinct connected components of $H - u$. Let H' be a connected component of $H - u$ that contains x. If x is not a pendant vertex of H then, by the minimality of H and Lemma 1, there exists a vertex $z \in V_{H'}$ that is adjacent to x and that is at distance 2 from u in H. Then, as every path from y to z in H contains u, we find that $yz \notin E_G$. This is a contradiction since x and y are true twins of G and $xz \in E_G$. We conclude that x is a pendant vertex of H.   □

The following definition plays a crucial role in our paper.

Definition 1. Let u be a cut vertex of a connected graph H. We say that

(i) u is *important* if $H - u$ has three vertices that belong to three distinct connected components of $H - u$ and that are each at distance at least 2 from u in H;

(ii) u is *essential* if $H - u$ has two vertices that belong to two distinct connected components of $H - u$ and that are both at distance at least 2 from u in H.

Definition 1(i) immediately implies the following lemma.

Lemma 3. *If u is an important cut vertex of a cactus root H of a graph G, then there are three vertices $x, y, z \in N_G(u)$ such that x, y and z are at distance at least 3 from each other in $G - u$.*

Although we have no implication in the opposite direction, we can show the following (which explains why we need the second and weaker part of Definition 1).

Lemma 4. *Let G be a graph with a cactus root H. If $u \in V_G$ has three neighbors x, y and z in G that are at distance at least 3 from each other in $G - u$, then u is an essential cut vertex of H. Moreover, at least two vertices of $\{x, y, z\}$ belong to distinct connected components of $H - u$.*

Proof. Assume that G has a cactus root H. Let $u \in V_G$ be such that u has three neighbors x, y and z in G that are at distance at least 3 from each other in $G - u$. Notice that because x, y and z are at distance at least 3 from each other in $G - u$, these vertices are all at distance 2 from u in H.

For contradiction, assume that u is not a cut vertex of H. Then u has at most two adjacent vertices in H, since H is a cactus (see Observation 1). Then at least two vertices of $\{x, y, z\}$ are adjacent to the same vertex of H (which is one of the two neighbors of u) implying that these two vertices of $\{x, y, z\}$ are adjacent in G and thus in $G - u$; a contradiction. Hence u is a cut vertex of H.

Now suppose that x, y and z are all in the same connected component H' of $H - u$. Since H is a cactus, we find, by Observation 1, that H' contains at most two vertices that are adjacent to u in H. Again, we obtain that at least two vertices of $\{x, y, z\}$ are adjacent to the same vertex of H; a contradiction. Hence, at least two vertices of $\{x, y, z\}$ belong to distinct connected components of $H - u$. Since x, y and z are at distance 2 from u in H, this implies that u is an essential cut vertex of H. $\qquad\square$

We now show that we can recognize edges of a cactus root that are incident to an essential cut vertex.

Lemma 5. *Let u be an essential cut vertex of a cactus root H of a graph G. Then for every $x \in N_G(u)$, it holds that $ux \notin E_H$ if and only if there exists a vertex $y \in N_G(u)$ such that x and y are at distance at least 3 in $G - u$.*

Proof. Let u be an essential cut vertex of a cactus root H of a graph G. Let $x \in N_G(u)$. First suppose that $ux \in E_H$. Let $y \in N_G(u)$. If $uy \in E_H$, then $xy \in E_G$. If $uy \notin E_H$, then there exist a vertex $z \in V_H$ and edges $uz, zy \in E_H$, as $y \in N_G(u)$. As $zy \in E_H$, we find that $zy \in E_G$. As $ux, uz \in E_H$, we also deduce that $xz \in E_G$. In both cases x and y are at distance at most 2 in $G - u$.

Now suppose that $ux \notin E_H$. Then, as $x \in N_G(u)$, we find that x is at distance 2 from u in H. Let H' be the connected component of $H - u$ containing x. Since u is an essential cut vertex of H, $H - u$ has another connected component H'' containing a vertex y at distance 2 from u in H. It remains to observe that $y \in N_G(u)$ and x and y are at distance 3 in $G - u$. $\qquad\square$

The next lemma is used to recognize vertices adjacent to an essential cut vertex that belong to the same block of a minimal cactus root.

Lemma 6. *Let H be a minimal cactus root of a graph G. For any $u \in V_H$, two distinct vertices $x, y \in N_H(u)$ are in the same block of H if and only if x and y are in the same connected component of $G' = G - E_{G[N_H(u)]} - u$.*

Proof. Let $x, y \in N_H(u)$. First suppose that x and y are in distinct blocks of H. Then x and y are readily seen to be in distinct connected components of G'. Now suppose that x and y are in the same block C of H. If $xy \in E_G$ then x and y are in the same connected component of G'. Suppose $xy \notin E_G$. Then C is a cycle by Observation 1. If C is not a triangle, then C has a unique (x, y)-path in H (avoiding u) of length at least 2. This path is an (x, y)-path in G' as well. Hence x and y are in the same connected component of G'. Suppose that C is a triangle. Then $xy \in E_H$. As H is a minimal cactus root, x or y has at least one neighbor $z \neq u$ in H due to Lemma 1. Assume without loss of generality that z is a neighbor of x. Then the edges $xy, xz \in E_H$ imply that $zy \in E_G$. We establish that xzy is an (x, y)-path in G', that is, also in this case x and y are in the same connected component of G'.   □

Finally we show how to determine which neighbors in G of an essential cut vertex u of a cactus root H are in the same connected component of $H - u$.

Lemma 7. *Let H be a minimal cactus root of a graph G. For any $u \in V_H$ and $x \in N_H(u)$, a vertex $y \in N_G(u)$ is in the same connected component of $H - u$ as x if and only if either $uy \in E_H$ and y in the same block of H as x, or $uy \notin E_H$ and there is a vertex $z \in N_H(u)$, such that z is in the same block of H as x and $yz \in E_G$.*

Proof. Let $y \in N_G(u)$. First suppose y is in the same connected component of $H - u$ as x. If $uy \in E_H$, then y is in the same block of H as x. Suppose $uy \notin E_H$. As $uy \in E_G$, there is a vertex $z \in N_H(u)$ such that $zy \in E_H$. Then z is in the same block of H as x, as x and y are in the same connected component of $H - u$.

To prove the reverse implication, if $uy \in E_H$ and x, y are in the same block of H, then x and y are in the same connected component of $H - u$. Suppose that $uy \notin E_H$ and there is a vertex $z \in N_H(u)$ such that z is in the same block of H as x and $yz \in E_G$. If $yz \in E_H$, then y and z are in the same connected component of $H - u$. If $yz \notin E_H$, then there is a $v \in V_G$ such that $yv, vz \in E_H$. Since $uy \notin E_H$, we obtain $v \neq u$. Therefore, y and z are in the same connected component of $H - u$. Because y and z are in the same connected component of $H - u$ and x, y are in the same block of H, we obtain that x, y are in the same connected component of $H - u$.   □

4   The Algorithm

In this section we use the structural results from the previous section to obtain a polynomial-time algorithm for CACTUS ROOT. The main idea is to reduce a given instance of CACTUS ROOT to a set of smaller instances of CACTUS ROOT WITH LABELS, each having bounded treewidth. We therefore need the following two lemmas which show, together with Observations 2 and 3, that we are done if we manage to achieve this goal. The first lemma is due to Bodlaender.

Lemma 8. ([2]) *For any fixed constant k, it is possible to decide in linear time whether the treewidth of a graph is at most k.*

Lemma 9. CACTUS ROOT WITH LABELS *can be solved in time $f(t) \cdot n$ for n-vertex graphs of treewidth at most t.*

Proof. It is not difficult to construct a dynamic programming algorithm for the problem, but for simplicity we give a non-constructive proof based on Courcelle's [5] theorem. By this theorem, it suffices to show that the existence of a cactus root can be expressed in monadic second-order logic.

Let (G, R, B) be an instance of CACTUS ROOT WITH LABELS. We observe that the existence of a cactus H such that $G = H^2$, $R \subseteq E_H$ and $B \cap E_H = \emptyset$ is equivalent to the existence of a subset $X \subseteq E_G$ such that the following four properties hold:

(i) $R \subseteq X$ and $B \cap X = \emptyset$;
(ii) for every $uv \in E_G$, $uv \in X$ or there exists a vertex w such that $uw, wv \in X$;
(iii) for every two distinct edges $uw, vw \in X$, $uv \in E_G$;
(iv) for every $uv \in X$ and for every two (u, v)-paths P_1 and P_2 in G such that $E_{P_1}, E_{P_2} \subseteq X \setminus \{uv\}$, it holds that $P_1 = P_2$.

Each of these properties can be expressed in monadic second-order logic. In particular, with respect to property (iv), expressing that a subgraph P of G is a (u, v)-path in G can be done in monadic second-order logic in a standard way (see, for example, [6]). Hence the lemma follows.   □

Now we are ready to prove the main result.

Theorem 1. CACTUS ROOT *can be solved in time $O(n^4)$ for n-vertex graphs.*

Proof. We first give an overview of our algorithm. As we can consider each connected component separately, we may assume without loss of generality that the input graph G is connected. First, we use Lemma 2 to recognize sets of pendant vertices in a (potential) cactus root adjacent to the same vertex that have size at least 7. For each of these sets, we show that it is safe to delete some vertices without changing the answer for the considered instance. After performing this step, we obtain a graph G' such that in any cactus root of G' each vertex is adjacent to at most six pendants. Further, we use Lemmas 3 and 4 to construct a set U of essential cut vertices in a (potential) cactus root such that U contains all important cut vertices. Next, we apply Lemma 5 to recognize which edges incident to the vertices of U are in any cactus root and which edges are not included in any cactus root. We label them *red* and *blue* respectively and obtain an instance of CACTUS ROOT WITH LABELS. Now we can use Lemmas 6 and 7 to determine for each $u \in U$, the partition of the set of vertices of $G - u$ into the sets of vertices of the connected components of $H - u$, where H is a cactus root of G'. This allows us to split G' via the vertices of U as shown in Fig. 2. Due to the presence of labeled edges incident to the vertices of U, we obtain an equivalent instance. Finally, we observe that the obtained graph has

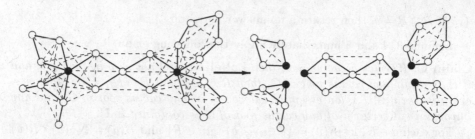

Fig. 2. Splitting of a graph; the vertices of U are black, the edges of a square root are shown by solid lines and the other edges are shown by dashed lines.

bounded treewidth using Observations 2 and 3, so we can use Lemmas 8 and 9 to solve the problem, as we pointed out already.

Now we formally explain the details of our algorithm. Let G be a connected graph. First, we preprocess G using Lemma 2 to reduce the number of pendant vertices adjacent to the same vertex in a (potential) cactus root of G. To do so, we exhaustively apply the following rule.

Pendants Reduction. If G has a set X of simplicial true twins of size at least 7, then delete an arbitrary $u \in X$ from G.

The following claim shows that this rule is safe.

Claim A. *If $G' = G - u$ is obtained from G by the application of* **Pendant reduction**, *then G has a cactus root if and only if G' has a cactus root.*

We prove Claim A as follows. Suppose that H is a minimal cactus root of G. By Lemma 2, H has a pendant vertex $u \in X$. It is easy to verify that $H' = H - u$ is a cactus root of G'. Assume now that H' is a minimal cactus root of G'. By Lemma 2, H has a pendant vertex $w \in X \setminus \{u\}$, since the vertices of $X \setminus \{u\}$ are simplicial true twins of G' and $|X \setminus \{u\}| \geq 6$. Let v be the unique neighbor of w in H'. We construct H from H' by adding u and making it adjacent to v. It is readily seen that H is a cactus root of G. This completes the proof of Claim A.

For simplicity, we call the graph obtained by exhaustive application of the pendants rule G again. The following property is important for us.

Claim B. *Every cactus root of G has at most six pendant vertices adjacent to the same vertex.*

Now we construct an instance of CACTUS ROOT WITH LABELS together with a set U of cut vertices of a (potential) cactus root.

Labeling. Set $U = \emptyset$, $R = \emptyset$ and $B = \emptyset$. For each $u \in V_G$ such that there are three distinct vertices $x, y, z \in N_G(u)$ that are at distance at least 3 from each other in $G - u$ do the following:

(i) set $U = U \cup \{u\}$,
(ii) set $B' = \{uv \in E_G \mid \exists w \in N_G(u) \text{ s.t. } \text{dist}_{G-u}(v, w) \geq 3\}$,
(iii) set $R' = \{uv \mid v \in N_G(u)\} \setminus B'$,
(iv) set $R = R \cup R'$ and $B = B \cup B'$,

(v) if $R \cap B \neq \emptyset$, then return a no-answer and stop.

Lemmas 3, 4 and 5 immediately imply the following claim.

Claim C. *If G has a cactus root, then* **Labeling** *does not stop in Step (v), and if H is a minimal cactus root of G, then $R \subseteq E_H$ and $B \cap E_H = \emptyset$. Moreover, every vertex $u \in U$ is an essential cut vertex of any cactus root of G, and any important cut vertex u of any cactus root of G is contained in U.*

For each $u \in U$, let $R(u) = \{v \in N_G(u) \mid uv \in R\}$ and $B(u) = N_G(u) \setminus R(u)$ and construct a partition $P(u) = \{S_1, S_2, \ldots, S_{k(u)}\}$ of $N_G(u)$ as follows.

Partition. For each $u \in U$,

(i) put $x, y \in R(u)$ in the same set of $P(u)$ if and only if x and y are in the same connected component of $G' = G - E_{G[R(u)]} - u$,
(ii) for each $x \in R(u)$, put $y \in B(y)$ in the same set with x if $xy \in E_G$,
(iii) if at least one of the following holds, then return a no-answer and stop:
  - $P(u)$ is not a partition of $N_G(u)$,
  - there is a set of $P(u)$ with at least three vertices of $R(u)$,
  - there is a vertex of $B(u)$ that is not in a set of $P(u)$ with a vertex of $R(u)$,
  - there are distinct $S, S' \in P(u)$ such that for some $x \in S$ and $y \in S'$, $xy \in R$,
  - there are distinct $S, S' \in P(u)$ such that for some $x \in S$ and $y \in S'$, $xy \in E_G$ but $ux \notin R$ or $uy \notin R$,
  - there are distinct $S, S' \in P(u)$ such that for some $x \in S$ and $y \in S'$, $xy \notin E_G$ but $ux \in R$ and $uy \in R$,
  - the graph $G - E_{G[R(u)]} - u$ has a path connecting vertices of distinct sets of $P(u)$.

By Lemmas 6, 7 and Claim C, we have the following.

Claim D. *If G has a cactus root, then* **Partition** *does not stop in Step (iii), and if H is a minimal cactus root of G, then*

(i) $R \subseteq E_H$ and $B \cap E_H = \emptyset$,
(ii) *every important cut vertex u of H is in U,*
(iii) *for any $u \in U$, $x, y \in N_G(u)$ are in the same connected component of $H - u$ if and only if x and y are in the same set of $P(u)$.*

Now we split the instance (G, R, B) of Cactus Root with Labels into several instances of the problem.

Splitting. For each $u \in U$, let $P(u) = \{S_1, \ldots, S_k\}$ and do the following:

(i) delete u and introduce k new vertices $u_1, \ldots, u_k$,
(ii) for each $i \in \{1, \ldots, k\}$, make u_i adjacent to all vertices of S_i,
(iii) for each $i \in \{1, \ldots, k\}$ and $v \in S_i$, if $uv \in R$, then replace uv by $u_i v$ in R, and if $uv \in B$, then replace uv by $u_i v$ in B,
(iv) for each $i, j \in \{1, \ldots, k\}$, $i \neq j$, delete the edges xy with $x \in S_i$ and $y \in S_j$,
(v) for each $i \in \{1, \ldots, k\}$ and $v \in S_i$, update $P(v)$ by replacing v by v_i in the sets and deleting the vertices of $N_G(u) \setminus S_i$ from the sets.

Let $G_1, \ldots, G_r$ be the connected components of the obtained graph. For $i \in \{1, \ldots, r\}$, let $R_i = R \cap E_{G_i}$ and $B_i = B \cap E_{G_i}$. By Claims B and D, we establish the following crucial claim.

Claim E. *The input graph G has a cactus root if and only if (G_i, R_i, B_i) is a yes-instance of* CACTUS ROOT WITH LABELS *for each $i \in \{1, \ldots, r\}$. Moreover, if (G_i, R_i, B_i) is a yes-instance, then G_i has a cactus root H with $R_i \subseteq E_H$ and $B_i \cap E_H = \emptyset$ such that every cut vertex of H belongs to at most eight blocks and to at most two blocks not being a K_2.*

By Claim E, if G has a cactus root, then $\Delta(G_i) \leq 10$ for $i \in \{1, \ldots, k\}$. By Observations 2 and 3, we obtain that $\mathbf{tw}(G_i) \leq 29$ in this case. We use Lemma 8 to check whether this holds for each $i \in \{1, \ldots, r\}$. If the algorithm reports that $\mathbf{tw}(G_i) \geq 30$ for some $i \in \{1, \ldots, r\}$, then we return a no-answer and stop. Otherwise, we solve CACTUS ROOT WITH LABELS for each instance (G_i, R_i, B_i) using Lemma 9 for $i \in \{1, \ldots, r\}$.

It remains to evaluate the running time of our algorithm. We can find all simplicial vertices and sort them into the equivalence classes with the true twin relation in time $O(n^3)$. This implies that the exhaustive application of the **Pendant reduction** rule can be done in time $O(n^3)$. For each vertex $u \in V_G$, we can compute the distances between the vertices of $G - u$ in time $O(n^3)$. Hence, the **Labeling** step can be done in time $O(n^4)$. For each $u \in U$ the sets $R(u)$ and $B(u)$ can be constructed in time $O(n^2)$. For each $u \in U$, we can construct $G' = G - E_{G[R(u)]}$ and find the connected components of G' in time $O(n^2)$. It follows, that the **Partition** step can be done in time $O(n^3)$. The **Splitting** step takes $O(n^3)$ time. The algorithm in Lemma 8 runs in $O(n)$ time. We conclude that the total running time is $O(n^4)$. □

5 Conclusions

We proved that the problem of testing whether a graph has a cactus root is $O(n^4)$-time solvable. In fact, our algorithm can be modified to find a cactus root in the same time (if it exists).

We recall that every cactus is planar and that the problem of settling the complexity of recognizing squares of planar graphs is open. We also recall that a cactus is a connected graph, in which each block is either a cycle or an edge. This leads to the following (known) generalization: a *cactus block graph* is a connected graph, in which each block is a cycle or a complete graph. Can we decide in polynomial time whether a given graph has a square root that is a cactus block graph? In order to answer this question, we need new arguments as our current proof for cactus roots does not carry over.

References

1. Adamaszek, A., Adamaszek, M.: Uniqueness of graph square roots of girth six. Electron. J. Comb. **18**, #P139 (2011)
2. Bodlaender, H.L.: A linear-time algorithm for finding tree-decompositions of small treewidth. SIAM J. Comput. **25**, 305–1317 (1996)
3. Cochefert, M., Couturier, J.-F., Golovach, P.A., Kratsch, D., Paulusma, D.: Sparse square roots. In: Brandstädt, A., Jansen, K., Reischuk, R. (eds.) WG 2013. LNCS, vol. 8165, pp. 177–188. Springer, Heidelberg (2013)
4. Cochefert, M., Couturier, J., Golovach, P.A., Kratsch, D., Paulusma, D.: Parameterized algorithms for finding square roots. Algorithmica **74**, 602–629 (2016)
5. Courcelle, B.: The monadic second-order logic of graphs III: tree-decompositions, minor and complexity issues. ITA **26**, 257–286 (1992)
6. Cygan, M., Fomin, F.V., Kowalik, L., Lokshtanov, D., Marx, D., Pilipczuk, M., Pilipczuk, M., Saurabh, S.: Parameterized Algorithms. Springer, Switzerland (2015)
7. Diestel, R.: Graph Theory, 4th edn. Graduate Texts in Mathematics, vol. 173. Springer, Heidelberg (2012)
8. Farzad, B., Karimi, M.: Square-root finding problem in graphs, a complete dichotomy theorem. CoRR, abs/1210.7684 (2012)
9. Farzad, B., Lau, L.C., Le, V.B., Tuy, N.N.: Complexity of finding graph roots with girth conditions. Algorithmica **62**, 38–53 (2012)
10. Golovach, P.A., Kratsch, D., Paulusma, D., Stewart, A.: A linear kernel for finding square roots of almost planar graphs. In: Proceedings of SWAT 2016, Leibniz International Proceedings in Informatics (2016, to appear)
11. Golovach, P.A., Kratsch, D., Paulusma, D., Stewart, A.: Squares of low clique number. In: Proceedings of CTW 2016, Electronic Notes in Discrete Mathematics (2016, to appear)
12. Lau, L.C.: Bipartite roots of graphs. ACM Trans. Algorithms **2**, 178–208 (2006)
13. Lau, L.C., Corneil, D.G.: Recognizing powers of proper interval, split, and chordal graphs. SIAM J. Discrete Math. **18**, 83–102 (2004)
14. Le, V.B., Oversberg, A., Schaudt, O.: Polynomial time recognition of squares of ptolemaic graphs and 3-sun-free split graphs. Theor. Comput. Sci. **602**, 39–49 (2015)
15. Le, V.B., Oversberg, A., Schaudt, O.: A unified approach for recognizing squares of split graphs. Manuscript (2015)
16. Le, V.B., Tuy, N.N.: The square of a block graph. Discrete Math. **310**, 734–741 (2010)
17. Le, V.B., Tuy, N.N.: A good characterization of squares of strongly chordal split graphs. Inf. Process. Lett. **111**, 120–123 (2011)
18. Lin, Y., Skiena, S.: Algorithms for square roots of graphs. SIAM J. Discrete Math. **8**, 99–118 (1995)
19. Milanic, M., Oversberg, A., Schaudt, O.: A characterization of line graphs that are squares of graphs. Discrete Appl. Math. **173**, 83–91 (2014)
20. Milanic, M., Schaudt, O.: Computing square roots of trivially perfect and threshold graphs. Discrete Appl. Math. **161**, 1538–1545 (2013)
21. Motwani, R., Sudan, M.: Computing roots of graphs is hard. Discrete Appl. Math. **54**, 81–88 (1994)
22. Mukhopadhyay, A.: The square root of a graph. J. Comb. Theor. **2**, 290–295 (1967)
23. Nestoridis, N.V., Thilikos, D.M.: Square roots of minor closed graph classes. Discrete Appl. Math. **168**, 34–39 (2014)
24. Ross, I.C., Harary, F.: The square of a tree. Bell Syst. Tech. J. **39**, 641–647 (1960)

Computing Giant Graph Diameters

Peter Damaschke[⊠]

Department of Computer Science and Engineering,
Chalmers University, 41296 Göteborg, Sweden
ptr@chalmers.se

Abstract. This paper is devoted to the fast and exact diameter computation in graphs with n vertices and m edges, if the diameter is a large fraction of n. We give an optimal $O(m+n)$ time algorithm for diameters above $n/2$. The problem changes its structure at diameter value $n/2$, as large cycles may be present. We propose a randomized $O(m + n \log n)$ time algorithm for diameters above $(1/3 + \epsilon)n$ for constant $\epsilon > 0$.

1 Introduction

Computing distances and shortest paths is one of the fundamental graph problems. The diameter of an undirected graph is the maximum distance between any two vertices. In a graph with n vertices and m edges of unit length, all distances from a single vertex (single-source shortest paths, SSSP) can be computed by breadth-first-search (BFS) in $O(n+m)$ time, and all pairwise distances (all-pairs shortest paths, APSP) can therefore be obtained in $O(nm)$ time by solving n times SSSP. Trivially, this also yields the diameter, but it was a longstanding open problem whether the diameter can be computed significantly faster than via APSP, see [1,14] for results. Many results are also known for diameter computation in special graph classes [5,6,8,13] and fast approximation of the diameter [2–4,11,13]. This bibliography is certainly far from being complete. Other related lines of research that we cannot survey here include faster APSP computation in special graph classes, and experimental studies of diameter computations in real-world graphs.

Instead of the graph structure one may also restrict the range of diameters. As discussed in [5], the problem of distinguishing between graphs of diameter 2 and 3, already for the special class of split graphs, is as hard as the disjoint sets problem (deciding whether a given set family contains two disjoint sets) and is therefore unlikely to have a subquadratic algorithm. In the present paper we look at the other end: graphs with "giant" diameters close to the number n of vertices. (The word is borrowed from the giant components of random graphs.) Whereas most real-world networks have small diameters, chain-like structures may appear as well in various contexts (chain molecules, connections between two fixed sites in a network, etc.).

Contributions. First we give an $O(m + n)$ time algorithm for diameters above $n/2$. One can think of different approaches, e.g., similar to diameter computation

© Springer International Publishing Switzerland 2016
V. Mäkinen et al. (Eds.): IWOCA 2016, LNCS 9843, pp. 373–384, 2016.
DOI: 10.1007/978-3-319-44543-4_29

in trees. Our approach is based on separators (articulation points in this case), and removal of irrelevant subgraphs. Moreover, it is not necessary to know in advance that the given graph has a large diameter. Admittedly we use quite a number of lemmas to prepare this result, but we want to point out all single steps, in the hope that future research can generalize them to larger separators and smaller diameters. We also show that the $O(n + m)$ time bound cannot be improved (say, to $O(n)$ time) under plausible assumptions on the graph representation. While our solution for diameters larger than $n/2$ works with articulation points, we observe some "phase transition" just below $n/2$: A graph with such a diameter may have a giant geodesic cycle, hence qualitatively different methods are needed to "choose the correct half cycle" that yields the diameter. For this purpose we define an auxiliary problem that might be of independent interest. Its solution is applied in a randomized $O(m + n \log n)$ time algorithm for diameters above $n/3$. Note that this bound is linear in the graph size if $m > n \log n$.

2   Preliminaries

Our graphs $G = (V, E)$ are undirected, unweighted, and connected, and have n vertices and m edges. A path joining two vertices u and v is denoted $u - v$, if the inner vertices are clear from context or irrelevant. The distance $d_G(u, v)$, or simply $d(u, v)$, is the length, i.e., number of edges, of a shortest $u - v$ path. A shortcut to a subgraph H of G is a path in G that connects two vertices u and v from H, but is shorter than $d_H(u, v)$. We call H a geodesic subgraph if H has no shortcuts in G. In particular, a geodesic path P is a shortest path connecting its end vertices. The diameter of G is $diam(G) := \max\{d(u, v)|\, u, v \in V\}$. Hence a longest geodesic path in G is a path of length $diam(G)$. We use the abbreviation $\delta := diam(G)/n$. Note that a cycle C is geodesic if, for any two vertices $u, v \in C$, their smaller distance (of at most $\frac{1}{2}|C|$) on C equals $d(u, v)$.

With respect to a root vertex r we refer to the sets $N_i(r) := \{v|\, d(u, v) = i\}$ as layers, and the depth of G is defined by $\max\{d(r, v)|\, v \in V\}$. Depth and layers can be computed using breadth-first-search (BFS).

To avoid heavy notation and technicalities we may neglect additive constants in arithmetic expressions, as well as rounding of fractional numbers to integers, as long as this does not affect asymptotic statements for large graphs.

For $U \subset V$ we denote by $G - U$ the graph that remains when the vertices of U and all incident edges are removed from G. If $U = \{u\}$, we write $G - u$ for $G - U$. A separator is a vertex set $S \subset V$ such that $G - S$ is disconnected. An articulation point is just a separator of size 1, that is, a vertex u such that $G - u$ is disconnected. A block is a biconnected graph, that is, a graph without articulation points. The block-cut tree of G has a node for every articulation point, and a node for every block (biconnected component) without the articulation points therein. The block-cut tree has edges between adjacent articulation points, and between those articulation points and blocks where G has edges.

A hair in a graph is a path H such that one end vertex of H has degree 1, all inner vertices have degree 2, and the other end vertex has degree larger than 2.

We can think of a hair as a simple path that is dangling at the rest of the graph. In particular, a hair is a geodesic path.

We say that a vertex v is between vertices u and w, in symbols $B(u, v, w)$, if the triangle inequality degenerates to the equation $d(u, w) = d(u, v) + d(v, w)$.

We tacitly use some elementary properties listed here: Any subpath of a geodesic path is geodesic. If we replace any subpath of a geodesic path with another geodesic (sub)path between the same two vertices, then the entire path remains geodesic. Any three vertices u, v, w that appear in this order on a geodesic path satisfy $B(u, v, w)$. Conversely, if $B(u, v, w)$ holds true, then any concatenation of two geodesic $u - v$ and $v - w$ paths is a geodesic $u - w$ path.

3 Diameters Larger than Half the Size

First we study the largest diameters, more precisely, the case $\delta > 1/2$. We show that this case can be solved in linear time. Our approach works with articulation points, and (in Lemma 3) pruning of irrelevant vertices. Lemma 2 below also holds for general graphs.

Lemma 1. *Suppose that $\delta = 1/2 + h$, and let P be any longest geodesic path. Then there exists a vertex $u \in P$ which is an articulation point of G and divides P in two subpaths of length at least hn each.*

Proof. Let v be an end vertex of P. Clearly, the number of vertices not in P is $(1/2 - h)n$. Hence at least $2hn$ of the layers $N_i(v)$ contain a single vertex. Since edges cannot skip layers, every such single vertex u (except for $i = 0$ and possibly the last layer) is an articulation point of G and an inner vertex of P.

Specifically, consider an articulation point u that belongs to P and is as close as possible to the center of P. In the worst case, only $2hn$ articulation points are on P, and they form two subpaths of equal lengths at the ends of P. Still, an innermost articulation point u divides P in two paths the shorter of which has length at least hn. □

Lemma 2. *Consider an articulation point u of G, a connected component C of $G - u$, and a longest geodesic path P in G. Define $C_u := C \cup \{u\}$. Then one of these three cases applies: (a) P does not intersect C. (b) P intersects both C and $G - C_u$. (c) P is entirely in C_u.*

In case (b), the subpath P_u of P in C_u is a geodesic path connecting u with some vertex of C at maximum distance from u. Moreover, any such geodesic path in C_u may replace P_u in P, and the resulting path is again a longest geodesic path in G. Case (c) can be true only if C_u has at least δn vertices.

Proof. The case distinction is evident, as well as the assertion about case (c). The assertion about case (b) follows from two facts: P has maximum length, and no edges join any vertices of C and $G - C_u$. Hence the new subpath cannot lead to shortcuts to vertices outside C. □

Lemma 3. *Consider an articulation point u of G, a connected component C of $G - u$, and a longest geodesic path P in G. Suppose that P is not entirely in C (for instance, because C has fewer than δn vertices). Then it is safe to keep only one geodesic path from u to a farthest vertex v (with maximum $d(u, v)$) in C and remove all other vertices of C. That is, this removal retains some longest geodesic path in G.*

Proof. By assumption, case (c) of Lemma 2 does not apply to C. If case (a) applies, then the assertion is vacuously true. If case (b) applies, then the assertion follows from the property mentioned in Lemma 2: Since any geodesic path from u to a farthest vertex v can be used, we need to keep only one.   □

As a consequence of the previous lemmas we can already settle one case:

Lemma 4. *Suppose that $\delta > 1/2$. Let u be an articulation point of G such that every connected component of $G - u$ has fewer than $n/2$ vertices. Then G has a longest geodesic path P composed of two subpaths that connect u with the farthest vertices in two distinct connected components of $G - u$ with the two largest depths. (Here, depth is understood with respect to the root u, and ties are broken arbitrarily if some depths are equal.)*

Proof. P has the claimed shape due to Lemma 3. Since P has the maximum length among all geodesic paths, the two connected components that intersect P must also have the largest depths.   □

The next lemma addresses some routine preprocessing.

Lemma 5. *Given a graph G, we can determine, in $O(n + m)$ time, the set A of all articulation points u, the block-cut tree of G, and the vertex numbers of all connected components of all graphs $G - u$ ($u \in A$).*

Proof. In $O(n + m)$ time one can find all articulation points of G [9,12], and furthermore construct the block-cut tree T straightforwardly. We declare an arbitrary node of T the root and compute, by bottom-up summation in the rooted tree T, the number of vertices (that is, original vertices of G) below every edge of T. From these numbers we get the vertex numbers of all connected components of $G - u$, for all articulation points u, in $O(n + m)$ time in total: In particular, note that one edge from any articulation point u except the root goes upwards in the rooted tree, and the size of the corresponding component is $n - 1$ minus the sum of sizes of all other connected components of $G - u$ being below u in the rooted tree.   □

Now we can either reduce an instance of our problem in linear time to an equivalent instance with a simple structure, or solve the problem.

Lemma 6. *In a graph G with $\delta > 1/2$ we can, in $O(n + m)$ time, either compute a longest geodesic path of G, or extract an induced subgraph of G that still contains a longest geodesic path of G and consists of only one block with hairs.*

Proof. We do computations as in Lemma 5. If, for an articulation point u, every connected component of $G-u$ has fewer than $n/2$ vertices, then we find a longest geodesic path by Lemma 4 in $O(n+m)$ time, by using BFS with root u.

The other case is that, for every articulation point u, one connected component of $G-u$ has at least $n/2$ vertices. Assume that the block-cut tree T has two or more blocks. Then there exists an articulation point u on the path of T between any two blocks. But now Lemma 3 applies to the connected components of $G-u$ except the largest one. Thus we can replace them all with one longest geodesic path from u into these components, ending now in a new leaf of T. In particular, we get rid of at least one block.

We repeat this procedure until only one block with hairs remains. The depths and hairs are computed by BFS, where we can append any previously computed hair as a whole, if BFS reaches its (non-leaf) start vertex. Thus all changes affect pairwise disjoint parts of T, thus the process costs $O(n+m)$ time in total. $\square$

In order to compute a longest geodesic path in arbitrary graphs with $\delta > 1/2$ it remains to treat the graphs as produced in Lemma 6, consisting of one block with hairs. Note that still $\delta > 1/2$, since the number of vertices has not increased. Now we also use the quantitative part of Lemma 1.

Lemma 7. *In a graph G with $\delta > 1/2$ consisting of one block with hairs, some longest geodesic path begins at one of the two longest hairs (where ties are broken arbitrarily if some hair lengths are equal).*

Proof. Lemma 1 implies for this special type of graph that any longest geodesic path P must begin with a hair of length at least hn. We define factors h_i such that $h_1 n \geq h_2 n \geq \ldots$ are the hair lengths in descending order, and we let $H_1, H_2 \ldots$ denote the hairs in this order (not including their last articulation points that belong to the block).

If P does not begin with H_1, then P is a longest geodesic path in the graph $G_1 := G - H_1$, thus in a graph with $n_1 := (1 - h_1)n$ vertices and with diameter $(\frac{1}{2} + h)n = (\frac{1}{2} - \frac{1}{2}h_1 + \frac{1}{2}h_1 + h)n = \frac{1}{2}n_1 + (\frac{1}{2}h_1 + h)n$. Since Lemma 1 also holds for G_1, we conclude that P must begin with a hair of length at least $(\frac{1}{2}h_1 + h)n$, thus $\frac{1}{2}h_1 + h \leq h_2$. If P does not begin with H_2 either, then P is a longest geodesic path in $G_2 := G_1 - H_2$, thus in a graph with $n_2 := (1 - h_1 - h_2)n$ vertices and, by a similar calculation, with diameter $\frac{1}{2}n_2 + (\frac{1}{2}(h_1 + h_2) + h)n$. The same reasoning as above implies $\frac{1}{2}(h_1 + h_2) + h \leq h_3$. This contradicts $h_1 \geq h_2 \geq h_3$. Thus, P must begin with H_1 or H_2. $\square$

This yields the final result of the section.

Theorem 1. *In a graph G with $\delta > 1/2$ we can find some longest geodesic path, and thus compute $diam(G)$, in $O(n+m)$ time.*

Proof. We run the procedure from Lemma 6. If it yields a subgraph of the special form mentioned there, we start BFS from the two longest hairs and output the longest of the two geodesic paths, which is correct by Lemma 7. $\square$

Corollary 1. *In a graph G we can decide whether $\delta > 1/2$, and in that case we can find some longest geodesic path, and thus compute $diam(G)$, altogether in $O(n+m)$ time.*

Proof. First we run an algorithm as in Theorem 1. (We remark that the following reasoning does not depend on the particular algorithm.) If it does not output a result, then $\delta \leq 1/2$. Otherwise, we test in $O(n+m)$ time whether the output path actually has a length above $n/2$ and is a geodesic path. This can be done by BFS from one end vertex, since BFS yields the distances from the root vertex. If the output passes the test, then $\delta > 1/2$. Conversely, if $\delta > 1/2$, then the test confirms it.   □

4   Optimality of Linear Time (in the Number of Edges)

Graphs with $\delta > 1/2$ can still have a quadratic number $m = O(n^2)$ of edges. For instance, consider a path of length δn with a clique of $(1-\delta)n$ vertices attached somewhere. One may suspect that we need not read all edges in dense subgraphs in order to compute $diam(G)$, since most of them cannot belong to a longest geodesic path. Therefore it is not obvious whether the time $O(n+m)$ is optimal. Perhaps one could solve the problem in $O(n)$ time? However, we will argue that $O(n+m)$ time is actually needed in the worst case, even for a good approximation, provided that graphs are given by adjacency lists where the vertices appear in no particular order. The idea is that $diam(G)$ can depend on the presence of single edges creating shortcuts, but they are hard to find between dense subgraphs. The crucial subproblem in pure form looks as follows.

CROSSING EDGE: Given is a graph on a vertex set $X \cup Y$, where $X \cap Y = \emptyset$. The graph is given by adjacency lists, where the vertices appear in no particular order, and the partitioning into X and Y is known. Find some edge xy with $x \in X$, $y \in Y$, or report that no such edge exist.

Note that the following lemma hinges on the cardinalities. It would not hold if, for instance, $|X| = k$ and $|Y| = 1$.

Lemma 8. *Any algorithm that solves CROSSING EDGE with $|X| = |Y| = k$ needs $\Omega(k^2)$ time in the worst case.*

Proof. We can think of any algorithm as a player that can only look up entries in the adjacency lists, whereas an adversary provides all information. This translates the problem into a game with the following rules. In each step, the player may choose an arbitrary vertex u, and the adversary returns one vertex v adjacent to u (meaning that the player reads v in u's adjacency list).

As we are proving a lower bound, we can give the player extra information: The adversary tells in advance that either none or two edges exist between X and Y, and all other edges are inside X or Y. The player also gets to know the degrees of all vertices, that is, the lengths of all adjacency lists. Now the player can examine the adjacency lists, thus learn the edges. After each step of the game, the adversary is even more helpful and removes not only v from u's adjacency list, but also u from v's adjacency list. Only the undetected edges are kept, and the degrees of u and v are reduced by 1.

It remains to specify an adversary strategy. Remember that the player's instantaneous knowledge is the degrees of all vertices of X and Y, respectively. We call a degree sequence (multiset of degrees) valid, if there exists a graph with that degree sequence. The adversary does not reveal the graph, but only valid degree sequences in both X and Y. Initially let all degrees be $k - 1$, thus we have roughly k^2 edges, and the degree sequences are valid, as both subgraphs can be cliques. As long as there remains at least one edge in both X and Y, the player cannot distinguish whether these edges in X and Y exist, or instead two edges between X and Y joining the same four vertices. Whenever the player has chosen a vertex u, the adversary takes a vertex v from the same set ($v \in X$ if $u \in X$, and $v \in Y$ if $u \in Y$) such that the resulting degree sequence after subtracting 1 remains valid. Such a vertex v does always exist: Since the current degree sequence is valid, there exists a graph realizing it, and in such a graph there exists an edge uv that can be removed.

This shows that the player must empty one of X and Y, and therefore see $\Omega(k^2)$ edges, in order to decide whether some edges join X and Y.   □

Proposition 1. Any algorithm that approximates the diameter of graphs with any fixed $\delta > 1/2$ within a factor better than 2 needs $\Omega(n+m)$ time in the worst case.

Proof. We construct a special graph G: We take a simple path P of length δn and attach two subgraphs with vertex sets X and Y at the ends of P, $|X| = |Y| = k := \frac{1}{2}(1 - \delta)n$. They are chosen as in Lemma 8; in particular, we have $m = \delta n + \Theta(k^2) = \Theta(n^2)$ edges. If X and Y are connected directly by some edge, then $diam(G) = \frac{1}{2}\delta n$ rather than $diam(G) = \delta n$ "as expected". By Lemma 8, a shortcut between X and Y cannot be recognized or excluded without reading $\Omega(k^2) = \Omega(n + m)$ edges, as this problem is an instance of CROSSING EDGE.   □

5   An Auxiliary Problem: Largest Mixed Sum

For $\delta \le 1/2$, diameter computation cannot be based on articulation points any more, for the trivial reason that there exist graphs with diameter about $n/2$ but without any articulation points, such as the chordless cycle. We argue that $\delta = 1/2$ is a barrier in the sense that already for δ slightly below $1/2$, due to the possibility of long geodesic cycles and the lack of articulation points, it is inevitable for diameter calculation to solve a specific new subproblem.

To introduce and motivate this problem, consider the following special case of graphs. Let $H = (V, E)$ and $H' = (V', E')$ be two vertex-disjoint graphs with distinguished vertices $u, v \in V$ and $u', v' \in V'$. We connect u and u' by a path of some length ℓ larger than the diameters of H and H'. Similarly we connect v and v' by another path of length ℓ, being vertex-disjoint to the first path. The graph G constructed in this way is, roughly speaking, a geodesic cycle with two subgraphs H and H' attached at diametral positions. For any two vertices $w \in V$

and $w' \in V'$ we have $d(w, w') = \ell + \min\{d(w, u) + d(w', u'), d(w, v) + d(w'v')\}$, since one of the paths $u - u'$ or $v - v'$ must be chosen. (Distances are meant with respect to G.) Define $s := d(u, v)$ and $s' := d(u', v')$. Note that G has a longest geodesic cycle (in general not uniquely determined) of length $2\ell + s + s'$. Any geodesic path that starts outside $V \cup V'$ is a subpath of some longest geodesic cycle and has therefore a length at most $\ell + \frac{1}{2}(s + s')$. Some geodesic path connecting H and H' can be longer, since a distance $d(w, w')$, as above, can be as large as $\ell + \frac{1}{2}(d(w, u) + d(w', u') + d(w, v) + d(w'v')) \geq \ell + \frac{1}{2}(s + s')$. (The two terms under "min" might be equal, and the triangle inequality holds.) Then we must find the maximum $d(w, w')$ to get the correct diameter. By abstracting from the graph problem and using the symbols

$$x := d(w, u), \ y := d(w, v), \ y' := d(w', u'), \ x' := d(w', v'),$$

we arrive at the following problem statement.

LARGEST MIXED SUM: We are given h pairs of numbers (x_i, y_i) and h' pairs of numbers (x'_j, y'_j), find two indices i and j so as to maximize $\min\{x_i + y'_j, y_i + x'_j\}$. We refer to the given pairs as h red and h' blue pairs, and we refer to the given numbers as coordinates. We can assume $h' \leq h$.

Observe that these values x, y and x', y' for all vertices w and w', respectively, can together be computed by four runs of BFS, in linear time in the number of edges of H and H'. From any identical pairs we keep only one copy. We say that a pair of numbers (a, b) is dominated by a pair (c, d) if $a \leq c$ and $b \leq d$. Within a given set of pairs, we call a pair non-dominated if that pair is not dominated by other pairs in the set.

Proposition 2. LARGEST MIXED SUM is solvable in $O(h \log h)$ time.

Proof. The subset of the non-dominated pairs in a set of h pairs, sorted by strictly ascending first coordinates (and thus by strictly descending second coordinates) can be computed in $O(h \log h)$ time: Sort the pairs by their first coordinates, scan this sequence, and maintain the sorted sequence of pairs being non-dominated so far. Since the second coordinates are decreasing there, for every new pair (a, b) we only have to find the correct place of b in the sequence by binary search, and then delete the current end of the sequence containing those pairs with second coordinates smaller than b.

An optimal solution to LARGEST MIXED SUM can always be formed by a red pair and a blue pair which are non-dominated in the set of red pairs and blue pairs, respectively. This is true by an obvious exchange argument. Thus, in order to solve the problem it suffices to take each red pair (x, y) and find an optimal partner (x', y') in the sorted sequence U of non-dominated blue pairs. Finally we take the best solution, with maximum $z := \min\{x + y', y + x'\}$.

We distinguish six cases regarding the relationships between the coordinates. In cases of equations, the equality signs $=$ can be arbitrarily replaced with the strict signs $<$ or $>$. (See the Fig. 1)

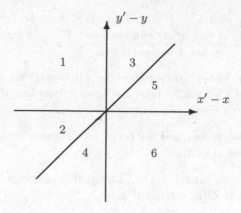

Fig. 1. These are the cases in the proof of Proposition 2.

(1) $x' < x$ and $y' > y$

(2) $x' < x$ and $y' < y$ and $x - x' > y - y'$

(3) $x' > x$ and $y' > y$ and $x' - x < y' - y$

(4) $x' < x$ and $y' < y$ and $x - x' < y - y'$

(5) $x' > x$ and $y' > y$ and $x' - x > y' - y$

(6) $x' > x$ and $y' < y$

Checking these cases one by one, we see that, if both x' increases y' decreases, then the objective z strictly increases in the regions (1)–(3) and strictly decreases in the regions (4)–(6). Moreover, the sorted sequence U first passes the regions (1)–(3) and then continues in the regions (4)–(6). Hence z is a unimodal discrete function on U, that is, z has only one local maximum which is therefore the global maximum. The maximum can be found by $O(\log h')$ look-ups of function values, by golden section search [10]. Since we have to do this at most h times (for every red pair), the time bound follows. □

Remark 1: Due to the search procedures, the log factor in Proposition 2 might be necessary for LARGEST MIXED SUM in general. Our particular objective function z is actually a "$\leftrightarrow\updownarrow$ unimodal 2D function" in the sense of [7], but we have used unimodality in only one direction. However, it is apparently unknown [7] whether this stronger property allows to find the global maximum in linear time. On another front, we have not established a linear-time reduction from LARGEST MIXED SUM to the diameter problem. The LARGEST MIXED SUM instances that can be realized by distances in graphs may have further properties that allow for linear time. We leave these questions open.

6   Diameters Larger than One Third of the Size

Generalizing Lemma 1 we can state, not surprisingly, that graphs with large diameter possess many small separators. We will use this version of the principle:

Lemma 9. *Suppose that* $\delta = \frac{1}{2} - h$, *where* $0 \leq h < \frac{1}{6}$. *Let* P *be any longest geodesic path, with* r *as one of its end vertices. Then at least* $(\frac{1}{4} - \frac{3}{2}h)n$ *of the layers* $N_i(r)$ *consist of at most two vertices.*

Proof. The $(\frac{1}{2} - h)n$ layers contain together all n vertices. Define x such that $x + 3(\frac{1}{2} - h - x) = 1$. Then at least xn layers have less than three vertices. Resolving the equation yields the claimed $x = \frac{1}{4} - \frac{3}{2}h$. $\qquad\square$

Based on this observation and the result of the previous section we will now propose a randomized algorithm.

Theorem 2. *For every fixed* $\delta > 1/3$, *a longest geodesic path can be computed with high probability in* $O(m + n \log n)$ *time.*

Proof. We attempt to construct a longest geodesic path by the following randomized procedure that we call a trial.

Trial, Preparation: Choosing Separator Vertices. We choose independently three random vertices u, v, w. The following happens with some guaranteed constant probability: (i) Each of u, v, w is in a layer of size at most 2, say $u \in N_i(r)$, $v \in N_j(r)$, $w \in N_k(r)$, where $i < j < k$, moreover, (ii) P goes through u and w. Note that constant probability for (i) holds due to Lemma 9, and for (ii) it follows from (i).

We can replace the subpath from u to w with any geodesic path Q between these vertices (if this geodesic path is not unique), as this yields another geodesic path between the end vertices of P. Thus, without loss of generality we may assume that some particular Q is a subpath of P.

Trial, Main Phase: Choosing a Geodesic Path. Observe the following:

(1) If $|N_j(r)| = 1$, then v is an articulation point, moreover, P also goes through v and intersects two different connected components of $G - v$.
(2) If $|N_j(r)| = 2$, then v is not on Q, with constant probability (since both vertices in $N_j(r)$ are proclaimed v with the same probability).

Now we "speculate" that our random u, v, w have properties (i) and (ii) above. Since we do not know which subcase appeared, we proceed as follows.

If v happens to be an articulation point, then situation (1) may be true. In order to capture this possible case, we apply Lemma 2 in order to determine, in $O(n + m)$ time, the longest geodesic path that intersects two different connected components of $G - v$. Since, in particular, P has this property in case (1), we find P or another longest geodesic path in this trial.

If v is not an articulation point, then we know that $|N_j(r)| = 2$, and we speculate that (2) is true. Since Q contains the other vertex of $N_j(r)$, and every layer is a separator S such that P intersects two connected components of $G - S$, we conclude that P also intersects two connected components of $G - (Q \cup \{v\})$. Furthermore, all vertices c and d of P in these two components satisfy $B(c, u, w)$ and $B(u, w, d)$, respectively. Defining the vertex sets $C := \{c| B(c, u, w)\}$ and $D := \{d| B(u, w, d)\}$, we can therefore set up an instance of LARGEST MIXED SUM, where the numbers x, x', y, y' are the distances of vertices in C and D to v and to some fixed reference vertex on Q. All these distances are computed by two runs of BFS, with roots u and v, in $O(n + m)$ time.

LARGEST MIXED SUM returns a path P' with the following properties: P' has its end vertices in C and D, its subpaths in C and D are geodesic, and either P' goes through Q and avoids v, or P' goes through v. (More precisely, only the end vertices of P' are returned, and the information whether P' uses Q or v, but this suffices to finally reconstruct a geodesic path between these ends.) If P' goes through Q (and hence is at most as long as the alternative path through v), we output P' in this trial, otherwise the trial has no output.

Analysis of a Trial. In case (1) we have already seen that a longest geodesic path is produced. In case (2), if P actually goes through Q as assumed, then we claim that the trial returns P (or another longest geodesic path). Assume for contradiction that some shorter path P' going through Q is returned. Let us divide P' in three subpaths: (A, Q, B). Since $A \subseteq C$, that is, A contains only vertices c with $B(c, u, w)$, it follows that the subpath (A, Q) is geodesic. By the symmetric argument, (Q, B) is geodesic. Hence, any shortcut on P' must connect A and B jumping over Q, and this is possible only by going through v, since $Q \cup \{v\}$ is a separator. However, by construction the alternative path through v was not shorter, hence P' has no shortcut at all, in other words, P' is geodesic. But since LARGEST MIXED SUM maximizes the minimum of the two lengths (of the paths through Q and v), it cannot yield a geodesic path shorter than P.

Conclusion. As shown above, our speculative assumptions are true with some guaranteed constant probability, and if they are, the path returned in the trial is in fact a longest geodesic path in the graph. As usual, one can amplify the probability of a correct result to any desired constant close to 1, by repeating the trial $O(1)$ times independently. $\qquad\square$

7   Further Research

Does a deterministic algorithm with the same time bound as in Theorem 2 exist? The difficulty is to hit a separator of two vertices that divides some (unknown!) longest geodesic path P. Alternatively we might use a version of Lemma 9 that guarantees a decrease of the largest connected component by a constant factor and thus enables divide-and-conquer, but now the catch is that a separator $S = \{u, v\}$ may have a large $d(u, v)$, and the long subpath $u - v$ of a solution may be in another connected component of $G - S$, such that the size of an instance to be solved recursively does not decrease enough.

The algorithm in Theorem 2 is Monte Carlo. It might be possible to turn it into a Las Vegas algorithm by verifying that the obtained geodesic path P is the longest one. This might be done by a technique as in Theorem 2, but now using the fact that P is already given. (Of course, the question becomes obsolete if a deterministic algorithm can be devised.)

Despite the mentioned difficulties we conjecture that the diameter can be found in nearly linear time for every fixed δ, by some smart use of $O(1/\delta)$ sized separators. By arguments similar to the case $\delta < 1/2$, this would also require a multi-dimensional generalization of LARGEST MIXED SUM.

Acknowledgment. The author would like to thank the anonymous referees for careful remarks which helped erase a number of small inaccuracies.

References

1. Abboud, A., Grandoni, F., Vassilevska Williams, V.: Subcubic equivalences between graph centrality problems, APSP and diameter. In: Indyk, P. (ed.) SODA 2015, pp. 1681–1697. ACM-SIAM (2015)
2. Abboud, A., Vassilevska Williams, V., Wang, J.R.: Approximation and fixed parameter subquadratic algorithms for radius and diameter in sparse graphs. In: Krauthgamer, R. (ed.) SODA 2016, pp. 377–391. ACM-SIAM (2016)
3. Boitmanis, K., Freivalds, K., Lediš, P., Opmanis, R.: Fast and simple approximation of the diameter and radius of a graph. In: Àlvarez, C., Serna, M. (eds.) WEA 2006. LNCS, vol. 4007, pp. 98–108. Springer, Heidelberg (2006)
4. Chechik, S., Larkin, D.H., Roditty, L., Schoenebeck, G., Tarjan, R.E., Vassilevska Williams, V.: Better approximation algorithms for the graph diameter. In: Chekuri, C. (ed.) SODA 2014, pp. 1041–1052. ACM-SIAM (2014)
5. Corneil, D.G., Dragan, F.F., Habib, M., Paul, C.: Diameter determination on restricted graph families. Discr. Appl. Math. **113**, 143–166 (2001)
6. Corneil, D.G., Dragan, F.F., Köhler, E.: On the power of BFS to determine a graph's diameter. Networks **42**, 209–222 (2003)
7. Demaine, E.D., Langerman, S.: Optimizing a 2D function satisfying unimodality properties. In: Brodal, G.S., Leonardi, S. (eds.) ESA 2005. LNCS, vol. 3669, pp. 887–898. Springer, Heidelberg (2005)
8. Dragan, F.F.: Almost diameter of a house-hole-free graph in linear time via LexBFS. Discr. Appl. Math. **95**, 223–239 (1999)
9. Hopcroft, J., Tarjan, R.E.: Algorithm 447: efficient algorithms for graph manipulation. Commun. ACM **16**, 372–378 (1973)
10. Kiefer, J.: Sequential minimax search for a maximum. Proc. Am. Math. Soc. **4**, 502–506 (1953)
11. Roditty, L., Vassilevska Williams, V.: Fast approximation algorithms for the diameter and radius of sparse graphs. In: Boneh, D., Roughgarden, T., Feigenbaum, J. (eds.) STOC 2013, pp. 515–524. ACM (2013)
12. Schmidt, J.M.: A simple test on 2-vertex- and 2-edge-connectivity. Inf. Proc. Lett. **113**, 241–244 (2013)
13. Weimann, O., Yuster, R.: Approximating the diameter of planar graphs in near linear time. ACM Trans. Algor. **12**(1), 12 (2016)
14. Yuster, R.: Computing the Diameter Polynomially Faster than APSP. CoRR abs/1011.6181 (2010)

Faster Computation of Path-Width

Martin Fürer[1,2]([✉])

[1] Department of Computer Science and Engineering, Pennsylvania State University,
University Park, PA 16802, USA
furer@cse.psu.edu
[2] Visiting Theoretical Computer Science, ETH Zürich, Zürich, Switzerland

Abstract. Tree-width and path-width are widely successful concepts. Many NP-hard problems have efficient solutions when restricted to graphs of bounded tree-width. Many efficient algorithms are based on a tree decomposition. Sometimes the more restricted path decomposition is required. The bottleneck for such algorithms is often the computation of the width and a corresponding tree or path decomposition. For graphs with n vertices and tree-width or path-width k, the standard linear time algorithm to compute these decompositions dates back to 1996. Its running time is linear in n and exponential in k^3 and not usable in practice. Here we present a more efficient algorithm to compute the path-width and provide a path decomposition. Its running time is $2^{O(k^2)}n$. In the classical algorithm of Bodlaender and Kloks, the path decomposition is computed from a tree decomposition. Here, an optimal path decomposition is computed from a path decomposition of about twice the width. The latter is computed from a constant factor smaller graph.

Keywords: Path-width · Tree-width · Bodlaender's algorithm · Path decomposition · FPT

1 Introduction

Tree-width and tree decompositions have been defined by Roberson and Seymour [15]. Independently, Arnborg and Proskurowski [2] introduced the equivalent concept of partial k-trees, as subgraphs of the previously known, simply structured k-trees. Many NP-hard graph problems have very efficient solutions when the graph is given with a tree decomposition of small width. Indeed, Courcelle's meta-theorem [6] says that all problems expressible in monadic second order logic have a linear time solution for graphs of bounded tree-width. Here, the dependence of the running time on the tree-width is allowed to be really bad. Theoretically this concept is captured by fixed parameter tractability (FPT). A parameterized problem is in FPT, if it can be solved by an algorithm with a running time of the form $O(f(k)n^c)$ for an arbitrary computable function $f(k)$ and some constant c, where n is the problem size and k is the parameter.

M. Fürer—Research supported in part by NSF Grant CCF-1320814.

© Springer International Publishing Switzerland 2016
V. Mäkinen et al. (Eds.): IWOCA 2016, LNCS 9843, pp. 385–396, 2016.
DOI: 10.1007/978-3-319-44543-4_30

Many faster solutions have been designed for specific problems. The goal is always to have efficient solutions for instances with small values of the parameter. For more background information on fixed parameter tractability, see e.g. [7–9,11].

Tree-width is an important parameter for enabling fast algorithms for interesting classes of graphs. But for some algorithms, the more restricted path-width parameter is of interest. (The Pathwidth entry on Wikipedia lists such applications in VLSI design, graph drawing, and computational linguistics.) The path-width is defined with tree-decompositions where the tree is a path.

Unfortunately, for graphs of small tree-width or path-width, it is not easy to find a corresponding tree decomposition of minimal width. Computing the tree-width is NP-hard [1]. For constant tree-width, a tree decomposition of minimal width can be computed in polynomial time [15]. The problem is even solvable by an FPT-algorithm [16]. But the only known linear time algorithms are variations of Bodlaender's algorithm [3]. Their running time is $2^{\Theta(k^3)}n$. This is too slow to be used in practice. Heuristic algorithms are used instead. Throughout this paper $n = |V|$ is the number of vertices of the graph in question, and k is a width parameter.

For the related notion of tree-depth [10], initially Bodlaender's algorithm provided the most efficient way to compute its value and to produce a corresponding tree decompositon. Recently, the exponent in the running time has been decreased from $O(k^3)$ to $O(k^2)$ [14]. We want to produce the same improvement for path-width, even though it seems to require a different method.

There have been many efforts to find better approximation algorithms for the tree-width. The main goal has been to achieve a small constant factor approximation with a running time $f(k)g(n)$, where $f(k)$ is $2^{O(k)}$ and $g(n)$ is polynomial, preferably linear. This combined goal has been achieved by the recent paper of Bodlaender et al. [4] producing a 5-approximation in time $O(c^k n)$. The authors write, "it would be very interesting to have an exact algorithm for testing if the treewidth of a given graph is at most k in $2^{o(k^3)}n^{O(1)}$ time." Downey and Fellows [7] remark that Bodlaender's Theorem, based on the algorithm of Bodlaender and Kloks is impractically exponential in k, namely 2^{ck^3} where $c \approx 32$, and they write, "It would be very interesting if this could be reduced to an exponential with exponent linear in k." We cannot get a linear exponent, but the improvement from $O(k^3)$ to $O(k^2)$ in the exponent for path-width is significant.

We will use some key ingredients of Bodlaender and Kloks [5] and its improved version of Perković and Reed [12]. First of all, it is the idea that a given tree decomposition is useful for the solution of all kinds of graph problems based on bottom-up dynamic programming. Even the problems of computing tree decompositions or path decompostions themselves are graph problems that can be solved this way. This makes sense, if one wants to compute a tree decomposition of width k, when one has available a tree decomposition of width linear in k. To obtain the needed constant factor approximation, one can use a top-down construction based on the repeated use of small vertex separators [13,16]. But such an FPT-algorithm runs in $O(n \log n)$ or even quadratic time.

Recently, a the 5-approximation has been obtained by Bodlaender et al. [4] running in time $O(c^k n)$.

For a theoretical result, this approximation would be sufficient, but the high approximation ratio makes the final step very expensive. Therefore, in order to have a chance of a practical algorithm, we show how to modify the Bodlaender and Kloks [5] method working with 2-approximations. Finally, we do the critical last step improving the constant factor approximation to an exact solution in time $2^{O(k^2)}$ instead of the previous $2^{O(k^3)}$ for graphs of path-width k. Here, the starting approximation ratio affects the constant factor in the exponent hidden by the O-notation.

The main idea of Bodlaender and Kloks [5] is the following. If a graph has a large matching, then a significantly smaller graph with the same or smaller tree-width is obtained by collapsing matched pairs of vertices into one. The smaller problem can be solved recursively, and expanding the collapsed pairs again results in a 2-approximation for the width and a corresponding tree decomposition of the original graph. On the other hand, if there is no large matching, then one can add more edges to the graph without increasing the tree-width. This in turn will create simplicial vertices, i.e., vertices whose neighborhood induces a clique. Simplicial vertices are easy to handle.

For computing the tree-width, these methods did not result in a practical algorithm, because of the cubic exponent and large constant factors. For path-width, with a quadratic exponent and much smaller constant factors, we could have a chance.

We use the standard notions of tree decomposition and a special notion of nice path decomposition.

Definition 1. A tree decomposition of a graph $G = (V, E)$ is a pair $(\{B_p : p \in I\}, T)$, where T is a tree, I is the node set of T, and the subsets $B_p \subseteq V$ have the following properties. (The set of vertices B_p associated with $p \in I$ is called the bag of p.)

1. $\bigcup_{p \in I} B_p = V$, i.e., each vertex belongs to at least one bag.
2. For all edges $e = \{u, v\} \in E$ there is at least one $p \in I$ with $\{u, v\} \subseteq B_p$, i.e., each edge is represented by at least one bag.
3. For every vertex $v \in V$, the set of indices p of bags containing v induces a subtree of T (i.e., a connected subgraph).

The tree-width of G is the smallest k such that G has a tree decomposition with largest bag size $k + 1$.

A rooted tree decomposition is a tree decomposition where T is a rooted tree. We assume all tree edges are oriented towards the root.

Definition 2. A nice tree decomposition is a rooted tree decomposition with the following four types of nodes.

Leaf node: p has no children, and $|B_p| = 1$.
Introduce node: p has one child q with $B_p = B_q \cup \{v\}$ for some vertex $v \notin B_q$.
Forget node: p has one child q with $B_p \cup \{v\} = B_q$ for some vertex $v \notin B_p$.

Join node: p *has 2 children* q *and* q' *with* $B_p = B_q = B_{q'}$.

Furthermore, the root is a forget node with an empty bag.

As an important concept, tree-width has several other equivalent definitions. A graph has tree-width at most k, if and only if it is a partial k-tree [2].

2   Path Decompositions

One can define a *path decomposition* of a graph $G = (V, E)$ to be a tree decomposing $(\{B_p : p \in I\}, T)$ where the tree $T = (I, F)$ is a path. A *rooted path decomposition* is a rooted tree decomposition where the root is an endpoint of the path.

We find it more convenient, to describe a nice path decomposition by the sequence of introduce and forget operations. Reminiscent of the traditional definition, we refer to the indices of the sequence as nodes. Every vertex has its introduce node before its forget node. The first node is the leaf, the last node is the root. It is a forget node. We consider the leaf node to be an introduce node too.

Definition 3. *A* path decomposition *of a graph* $G = (V, E)$ *with* $|V| = n$ *is a sequence of triples* $P = ((p_1, t_1, w_1), \ldots, (p_{2n}, t_{2n}, w_{2n}))$ *with the following properties.*

- *Every vertex* $v \in V$ *occurs exactly twice in the sequence* $(p_1, \ldots, p_{2n})$, *first,* $p_j = v$ *with* $t_j = +1$ *indicating* j *being the introduce node for* v, *then,* $p_{j'} = v$ *with* $t_{j'} = -1$ *indicating* j' *with* $j' > j$ *being the forget node for* v.
- *The sequence* $(w_1, \ldots, w_{2n})$ *is defined by*

$$
w_j = \begin{cases} 0 & \text{if } j = 1 \\ w_{j-1} + 1 & \text{if } 1 < j \le 2n \ \text{and} \ t_j = +1 \ (\text{introduce node}) \\ w_{j-1} - 1 & \text{if } 1 < j \le 2n \ \text{and} \ t_j = -1 \ (\text{forget node}). \end{cases}
$$

Bags have the traditional meaning and can easily be defined recursively.

$$
B_j = \begin{cases} \{p_1\} & \text{if } j = 1 \\ B_{j-1} \cup \{p_j\} & \text{if } 1 < j \le 2n \text{ and } t_j = +1 \ (\text{introduce node}) \\ B_{j-1} \setminus \{p_j\} & \text{if } 1 < j \le 2n \text{ and } t_j = -1 \ (\text{forget node}). \end{cases}
$$

The width of a node j is defined to be 1 less than the number of vertices in its bag B_j. Thus, w_j is the width of the node j.

Definition 4. *The* width of a path decomposition *is the maximum width of any of its nodes.*

The path-width *of a graph is the minimum width of any of its path decompositions.*

We use the double factorial $(2n-1)!! = (2n-1)(2n-3)\ldots3\,1 = (2n)!/(n!2^n)$. We count the number of path decompositions.

Proposition 1. *The number of path decompositions of an n-vertex graph is*

$$n!(2n-1)!! = \frac{(2n)!}{2^n} \sim \sqrt{\pi n}\left(\frac{n}{e}\right)^{2n}2^{n+1}.$$

Proof. Induction on n shows that the number of path decompositions for n' vertices which are forgotten in a fixed order is $(2n-1)!!$, as there are $2n-1$ places to introduce the last forgotten vertex. Considering all $n!$ permutations of the forgetting order of the vertices proves the result. Finally Stirling's approximation is used.   □

We now assume, we are given a path decomposition of width ℓ, and we want to produce a path decomposition of width $k < \ell$ or conclude that no such decomposition exists.

Definition 5. *The* full skeleton $Q_P(U)$ *induced by a non-empty subset $U \subseteq V$ of a path decomposition $P = ((p_1, t_1, w_1), \ldots, (p_{2n}, t_{2n}, w_{2n}))$ of $G = (V, E)$ is obtained from P by replacing all $p_j \in V \setminus U$ by 0.*

Definition 6. *The* skeleton Q *induced by a non-empty subset $U \subseteq V$ of a path decomposition $P = ((p_1, t_1, w_1), \ldots, (p_{2n}, t_{2n}, w_{2n}))$ of $G = (V, E)$ is obtained from $Q_P(U)$ by repeatedly deleting maximal length intervals of the form $((p_{j+1}, t_{j+1}, w_{j+1}), \ldots, (p_{j'-1}, t_{j'-1}, w_{j'-1}))$ with $p_{j+1} = p_{j+2} = \cdots = p_{j'-1} = 0$ and*

$$\min\{w_j, w_{j'}\} \le w_{j''} \le \max\{w_j, w_{j'}\}$$

for all j'' with $j < j'' < j'$. We refer to this step as simplifying.

Note that every deleted node has a width between the width of the remaining node immediately before it and the width of the remaining node immediately after it. w_j is the width of the jth node. Thus, if the jth node is an introduce node for vertex v, then w_j is the width just after the insertion of vertex v, and if the jth node is a forget node for vertex v, then w_j is the width just after the deletion of vertex v.

Definition 7. *The* width of a skeleton *is its maximum w_j entry.*

Proposition 2. *The width of a path decomposition is equal to the width of any of its skeletons.*

Proof. Sequences of deleted nodes are always next to a node whose width is at least equal to the width of any node in the deleted sequence.   □

3   Overview

The fastest published linear time algorithm to decide whether the path-width of a graph G is at most k, and to produce a width k path decomposition is obtained in two steps. The first step uses a version of Bodlaender's algorithm [3] to compute a tree decomposition of width $\ell = O(k)$ (or show that none exists). The second step uses the method of Bodlaender and Kloks [5] to produce a path decomposition of width k (or show that none exists) from the tree decomposition of width ℓ. Also the first step uses the method of Bodlaender and Kloks in recursive calls, to compute tree decompositions of smaller width from tree decompositions of roughly twice the width for smaller graphs. The improved version of Bodlaender's algorithm by Perković and Reed [12] computes the small width tree decomposition much faster, but like the original version, its running time has an exponent of order k^3. A theoretical alternative would be to start with the recent 5-approximation algorithm [4], but if an exact solution is desired, the second step would be significantly more expensive due to the higher approximation factor.

We propose a linear time path-width and path decomposition algorithm which recurses on path decompositions rather than the more costly tree decompositions. The exponent is only quadratic in k, and there are no large hidden constants. Note that we concentrate on the worst case in terms of the path-width. It is possible that the tree-width is significantly smaller than the path-width (by a factor of up to $\log n$). In this special case, the traditional approach can be faster.

4   The Efficient Algorithm

The crucial step of producing our faster path decomposition is to produce a small width path decomposition from one with a constant factor bigger width.

We are given a path decomposition P. Let

$$B_j^* = \bigcup_{i=1}^{j} B_i.$$

Let $G_j = G[B_j^*]$ be the subgraph of G induced by B_j^*.

We want to construct a minimum width path decomposition P' of G. As for most efficient algorithms based on small tree-width, our path decomposition algorithm uses a bottom-up dynamic programming approach. Any optimal path decomposition P' of G contains a path decomposition P_j' of the subgraph G_j of G.

It is sufficient to try for all small k whether there is a path decomposition of width k, and pick one in the affirmative case. For simplicity, we just describe the decision algorithm, because a solution can be found by standard back tracing of the dynamic programming solution. The basic idea of the algorithm is to produce the skeleton Q_j of an optimal P_j' induced by B_j for $j = 1, \ldots, 2n$. Then the path-width of Q_{2n} is the path-width of G.

Naturally, the problem with this basic idea is that the optimal path decomposition P' is unknown. A pessimistic approach would be to compute all skeletons obtained from all possible path decompositions of G. Fortunately, a good compromise is possible. Instead of computing all skeletons Q_j, we compute a set $\mathcal{Q}_j$ of skeletons, with the assurance that $\mathcal{Q}_j$ contains at least one skeleton Q_j of an optimal path decomposition.

Theorem 1. *Given a graph G, a number k, and a path decomposition P of G of width $\ell = 2^{O(k)}$, one can decide whether the path-width of G is at most k in time $2^{O(\ell k)} n$. A corresponding path decomposition can be computed with the same time bound.*

Proof. Assume G and its path decomposition P of width ℓ is given. Each bag B_j is a set of vertices of G with $|B_j| \leq \ell + 1$. We define $\mathcal{A}_j$ to be the set of all path decompositions of $G_j = G[B_j^*]$ of width at most k.

We will now define an algorithm that visits the nodes of G in the order given by P. For more details see Fig. 1. The algorithm will have the following property.

Claim. During the visit of node j, the algorithm computes a set of skeletons $\mathcal{Q}_j$ of $\mathcal{A}_j$, induced by B_j. This set $\mathcal{Q}_j$ includes at least one of minimal width.

Each set of skeletons is computed from the previously computed set of skeletons of the predecessor $j-1$ in P. We will show that if G_j has a path decomposition P'_j of width at most $k' \leq k$, then $\mathcal{Q}_j$ contains at least one skeleton of G_j (induced by B_j) of width at most k'.

The algorithm Decrease Path-Width depends on the type of node j in the path decomposition P. For every type of node, we now describe the action of the algorithm and prove the claim inductively.

Leaf node 1: The bag of the leaf node 1 contains just one vertex $v = p_1$, the only skeleton in $\mathcal{Q}_1$ is a two node skeleton $Q = (v, +1, 0), (v, -1, -1)$ (v is introduced and then forgotten). The leaf node 1 has bag $B_1 = \{v\}$, the root node 2 has bag $B_2 = \emptyset$. The claim is satisfied, because there is only one path decomposition P' of the one vertex graph G_1. Thus Q is the skeleton of the minimal path decomposition P'.

Introduce node j: If j is an introduce node, then the bag B_j of node j of P contains a new vertex v not in the bag B_{j-1} of the predecessor node. More precisely, $B_j = B_{j-1} \cup \{v\}$ with $v \notin B_{j-1}$.

The algorithm goes through all skeletons Q of $\mathcal{Q}_{j-1}$. For each Q, it creates various skeletons of $\mathcal{Q}_j$ by inserting an introduce node and a forget node for the new vertex v. All suitable places are tried for the insertion of v. A place is suitable if it is before the forget nodes of all neighbors u of v in G. Now the forget node of v is inserted somewhere after the introduce node of v. This includes the option immediately after the introduce node of v if that place is suitable. Again, all suitable places are tried. A place is suitable if it is after the introduce nodes of v and all its neighbors u in G. A newly created skeleton is discarded rather than put into $\mathcal{Q}_j$, if its width is greater than k.

Algorithm Decrease Path-Width:

Input: A graph G, widths ℓ and k, and a path decomposition
$P = ((p_1, t_1, w_1), \ldots, (p_{2n}, t_{2n}, w_{2n}))$ of G of width ℓ.
Task: Decide whether G has path-width at most k.
Comment: Will be used with $\ell \leq 2k + 1$.
Comment: Compute a sequence of sets of skeletons $\mathcal{Q}_j$ of path decompositions of
$G_j = G[B_j^*]$ induced by the bag B_j of node j, for $j = 1, \ldots, 2n$.
Comment: For some optimal path decomposition P' of G, each $\mathcal{Q}_j$ contains a skeleton of $P'[Q_j]$, the restriction of P' to B_j^*.

// Start Node
$w_1' = 0; w_2' = -1; \mathcal{Q}_1 = \{((p_1, +1, w_1'), (p_1, -1, w_2'))\}$
for $j = 2$ **to** $2n$ **do**
  // Introduce Node
  if $t_j = +1$ **then**
    for each skeleton Q in $\mathcal{Q}_{j-1}$ create skeletons for $\mathcal{Q}_j$ by
      inserting $(p_j, +1, w_i' + 1)$ in Q after any position i,
      inserting $(p_j, -1, w_{i'}' - 1)$ in Q after any position i' with $i' > i$, and
      incrementing $w_{i''}'$ for all positions i'' in-between,
      as long as it does not increase the width above k,
      i is before the forget node in Q of any neighbor u of v, and
      i' is after the introduce node in Q of any neighbor u of v.
      (Positions include position 0 meaning insertion to the left of Q.)
    if $\mathcal{Q}_j = \emptyset$ **then** Return("tree-width $> k$").
  // Forget Node
  if $t_j = -1$ **then**
    for each skeleton Q in $\mathcal{Q}_{j-1}$ create a skeleton for $\mathcal{Q}_j$ by
      replacing p_j both times by 0, and simplifying as specified in Def. 6.
// Success
Return($\min\{$width of $Q : Q \in \mathcal{Q}_{2n}\}$)

Fig. 1. The algorithm Decrease path-width

The algorithm would certainly be correct, if the insertions were tried in all positions of the full skeletons. This would be an extremely slow brute force algorithm. Thus it is crucial to argue that there is no benefit in trying those intervals I of positions in the full skeletons that have been deleted in the proper skeletons. Such an interval I consists of the nodes j'' between two positions j and j' with $\min\{w_j', w_{j'}'\} \leq w_{j''}' \leq \max\{w_j', w_{j'}'\}$. All the vertices introduced and deleted in these intervals are from B_{i-1}^*, while all vertices that still have to be included in the path decomposition are from bags after i. There are no edges between vertices of these intervals and later introduced vertices. Thus the only concern is the width caused by insertions between j and j'.

The widths between positions j and j' can be viewed as a mountain range with the height at j'' being the current width $w_{j''}'$ of the bag $B_{j''}'$. If an insert or forget node of a later vertex is placed between positions j and j', it is not always an advantage to place these nodes in the deepest valley, because the width is also affected by mountain tops between the insertion and the deletion of a vertex.

Nevertheless, it is never an advantage to place insertions or deletions of later nodes along the slope of a mountain. All the later nodes can just as well, and often with an advantage, be placed at the bottom of a valley. From there, still the same mountain tops have to be crossed, but it can only be an advantage if the new width caused by later placed vertices is added to a smaller width (from a valley) of the earlier placed vertices. More precisely, if a highest and a lowest point of an interval I are at its boundaries, and all the intermediate nodes insert and forget vertices of B_{j-1}^*, then there is never an advantage to insert or forget a new vertex at any other place on I than at the lowest point.

To prove the claim by induction, we assume that $\mathcal{Q}_{j-1}$ contains a skeleton Q_{j-1} of width at most k induced by B_{j-1} corresponding to an optimal path decomposition $P^{(j-1)}$ of G. If P^{j-1} has any vertices introduced or forgotten on the slopes of a skeleton Q_{j-1}, then $P^{(j-1)}$ is modified to a path decomposition $P^{(j)}$ by sliding all these vertices down the slope to the lowest valley of its interval. The width of $P^{(j)}$ is also optimal, because it is not more than the width of $P^{(j-1)}$. The skeleton Q_j of this $P^{(j)}$ is in $\mathcal{Q}_j$ proving the induction step of the claim.

Forget node j: If j is a forget node, then the bag B_j of node j of P contains a new vertex v not in the bag B_{j-1} of the predecessor node. More precisely, $B_j = B_{j-1} \setminus \{v\}$ with $v \in B_{j-1}$. Now, $\mathcal{Q}_j$ is obtained from $\mathcal{Q}_{j-1}$ simply by restricting to the smaller set B_j, i.e., by replacing both occurrences of v by 0. This shows the correctness of the claim.

Running time: The running time of the algorithm is mainly determined by the number of skeletons used. We have $O(n)$ nodes j in the path decomposition P. For each node, we consider skeletons induced by the $\ell + 1$ vertices from the bag B_j. By Proposition 1, we have $2^{O(\ell \log \ell)}$ path decompositions with $\ell + 1$ vertices. For each of these path decompositions, we have $2\ell + 1$ intervals between the nodes where the vertices of the bag B_j are inserted and deleted. These intervals are determined by their sequence of widths.

The lengths of these intervals between two nodes involving vertices of B_j in any skeleton are at most $2k + 1$. This is so, because the worst width sequences in such a interval is $\ldots, k-2, 2, k-1, 1, k, 0, k, 1, k-1, 2, k-2, \ldots$ and $\ldots, k-2, 1, k-1, 0, k, 0, k-1, 1, k-2, \ldots$. More importantly, there are only $2^{O(k)}$ such sequences [5, Lemma 3.5].

In summary, when handling bag B_j for $j = 1, \ldots, 2n$, we have $2^{O(\ell \log \ell)}$ path decompositions of B_j. Each has $2\ell + 1$ intervals with $2^{O(k)}$ possible sequences, resulting in $2^{O(\ell \log \ell)} 2^{O(\ell k)} = 2^{O(\ell k)}$ skeletons. Thus for all $2n$ nodes of P together, there are $2^{O(\ell k)} n$ possible skeletons. As the algorithm only makes polynomial time (in k) manipulations on each skeleton, the total running time is still $2^{O(\ell k)} n$.

It is standard for dynamic programming algorithms to actually recover the structure (the path decomposition in our case) that has produced the minimum. □

One should notice that there are no large hidden constants involved in the time analysis.

Corollary 1. *The path-width of a graph can be computed in time $2^{O(k^2)} n$ for graphs of path-width k if a path decomposition of width $O(k)$ is provided.*

Proof. Theoretically, one can just try $k = 1, 2, \ldots$ until one succeeds. Naturally, most of the work for $k - 1$ could be used for k. $\square$

5   Computing a Path-Width Approximation

We use the well known result that in any tree decomposition of any graph containing the complete bipartite graph $K_{p,q}$, there is a bag containing either all the p vertices of one side or all the q vertices of the other side and an additional vertex. Thus if p is greater than the tree-width k, then the addition of all edges between the q vertices on the other side does not increase the tree-width or path-width. The graph obtained from G by adding all such forced edges is called the augmented graph.

A vertex is simplicial, if its neighbors form a clique.

Theorem 2. *The path-width of a graph can be computed and a corresponding path decomposition can be found in time $2^{O(k^2)}n$ for graphs of path-width k.*

Proof. We use the results of [5,12] that for any graph G of tree-width k one can quickly augment G and find a linear size matching M or a linear size subset V' of simplicial vertices in the augmented graph.

If M is large, then one recursively computes an optimal path decomposition of width k of the graph with the vertices of M merged. It implies a path decomposition of the original graph G of width at most $2k + 1$. It can be improved to a path decomposition of width k as seen in Theorem 1.

If there is a large set V' of simplicial vertices, then one recursively computes an optimal path decomposition of width k of the graph with the vertices V' removed. If a tree decomposition of width k is found for the graph with the simplicial vertices removed, then immediately such a decomposition can be obtained for the given graph. This does not work for path decompositions. One obtains a caterpillar graph instead. Then, we just change the caterpillar decomposition into a path decomposition of width 1 more. This width might not be optimal, but it is a very good approximation, that can be improved to an optimal path decomposition as before. $\square$

The reason why [5] works with tree decompositions, even when computing path decompositions, might be because the intermediate caterpillar decompositions prevent a direct production of path decompositions. Another reason is that for some graphs the tree-width is smaller than the path-width by a $\log n$ factor. In such situations, it could be better to work with tree-decompositions.

Corollary 2. *There is an algorithm computing the path-width, and outputting a corresponding path decomposition in time $2^{O(k^2)}n$, where k is the path-width.*

Proof. As is usual for dynamic programming algorithms doing some minimization, whenever the algorithm makes a choice, computing the minimum width of two skeletons, it can record which option provided the minimum (an arbitrary choice is sufficient in case of a tie). Then it is easy to trace back to find an actual path decomposition once the global minimum width has been determined. $\square$

6   Conclusion

Path-width is an important width parameter. The worst case linear running time for its computation has not been improved over the last two decades.

Our algorithm is significantly faster for path decomposition than the fastest linear time algorithms for tree decomposition. Furthermore, there are no large hidden constant factors in the expressions for the running time. We conjecture that this algorithm can be implemented to run satisfactory for small path-width.

The main open problem in this area is to get an improvement of Bodlaender's algorithms of tree-width computation and the production of tree-decompositions of small width. In particular, one would like to know whether an $O(c^k n)$ time algorithm is possible for some constant c. In a wider context the open question is whether similar results are possible for other important width parameters, in particular for clique-width.

References

1. Arnborg, S., Corneil, D.G., Proskurowski, A.: Complexity of finding embeddings in a k-tree. SIAM J. Algebraic Discrete Methods **8**(2), 277–284 (1987). http://dx.doi.org/10.1137/0608024
2. Arnborg, S., Proskurowski, A.: Linear time algorithms for NP-hard problems restricted to partial k-trees. Discrete Appl. Math. **23**, 11–24 (1989)
3. Bodlaender, H.L.: A linear-time algorithm for finding tree-decompositions of small treewidth. SIAM J. Comput. **25**(6), 1305–1317 (1996)
4. Bodlaender, H.L., Drange, P.G., Dregi, M.S., Fomin, F.V., Lokshtanov, D., Pilipczuk, M.: An $o(c^k n)$ 5-approximation algorithm for treewidth. In: 54th Annual IEEE Symposium on Foundations of Computer Science, FOCS 2013, 26–29 October 2013, Berkeley, CA, USA, pp. 499–508. IEEE Computer Society (2013). http://doi.ieeecomputersociety.org/10.1109/FOCS.2013.60
5. Bodlaender, H.L., Kloks, T.: Efficient and constructive algorithms for the pathwidth and treewidth of graphs. J. Algorithms **21**(2), 358–402 (1996). http://dx.doi.org/10.1006/jagm.1996.0049
6. Courcelle, B.: Graph rewriting: an algebraic and logic approach. In: van Leeuwen, J. (ed.) Handbook of Theoretical Computer Science, Volume B: Formal Models and Sematics (B), pp. 193–242. Elsevier/MIT Press, Cambridge (1990)
7. Downey, R.G., Fellows, M.R.: Parameterized Complexity. Springer, New York (1999)
8. Downey, R.G., Fellows, M.R.: Fundamentals of Parameterized Complexity. Springer, London (2013)
9. Flum, J., Grohe, M.: Parameterized Complexity Theory. Text in Theoretical Computer Science. Springer, Heidelberg (2006)
10. Nešetřil, J., de Mendez, P.O.: Tree-depth, subgraph coloring and homomorphism bounds. Eur. J. Comb. **27**(6), 1022–1041 (2006). http://dx.doi.org/10.1016/j.ejc.2005.01.010
11. Niedermeier, R.: Invitation to Fixed-Parameter Algorithms. Oxford University Press, Oxford (2006)
12. Perković, L., Reed, B.: An improved algorithm for finding tree decompositions of small width. Int. J. Found. Comput. Sci **11**(3), 365–371 (2000). http://dx.doi.org/10.1142/S0129054100000247

13. Reed, B.A.: Finding approximate separators and computing tree width quickly. In: Proceedings of the Twenty-Fourth Annual ACM Symposium on the Theory of Computing, Victoria, British Columbia, Canada, pp. 221–228, 4–6 May 1992

14. Reidl, F., Rossmanith, P., Villaamil, F.S., Sikdar, S.: A faster parameterized algorithm for treedepth. In: Esparza, J., Fraigniaud, P., Husfeldt, T., Koutsoupias, E. (eds.) ICALP 2014. LNCS, vol. 8572, pp. 931–942. Springer, Heidelberg (2014). http://dx.doi.org/10.1007/978-3-662-43948-7

15. Robertson, N., Seymour, P.D.: Graph minors II. Algorithmic aspects of tree-width. J. Algorithms **7**(3), 309–322 (1986). http://dx.doi.org/10.1016/0196-6774(86)90023-4

16. Robertson, N., Seymour, P.D.: Graph minors XIII. The disjoint paths problem. J. Comb. Theory Ser. B **63**, 65–110 (1995)

The Solution Space of Sorting with Recurring Comparison Faults

Peter Damaschke[✉]

Department of Computer Science and Engineering,
Chalmers University, 41296 Göteborg, Sweden
ptr@chalmers.se

Abstract. Suppose that n elements shall be sorted by comparisons, but an unknown subset of at most k pairs systematically returns false comparison results. Using a known connection with feedback arc sets in tournaments (FAST), we characterize the solution space of sorting with recurring comparison faults by a FAST enumeration, which represents all information about the order that can be obtained by doing all $\binom{n}{2}$ comparisons. An optimal parameterized enumeration algorithm for FAST also works for the more general chordal graphs, and this fact contributes to the efficiency of our representation. Then, we compute the solution space more efficiently, by fault-tolerant versions of Treesort and Quicksort. We need $O(n \log n + kn + k^2 \log n)$ comparisons and $O(n \log n + kn + k^2 \log n + kF(k^2, k))$ time, where $F(n, k)$ is any parameterized time bound for finding a FAST with at most k arcs. Thus, for rare faults the complexity is close to optimal.

1 Introduction

In the model of recurring faults in computations as introduced in [10], operations on certain items yield false results even when repeated. As opposed to transient or probabilistic failures, this model accounts for systematic errors. One of the problems investigated in [10] is to sort a set of n elements by comparisons, where at most k pairs return false comparison results; let us denote this assumption A_k. Recurring comparison faults can result from software bugs. One can also think of applications where the elements are real entities rather than data items in computer memory. For instance, archaeological finds or historical events may be brought into chronological order by pairwise comparisons, say by comparing style characteristics or by causal dependencies, respectively, but for a few pairs the comparison criteria may be misleading.

It is impossible to verify A_k from the comparison results only, since false but consistent answers might pretend any order. The best we can do is to determine all orders compatible with A_k, and then we know: *If A_k holds true, then these are the possible orders.* Only if no compatible order exists, we recognize that A_k is false. Hence the problem belongs to the category of promise problems: We must know in advance that comparisons are reliable, subject to a certain "small" number of at most k false pairs.

© Springer International Publishing Switzerland 2016
V. Mäkinen et al. (Eds.): IWOCA 2016, LNCS 9843, pp. 397–408, 2016.
DOI: 10.1007/978-3-319-44543-4_31

In [10], quality measures for alleged sorted sequences are defined and related to each other. This is done from the approximation point of view, asking: How much does an order obtained by doing *all comparisons* and some postprocessing differ from the unknown sorted order? What is not considered is the full solution space obtained from the comparisons, and the number of comparisons actually needed. A fault-tolerant search for the minimum element is provided, which returns an element of rank $O(k)$ by using $O(\sqrt{k}n)$ comparisons and time. Here we aim at similar results for the sorting problem. We separate the number of comparisons and auxiliary computations, as comparisons may be more expensive, depending on the nature of elements to compare.

Our contributions. We answer two different questions: 1. What can we learn at all about an unknown order by faulty comparisons? 2. How can we efficiently extract this entire information? Specifically, how can we infer all comparison results by doing only a minority of them, ideally in a time close to $O(n \log n)$?

Starting from a version of the reversal lemma for minimal feedback arc sets (MFAS), we enumerate in $O(3^k k(n + m))$ time all MFAS with at most k arcs in a directed graph of n vertices and m edges whose underlying undirected graph is chordal. This extends an early algorithm [11] for finding smallest MFAS in tournaments, called FAST. While a single minimum FAST can be computed faster, base 3 is optimal for explicit enumerations. Next, the MFAS enumeration characterizes the solution space, i.e., the orders compatible with all comparisons. While there can be $n^{O(k)}$ such orders, it suffices to know at most 3^k MFAS, as all other compatible orders are obtained from them by simple transposition sequences. Next we observe: If we know already a compatible order, we can certify it with only $O(kn)$ comparisons that form a chordal graph, hence the MFAS that describe all compatible orders can be enumerated in $O(3^k k^2 n)$ time. Finally we give efficient algorithms that actually find a compatible order and the information needed to reconstruct the solution space. A building block is a procedure to insert another vertex in an existing order with a minimum number of backward arcs. This leads to fault-tolerant sorting algorithms based on Treesort and Quicksort, that essentially need $O(n \log n)$ comparisons for fixed k, which is optimal in a sense. The time is larger by just some "FPT term" in the parameter k. These are the first subquadratic algorithms for sorting with recurring comparison faults.

Other related literature. As much work exists on fault-tolerant searching and sorting (see the survey [5]), it is important to pay attention to similarities. Liar models are also deterministic fault models with a maximum number of false answers, but they count repeated false answers, and the searcher can reconstruct the true results. Sorting in a model where some elements can be corrupted (but comparisons are correct) is considered in [8], where the goal is to sort the uncorrupted elements. Sorting under probabilistic errors is studied in [3,4]. Some steps of our insertion procedure resemble some of their lemmas, as well as arguments from the kernelization of FAST [1]. Enumeration problems find attention in various fields (see, e.g. [2]). The number of comparisons needed to decide properties of partial orders is studied in [7].

2 Preliminaries

Orders are ascending from left to right. We use the terms *vertex* and *element* interchangeably. Suppose that k is fixed. In a directed *comparison graph* $D = (V, A)$, where V is the set of the n elements to be sorted, every arc (u, v) indicates a comparison that claimed $u < v$. We call the arc (u, v) *true* if actually $u < v$, and *false* if $v < u$. With respect to an order σ of V, an arc is *forward (backward)* if it points to the right (left). We denote the set of backward arcs $B(\sigma)$. The *length* of an arc (u, v) is the absolute difference of the positions of u and v in σ. Provided that at most k comparisons are false, clearly, an order σ is a candidate for the correctly sorted sequence if and only if $|B(\sigma)| \leq k$. As in [10] we call such σ *compatible*.

A *transposition* flips the positions of two neighbored vertices u, v in an order. It turns the arc (u, v), if there is one, from forward to backward or vice versa, while all other arcs are not affected. For two orders π and σ of the same set, an *inversion* is a pair of elements u, v such that u is to the left of v in π but v is to the left of u in σ. The *Kemeny distance* $d(\pi, \sigma)$ is the number of inversions. Starting from π, consider any sequence of transpositions with the property that each transposition removes an inversion. Every maximal sequence of this kind has length $d(\pi, \sigma)$ and ends in σ.

As usual, n and m denotes the number of vertices and arcs of a graph. A directed graph $D = (V, A)$ is *acyclic* if it has no directed cycles. As is well known, a directed graph is acyclic if and only if it admits an order without backward arcs, called *topological order*, and one can construct a topological order or output a directed cycle in $O(n + m)$ time. For general $D = (V, A)$ we call an order σ of V a *minimal backward order* if no other order τ has $B(\tau) \subset B(\sigma)$. Hence, in acyclic graphs, minimal backward and topological orders are the same. A *minimum backward* order also has a minimum number of backward arcs.

A computational problem is fixed-parameter tractable (FPT) if instances of size n and with an additional input parameter k can be solved in $f(k) \cdot n^{O(1)}$ time, with some computable function f. A *feedback arc set (FAS)* is a subset of arcs whose removal makes the graph acyclic, and a *minimal FAS (MFAS)* is a FAS such that no proper subset of it is a FAS, too. A *tournament* is a complete directed graph $D = (V, A)$. A (directed) *triangle* is a (directed) cycle of three vertices. The FAST problem requires to find a minimum FAS in a tournament. Let $F(n, k)$ be a time bound of an FPT algorithm for FAST, for graphs with $O(n)$ vertices and solution size k. Note that $F(n, k)$ is well-defined for any FPT algorithm: Since the dependency of the time bound on n is polynomial, a constant factor in n only affects the constant factor in $F(n, k)$. We can use $F(n, k) = 2^{O(\sqrt{k})} n^{O(1)}$ [6,9]. We will give time bounds in terms of $F(n, k)$, not in order to hide the exponential part, but in order to state the bounds in a generic way, independently of the current state of FAST.

The undirected *underlying graph* of a directed graph is obtained by ignoring the orientations of arcs. An undirected graph is *chordal* if every cycle C is a triangle or has a *chord*, that is, an edge joining two non-consecutive vertices in C. Every chordal graph has a *perfect elimination order (PEO)*, defined by the following property: If u is the first of u, v, w in the order, and uv and uw are edges, then vw is an edge, too. A PEO is constructed in $O(n + m)$ time [12].

3   Characterizing and Enumerating MFAS

The "reversal lemma" was used in [11] and already discovered several times in the 1960s. It states that reversing the arcs of an MFAS makes a directed graph acyclic. The following extended version also considers orders.

Lemma 1. *An arc set $F \subseteq A$ is an MFAS in a directed graph $D = (V, A)$, if and only if $F = B(\sigma)$ for some minimal backward order σ. Moreover, the possible σ are exactly the topological orders of $(V, A \setminus F)$.*

Proof. For any order σ, trivially, $B(\sigma)$ is a FAS. Let F be any FAS. Then $(V, A \setminus F)$ is acyclic. We take any topological order σ and re-insert the arcs of F. Clearly, $B(\sigma) \subseteq F$. If F is an MFAS then, since $B(\sigma)$ is a FAS, it also follows $B(\sigma) = F$. Now assume that σ is not minimal backward. Then there exists another σ' with $B(\sigma') \subset B(\sigma)$. But $F' := B(\sigma')$ is also a FAS, and $F' \subset F$ contradicts the minimality of F. Thus, every topological order of $(V, A \setminus F)$ is minimal backward.

Conversely, let σ be any minimal backward order, and $F := B(\sigma)$. Then F is a FAS. Assume that a smaller FAS $F' \subset F$ exists. As we saw above, there exists a topological order σ' such that $B(\sigma') \subseteq F' \subset F = B(\sigma)$, which contradicts the assumed backward minimality of σ.   □

Lemma 2. *A directed graph with an underlying chordal graph is acyclic if and only if it has no directed triangle. Furthermore, we can confirm that the graph is acyclic or find a triangle in $O(n + m)$ time.*

Proof. We run a standard $O(n+m)$ time algorithm that constructs a topological order or outputs a directed cycle. If the graph is acyclic, trivially it has no directed triangle. If we get a directed cycle C, represented as a doubly linked circular list, it remains to find a directed triangle in $O(n + m)$ time. To this end we construct in $O(n + m)$ time a PEO of the underlying chordal graph and mark the vertices of C therein. We scan the PEO from left to right until we find the first vertex $u \in C$. Let v and w be its neighbors in C (in the circular list). Then u, v, w form a triangle, due to the PEO. If this triangle is directed, we can stop. If not, then we update C by removing u and its two incident arcs, and inserting the arc (v, w) or (w, v) instead. The shortened cycle is still directed, and the update is done in $O(1)$ time. We keep on scanning the PEO until the next vertex of C is found. Since the cycle is shortened each time and remains directed, eventually we get a directed triangle.   □

Theorem 1. *In a directed graph with chordal underlying graph, at most 3^k MFAS of at most k arcs exist, and they are enumerated in $O(3^k k(n + m))$ time.*

Proof. We pick any directed triangle T and branch on it. That means, we generate at most three sub-instances of the problem as follows: In every branch we choose one arc of T, reverse it and mark it. Marked arcs are not reversed again in later steps (dealing with other triangles). If all three arcs in T were already marked, then the sub-instance is discarded. Each of the, at most 3^k, paths of branching steps is followed until k steps are done or the obtained directed graph is free of directed triangles. We collect the latter graphs. By Lemma 2, each of them is acyclic, hence the reversed arcs form a FAS. Eventually we throw out all FAS that are not MFAS or are duplicates of other MFAS.

For correctness it remains to show that every MFAS F with at most k arcs is found in this collection. We use Lemma 1 and fix an order σ where $F = B(\sigma)$. We follow a path of reversals where only arcs of F get reversed. As long as the obtained graph is not acyclic, by Lemma 2, it retains a directed triangle. The algorithm picks some; let us call it T. Clearly, some of the three arcs in T is still backward in σ, thus the arc is in F and not yet reversed and marked, and one of the branches reverses just this arc. As soon as the obtained graph is acyclic, the graph without the reversed arcs is acyclic, too, but since F is an MFAS, it follows that all arcs of F have already been reversed. These two cases show that our path never gets stuck with a proper subset of F reversed.

We have $O(3^k)$ branching steps, and the main work in each of them is to find a directed triangle. By Lemma 2 this us done in $O(n + m)$ time. Every FAS F not being an MFAS is detected easily: For every arc e we check whether $F \setminus \{e\}$ is still a FAS, in $O(n + m)$ time. This costs $O(3^k k(n + m))$ time for all collected FAS. Duplicates are recognized by bucketsorting.   □

One could also make the enumeration repetition-free by sorting the edges in each triangle and marking the reversed arc and the preceding arcs, but we remove duplicates anyhow, and the base in the 3^k factor is optimal, even for tournaments. To see this, take for instance k disjoint directed triangles, arrange them in an order, and insert all possible forward edges between vertices of different triangles. Then each of the 3^k selections of one arc from each triangle is an MFAS. By this 3^k lower bound, none of the faster algorithms that compute a minimum FAS can be turned into a faster algorithm that enumerates all MFAS as an explicit list.

4   MFAS and the Solution Space of Faulty Sorting

In this section we describe the family of all orders of a set being compatible with a comparison graph, by virtue of an MFAS enumeration and transpositions.

Lemma 3. *An order σ of the vertex set of a directed tournament $D = (V, A)$ is minimal backward if and only if no backward arc has length 1.*

Proof. One direction is trivial: If some backward arc has length 1, then a transposition makes it a forward arc, hence σ is not minimal backward.

Conversely, assume for contradiction that no backward arc in σ has length 1, but there exists an order τ with $B(\tau) \subset B(\sigma)$. Consider an arc $(u, v) \in B(\sigma) \setminus B(\tau)$ that has minimum length in σ, among all arcs in this set difference. Since this length is not 1, there exists a vertex $w \in V$ such that v, w, u appear in this order in σ. Clearly, u appears before v in τ, hence w swaps its position relative to u or v or both. We look at the conceivable cases:

Assume that w appears before v in τ; the other case is symmetric. If $(v, w) \in A$ then $(v, w) \notin B(\sigma)$ and $(v, w) \in B(\tau)$, which contradicts the choice of τ. If $(w, v) \in A$ then $(w, v) \in B(\sigma)$ and $(w, v) \notin B(\tau)$, but since (w, v) is shorter than (u, v), this contradicts the choice of (u, v). $\square$

Theorem 2. *For a tournament $D = (V, A)$ and an integer k, every compatible order can be obtained from a compatible, minimal backward order by a sequence of transpositions, each turning a (current) forward arc into a backward arc.*

Proof. Consider any compatible order σ. If σ is not minimal backward, then, by Lemma 3, it has a backward arc of length 1. A transposition at this place removes exactly this arc from $B(\sigma)$. By an inductive argument, a sequence of such transpositions ends in some minimal backward order. (We remark that this final order is not unique.) Trivially, this order is compatible, too. The assertion follows by reversing the sequence of transpositions. $\square$

Suppose that we have done all $\binom{n}{2}$ comparisons, that is, the comparison graph is a tournament. Then, the results provide a simple implicit description of all compatible orders, which is also practical for rare faults, that is, for small k:

Enumerate all MFAS with most k arcs in the comparison graph, in $O^*(3^k)$ time (as in Theorem 1). For any solution, reverse the arcs in the MFAS, and output the resulting order, which is a compatible, minimal backward order (by Lemma 1). If the number of backward arcs is $b < k$, all other compatible orders are obtained by up to $k - b$ transpositions that preserve the backward arcs, but are arbitrary, subject to this condition. Equivalently, these orders have Kemeny distance at most $k - b$ from the minimal backward orders. We comment that Theorem 2 is not an isolated observation but an integral part of the characterization of the solution space. It implies that algorithms for fault-tolerant sorting need to care about minimal backward orders only.

5   A Certificate for Sorting with Recurring Faults

The next natural question is whether we can get the solution space without doing all $O(n^2)$ comparisons when k is small, in view of the fact that sorting without faults (the case $k = 0$) needs only $O(n \log n)$ comparisons. Intuitively, we could first apply any $O(n \log n)$-time sort and then check the result and fix errors. However, usual sorting algorithms would not notice comparison faults, as they do not cause inconsistencies. They just continue and output a possibly false order σ. Another observation is that for any two neighbored elements u and v in σ we must actually do the comparison between u and v, since otherwise one could not tell whether $u < v$ or $v < u$. In order to spot errors we insert some redundancy, namely all arcs of length at most $2k+1$ in σ, and we do these $O(kn)$ comparisons. We do not take advantage of longer arcs from other comparisons possibly made before.

Definition 1. *Given an order σ of the vertex set V of a comparison graph, let $D(\sigma)$ be the subgraph consisting of V and all arcs of length at most $2k + 1$.*

Suppose that $D(\sigma)$ has at most k backward arcs. In this case we are in a good position, as the next theorem says that further comparisons would not add more information, thus the instance of the sorting problem is then solved after $O(n \log n + kn)$ comparisons.

Theorem 3. *Consider an ordered comparison graph that contains all arcs of length at most $2k + 1$, and at most k of them are backward arcs. Let u and v be any two vertices such that v appears more than $2k + 1$ positions to the right of u. Then we can safely conclude $u < v$.*

Proof. We use induction on the distance d between u and v in the order. Suppose that the assertion holds for all distances between $2k + 1$ and d. Let w be any of the $d - 1$ vertices between u and v. We have either (1) $u < w$ by the induction hypothesis, or (2) (u, w) is a forward arc, or (3) (w, u) is a backward arc. Similarly, we have either (1) $w < v$ by the induction hypothesis, or (2) (w, v) is a forward arc, or (3) (v, w) is a backward arc.

Since at most k backward arcs exist, for at least $d - 1 - k$ of the vertices w, only cases (1) and (2) apply, with respect to both u and v. Since at most k arcs are false, for at least $d - 1 - 2k \geq 1$ of the vertices w, we have both $u < w$ and $w < v$, where each of the two inequalities holds either by the induction hypothesis or since the forward arc, (u, w) or (w, v), is true. Note that we do not know which forward arcs are true, yet we can infer the existence of a vertex w with $u < w < v$. This concludes the induction step and the proof. $\qquad \square$

By Theorem 3, a graph $D(\sigma)$ with at most k backward arcs is a certificate that all other arcs are forward. Thus, the solution space description from Sect. 4 can be based on $D(\sigma)$, as we know the other arcs without testing them. Since $D(\sigma)$ has $m = O(kn)$ edges and is chordal, by Theorem 1 we can enumerate its MFAS already in $O(3^k k^2 n)$ time, which is $O(n)$ for any fixed k. However, in general we cannot expect to be lucky and get $D(\sigma)$ with at most k backward arcs already in one pass of a usual sorting algorithm. The following sections deal with the actual construction of an order that satisfies the condition in Theorem 3. We conclude this section with another structural property that will be needed.

Definition 2. *Consider a tournament and an order of its vertices. We partition it into components with the following properties: every component is a consecutive set of vertices; every backward arc is within a component; and for every point between two vertices in a component there exists a backward arc from a vertex on the right side to a vertex on the left side of this point. A trivial component has only one vertex, and a nontrivial component has more than one vertex.*

The components are uniquely determined. We index them from left to right by C_1, C_2, C_3, and so on. Let b_i denote the minimum number of backward arcs, in an optimal order of C_i, and $b := \sum_i b_i$. We define the following routine:

Procedure MB. In every nontrivial component C_i we compute a minimum FAS. Due to Lemma 1, topological sorting then yields a minimal backward order of C_i. We rearrange the vertices in every C_i accordingly.

Lemma 4. *The order from MB has exactly b backward arcs, which is optimal.*

Proof. The minimal backward order of every C_i has b_i backward arcs. Since we keep the order of components, and there exist no backward arcs between components, the number b is evident. To show optimality, consider any order of the whole set. The order induced on every C_i still has at least b_i backward arcs, since b_i is optimal in C_i. Since the components are disjoint, no backward arcs are counted twice. It follows that at least b backward arcs are needed.   □

6   Insertion in a Compatible Minimum Backward Order

Suppose that we have already found an order σ of a subset $U \subset V$, such that $D(\sigma)$ exhibits at most k backward arcs. Due to Theorem 3 this also implies that all longer arcs are forward. We can further suppose that the number of backward arcs in σ, or equivalently, in every component, is minimized (see Lemma 4). Let us store the sequence σ in an array indexed with consecutive integers. Now we want to insert another vertex $v \notin U$ and find an order τ of $U \cup \{v\}$ that still enjoys the same properties. Such an order must exist, if at most k comparison faults are present, but it is not obvious how to get τ efficiently from σ. We begin with a transitivity lemma and then establish a fault-tolerant binary search that runs, so to speak, on an almost sorted set blurred by comparison faults.

Lemma 5. *Suppose that u' stands to the left of u, at a distance larger than $2k + 1$. If (u, v) is true, then $u' < v$. A similar assertion holds in the symmetric case.*

Proof. By the assumed distance and Theorem 3, we have $u' < u$. Since (u, v) is true, we also have $u < v$, hence $u' < v$.   □

Lemma 6. *We can find elements ℓ and r with distance $O(k)$ in σ and $\ell < v < r$, by using $O(k \log n)$ comparisons of elements of U with v, in $O(k \log n)$ time.*

Proof. Let us append dummy vertices to σ: one at the left end which is smaller than all real elements, and one at the right end which is larger than all real elements. Initially let ℓ and r be these dummy elements, hence $\ell < v < r$ is true. To query a vertex means to compare it to v. Since σ is stored as an array, we have access to the indices and can find the center of an interval in $O(1)$ time.

We query consecutive vertices u around the center of the interval $[\ell, r]$, until $k+1$ of them give the same answer, say $u < v$. Clearly, this happens after at most $2k + 1$ comparisons. Since at most k comparisons are false, we know that $u < v$ is true for some queried vertex u, but we cannot say which. However, Lemma 5 ensures $u' < v$ for all u' more than $2k + 1$ positions to the left of u. Thus it is safe to update ℓ to the vertex at distance $2k + 2$ to the left of the leftmost queried vertex. Similarly we proceed with r in the symmetric case. Thus, the property $\ell < v < r$ is preserved. In each step we halve the interval $[\ell, r]$ and add an offset of $O(k)$. Clearly, after $O(\log n)$ such steps with $O(k \log n)$ comparisons, the length of $[\ell, r]$ is reduced down to $O(k)$. □

Next we finalize the procedure. Recall the FAST time bound $F(n, k)$ from Sect. 2, the notion of components in Definition 2, and note that a component has $O(k^2)$ vertices, since at most k backward arcs exist, all of length $O(k)$.

Lemma 7. *Given an order σ of U such that $D(\sigma)$ has a minimum number of backwards arcs, bounded by k, we can get an order τ of $U \cup \{v\}$ with the same properties, by $O(k \log n)$ comparisons in $O(k \log n + F(k^2, k) + n)$ time.*

Proof. After running the procedure in Lemma 6 we insert v anywhere in $[\ell, r]$ and denote by σ' the resulting order of $U \cup \{v\}$. By Theorem 3, ℓ is larger than all vertices to the left, and r is smaller than all vertices to the right, with the exception of at most $2k + 1$ vertices next to ℓ and r. From Lemma 6 we have $|[\ell, r]| = O(k)$ and $\ell < v < r$. Thus v is only involved in backward arcs of length $O(k)$ in σ'. (Longer backward arcs from comparisons with v that contradict these relations are now recognized as false and can be reversed.) The backward arcs with v create a new component that may incorporate some components from σ and contains only $O(k^2)$ vertices.

We have now learned the complete comparison graph of $U \cup \{v\}$ by actually doing only $O(k \log n)$ comparisons. By Lemma 4 it remains to apply MB to σ', and to take the resulting order τ. Actually it suffices to optimize the component including v, since all other components were already optimal in σ and have not changed. At this point, $D(\tau)$ has at most k backward arcs (otherwise more than k faults exist, and we can stop), these are the only true backward arcs in τ, and their number is minimized. This establishes correctness.

In addition we need $F(k^2, k)$ time to optimize the new component of length $O(k^2)$, and $O(n)$ time to update the indices, due to the insertion of v. □

Lemma 7 is complemented with a simpler and faster procedure for the special case when backwards arcs did not yet appear. Then we can either insert another vertex as above, or recognize a fault.

Lemma 8. *Given an order σ of U such that $D(\sigma)$ has no backwards arcs, we can construct an order τ with the same property, such that either τ is an order of $U \cup \{v\}$, or τ is an order of $U \setminus \{u, u'\}$ for some $u, u' \in U$ where some comparison among u, u', v is false, by using $O(\log n + k)$ comparisons and $O(n)$ time.*

Proof. First we do usual binary search and temporarily believe the results. We insert v at the resulting position in σ. Note that all arcs of length 1 are forward. Only now we compare v to all vertices at distance at most $2k + 1$. If we get only forward arcs, then τ is the obtained order of $U \cup \{v\}$. Otherwise we take some shortest backward arc. Since its length is not 1, it forms a directed triangle with two forward arcs. We remove the three involved vertices and let τ be the resulting order. Trivially, some of the arcs in the directed triangle is false.

The number of $O(\log n + k)$ comparisons is obvious. We need $O(n)$ time to update the indices, and the time for all other operations is no larger.   □

Theorem 4. *For $k < \sqrt{n}$, sorting with at most k recurring comparison faults can be accomplished with $O(n \log n + kn + k^2 \log n)$ comparisons.*

Proof. We do fault-tolerant Insertion Sort, that is, beginning with the empty order we insert all n elements one by one in a minimum backward order. If k is small compared to n, actually the special case of no backward arcs is the more frequent one. In detail:

Phase 1: We apply Lemma 8 as long as possible. Since in total at most k comparisons are false and the removed triples are disjoint, at most $3k$ vertices are removed from the order. We put these vertices aside. This needs $O(n \log n + kn)$ comparisons and $O(n^2)$ time.

Phase 2: We switch to Lemma 7 and insert the remaining $O(k)$ vertices. This needs $O(k^2 \log n)$ comparisons and $O(k^2 \log n + kn + kF(k^2, k))$ time.   □

While the number of comparisons is already pleasant, this method would need $O(n^2 + k^2 \log n + kF(k^2, k))$ computations. As a final step we will do the insertion procedures in a more economic way, to get rid of the $O(n^2)$ term.

7   Fault-Tolerant Treesort and Quicksort

For ease of presentation we did not pay much attention to the data structures so far. The catch with the use of an array for the order is that $O(n^2)$ time is needed only for updating the indices n times. Of course, fault-tolerant Insertion Sort cannot be faster than the error-free case. But we can also maintain the order and at the same time use a balanced search tree for the comparisons. This does not affect the comparison graphs $D(\sigma)$ and accelerates the updates.

Theorem 5. *Sorting with at most k recurring comparison faults can be accomplished with $O(n \log n + kn + k^2 \log n)$ comparisons and in $O(n \log n + kn + k^2 \log n + kF(k^2, k))$ time.*

Proof. We explain the modifications of the method from Theorem 4

We maintain a partitioning of σ into buckets, which are sets of at least $2k + 2$ but at most $4k + 3$ consecutive vertices. The leftmost vertex of each bucket is the leading vertex. Since the leading vertices have distances larger than $2k + 1$, by Theorem 3, they are in the correct order.

We store the leading vertices in a balanced binary search tree. Instead of using indices for the positions of vertices in σ we use the search tree to find the appropriate position for insertion of the new vertex v. During Phase 1, in every node of the search tree we compare v to the leading vertex only. During Phase 2, in every node of the search tree we compare v to the leading vertex and its entire bucket. Since the buckets are larger than $2k + 1$, majority vote sends v in the correct direction (as we have seen before), and since the buckets have size $O(k)$, also the last comparisons during this search cost only $O(k)$ time. For every vertex we used $O(\log n + k)$ comparisons and $O(\log n + k)$ time in Phase 1, and $O(k \log n)$ comparisons and $O(k \log n + F(k^2, k))$ time in Phase 2.

Optimizing and re-ordering the $O(k)$-sized component of v affects only $O(1)$ buckets. If the new vertex v exceeds the size limit of buckets, we also split one bucket in two smaller ones. Then we update the search tree in $O(\log n)$ time. Altogether we get the claimed complexity bounds.   □

A drawback of Treesort is the overhead for tree manipulations which deteriorates the constant factor in the time bound. Therefore we also present an alternative: to equip Quicksort with fault tolerance. Interestingly enough, it is possible to invoke our insertion procedure from Lemma 7 also there. The reason why it works is that Quicksort divides an instance recursively in two smaller instances that are independent in the error-free setting and interact only a little in the case of a few faults. We formally state the theorem as follows, although the (expected) complexity in O-notation is the same as for the deterministic algorithm. We remark that the expected $O(n \log n)$ bound for Quicksort holds for every instance, and the only randomness is in the choice of pivots; loosely speaking, there is no "interference" with our comparison faults.

Theorem 6. *Sorting with at most k recurring comparison faults can be accomplished with $O(n \log n + kn + k^2 \log n)$ expected comparisons in $O(n \log n + kn + k^2 \log n + kF(k^2, k))$ expected time.*

Proof. First remember how Quicksort works. A random pivot element p is compared to all other elements. A set L (R) collects all elements smaller (larger) than p, then L and R are sorted recursively, and L, p, R is the sorted order. In expectation this costs $O(n \log n)$ comparisons and time. Now, some extra work is needed due to possible comparison faults. Instead of sorting L and R completely, we only produce minimum backward orders recursively. Since some comparisons with p may be false, some vertices in L should actually be in R and vice versa. We call them the dislocated vertices. Due to Theorem 3, dislocated vertices can only exist in a segment of length $O(k)$ at the right end of L and at the left end of R. Each of the $O(k)$ candidates v for a dislocated vertex in L is compared to the first $2k + 1$ vertices in R. If the majority claims that v is smaller, then Lemma 5 yields that v is actually smaller than all vertices of R, with $O(k)$ exceptions at the left end. In the other case we insert v in R as in Lemma 7. We proceed similarly with dislocated vertices in R. To turn the concatenation L, p, R into a minimum backward order, it remains to optimize the component of p.

Only $O(n/k)$ pivots are considered, because segments of length $O(k)$ are not further split recursively. The dislocation tests require $O(k^2)$ comparisons for every pivot, in total $O(kn)$. Since at most k vertices are dislocated in total (not only per pivot), all insertions together are done in $O(k^2 \log n + kF(k^2, k) + kn)$ time. For every pivot p, the component of p has length $O(k^2)$, thus in can be optimized in $F(k^2, k)$ time. We need to call an FPT algorithm at most k times, since every nontrivial component exists due to a comparison fault. Altogether the asserted expected complexity follows.   □

8   Conclusions and Further Work

We presented the first efficient algorithms for sorting with recurring faults. The methods are elementary but not obvious. It is unclear whether the approach of error detection and correction by majority voting would also work with Merge-sort. (It works in [8], but in a different error model.) Simplicity should make it possible to implement the proposed algorithms, which was outside the scope of this study. We assumed small k, that is, applications with exceptional faults. For growing k, the dependency on k, which is subexponential but still has k in the exponent, becomes an issue. Further research may find improved bounds, e.g., by more sophisticated Quicksort versions. Our aim was mainly to explore the structure of the solution space. In practice this does not necessarily mean that one must explicitly enumerate all compatible orders. Faults may appear independently in different segments, and then succinct enumerations of MFAS using binary decision diagrams can be much smaller. Some experimentation is needed to study the practicality.

References

1. Bessy, S., Fomin, F.V., Gaspers, S., Paul, C., Perez, A., Saurabh, S., Thomassé, S.: Kernels for feedback arc set in tournaments. J. Comput. Syst. Sci. **77**, 1071–1078 (2011)
2. Bodlaender, H.L., Boros, E., Heggernes, P., Kratsch, D.: Open problems of the lorentz workshop. In: Enumeration Algorithms using Structure. Technical Report UU-CS-2015-016, Utrecht Univ. (2015)
3. Braverman, M., Mossel, E.: Noisy sorting without resampling. SODA **2008**, 268–276 (2008)
4. Braverman, M., Mossel, E.: Sorting from Noisy Information (2009). CoRR abs/0910.1191
5. Cicalese, F.: Fault-Tolerant Search Algorithms. Monographs in Theoretical Computer Science. An EATCS Series. Springer, Heidelberg (2013)
6. Feige, U.: Faster FAST (Feedback Arc Set in Tournaments) (2009). CoRR abs/0911.5094
7. Felsner, S., Kant, R., Pandu Rangan, C., Wagner, D.: On the complexity of partial order properties. Order **17**, 179–193 (2000)
8. Finocchi, I., Grandoni, F., Italiano, G.F.: Optimal resilient sorting and searching in the presence of memory faults. In: Bugliesi, M., Preneel, B., Sassone, V., Wegener, I. (eds.) ICALP 2006. LNCS, vol. 4051, pp. 286–298. Springer, Heidelberg (2006)
9. Fomin, F.V., Pilipczuk, M.: Subexponential parameterized algorithm for computing the cutwidth of a semi-complete digraph. In: Bodlaender, H.L., Italiano, G.F. (eds.) ESA 2013. LNCS, vol. 8125, pp. 505–516. Springer, Heidelberg (2013)
10. Geissmann, B., Mihalák, M., Widmayer, P.: Recurring comparison faults: sorting and finding the minimum. In: Kosowski, A., Walukiewicz, I. (eds.) FCT 2015. LNCS, vol. 9210, pp. 227–239. Springer, Heidelberg (2015)
11. Raman, V., Saurabh, S.: Parameterized algorithms for feedback set problems and their duals in tournaments. Theor. Comp. Sci. **351**, 446–458 (2006)
12. Rose, D., Lueker, G., Tarjan, R.E.: Algorithmic aspects of vertex elimination on graphs. SIAM J. Comp. **5**, 266–283 (1976)

Combinatorics

Monotone Paths in Geometric Triangulations

Adrian Dumitrescu[1], Ritankar Mandal[1(✉)], and Csaba D. Tóth[2,3]

[1] University of Wisconsin-Milwaukee, Milwaukee, WI, USA
{dumitres,rmandal}@uwm.edu
[2] California State University Northridge, Los Angeles, CA, USA
[3] Tufts University, Medford, MA, USA
cdtoth@acm.org

Abstract. (I) We prove that the (maximum) number of monotone paths in a triangulation of n points in the plane is $O(1.8027^n)$. This improves an earlier upper bound of $O(1.8393^n)$; the current best lower bound is $\Omega(1.7034^n)$. (II) Given a planar straight-line graph G with n vertices, we show that the number of monotone paths in G can be computed in $O(n^2)$ time.

Keywords: Monotone path · Triangulation · Counting algorithm

1 Introduction

A directed polygonal path ξ in $\mathbb{R}^d$ is *monotone* if there exists a nonzero vector $\mathbf{u} \in \mathbb{R}^d$ that has a positive inner product with every directed edge of ξ. The classical simplex algorithm in linear programming produces a monotone path on the 1-skeleton of a d-dimensional polytope of feasible solutions. According to the *monotone Hirsch conjecture* [26], for every $\mathbf{u} \in \mathbb{R}^2 \setminus \{\mathbf{0}\}$, the 1-skeleton of every d-dimensional polytope with n facets contains a $\mathbf{u}$-monotone path with at most $n - d$ edges from any vertex to a $\mathbf{u}$-maximal vertex. Klee [14] verified the conjecture for 3-dimensional polytopes, but counterexamples have been found in dimensions $d \geq 4$ [25]; see also [19]. Kalai [12,13] gave a subexponential upper bound for the length of a *shortest* monotone path between any two vertices. However, even in 3 dimensions, no deterministic pivot rule is known to find a monotone path of length $n - 3$ [11], and the expected length of a path found by randomized pivot rules requires averaging over all $\mathbf{u}$-monotone paths [10,16]. This motivates the study of the maximum number of monotone paths in geometric graphs on n vertices.

Our Results. We first show that the number of monotone paths in a triangulation of n points in the plane is $O(1.8193^n)$, using a fingerprinting technique in which incidence patterns of groups of size 8 are analyzed. We then give a sharper bound of $O(1.8027^n)$ using the same strategy, by enumerating incidence patterns of groups of size 9 by a computer program.

C.D. Tóth—Supported in part by the NSF awards CCF-1422311 and CCF-1423615.

© Springer International Publishing Switzerland 2016
V. Mäkinen et al. (Eds.): IWOCA 2016, LNCS 9843, pp. 411–422, 2016.
DOI: 10.1007/978-3-319-44543-4_32

Theorem 1. *The number of monotone paths in a triangulation on n vertices in the plane is $O(1.8027^n)$.*

The number of crossing-free structures (matchings, spanning trees, spanning cycles, triangulations) on a set of n points in the plane is known to be exponential [1,6,9,18,20–23]; see also [7,24]. Early upper bounds in this area were obtained by multiplying an upper bound on the maximum number of triangulations on n points with an upper bound on the maximum number of desired configurations in an n-vertex triangulation; valid upper bounds result since every planar straight-line graph can be augmented into a triangulation.

It is often challenging to determine the number of configurations (i.e., count) faster than listing all such configurations (i.e., enumerate). In Sect. 5 we show that monotone paths can be counted in polynomial time in plane graphs.

Theorem 2. *Given a plane straight-line graph G with n vertices, the number of monotone paths in G can be computed in $O(n^2)$ time. The monotone paths can be enumerated in an additional $O(1)$-time per edge.*

Related Previous Work. We derive a new upper bound on the maximum number of monotone paths in straight-line triangulations of n points in the plane. Analogous problems have been studied for cycles, spanning cycles, spanning trees, and matchings [3] in n-vertex edge-maximal planar graphs, which are defined in purely graph theoretic terms. In contrast, the monotonicity of a path depends on the embedding of the point set in the plane. The number of *geometric* configurations contained (as a subgraph) in a triangulation of n points have been considered only recently. The *maximum* number of convex polygons is known to be between $\Omega(1.5028^n)$ and $O(1.5029^n)$ [8,15]; while the *minimum* number of monotone paths in an n-vertex triangulation lies between $\Omega(n^2)$ and $O(n^{3.17})$ [4].

2 Preliminaries

A polygonal path $\xi = (v_1, v_2, \ldots, v_t)$ is *monotone in direction* $\mathbf{u} \in \mathbb{R}^2 \setminus \{\mathbf{0}\}$ if every directed edge of ξ has a positive inner product with $\mathbf{u}$, that is, $\langle \overrightarrow{v_i v_{i+1}}, \mathbf{u} \rangle > 0$ for $i = 1, \ldots, t-1$; here $\mathbf{0} = (0,0)$. A path $\xi = (v_1, v_2, \ldots, v_t)$ is *monotone* if it is monotone in some direction $\mathbf{u} \in \mathbb{R}^2 \setminus \{\mathbf{0}\}$. Every triangulation contains at least $\Omega(n^2)$ monotone paths, since there is a monotone path between any two vertices (by a straightforward adaptation of [5, Lemma 1] from convex subdivisions to triangulations).

Let $S \subset \mathbb{R}^2$ be an n-element point set in the plane with no three points collinear. A (geometric) *triangulation* of S is a plane straight-line graph with vertex set S such that the bounded faces are triangles that jointly tile of the convex hull of S. Since a triangulation has at most $3n - 6$ edges, it is enough to consider monotone paths in $O(n)$ directions parallel to the edges. In the remainder of the paper, we fix a direction, which we may assume to be the x-axis after a suitable rotation.

We prove the upper bound for a broader class of graphs, *plane monotone graphs*, in which every edge is an x-monotone Jordan arc. Consider a plane monotone graph G on n vertices with a maximum number of x-monotone paths. We may assume that the vertices have distinct x-coordinates; otherwise we can perturb the vertices without decreasing the number of x-monotone paths. Since adding extra edges to G can only increase the number of x-monotone paths, we may also assume that G is fully triangulated [17, Lemma 3.1], i.e., it is an edge-maximal planar graph. Denote the vertex set by $W = \{w_1, w_2, \ldots, w_n\}$, ordered by increasing x-coordinates; and orient each edge $w_i w_j \in E(G)$ from w_i to w_j if $i < j$; we thereby obtain a directed graph G.

By [4, Lemma 3], all edges $w_i w_{i+1}$ must be present, i.e., $\xi_0 = (w_1, w_2, \ldots, w_n)$ is a Hamiltonian path in G. The recurrence $T(i) = T(i-1) + T(i-2) + T(i-3)$ for $i \geq 4$, where $T(i)$ denotes the number of x-monotone paths that start at vertex w_{n-i+1}, was established in [4]. The recurrence solves to $T(n) = O(1.8393^n)$.

Proof Technique. An x-monotone path can be represented uniquely by the subset of visited vertices. This unique representation gives the trivial upper bound of 2^n for the number of x-monotone paths. For a set of k vertices $V \subseteq W$, an *incidence pattern* of V (*pattern*, for short) is a subset of V that appears in a monotone path ξ (i.e., the intersection between V and a monotone path ξ). Denote by $I(V)$ the set of all incidence patterns of V; see Fig. 1. For instance, $v_1 v_3 \in I(V)$ implies that there exists a monotone path ξ in G that is incident to v_1 and v_3 in V, but no other vertices in V. The incidence pattern $\emptyset \in I(V)$ denotes an empty intersection between ξ and V, i.e., a monotone path that has no vertices in V.

We now describe a *divide & conquer* application of the fingerprinting technique we use in our proof. Assuming that n is a multiple of 4, partition the path ξ_0 into groups of 4 consecutive vertices. We show that such a group can have at most 13 patterns (Sect. 3). By the product rule, we can deduce an upper bound of $13^{n/4} < 1.8989^n$ on the number of x-monotone paths in G. A careful analysis of the edges between two consecutive groups of 4 shows that, for 8 consecutive vertices, at most 120 out of the $13^2 = 169$ incidence patterns are possible (Lemma 6). It follows that the maximum number of x-monotone paths is $120^{n/8} < 1.8193^n$ if n is a multiple of 8, and $O(1.8193^n)$ in general (Sect. 4).

Computer search reveals that a group of 9 consecutive vertices admits 201 patterns, and so the number of x-monotone paths is $O(201^{n/9}) = O(1.8027^n)$; see Sect. 4. The analysis of larger groups, using the same technique is expected to yield further improvements. Handling groups of 10 or 11 is still realistic (although time consuming), but working with larger groups is currently prohibitive, both by analytic methods and with computer search. Significant improvement over our results will likely require new ideas.

Definitions and Notations. Let G be a directed plane monotone triangulation that contains a Hamiltonian path $\xi_0 = (w_1, w_2, \ldots, w_n)$. Denote by G^- (resp., G^+) the path ξ_0 together with all edges below (resp., above) ξ_0. Let $V = \{v_1, \ldots, v_k\}$ be a set of k consecutive vertices of ξ_0. We wish to identify the edges relevant for the incidence patterns of V. For this purpose, the edges between a vertex $v_i \in V$ and any vertex preceding V (resp., succeeding V) are equivalent. We apply a graph homomorphism φ on G^- and G^+, respectively, that maps all vertices preceding V to a new node v_0, and all vertices succeeding V to a new node v_{k+1}. The path ξ_0 is mapped to a new path $(v_0, v_1, \ldots, v_k, v_{k+1})$. Denote the edges in $\varphi(G^- \setminus \xi_0)$ and $\varphi(G^+ \setminus \xi_0)$, respectively, by $E^-(V)$ and $E^+(V)$; they are referred to as the *upper side* and the *lower side*; and let $E(V) = E^-(V) \cup E^+(V)$. The incidence pattern of the vertex set V is determined by the triple $(V, E^-(V), E^+(V))$. We call this triple the *group induced* by V.

Fig. 1. Left: a group U with incidence patterns $I(U) = \{\emptyset, u_1u_2, u_1u_2u_3, u_1u_2u_3u_4, u_1u_2u_4, u_2, u_2u_3, u_2u_3u_4, u_2u_4, u_3, u_3u_4\}$. Right: a group V with $I(V) = \{\emptyset, v_1v_2, v_1v_2v_3, v_1v_2v_3v_4, v_1v_2v_4, v_1v_3, v_1v_3v_4, v_2, v_2v_3, v_2v_3v_4, v_2v_4, v_3, v_3v_4\}$.

The edges $v_iv_j \in E(V)$, $1 \leq i < j \leq k$, are called *inner edges*. The edges v_0v_i, $1 \leq i \leq k$, are called *incoming edges* of $v_i \in V$; and the edges v_iv_{k+1}, $1 \leq i \leq k$, are *outgoing edges* of $v_i \in V$ (note that v_0 and v_{k+1} are not in V). An incoming edge v_0v_i for $1 < i \leq k$ (resp., and outgoing edge v_iv_{k+1} for $1 \leq i < k$) may be present in both $E^-(V)$ and $E^+(V)$. Denote by $\mathrm{In}(v)$ and $\mathrm{Out}(v)$, respectively, the number of incoming and outgoing edges of a vertex $v \in V$; and note that $\mathrm{In}(v)$ and $\mathrm{Out}(v)$ can be 0, 1 or 2.

For $1 \leq i \leq k$, let V_{*i} denote the set of incidence patterns in the group V ending at i (i.e., leaving the group at v_i). For example in Fig. 1 (right), $V_{*3} = \{v_1v_2v_3, v_1v_3, v_2v_3, v_3\}$. By definition we have $|V_{*i}| \leq 2^{i-1}$. Similarly V_{i*} denotes the set of incidence patterns in the group V starting at i (i.e., entering the group at v_i). In Fig. 1 (left), $U_{2*} = \{u_2, u_2u_3, u_2u_3u_4, u_2u_4\}$. Observe that $|V_{i*}| \leq 2^{k-i}$. Note that

$$|I(V)| = 1 + \sum_{i=1}^{k} |V_{*i}| \qquad \text{and} \qquad |I(V)| = 1 + \sum_{i=1}^{k} |V_{i*}|. \qquad (1)$$

Reflecting all components of a triple $(V, E^-(V), E^+(V))$ with respect to the x-axis generates a new group denoted by V^R. By definition, both V and V^R have the same set of incidence patterns.

3   Groups of 4 or 8 Vertices

Lemma 1. *Let V be a group of 4 vertices with at least 10 incidence patterns. Then there is: (i) an outgoing edge from v_2 or v_3; and (ii) an incoming edge into v_2 or v_3.*

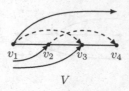

$$V$$

Fig. 2. v_1 cannot be the last vertex with an outgoing edge from a group $V = \{v_1, v_2, v_3, v_4\}$ with at least 10 incidence patterns.

Proof. (i) There is at least one outgoing edge from $\{v_1, v_2, v_3\}$, since otherwise $V_{*1} = V_{*2} = V_{*3} = \emptyset$ implying $|I(V)| = |V_{*4}| + 1 \leq 9$. Assume there is no outgoing edge from v_2 and v_3; then $V_{*1} = \{v_1\}$ and $V_{*2} = V_{*3} = \emptyset$. From (1), we have $|V_{*4}| = 8$ and this implies $\{v_1v_3v_4, v_2v_4, v_3v_4\} \subset V_{*4}$. The patterns $v_1v_3v_4$ and v_2v_4, respectively, imply that $v_1v_3, v_2v_4 \in E(V)$. The patterns v_2v_4 and v_3v_4, respectively, imply there are incoming edges into v_2 and v_3. Refer to Fig. 2. Without loss of generality, an outgoing edge from v_1 is in $E^+(V)$. By planarity, an incoming edges into v_2 and v_3 have to be in $E^-(V)$. Then v_1v_3 and v_2v_4 both have to be in $E^+(V)$ which by planarity is impossible.

(ii) By symmetry in a vertical axis, there is an incoming edge into v_2 or v_3. □

Lemma 2. *Let V be a group of 4 vertices with at least 11 incidence patterns. Then there is:* (i) *an incoming edge into v_2; and* (ii) *an outgoing edge from v_3.*

Proof. (i) Assume $\text{In}(v_2) = 0$. Hence $|V_{2*}| = 0$. By Lemma 1 (ii), we have $\text{In}(v_3) > 0$. By definition $|V_{3*}| \leq 2$. We distinguish two cases.

Case 1: $\text{In}(v_4) = 0$. In this case, $|V_{4*}| = 0$. Refer to Fig. 3 (left). By planarity, the edge v_1v_4 and an outgoing edge from v_2 cannot coexist with an incoming edge into v_3. So either v_1v_4 or v_1v_2 is not in V_{1*}, which implies $|V_{1*}| < 8$. Therefore, (1) yields $|I(V)| = |V_{1*}| + |V_{3*}| + 1 < 8 + 2 + 1 = 11$, which is a contradiction.

Case 2: $\text{In}(v_4) > 0$. In this case, $|V_{4*}| = 1$. If the incoming edges into v_3 and v_4 are on opposite sides (see Fig. 3 (center)), then by planarity there are outgoing edges from neither v_1 nor v_2, which implies that the patterns v_1 and v_1v_2 are not in V_{1*}, and so $|V_{1*}| \leq 8 - 2 = 6$. If the incoming edges into v_3 and v_4 are on the same side (see Fig. 3 (right)), then by planarity either the edges v_1v_4 and v_2v_4 or an outgoing edge from v_3 cannot exist, which implies that either v_1v_4 and $v_1v_2v_4$ are not in V_{1*} or v_1v_3 and $v_1v_2v_3$ are not in V_{1*}. In either case, $|V_{1*}| \leq 8 - 2 = 6$.

Therefore, irrespective of the relative position of the incoming edges into v_3 and v_4, (1) yields $|I(V)| = |V_{1*}| + |V_{3*}| + |V_{4*}| + 1 \leq 6 + 2 + 1 + 1 = 10$, which is a contradiction.

(ii) By symmetry in a vertical axis, $\text{Out}(v_3) > 0$. □

Lemma 3. *Let V be a group of 4 vertices with exactly 11 incidence patterns. Then the following hold.*

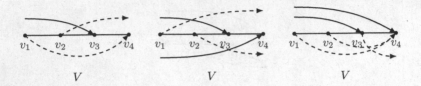

Fig. 3. Left: an incoming edge arrives into v_3, but not into v_4. Center and right: incoming edges arrive into both v_3 and v_4; either on the same or on opposite sides of ξ_0.

(i) If $\text{In}(v_3) = 0$, then all the incoming edges into v_2 are on the same side, $|V_{1*}| \geq 5$ and $|V_{2*}| \geq 3$.

(ii) If $\text{In}(v_3) > 0$, then all the incoming edges into v_3 are on the same side, $|V_{1*}| \geq 4$, $|V_{2*}| \geq 2$ and $|V_{3*}| = 2$.

Proof. By Lemma 2, $\text{In}(v_2) \neq 0$ and $\text{Out}(v_3) \neq 0$. Therefore $\{v_2 v_3, v_2 v_3 v_4\} \subseteq V_{2*}$, implying $|V_{2*}| \geq 2$. By definition $|V_{4*}| \leq 1$.

(i) Assume $\text{In}(v_3) = 0$. Then we have $|V_{3*}| = 0$. Using (1), $|V_{1*}| + |V_{2*}| \geq 9$. By definition $|V_{2*}| \leq 4$, implying $|V_{1*}| \geq 5$. All incoming edges into v_2 are on the same side, otherwise the patterns $\{v_1, v_1 v_3, v_1 v_3 v_4, v_1 v_4\}$ cannot exist, which would imply $|V_{1*}| < 5$. If $|V_{2*}| < 3$, then v_2 and $v_2 v_4$ are not in V_{2*} implying that $v_1 v_2$ and $v_1 v_2 v_4$ are not in V_{1*}; hence $|V_{1*}| \leq 6$ and thus $|V_{1*}| + |V_{2*}| < 9$, which is a contradiction. We conclude that $|V_{2*}| \geq 3$.

(ii) Assume $\text{In}(v_3) > 0$. Then we have $\{v_3, v_3 v_4\} \subseteq V_{3*}$, hence $|V_{3*}| = 2$. By (1), we obtain $|V_{1*}| + |V_{2*}| \geq 7$. If $|V_{1*}| < 4$, then $|V_{2*}| \geq 4$ and so $\{v_2, v_2 v_3, v_2 v_4, v_2 v_3 v_4\} \subseteq V_{2*}$. This implies $\{v_1 v_2, v_1 v_2 v_3, v_1 v_2 v_4, v_1 v_2 v_3 v_4\} \subseteq V_{1*}$, hence $|V_{1*}| \geq 4$ and $|V_{1*}| + |V_{2*}| \geq 4 + 4 = 8$ which is a contradiction. We conclude $|V_{1*}| \geq 4$. All incoming edges into v_3 are on the same side, otherwise the patterns $\{v_1, v_1 v_2, v_1 v_2 v_4, v_1 v_4, v_2, v_2 v_4\}$ cannot exist, and thus $|I(V)| \leq 10$ which is a contradiction.   □

Lemma 4. *Let V be a group of 4 vertices with exactly 12 incidence patterns. Then the following hold.*

(i) *For $i = 1, 2, 3$, all outgoing edges from v_i, if any, are on one side of ξ_0.*

(ii) *If V has outgoing edges from exactly one vertex, then this vertex is v_3 and we have $|V_{*3}| = 4$ and $|V_{*4}| = 7$. Otherwise there are outgoing edges from v_2 and v_3, and we have $|V_{*2}| = 2$, $|V_{*3}| \geq 3$ and $|V_{*4}| \geq 5$.*

(iii) *For $i = 2, 3, 4$, all incoming edges into v_i, if any, are on one side of ξ_0.*

(iv) *If V has incoming edges into exactly one vertex, then they are into v_2 and we have $|V_{2*}| = 4$ and $|V_{1*}| = 7$. Otherwise there are incoming edges into v_3 and v_2, and we have $|V_{3*}| = 2$, $|V_{2*}| \geq 3$ and $|V_{1*}| \geq 5$.*

Proof. (i) By Lemma 2 (i), there is an incoming edge into v_2. So by planarity, all outgoing edges from v_1, if any, are on one side of ξ_0.

If there are outgoing edges from v_2 on both sides, then by planarity the edges $v_1 v_3$, $v_1 v_4$ and any incoming edge into v_3 cannot exist, implying the five patterns

$\{v_1v_3, v_1v_3v_4, v_1v_4, v_3, v_3v_4\}$ are not in $I(V)$ and thus $|I(V)| \leq 16-5 = 11$, which is a contradiction.

If there are outgoing edges from v_3 on both sides (see Fig. 4 (a)) then by planarity the edges v_1v_4, v_2v_4 and an incoming edge into v_4 cannot exist which implies that the four patterns $\{v_1v_2v_4, v_1v_4, v_2v_4, v_4\}$ are not in $I(V)$. Without loss of generality, an incoming edge into v_2 is in $E^+(V)$. Then by planarity, any outgoing edge of v_1 and the edge v_1v_3 (which must be present) are in $E^-(V)$. Then by planarity either an incoming edge into v_3 or an outgoing edge from v_2 cannot exist. So either the patterns $\{v_3, v_3v_4\}$ or the patterns $\{v_1v_2, v_2\}$ are not in $I(V)$. Hence $|I(V)| \leq 16-(4+2) = 10$, which is a contradiction. Consequently, all outgoing edges of v_i are on the same side of ξ_0, for $i = 1, 2, 3$.

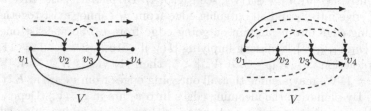

Fig. 4. (a) Having outgoing edges from v_3 on both sides is impossible. (b) Existence of outgoing edges only from $\{v_1v_3\}$ is impossible.

(ii) If V has outgoing edges from exactly one vertex, then by Lemma 2 (ii), this vertex is v_3. Consequently, $V_{*1} = V_{*2} = \emptyset$. Using (1), $|V_{*3}| + |V_{*4}| = 11$. Therefore $|V_{*4}| \geq 7$, since by definition $|V_{*3}| \leq 4$. If $|V_{*4}| = 8$, then we have $\{v_1v_2v_3v_4, v_1v_3v_4, v_2v_3v_4, v_3v_4\} \subset V_{*4}$. Existence of these four patterns along with an outgoing edge from v_3 implies $\{v_1v_2v_3, v_1v_3, v_2v_3, v_3\} \subseteq V_{*3}$ and thus $|V_{*3}| + |V_{*4}| = 4 + 8 = 12$, which is a contradiction. Therefore $|V_{*4}| = 7$ and $|V_{*3}| = 4$.

If V has outgoing edges from more than one vertex, the the possible vertex sets with outgoing edges are $\{v_1, v_3\}$, $\{v_2, v_3\}$, and $\{v_1, v_2, v_3\}$. We show that it is impossible that all outgoing edges are from $\{v_1, v_3\}$, which will imply that there are outgoing edges from both v_2 and v_3.

If there are outgoing edges from $\{v_1, v_3\}$ only, we may assume the ones from v_1 are in $E^+(V)$ and then by planarity all incoming edges into v_2 are in $E^-(V)$, see Fig. 4 (b). Then by planarity, either v_1v_3 or v_2v_4 or an incoming edge into v_3 cannot exist implying that $\{v_1v_3, v_1v_3v_4\}$ or $\{v_1v_2v_4, v_2v_4\}$ or $\{v_3, v_3v_4\}$ is not in $I(V)$. By the same token, depending on the side the outgoing edges from v_3 are on, either the edge v_1v_4 or an incoming edge into v_4 cannot exist, implying that either v_1v_4 or v_4 is not in $I(V)$. Since $V_{*2} = \emptyset$, $\{v_1v_2, v_2\}$ are not in $I(V)$. So $|I(V)| \leq 16 - (2 + 1 + 2) = 11$, which is a contradiction. Therefore the existence of outgoing edges only from v_1 and v_3 is impossible.

If there are outgoing edges from only $\{v_2, v_3\}$ or only $\{v_1, v_2, v_3\}$, then we have $\{v_1v_2, v_2\} \subseteq V_{*2}$ and $\{v_1v_2v_3, v_2v_3\} \subseteq V_{*3}$, since $\text{In}(v_2) \neq 0$ and $\text{Out}(v_3) \neq 0$

Fig. 5. (a) $|V_{*3}| \geq 3$. (b) $|V_{*4}| \geq 5$.

by Lemma 2. Therefore $|V_{*2}| = 2$ and $|V_{*3}| \geq 2$. If $|V_{*3}| < 3$, then $v_1 v_3, v_3 \notin V_{*3}$, which implies $v_1 v_3$ and that an incoming edge into v_3 are not in $E(V)$. Consequently, $v_1 v_3 v_4, v_3 v_4 \notin I(V)$; see Fig. 5(a). By planarity the edge $v_1 v_4$, an incoming edge into v_4 and an outgoing edge from v_1 cannot exist together with an incoming edge into v_2 and an outgoing edge from v_3. So at least one of the patterns $\{v_1, v_1 v_4, v_4\}$ is missing implying $|I(V)| \leq 16 - (2 + 2 + 1) = 11$, which is a contradiction. So $|V_{*3}| \geq 3$. If $|V_{*4}| < 5$, then (1) yields $|V_{*3}| = 4$, $|V_{*2}| = 2$ and $|V_{*1}| = 1$. We may assume that all outgoing edges from v_1 are in $E^+(V)$; see Fig. 5(b). By planarity, the incoming edges into v_2 are in $E^-(V)$. Depending on the side the outgoing edges from v_2 are on, either $v_1 v_3$ or an incoming edge into v_3 cannot exist, implying that either $v_1 v_3$ or v_3 is not in V_{*3}, therefore $|V_{*3}| < 4$, creating a contradiction. We conclude that $|V_{*4}| \geq 5$.

(iii) and (iv) follow from (i) and (ii), respectively, by symmetry.   □

Lemma 5. *Let V be a group of 4 vertices. Then V has at most 13 incidence patterns. If V has 13 incidence patterns, then V is either A or A^R in Fig. 6.*

Fig. 6. $I(A) = I(A^R) = \emptyset, 12, 123, 1234, 124, 13, 134, 2, 23, 234, 24, 3, 34$. A and A^R are the only groups with 13 incidence patterns.

Proof. Observe that group A in Fig. 6 has 13 patterns. Let V be a group of 4 vertices with at least 13 patterns. We first prove that V has an incoming edge into v_3 and an outgoing edge from v_2. Their existence combined with Lemma 2 implies that $\{v_3 v_4, v_3\} \subset I(V)$ and $\{v_1 v_2, v_2\} \subset I(V)$, respectively. At least one of these two edges has to be in $E(V)$, otherwise V has at most $16 - (2 + 2) = 12$ patterns. Assume that one of the two, without loss of generality, the outgoing edge from v_2 is not in $E(V)$. Then $\{v_1 v_3, v_2 v_4\} \subseteq E(V)$, otherwise either patterns $\{v_1v_3, v_1v_3v_4\}$ or $\{v_1v_2v_4, v_2v_4\}$ are not in $I(V)$ and there are at most $16 - (2+2) = 12$ patterns. By Lemma 2, there is an incoming edge into v_2 and an outgoing edge from v_3. Without loss of generality, the outgoing edge from v_3 is in $E^-(V)$. So by planarity v_2v_4 is in $E^+(V)$ which implies that v_1v_3 and the incoming edge into v_3 are in $E^-(V)$. By the same token, the incoming edge into v_2 is in $E^+(V)$. So by planarity the edge v_1v_4 and an outgoing edge from v_1 cannot be in $E(V)$. Then the patterns $\{v_1v_4, v_1\}$ are not in $I(V)$, thus V has at most $16 - (2+2) = 12$ patterns which is a contradiction.

We may assume that the incoming edge into v_3 is in $E^-(V)$. By planarity, the outgoing edge from v_2 is in $E^+(V)$. If the outgoing edge from v_1 is in $E(V)$, then by planarity it has to be in $E^+(V)$, which implies incoming edge into v_2 is in $E^-(V)$ and the edge v_1v_3 is not in $E(V)$. Since outgoing edge from v_1 implies only one pattern v_1 where the edge v_1v_3 implies two patterns $\{v_1v_3, v_1v_3v_4\}$, outgoing edge from v_1 cannot be in $E(V)$ but the edge v_1v_3 is in $E^-(V)$. By a similar argument we show that the incoming edge into v_4 cannot be in $E(V)$ and the edge v_2v_4 is in $E^+(V)$. Therefore V is A and it has 13 patterns.

If the incoming edge into v_3 is in $E^+(V)$, then V is A^R (again with 13 patterns). $\qquad\square$

Lemmas 1–5 together with other arguments omitted here give the following.

Lemma 6. *Consider a group UV consisting of two consecutive groups of 4 vertices, where $10 \leq |I(U)| \leq 13$ or $10 \leq |I(V)| \leq 13$. Then UV allows at most 120 incidence patterns.*

4   Proof of Theorem 1

Partition the path ξ_0 into groups of 8 consecutive vertices. A group of 8, denoted by UV, where U and V are the groups induced by the first and last four vertices of UV, respectively. If $|I(U)| \leq 9$ or $|I(V)| \leq 9$, then $|I(UV)| \leq |I(U)| \cdot |I(V)| \leq 9 \times 13 = 117$ by Lemma 5. Otherwise, Lemma 6 shows that $|I(UV)| \leq 120$. It follows by the product rule that the number of x-monotone paths is bounded above by $120^{n/8} < 1.8193^n$ if n is a multiple of 8, and by $O(1.8193^n)$ in general. Consequently, the number of monotone paths in any direction is $O(n\,120^{n/8}) = O(1.8193^n)$.

To verify the tightness of the upper bound of our analysis, consider the group $(U, E^-(U), E^+(U))$ of 8 vertices depicted in Fig. 7 (right). The first and second half of U are the groups $B2$ and $B3$, each with 12 patterns. Observe that exactly 24 patterns are incompatible, thus U has exactly $|I(B2)| \cdot |I(B3)| - 24 = 12 \cdot 12 - 24 = 120$ patterns.

The application of the same fingerprinting technique to groups of 9 vertices via a computer program[1] shows that a group of 9 allows at most 201 incidence patterns; the extremal configuration appears in Fig. 8. This yields a sharper upper bound of $O(n\,201^{n/9}) = O(1.8027^n)$ for the number of monotone paths in an n-vertex triangulation, as given in Theorem 1.

---

[1] The program will be made available on the arXiv in due time.

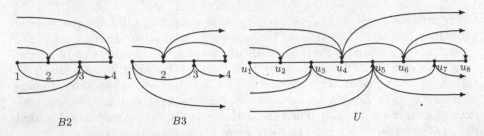

Fig. 7. U has 120 patterns. The 24 missing patterns are 123678, 12367, 12368, 1236, 123,13678,1367,1368,136,13,23678,2367, 2368,236,23,3678,367,368,36,3,678,67,68,6.

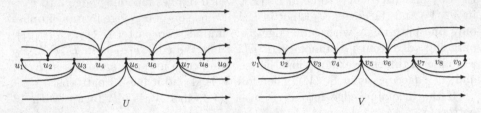

Fig. 8. Groups U and V (and their reflections in the x-axis) are the only groups of 9 vertices that have 201 patterns. Observe that V is the reflection of U in the y-axis.

To generate all groups of k vertices, the program first generates all possible *sides* of k vertices, essentially by brute force. A side of k vertices $V = \{v_1, \ldots v_k\}$ is represented by a directed planar graph with $k + 2$ vertices, where the edges $v_0 v_i$ and $v_i v_{k+1}$, for $1 \le i \le k$, denote an incoming edge into v_i and an outgoing edge from v_i, respectively. The edge $v_0 v_{k+1}$ represents the $\emptyset$ pattern. Note that $\xi_0 \cup v_0 v_{k+1}$ forms a plane cycle on $k + 2$ vertices in the underlying undirected graph. Therefore, $E^+(V)$ and $E^-(V)$ can each have at most $(k + 2) - 3 = k - 1$ edges. After all possible sides are generated, the program combines all pairs of sides with no common inner edge to generate a group $(V, E^-(V), E^+(V))$. For each generated group, the program calculates the corresponding number of patterns and in the end returns the group with the maximum number of patterns. $\square$

5   Algorithm for Counting Monotone Paths

Let $G = (V, E)$ be a planar straight-line graph with n vertices. We first observe that the number of x-monotone paths in G can be easily computed by a sweep-line algorithm (and similarly the number of **u**-monotone paths in G for any direction $\mathbf{u} \in \mathbb{R}^2 \setminus \{\mathbf{0}\}$). For every vertex $v \in V$, denote by $m(v)$ the number of (directed) nonempty x-monotone paths that end at v.

Sweep a vertical line ℓ from left to right, and whenever ℓ reaches a vertex v, we compute $m(v)$ according to the relation $m(v) = \sum_{q \in L^-(v)} [m(q) + 1]$, where $L^-(v)$ denotes the set of neighbors of vertex v in G that lie to the left of v.

The total number of x-monotone paths in G is $\sum_{v \in V} m(v)$. For every $v \in V$, the computation of $m(v)$ takes $O(\deg(v))$ time, thus computing $m(v)$ for $v \in V$ takes $\sum_{v \in V} O(\deg(v)) = O(n)$ time. Together with the sorting step, the algorithm for computing the number of x-monotone paths in G takes $O(n \log n)$ time.

For computing the total number of monotone paths over all directions $\mathbf{u} \in \mathbb{R}^2 \setminus \{\mathbf{0}\}$, some care is required. Note that it is enough to consider the $|E|$ directions parallel to the edges of G. However, we cannot simply sum up the number of monotone paths for all $|E|$ directions, since a monotone path in G may be monotone in several of these directions. Instead, for each new direction, we compute the number of *new* paths. In fact, we count *directed* monotone paths, that is, each path will be counted twice, as traversed in two opposite directions.

Sort the edges of G by their slopes. To avoid ties (corresponding to parallel edges), perturb the slopes arbitrarily to obtain a set $\mathcal{U}$ of $|E| \leq 3n - 6$ distinct directions. Let $\mathbf{u}_0 \in \mathcal{U}$ denote the direction of the minimum slope in E. We first compute the number of $\mathbf{u}_0$-monotone directed paths in G by the sweep-line algorithm described above in $O(n)$ time. Consider the directions $\mathbf{u} \in \mathcal{U} \setminus \{\mathbf{u}_0\}$, in increasing slope order. For each $\mathbf{u}$, we maintain the number of directed paths in G that are monotone in some direction between $\mathbf{u}_0$ and $\mathbf{u}$. For each new direction $\mathbf{u}$, exactly one directed edge, say (a, b) becomes $\mathbf{u}$-monotone. Therefore, it is enough to count the number of $\mathbf{u}$-monotone paths that traverse edge (a, b). These paths can easily be counted by sweeping G with a line ℓ orthogonal to $\mathbf{u}$: Sort the vertices in direction $\mathbf{u}$, and then count all $\mathbf{u}$-monotone paths that end at a, and all $\mathbf{u}$-monotone paths that start at b, both in $O(n)$ time. The total number of $\mathbf{u}$ monotone paths traversing (a, b) is the product of these two counts plus one (for the 1-edge path (a, b)). Note that the sorted order of vertices in direction $\mathbf{u}$ can be maintained in $O(n^2)$ time over *all* $|E| = O(n)$ directions [2, Ch. 8]. Consequently, the total running time of the algorithm is $O(n^2)$, as claimed. $\square$

References

1. Ajtai, M., Chvàtal, V., Newborn, M., Szemerèdi, E.: Crossing-free subgraphs. Ann. Discrete Math. **12**, 9–12 (1982)
2. de Berg, M., Cheong, O., van Kreveld, M., Overmars, M.: Computational Geometry: Algorithms and Applications, 3rd edn. Springer, Heidelberg (2008)
3. Buchin, K., Knauer, C., Kriegel, K., Schulz, A., Seidel, R.: On the number of cycles in planar graphs. COCOON 2007. LNCS, vol. 4598, pp. 97–107. Springer, Heidelberg (2007)
4. Dumitrescu, A., Löffler, M., Schulz, A., Tóth, C.D.: Counting carambolas. Graphs Combin. **32**(3), 923–942 (2016)
5. Dumitrescu, A., Rote, G., Tóth, C.D.: Monotone paths in planar convex subdivisions and polytopes. In: Bezdek, K., Deza, A., Ye, Y. (eds.) Discrete Geometry and Optimization. Fields Institute Communications, vol. 69, pp. 79–104. Springer, Cham (2013)
6. Dumitrescu, A., Schulz, A., Sheffer, A., Tóth, C.D.: Bounds on the maximum multiplicity of some common geometric graphs. SIAM J. Discrete Math. **27**(2), 802–826 (2013)

7. Dumitrescu, A., Tóth, C.D.: Computational geometry column 54. SIGACT News Bull. **43**(4), 90–97 (2012)
8. Dumitrescu, A., Tóth, C.D.: Convex polygons in geometric triangulations. In: Dehne, F., Sack, J.R., Stege, U. (eds.) WADS 2015. LNCS, vol. 9214, pp. 289–300. Springer, Cham (2015)
9. García, A., Noy, M., Tejel, A.: Lower bounds on the number of crossing-free subgraphs of K_N. Comput. Geom. **16**(4), 211–221 (2000)
10. Gärtner, B., Kaibel, V.: Two new bounds for the random-edge simplex-algorithm. SIAM J. Discrete Math. **21**(1), 178–190 (2007)
11. Kaibel, V., Mechtel, R., Sharir, M., Ziegler, G.M.: The simplex algorithm in dimension three. SIAM J. Comput. **34**(2), 475–497 (2005)
12. Kalai, G.: Upper bounds for the diameter and height of graphs of convex polyhedra. Discrete Comput. Geom. **8**(4), 363–372 (1992)
13. Kalai, G.: Polytope skeletons and paths. In: O'Rourke, J., Goodman, J.E. (eds.) Handbook of Discrete and Computational Geometry, pp. 455–476. CRC Press, Boca Raton (2004)
14. Klee, V.: Paths on polyhedra I. J. SIAM **13**(4), 946–956 (1965)
15. van Kreveld, M., Löffler, M., Pach, J.: How many potatoes are in a mesh? ISAAC 2012. LNCS, vol. 7676, pp. 166–176. Springer, Heidelberg (2012)
16. Matoušek, J., Szabó, T.: RANDOM EDGE can be exponential on abstract cubes. Adv. Math. **204**(1), 262–277 (2006)
17. Pach, J., Tóth, G.: Monotone drawings of planar graphs. J. Graph Theor. **46**, 39–47 (2004). Corrected version: arXiv:1101.0967 (2011)
18. Razen, A., Snoeyink, J., Welzl, E.: Number of crossing-free geometric graphs vs. triangulations. Electron. Notes Discrete Math. **31**, 195–200 (2008)
19. Santos, F.: A counterexample to the Hirsch conjecture. Ann. Math. **176**(1), 383–412 (2012)
20. Sharir, M., Sheffer, A.: Counting triangulations of planar point sets. Electron. J. Combin. **18**, P70 (2011)
21. Sharir, M., Sheffer, A.: Counting plane graphs: cross-graph charging schemes. Combin. Probab. Comput. **22**(6), 935–954 (2013)
22. Sharir, M., Sheffer, A., Welzl, E.: Counting plane graphs: perfect matchings, spanning cycles, and Kasteleyn's technique. J. Combin. Theor. Ser. A **120**(4), 777–794 (2013)
23. Sharir, M., Welzl, E.: On the number of crossing-free matchings, cycles, and partitions. SIAM J. Comput. **36**(3), 695–720 (2006)
24. Sheffer, A.: Numbers of plane graphs (2016). https://adamsheffer.wordpress.com/numbers-of-plane-graphs/
25. Todd, M.J.: The monotonic bounded Hirsch conjecture is false for dimension at least 4. Math. Oper. Res. **5**(4), 599–601 (1980)
26. Ziegler, G.M.: Lectures on Polytopes. GTM, vol. 152, pp. 83–93. Springer, New York (1994)

On Computing the Total Displacement Number via Weighted Motzkin Paths

Andreas Bärtschi[✉], Barbara Geissmann[✉], Daniel Graf, Tomas Hruz[✉], Paolo Penna[✉], and Thomas Tschager[✉]

Department of Computer Science, ETH Zurich, Zürich, Switzerland
{andreas.baertschi,barbara.geissmann,daniel.graf,tomas.hruz,
paolo.penna,thomas.tschager}@inf.ethz.ch

Abstract. Counting the number of permutations of a given total displacement is equivalent to counting weighted Motzkin paths of a given area (Guay-Paquet and Petersen [11]). The former combinatorial problem is still open. In this work we show that this connection allows to construct efficient algorithms for counting and for sampling such permutations. These algorithms provide a tool to better understand the original combinatorial problem. A by-product of our approach is a different way of counting based on certain "building sequences" for Motzkin paths, which may be of independent interest.

1 Introduction

Consider the set $\mathcal{S}_n$ of all permutations over n elements $\{1, 2, \ldots, n\}$. Diaconis and Graham [6] studied the *disarray* statistic of permutations, also called *total displacement* by Knuth [14, Problem 5.1.1.28], defined as follows. For any permutation π define its distance to the identity permutation as the sum of the displacements of all elements:

$$D(\pi) := \sum_{i=1}^{n} |i - \pi(i)| = 2 \sum_{\pi(i) > i} (\pi(i) - i).$$

Note that this distance is always *even*. The following natural question is still unresolved:

> *How many permutations at a given distance $2d$ from the identity permutation are there?*

That is, one would like to know the following *total displacement* number:

$$D(n, d) := |\{\pi \in \mathcal{S}_n \mid D(\pi) = 2d\}|,$$

that is the number of permutations of total displacement equal to $2d$. So far, a closed formula for arbitrary n is only known for fixed d *up to seven* ($d \leq 7$) [11]. Entry A062869 [8] of the OEIS previously reported values of $D(n, d)$ for small n and d ($n \leq 30$).

© Springer International Publishing Switzerland 2016
V. Mäkinen et al. (Eds.): IWOCA 2016, LNCS 9843, pp. 423–434, 2016.
DOI: 10.1007/978-3-319-44543-4_33

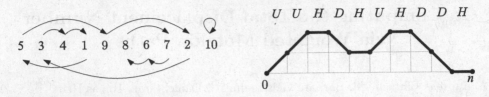

Fig. 1. A permutation and its Motzkin path of width 10 and area 12.

Guay-Paquet and Petersen [11] made recently significant progress in this question by showing that these permutations are in correspondence to *Motzkin paths* whose *area* is exactly the distance d under consideration. Their result shows that, for any Motzkin path (see below) of area d, one can easily calculate the number of permutations that correspond to this specific path. Therefore the problem above translates into the problem of counting *weighted* Motzkin paths of a given area.

A Motzkin path consists of a sequence of U (Up-right), H (Horizontal-right), and D (Down-right) moves over the two-dimensional lattice starting at coordinate $(0,0)$ and such that the path never goes below the x-axis and ends on the x-axis (see Fig. 1 (right) for an example). For any such path, one can consider its *width* and its *area* defined as the number of moves and the area of the region between the $y = 0$ axis and the path. The permutations over n elements with total displacement $2d$ map into Motzkin paths of width n and area $A = d$.

For instance, the permutation in Fig. 1 is mapped into a Motzkin path according to the following rule. The first element $\pi(1) = 5$ is mapped into a U because the element at position 1 goes to a higher position (right) and also the number coming into position 1 is higher than 1: $\pi(1) > 1 < \pi^{-1}(1)$. The fourth element is mapped into D because the opposite happens: $\pi(4) = 1 < 4 > 3 = \pi^{-1}(4)$. Finally, elements $3, 5, 7, 10$ are mapped into H because neither of the previous cases apply.

Let h_i denote the maximum height of the path during move i (for U: after the move, for D: before the move, and anytime for H). Then the number $\omega(mz)$ of permutations that map to a certain Motzkin path mz is [11]

$$\omega(mz) = \prod_i^n \omega_i \text{ where } \omega_i = \begin{cases} h_i & \text{if } mz_i = U \quad \text{or} \quad mz_i = D, \\ 2h_i + 1 & \text{if } mz_i = H. \end{cases} \tag{1}$$

We also refer to $\omega(mz)$ as the *weight* of mz. In the example in Fig. 1 this gives $1 \cdot 2 \cdot 5 \cdot 2 \cdot 3 \cdot 2 \cdot 5 \cdot 2 \cdot 1 \cdot 1 = 1200$. Note how this formula separates over the moves of the Motzkin path. This independence is what we will exploit in this article.

Theorem 1 ([11]). *For any n and d, let $MZ(n, A)$ be the set of all Motzkin paths of width n and area $A = d$. Then it holds that*

$$D(n, d) = \sum_{mz \in MZ(n,A)} \omega(mz). \tag{2}$$

Corollary 1 ([2]). *Given a Motzkin path mz of length n, we can sample uniformly at random one of the ω(mz) many permutations mapping into mz in time $\mathcal{O}(n)$.*

Our contribution. In this work, we address counting and sampling of permutations from both a combinatorial and computational point of view. Specifically:

- On the computational side, we show that the total displacement number $D(n, d)$ can be computed efficiently, namely, in time $\mathcal{O}(n^4)$ and space $\mathcal{O}(n^3)$.
- On the combinatorial side, we introduce sequences of certain *building blocks* which provide a different perspective on the problem structure. Moreover, this is a crucial part of a Markov chain sampling method which constitutes the third contribution of this paper.
- Finally, we consider the task of *sampling* permutations of a given total displacement with uniform distribution.

To compute the number of permutations efficiently, we look at the paths from left to right. Building on an operation introduced by Barcucci et al. [1], we can provide an elegant dynamic programming formulation which achieves a running time of $\mathcal{O}(n^4)$ and needs space $\mathcal{O}(n^3)$. Consequently, we can compute the sequences A062869 [8] and A129181 [4] to much higher values of n and d than was possible before.

Considering the combinatorial aspects, we show that every Motzkin path comes from a sequence a describing its building blocks. We provide an explicit formula for the number $m(a)$ of paths that these building blocks can form. The weights in Eq. (1) are preserved in the sense that the weight of a path depends only on its building sequence.

Since the exact formula seems to be currently out of reach, we contribute a dynamic programming algorithm which computes $D(n, d)$ for large n and d. Given the dynamic programming table one can efficiently sample permutations of total displacement $2d$ in linear time $\mathcal{O}(n)$. Further, we show that sampling sequences of building blocks with the appropriate distribution automatically gives a sampler for the permutations. One application of the latter result is a Monte Carlo Markov chain (MCMC) method which gives an alternative approach to the dynamic programming. The computational experiments with the MCMC method show a promising convergence speed leading to a sampler for very high values of n and d. The experimental results support a hypothesis that the MCMC method is faster than the method based on dynamic programming and runs in $\mathcal{O}(n^3)$ time.

Omitted proofs and an extensive experimental evaluation of our MCMC can be found in the full version of the paper [2].

Related Work. Different metrics on permutations have been studied, for a survey see [5]. Sampling and counting of permutations of a fixed distance was studied for several metrics [13] but not for total displacement.

The number of Motzkin paths under various conditions were also studied in a more general frame of enumeration of lattice paths [9,12]. Motzkin numbers play a role in many combinatorial problems as is illustrated for example in [7].

The total area under a set of generalized Motzkin paths, where the horizontal segments have a constant length k ($k \geq 0$) have been studied in [16] and [15]. Moreover, the author in [17] studies the moments of generalized Motzkin paths where the first moment describes the area under a Motzkin path. Heinz [4] describes a different algorithm for enumerating unweighted Motzkin paths with a given area, cf. Remark 2 in Sect. 2.1.

The use of Markov chains for sampling and counting combinatorial objects is a very active research area (see e.g. the book [3]), and some works exploit the connection between combinatorial structures and paths of a certain type to accomplish this task (see e.g. [10]).

Paper Organization. Section 2 describes the dynamic programming algorithm. Section 3 describes how weighted Motzkin paths can be counted via building block sequences. Section 4 provides a Markov chain sampling algorithm as well as its experimental evaluation.

2   Weighted Motzkin Paths Using Dynamic Programming

Recall that we denote by $D(n, d)$ the number of permutations on n elements with total displacement $2d$ (OEIS A062869 [8]). Let $M(n, A)$ denote the number of Motzkin paths of width n and area A (OEIS A129181 [4]).

2.1   Dynamic Program for Counting Weighted Motzkin Paths

Theorem 2. *Computing $M(n, A)$ and $D(n, d)$ can be done in time $\mathcal{O}(n^4)$ and space $\mathcal{O}(n^3)$.*

Proof. The key ingredient is a construction by Barcucci et al. [1] that produces every possible Motzkin path through a unique sequence of insertion steps. Let us look at the last fall of a given Motzkin path, i.e., its suffix of Down-right moves. At one of the positions before or after any of these fall moves, we insert a new *peak* (a U and a D) or we insert a new *flat* (an H). Repeatedly inserting peaks and flats this way along the last fall will create our path. See Fig. 2 for an example. This construction is complete and unique [1], meaning that every Motzkin path can be created through a unique sequence of such insertions.

This allows us to derive a dynamic programming formulation for counting $M(n, A)$. We add the last fall length l to our state and write it as $M(n, A, l)$. So how can we recursively express $M(n, A, l)$? We undo the last insertion step. If we inserted a flat last, then we were at $M(n-1, A-l, l')$ before the insertion, for some $l' \geq l$, because the last fall was at least as long before the insert. When inserting a peak, we might increase the last fall length by one, but not more. So $M(n-2, A-(2l-1), l')$ for all $l' \geq l-1$ are also possible predecessor states. Together with the base case $M(0, 0, 0) = 1$ this gives the recurrence

$$M(n, A, l) = \sum_{l' \geq l}^{n/2} M(n-1, A-l, l') + \sum_{l' \geq l-1}^{n/2} M(n-2, A-(2l-1), l'), \quad (3)$$

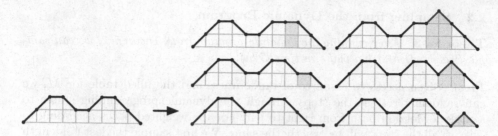

Fig. 2. All six possible flat- and peak-extensions of the last fall of length 2.

which allows for $\mathcal{O}(n^4)$ many states as $A \leq n^2$ and $l \leq n$. Hence we immediately get an $\mathcal{O}(n^5)$ time algorithm with $\mathcal{O}(n^4)$ space. We can shave off one factor of n in both time and space as follows: We first note, that we can compute the two sums in constant time if we precompute the prefix sums over the last variable l'. Let us denote these prefix sums as $SM(n, A, l) = \sum_{l'=0}^{l} M(n, A, l') = SM(n, A, l-1) + M(n, A, l)$. This allows us to compute every value of $M(n, A, l)$ in amortized constant time, so in time $\mathcal{O}(n^4)$ overall. Finally, our recurrence only relies on the last two values of n, so when computing $M(n, \cdot, \cdot)$ only the $\mathcal{O}(n^3)$ many values for $M(n-1, \cdot, \cdot)$ and $M(n-2, \cdot, \cdot)$ need to be stored. The values $M(n, A)$ are then simply the marginals of $M(n, A, l)$ over all last fall lengths l.

We can extend this recurrence to the weighted case which by Corollary 1 gives rise to the total displacement count: We distribute the factors of the weight $\omega(mz)$ (Eq. (1)) over the steps of the dynamic program. As l denotes the height of the last flat or peak that we add, we have factors $2l + 1$ or l^2:

$$D(n, d, l) = (2l + 1) \sum_{l' \geq l}^{n/2} D(n-1, d-l, l') + l^2 \sum_{l' \geq l-1}^{n/2} D(n-2, d-(2l-1), l'). \quad \square$$

Remark 1. The bounds in Theorem 2 assumed that basic operations have unit-cost. The numbers involved can be exponential in n however. We can easily bound $M(n, A) \leq 3^n$ and $D(n, d) \leq n!$ showing that their bit-representations are at most of length $\mathcal{O}(n \log n)$. Our dynamic programs only use multiplication with small numbers of size $\mathcal{O}(\log n)$ and addition. So one can consider a refined analysis by multiplying both the time and space bounds of Theorem 2 by $\Theta(n \operatorname{polylog} n)$. Finally, as suggested by an anonymous reviewer, the space could be further improved by counting modulo small primes and using the Chinese Reminder Theorem.

Remark 2. For computing $M(n, A)$, the OEIS contains a dynamic program by Heinz [4]. It is stated as a *Maple* code snippet without any further comment or reference. It uses a different state and might have the same time complexity as ours. We believe that our extension to the weights of $D(n, d)$ can also be applied.

2.2 Sampling from the Dynamic Program

Theorem 3. *After running the dynamic program from Theorem 2, we can sample (weighted) Motzkin paths in time $\mathcal{O}(n)$.*

Proof. Given access to a source of randomness and the filled table for M, we can randomly retrace the steps through the dynamic programming states to sample a Motzkin path from right to left. For the weighted paths according to $D(n, d)$ all the steps will be exactly the same. We first sample the last fall length by picking a random number $x \in_{\text{u.a.r.}} \{0, \ldots, M(n, A) - 1\}$ and then finding the smallest l such that its prefix sum $SM(n, A, l)$ is larger than x. We continue with $x - SM(n, A, l - 1)$, the offset within the class of paths with last fall length l. For each step, we first decide whether we are in the flat-case or in the peak-case of the recurrence by comparing x to the left summand of (3). We then know whether the move before the last fall was an H or a U. We increment l' until we find the last fall length of the previous state. We adapt x and recurse until we end at $M(0, 0, 0)$ with $x = 0$. Note that the search for the initial l takes linear time. After that, every time we compare x to a value of M, we fix at least one move of the sampled Motzkin path, so sampling takes $\mathcal{O}(n)$ time overall. □

Remark 3. This sampling procedure requires the full table of the dynamic program to be stored. Hence the memory optimization from $\mathcal{O}(n^4)$ to $\mathcal{O}(n^3)$ in Theorem 2 can not be used simultaneously.

Remark 4. A C++ implementation of our counting and sampling approaches by Theorems 2 and 3 is available at http://people.inf.ethz.ch/grafdan/motzkin/. With our code, we can quickly compute for n up to 100 (and all d) the integer sequences A062869 [8] and A129181 [4] of which the former was only known up to $n \leq 30$ before.

3   Combinatorial Structure of Motzkin Paths

In this section, we look at the combinatorial structure of Motzkin paths: There is a natural decomposition of any Motzkin path into "building blocks", already hinted at in the last section. For each height i of the Motzkin path we count the number of flats f_i and peaks p_i.

Definition 1 (building sequence). *For given positive integers n and A, a finite sequence of non-negative integers $a = (f_0, p_1, f_1, p_2, \ldots, p_h, f_h)$ is a building sequence if all p-entries are non-zero, $p_1, p_2, \ldots, p_h > 0$, and the following two conditions hold:*

$$(f_0 + f_1 + \ldots + f_h) + 2(p_1 + p_2 + \ldots + p_h) = n, \qquad (4)$$
$$(0f_0 + 1f_1 + \ldots + hf_h) + (1p_1 + 3p_2 + \ldots + (2h - 1)p_h) = A. \qquad (5)$$

The set of all building sequences satisfying (4)–(5) is denoted as $S(n, A)$.

7

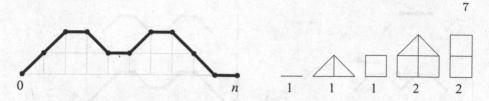

Fig. 3. The Motzkin path on the left can be obtained from its building blocks.

Such a sequence has a natural interpretation as a set of "building blocks" that generate a number of Motzkin paths of width n and area A (see Fig. 3): We have f_i flats and p_i peaks of height i which can be split into pieces of width 1 and then rearranged into a Motzkin path.

Proposition 1. *For any Motzkin path mz of width n and area A there exists a unique building sequence $a^{(mz)} \in S(n, A)$ such that mz can be obtained by splitting and rearranging the blocks of this sequence.*

Theorem 1 gives a surjective mapping from permutations into Motzkin paths. It is easy to see that the number of permutations $\omega(mz)$ mapping into the same path mz, given by Eq. (1), is uniquely determined by the building block sequence $a = a^{(mz)}$, since we have

$$perm(a) := \prod_{f_i} (2i+1)^{f_i} \prod_{p_i} i^{2p_i} = \omega(mz). \tag{6}$$

Hence $\omega(mz)$ is independent of the actual Motzkin path and only depends on its combinatorial structure. This raises the question of whether also the number of Motzkin paths which share a common building sequence a is solely determined by a. We answer this in the positive, deriving a formula for this number, denoted by $m(a)$. We proceed in a top-down fashion by looking at the number of peaks and flats in the highest level and how these can be rearranged. Once a level is fixed, we proceed recursively by arranging the blocks one level below.

Theorem 4. *For any building sequence $a = (f_0, p_1, f_1, \ldots, p_h, f_h) \in S(n, A)$, the number of Motzkin paths of width n and area A that can be constructed out of the building sequence a is exactly*

$$m(a) = \binom{f_h + p_h - 1}{p_h - 1}\binom{p_h + f_{h-1}}{f_{h-1}}\binom{p_h + f_{h-1} + p_{h-1} - 1}{p_{h-1} - 1}\binom{p_{h-1} + f_{h-2}}{f_{h-2}}\cdots$$

$$\cdots\binom{p_3 + f_2 + p_2 - 1}{p_2 - 1}\binom{p_2 + f_1}{f_1}\binom{p_2 + f_1 + p_1 - 1}{p_1 - 1}\binom{p_1 + f_0}{f_0}. \tag{7}$$

Proof. We start with the highest flats of the sequence a. There are f_h of those flats. Two (or more) such flats can either lie directly next to each other, or they

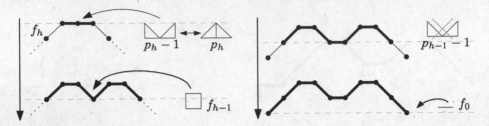

Fig. 4. The top down construction of paths from the given sequence $(1, 1, 1, 2, 2)$. Note that $p_1 - 1 = 0$, and thus no DU valley is inserted at height 1.

might be separated by a *Down-right* move followed at some point by an *Up-right* move. We call this setting a DU valley; we get such valleys by splitting peaks of height h and reassembling them the other way round, see Fig. 4. A feasible Motzkin path has to have a U slope at the very left and a D slope at the very right of all height h pieces. The remaining $p_h - 1$ DU valleys can be freely placed around the f_h flats, that is we choose their places from $f_h + p_h - 1$ available positions. The number of ways to do this is

$$\binom{f_h + p_h - 1}{p_h - 1} \tag{8}$$

Now we continue on the second highest level $h - 1$. Naturally, the number of times that our Motzkin path rises above level $h - 1$ is exactly the number p_h of peaks of height h. We can distribute our f_{h-1} flats of height $h - 1$ around those peaks, i.e. pick from $p_h + f_{h-1}$ many positions, hence we can choose from

$$\binom{p_h + f_{h-1}}{f_{h-1}} \tag{9}$$

many possibilities. After placing the flats, we will have to place new valleys down to the next lower level around the existing p_h peaks and f_{h-1} flats. As before, the leftmost up and down slopes are fixed, hence the number of ways to distribute $p_{h-1} - 1$ valleys is given by the third factor in Eq. (7). Since the choices in different levels are independent, we can iterate this reasoning until we include flats of height 0.   □

We conclude with a corollary of Theorems 1 and 4:

Corollary 2. *There exists a surjective mapping from permutations over n elements into* building sequences *satisfying the following condition: For any building sequence $a \in S(n, A)$, the number of permutations π which are at distance $D(\pi) = 2d = 2A$ from the identity permutation and that are mapped into this building sequence a is precisely*

$$P(a) := m(a) \cdot perm(a), \tag{10}$$

where $m(a)$ is given by Eq. (7) and $perm(a)$ by Eq. (6). Therefore the total number of permutations at distance $2d = 2A$ from the identity permutation satisfies

$$D(n, d) \overset{(2)}{=} \sum_{mz \in MZ(n,A)} \omega(mz) = \sum_{a \in S(n,A)} P(a). \tag{11}$$

Example 1. The building blocks in Fig. 3 yield $\binom{3}{1}\binom{3}{1}\binom{3}{0}\binom{2}{1} = 18$ Motzkin paths, and each path corresponds to 1200 permutations. So, there are $1200 \cdot 18 = 21\,600$ permutations mapping into the building sequence $a = (1, 1, 1, 2, 2)$.

Remark 5. Theorem 4 and Corollary 2 allow for a dynamic program for counting and sampling weighted Motzkin paths, similar to Sects. 2.1 and 2.2. Additionally, we can easily sample paths with a fixed number of highest peaks and flats, at the cost of an additional $O(n^3)$-factor in the running time, see [2].

4   Sampling Weighted Motzkin Paths by Length and Area

In this section, we consider the task of selecting (sampling) permutations with uniform distribution over all permutations of a given total displacement. By Corollary 1 it is enough to sample Motzkin paths with the proper weights. We have already seen in Sect. 2.2 that we can sample such weighted Motzkin paths using dynamic programming at the cost of large memory consumption.

We will show in Sect. 4.1 an approach to sample weighted Motzkin paths based on the building sequences introduced in Sect. 3 that requires only $\mathcal{O}(n)$ memory. In general, observe that sampling permutations can be accomplished efficiently if we can sample building sequences with a probability proportional to $P(a) = m(a) \cdot perm(a)$ in polynomial time:

Theorem 5. *Every polynomial-time algorithm that samples sequences in $S(n, A)$ with probability $\pi(a) \propto P(a)$ can be turned into a polynomial-time algorithm for sampling permutations uniformly at random among the permutations over n elements and of total displacement $2d = 2A$.*

Proof (Sketch). The sampler maps the sequence into a random Motzkin path, and then into a random permutation as follows: (1) Pick a Motzkin path mz uniformly at random among those that can be created with a, that is, with probability $1/m(a)$; (2) Pick a permutation uniformly at random among those that map into the Motzkin path mz, that is, with probability $1/perm(a)$, according to Corollary 1. Using Corollary 2 it is immediate to see that every permutation has probability of being selected equal to $1/D(n, d)$.

4.1   A Markov Chain Sampler

Suppose we have a set of k possible local changes transforming any sequence a into another sequence a' such that all sequences can be obtained by applying a certain number of such operations. Then the following standard Metropolis chain samples sequences with the desired distribution:

1. With probability $\frac{1}{2}$ do nothing. Otherwise,
2. Select one of the k local operations u.a.r. If this operation cannot be applied to the current sequence a (the new sequence is unfeasible) do nothing; Otherwise, let a' be the sequence obtained from a by applying this operation;
3. Accept the operation transforming a to a' with probability

$$A(a, a') := \min\left\{1, \frac{P(a')}{P(a)}\right\} = \min\left\{1, \frac{m(a')}{m(a)} \cdot \frac{perm(a')}{perm(a)}\right\}, \qquad (12)$$

and do nothing with remaining probability $1 - A(a, a')$.

Local operations over the sequences. We define our Metropolis chain $\mathcal{M}_{blocks}$ through four types of operations: Peak to Flat (PF), Flat to Valley (FV), Flat to Flat (FF), and Peak into Valley (PV). We formally define them as:

$$PF(i,j) := \begin{cases} p_i & \leftarrow p_i - 1 \\ f_{i-1} & \leftarrow f_{i-1} + 2 \\ f_j & \leftarrow f_j - 1 \\ f_{j+1} & \leftarrow f_{j+1} + 1 \end{cases}, \qquad FV(i,j) := \begin{cases} f_i & \leftarrow f_i - 2 \\ p_i & \leftarrow p_i + 1 \\ f_j & \leftarrow f_j - 1 \\ f_{j+1} & \leftarrow f_{j+1} + 1 \end{cases}$$

$$FF(i,j) := \begin{cases} f_i & \leftarrow f_i - 1 \\ f_{i+1} & \leftarrow f_{i+1} + 1 \\ f_j & \leftarrow f_j - 1 \\ f_{j-1} & \leftarrow f_{j-1} + 1 \end{cases} \qquad PV(i,j) := \begin{cases} p_i & \leftarrow p_i - 1 \\ f_{i-1} & \leftarrow f_{i-1} + 2 \\ p_j & \leftarrow p_j - 1 \\ f_j & \leftarrow f_j + 2 \end{cases}$$

Note that each type of operation applies to two indices i and j, and we also implicitly consider the reversed operations which "undo" the changes. We now explain step 4.1 of the chain $\mathcal{M}_{blocks}$ in more detail: The Markov chain $\mathcal{M}_{blocks}$ picks two indices i and j at random, then picks one of the four operations above, and decides with probability $1/2$ whether to choose the operation or its reversed version. As for step 4.1, computing the transitional probability $A(a, a')$ can be done in constant time as only a few of the factors in Eqs. (6) and (7) change.

Theorem 6. *The Markov chain $\mathcal{M}_{blocks}$ defined above is ergodic and its unique stationary distribution satisfies $\pi(a) \propto P(a)$ for every $a \in S(n, A)$.*

Experimental Evaluation of $\mathcal{M}_{blocks}$. We are interested in the required number of steps until the distribution of $\mathcal{M}_{blocks}$ is sufficiently close to its stationary distribution. We measure the distance between two distributions by the *total variation distance*. The *mixing time* of a Markov chain is the smallest time t such that the total variation distance between the stationary distribution and the distribution after t steps, starting from any state, is smaller than some small $\epsilon > 0$.

We study the mixing time of $\mathcal{M}_{blocks}$ for a given area A and a given width n by running the following experiment. We estimate the distribution after a given number of steps by repeatedly running $\mathcal{M}_{blocks}$ with an initial state a_0 defined as follows: The building block sequence consists of one peak of height

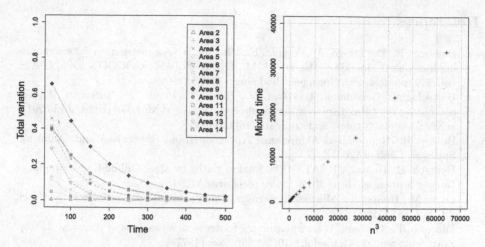

Fig. 5. (left) Total variation for $n = 8$ and all $A \leq (n/2)^2$ with $|S(8, A)| > 1$. (right) Maximal mixing time for given widths ($\bullet$), mixing time for areas A_n^* (+).

h for every $h \leq \lfloor \sqrt{A} \rfloor$ and the remaining area and width is filled greedily with flats of maximal possible height. The total variation distance of the distribution of $\mathcal{M}_{blocks}$ after some number of steps t from its stationary distribution π is

$$d_{TV}(P^t(a_0, \cdot), \pi) = \frac{1}{2} \cdot \sum_{a \in S(n, A)} \left| P^t(a_0, a) - \pi(a) \right|.$$

We estimate the mixing time for a given area A and a given width w by computing the total variation distance for increasing t until the total variation distance is below 0.05.

Figure 5 (left) illustrates the mixing time for width 8 and every area A with more than one possible building block sequence. The maximal mixing time (400 steps) is necessary for area 9. In fact, for every width smaller than 13, the mixing time is maximal for area $A_n^* = ((n - 2)/2)^2$ if n is even and $A_n^* = ((n - 1)/2)^2$ otherwise. This is due to our choice of the initial state of $\mathcal{M}_{blocks}$. We estimate the maximal mixing time for widths larger than 12 by computing the mixing time for A_n^* *only*, as the number of repeats necessary to estimate the distribution of $\mathcal{M}_{blocks}$ after t steps depends on the number of possible building block sequences, which grows exponentially depending on n. Figure 5 (right) shows the maximal mixing time up to width 40. The plot suggests that the number of steps necessary to approximate the stationary distribution does not grow exponentially depending on the width n, the algorithm is probably faster than the sampler based on dynamic programming and the results suggest that the MCMC sampler achieves the mixing time $\mathcal{O}(n^3)$.

Conjecture 1. $\mathcal{M}_{blocks}$ mixes in time $\mathcal{O}(n^3)$.

Remark 6. The implementation of $\mathcal{M}_{blocks}$ is available at http://people.inf.ethz.ch/grafdan/motzkin/.

References

1. Barcucci, E., Del Lungo, A., Pergola, E., Pinzani, R.: A construction for enumerating k-coloured Motzkin paths. In: Li, M., Du, D.-Z. (eds.) COCOON 1995. LNCS, vol. 959, pp. 254–263. Springer, Heidelberg (1995)
2. Bärtschi, A., Geissmann, B., Graf, D., Hruz, T., Penna, P., Tschager, T.: On computing the total displacement number via weighted Motzkin paths, June 2016, arXiv preprint. https://arxiv.org/abs/1606.05538
3. Bubley, R.: Randomized Algorithms: Approximation, Generation and Counting. Springer, London (2001)
4. Deutsch, E., Heinz, A.P.: A129181 Motzkin paths by area, Online Encyclopedia of Integer Sequences, June 2012. http://oeis.org/A129181
5. Deza, M., Huang, T.: Metrics on permutations, a survey. J. Comb. Inf. Syst. Sci. **23**, 173–185 (1998)
6. Diaconis, P., Graham, R.L.: Spearman's footrule as a measure of disarray. J. Roy. Stat. Soc.: Ser. B (Methodol.) **39**(2), 262–268 (1977)
7. Donaghey, R., Shapiro, L.W.: Motzkin numbers. J. Comb. Theory: Ser. A **23**(3), 291–301 (1977)
8. Gérard, O., Guay-Paquet, M., Heinz, A.P.: A062869 permutation with fixed total displacement, Online Encyclopedia of Integer Sequences, May 2014. https://oeis.org/A062869
9. Goulden, I.P., Jackson, D.M.: Combinatorial Enumeration. Dover Publications, Mineola (2004)
10. Greenberg, S., Pascoe, A., Randall, D.: Sampling biased lattice configurations using exponential metrics. In: 20th ACM-SIAM Symposium on Discrete Algorithms SODA 2009, pp. 76–85 (2009)
11. Guay-Paquet, M., Petersen, K.: The generating function for total displacement. Electron. J. Comb. **21**(3), P3–37 (2014)
12. Humphreys, K.: A history and a survey of lattice path enumeration. J. Stat. Plan. Infer. **140**(8), 2237–2254 (2010)
13. Irurozki, E.: Sampling and learning distance-based probability models for permutation spaces. Ph.D. thesis, University of the Basque Country, Donostia - San Sebastián, July 2014
14. Knuth, D.E.: The art of computer programming. Sorting Search. **3**, 426–458 (1999)
15. Merlini, D.: Generating functions for the area below some lattice paths. In: Discrete Random Walks, DRW 2003, pp. 217–228 (2003)
16. Pergola, E., Pinzani, R., Rinaldi, S., Sulanke, R.: A bijective approach to the area of generalized Motzkin paths. Adv. Appl. Math. **28**(3), 580–591 (2002)
17. Sulanke, R.A.: Moments of generalized Motzkin paths. J. Integer Sequences **3**(00.1), 1–14 (2000)

Probabilistics

Partial Covering Arrays: Algorithms and Asymptotics

Kaushik Sarkar[1]([✉]), Charles J. Colbourn[1], Annalisa de Bonis[2], and Ugo Vaccaro[2]

[1] CIDSE, Arizona State University, Tempe, USA
ksarkar1@asu.edu
[2] Dipartimento di Informatica, University of Salerno, Fisciano, Italy

Abstract. A covering array $CA(N; t, k, v)$ is an $N \times k$ array with entries in $\{1, 2, \ldots, v\}$, for which *every* $N \times t$ subarray contains *each* t-tuple of $\{1, 2, \ldots, v\}^t$ among its rows. Covering arrays find application in interaction testing, including software and hardware testing, advanced materials development, and biological systems. A central question is to determine or bound $CAN(t, k, v)$, the minimum number N of rows of a $CA(N; t, k, v)$. The well known bound $CAN(t, k, v) = O((t - 1)v^t \log k)$ is not too far from being asymptotically optimal. Sensible relaxations of the covering requirement arise when (1) the set $\{1, 2, \ldots, v\}^t$ need only be contained among the rows of *at least* $(1 - \epsilon)\binom{k}{t}$ of the $N \times t$ subarrays and (2) the rows of *every* $N \times t$ subarray need only contain a (large) *subset* of $\{1, 2, \ldots, v\}^t$. In this paper, using probabilistic methods, significant improvements on the covering array upper bound are established for both relaxations, and for the conjunction of the two. In each case, a randomized algorithm constructs such arrays in expected polynomial time.

1 Introduction

Let $[n]$ denote the set $\{1, 2, \ldots, n\}$. Let $N, t, k,$ and v be integers such that $k \geq t \geq 2$ and $v \geq 2$. Let A be an $N \times k$ array where each entry is from the set $[v]$. For $I = \{j_1, \ldots, j_\rho\} \subseteq [k]$ where $j_1 < \ldots < j_\rho$, let A_I denote the $N \times \rho$ array in which $A_I(i, \ell) = A(i, j_\ell)$ for $1 \leq i \leq N$ and $1 \leq \ell \leq \rho$; A_I is the projection of A onto the columns in I.

A *covering array* $CA(N; t, k, v)$ is an $N \times k$ array A with each entry from $[v]$ so that for each t-set of columns $C \in \binom{[k]}{t}$, each t-tuple $x \in [v]^t$ appears as a row in A_C. The smallest N for which a $CA(N; t, k, v)$ exists is denoted by $CAN(t, k, v)$.

Covering arrays find important application in software and hardware testing (see [22] and references therein). Applications of covering arrays also arise in experimental testing for advanced materials [4], inference of interactions that regulate gene expression [29], fault-tolerance of parallel architectures [15], synchronization of robot behavior [17], drug screening [30], and learning of boolean functions [11]. Covering arrays have been studied using different nomenclature,

© Springer International Publishing Switzerland 2016
V. Mäkinen et al. (Eds.): IWOCA 2016, LNCS 9843, pp. 437–448, 2016.
DOI: 10.1007/978-3-319-44543-4_34

as qualitatively independent partitions [13], t-surjective arrays [5], and (k,t)-universal sets [19], among others. Covering arrays are closely related to hash families [10] and orthogonal arrays [8].

2 Background and Motivation

The exact or approximate determination of $\mathsf{CAN}(t,k,v)$ is central in applications of covering arrays, but remains an open problem. For fixed t and v, only when $t=v=2$ is $\mathsf{CAN}(t,k,v)$ known precisely for infinitely many values of k. Kleitman and Spencer [21] and Katona [20] independently proved that the largest k for which a $\mathsf{CA}(N;2,k,2)$ exists satisfies $k = \binom{N-1}{\lceil N/2 \rceil}$. When $t=2$, Gargano, Körner, and Vaccaro [13] establish that

$$\mathsf{CAN}(2,k,v) = \frac{v}{2}\log k(1+\mathrm{o}(1)). \tag{1}$$

(We write log for logarithms base 2, and ln for natural logarithms.) Several researchers [2,5,14,16] establish a general asymptotic upper bound on $\mathsf{CAN}(t,k,v)$:

$$\mathsf{CAN}(t,k,v) \le \frac{t-1}{\log \frac{v^t}{v^t-1}}\log k(1+\mathrm{o}(1)). \tag{2}$$

A slight improvement on (2) has recently been proved [12,28]. An (essentially) equivalent but more convenient form of (2) is:

$$\mathsf{CAN}(t,k,v) \le (t-1)v^t \log k(1+o(1)). \tag{3}$$

A lower bound on $\mathsf{CAN}(t,k,v)$ results from the inequality $\mathsf{CAN}(t,k,v) \ge v \cdot \mathsf{CAN}(t-1,k-1,v)$ obtained by derivation, together with (1), to establish that $\mathsf{CAN}(t,k,v) \ge v^{t-2} \cdot \mathsf{CAN}(2,k-t+2,v) = v^{t-2} \cdot \frac{v}{2}\log(k-t+2)(1+\mathrm{o}(1))$. When $\frac{t}{k} < 1$, we obtain:

$$\mathsf{CAN}(t,k,v) = \Omega(v^{t-1}\log k). \tag{4}$$

Because (4) ensures that the number of rows in covering arrays can be considerable, researchers have suggested the need for relaxations in which not all interactions must be covered [7,18,23,24] in order to reduce the number of rows. The practical relevance is that each row corresponds to a test to be performed, adding to the cost of testing.

For example, an array *covers a t-set of columns* when it covers each of the v^t interactions on this t-set. Hartman and Raskin [18] consider arrays with a fixed number of rows that cover the *maximum* number of t-sets of columns. A similar question was also considered in [24]. In [23,24] a more refined measure of the (partial) coverage of an $N \times k$ array A is introduced. For a given $q \in [0,1]$, let $\alpha(A,q)$ be the number of $N \times t$ submatrices of A with the property that at least qv^t elements of $[v]^t$ appear in their set of rows; the (q,t)-*completeness* of A is $\alpha(A,q)/\binom{k}{t}$. Then for practical purposes one wants "high"(q,t)-completeness with few rows.

In these works, no theoretical results on partial coverage appear to have been stated; earlier contributions focus on experimental investigations of heuristic construction methods. Our purpose is to initiate a mathematical investigation of arrays offering "partial" coverage. More precisely, we address:

- Can one obtain a significant improvement on the upper bound (3) if the set $[v]^t$ is only required to be contained among the rows of *at least* $(1 - \epsilon)\binom{k}{t}$ subarrays of A of dimension $N \times t$?
- Can one obtain a significant improvement if, among the rows of *every* $N \times t$ subarray of A, only a (large) *subset* of $[v]^t$ is required to be contained?
- Can one obtain a significant improvement if the set $[v]^t$ is only required to be contained among the rows of *at least* $(1 - \epsilon)\binom{k}{t}$ subarrays of A of dimension $N \times t$, **and** among the rows of each of the $\epsilon\binom{k}{t}$ subarrays that remain, a (large) *subset* of $[v]^t$ is required to be contained?

We answer these questions both theoretically and algorithmically in the following sections.

3   Partial Covering Arrays

When $1 \le m \le v^t$, a *partial m-covering array*, $\mathsf{PCA}(N; t, k, v, m)$, is an $N \times k$ array A with each entry from $[v]$ so that for each t-set of columns $C \in \binom{[k]}{t}$, at least m distinct tuples $x \in [v]^t$ appear as rows in A_C. Hence a covering array $\mathsf{CA}(N; t, k, v)$ is precisely a partial v^t-covering array $\mathsf{PCA}(N; t, k, v, v^t)$.

Theorem 1. *For integers t, k, v, and m where $k \ge t \ge 2$, $v \ge 2$ and $1 \le m \le v^t$ there exists a $\mathsf{PCA}(N; t, k, v, m)$ with*

$$N \le \frac{\ln\left\{\binom{k}{t}\binom{v^t}{m-1}\right\}}{\ln\left(\frac{v^t}{m-1}\right)}. \tag{5}$$

Proof. Let $r = v^t - m + 1$, and A be a random $N \times k$ array where each entry is chosen independently from $[v]$ with uniform probability. For $C \in \binom{[k]}{t}$, let B_C denote the event that at least r tuples from $[v]^t$ are missing in A_C. The probability that a particular r-set of tuples from $[v]^t$ is missing in A_C is $\left(1 - \frac{r}{v^t}\right)^N$. Applying the union bound to all r-sets of tuples from $[v]^t$, we obtain $\Pr[B_C] \le \binom{v^t}{r}\left(1 - \frac{r}{v^t}\right)^N$. By linearity of expectation, the expected number of t-sets C for which A_C misses at least r tuples from $[v]^t$ is at most $\binom{k}{t}\binom{v^t}{r}\left(1 - \frac{r}{v^t}\right)^N$. When A has at least $\frac{\ln\left\{\binom{k}{t}\binom{v^t}{m-1}\right\}}{\ln\left(\frac{v^t}{m-1}\right)}$ rows this expected number is less than 1. Therefore, an array A exists with the required number of rows such that for all $C \in \binom{[k]}{t}$, A_C misses at most $r - 1$ tuples from $[v]^t$, i.e. A_C covers at least m tuples from $[v]^t$. $\square$

Theorem 1 can be improved upon using the Lovász local lemma.

Lemma 1. *(Lovász local lemma; symmetric case) (see [1]) Let $A_1, A_2, \ldots, A_n$ be events in an arbitrary probability space. Suppose that each event A_i is mutually independent of a set of all other events A_j except for at most d, and that $\Pr[A_i] \leq p$ for all $1 \leq i \leq n$. If $ep(d+1) \leq 1$, then $\Pr[\cap_{i=1}^n \bar{A}_i] > 0$.*

Lemma 1 provides an upper bound on the probability of a "bad" event in terms of the dependence structure among such bad events, so that there is a guaranteed outcome in which all "bad" events are avoided. This lemma is most useful when there is limited dependence among the "bad" events, as in the following:

Theorem 2. *For integers t, k, v and m where $v, t \geq 2$, $k \geq 2t$ and $1 \leq m \leq v^t$ there exists a $\mathrm{PCA}(N; t, k, v, m)$ with*

$$N \leq \frac{1 + \ln\left\{t\binom{k}{t-1}\binom{v^t}{m-1}\right\}}{\ln\left(\frac{v^t}{m-1}\right)}. \tag{6}$$

Proof. When $k \geq 2t$, each event B_C with $C \in \binom{[k]}{t}$ (that is, at least $v^t - m + 1$ tuples are missing in A_C) is independent of all but at most $\binom{t}{1}\binom{k-1}{t-1} < t\binom{k}{t-1}$ events in $\{B_{C'} : C' \in \binom{[k]}{t} \setminus \{C\}\}$. Applying Lemma 1, $\Pr[\wedge_{C \in \binom{[k]}{t}} \overline{B_C}] > 0$ when

$$e\binom{v^t}{r}\left(1 - \frac{r}{v^t}\right)^N t\binom{k}{t-1} \leq 1. \tag{7}$$

Solve (7) to obtain the required upper bound on N.   □

When $m = v^t$, apply the Taylor series expansion to obtain $\ln\left(\frac{v^t}{m-1}\right) \geq \frac{1}{v^t}$, and thereby recover the upper bound (3). Theorem 2 implies:

Corollary 1. *Given $q \in [0, 1]$ and integers $2 \leq t \leq k$, $v \geq 2$, there exists an $N \times k$ array on $[v]$ with (q, t)-completeness equal to 1 (i.e., maximal), whose number of rows, N satisfies*

$$N \leq \frac{1 + \ln\left\{t\binom{k}{t-1}\binom{v^t}{qv^t-1}\right\}}{\ln\left(\frac{v^t}{qv^t-1}\right)}.$$

Rewriting (6), setting $r = v^t - m + 1$, and using the Taylor series expansion of $\ln\left(1 - \frac{r}{v^t}\right)$, we get

$$N \leq \frac{1 + \ln\left\{t\binom{k}{t-1}\binom{v^t}{r}\right\}}{\ln\left(\frac{v^t}{v^t-r}\right)} \leq \frac{v^t(t-1)\ln k}{r}\left\{1 - \frac{\ln r}{\ln k} + o(1)\right\}. \tag{8}$$

Hence when $r = v(t-1)$ (or equivalently, $m = v^t - v(t-1) + 1$), there is a partial m-covering array with $\Theta(v^{t-1} \ln k)$ rows. This matches the lower bound (4)

Algorithm 1. Moser-Tardos type algorithm for partial m-covering arrays.

Input: Integers N, t, k, v and m where $v, t \geq 2$, $k \geq 2t$ and $1 \leq m \leq v^t$

Output: A : a PCA$(N; t, k, v, m)$

1 Let $N := \dfrac{1 + \ln\left\{t\binom{k}{t-1}\binom{v^t}{m-1}\right\}}{\ln\left(\frac{v^t}{m-1}\right)}$;

2 Construct an $N \times k$ array A where each entry is chosen independently and uniformly at random from $[v]$;

3 **repeat**

4   Set *covered*:= true;

5   **for** *each column t-set* $C \in \binom{[k]}{t}$ **do**

6   **if** A_C *does not cover at least m distinct t-tuples* $x \in [v]^t$ **then**

7   Set *covered*:= false;

8   Set *missing-column-set* := C;

9   **break** ;

10   **end**

11   **end**

12   **if** *covered* = *false* **then**

13   Choose all the entries in the t columns of *missing-column-set* independently and uniformly at random from $[v]$;

14   **end**

15 **until** *covered* = *true*;

16 Output A;

---

asymptotically for covering arrays by missing, in each t-set of columns, *no more than* $v(t-1) - 1$ of the v^t possible rows.

The dependence of the bound (6) on the number of v-ary t-vectors that must appear in the t-tuples of columns is particularly of interest when test suites are run sequentially until a fault is revealed, as in [3]. Indeed the arguments here may have useful consequences for the rate of fault detection.

Lemma 1 and hence Theorem 2 have proofs that are non-constructive in nature. Nevertheless, Moser and Tardos [26] provide a randomized algorithm with the same guarantee. Patterned on their method, Algorithm 1 constructs a partial m-covering array with exactly the same number of rows as (6) in expected polynomial time. Indeed, for fixed t, the expected number of times the resampling step (line 13) is repeated is linear in k (see [26] for more details).

4 Almost Partial Covering Arrays

For $0 < \epsilon < 1$, an ϵ-*almost partial m-covering array*, APCA$(N; t, k, v, m, \epsilon)$, is an $N \times k$ array A with each entry from $[v]$ so that for at least $(1 - \epsilon)\binom{k}{t}$ column t-sets $C \in \binom{[k]}{t}$, A_C covers at least m distinct tuples $x \in [v]^t$. Again, a covering array CA$(N; t, k, v)$ is precisely an APCA$(N; t, k, v, v^t, \epsilon)$ when $\epsilon < 1/\binom{k}{t}$. Our first result on ϵ-*almost* partial m-covering arrays is the following.

Theorem 3. *For integers t, k, v, m and real ϵ where $k \geq t \geq 2$, $v \geq 2$, $1 \leq m \leq v^t$ and $0 \leq \epsilon \leq 1$, there exists an APCA$(N; t, k, v, m, \epsilon)$ with*

$$N \leq \frac{\ln \left\{ \binom{v^t}{m-1} / \epsilon \right\}}{\ln \left(\frac{v^t}{m-1} \right)}. \tag{9}$$

Proof. Parallelling the proof of Theorem 1 we compute an upper bound on the expected number of t-sets $C \in \binom{[k]}{t}$ for which A_C misses at least r tuples $x \in [v]^t$. When this expected number is at most $\epsilon \binom{k}{t}$, an array A is guaranteed to exist with at least $(1 - \epsilon) \binom{k}{t}$ t-sets of columns $C \in \binom{[k]}{t}$ such that A_C misses at most $r - 1$ distinct tuples $x \in [v]^t$. Thus A is an APCA$(N; t, k, v, m, \epsilon)$. To establish the theorem, solve the following for N:

$$\binom{k}{t} \binom{v^t}{r} \left(1 - \frac{r}{v^t} \right)^N \leq \epsilon \binom{k}{t}. \qquad \square$$

When $\epsilon < 1/\binom{k}{t}$ we recover the bound from Theorem 1 for partial m-covering arrays. In terms of (q, t)-completeness, Theorem 3 yields the following.

Corollary 2. *For $q \in [0, 1]$ and integers $2 \leq t \leq k$, $v \geq 2$, there exists an $N \times k$ array on $[v]$ with (q, t)-completeness equal to $1 - \epsilon$, with*

$$N \leq \frac{\ln \left\{ \binom{v^t}{m-1} / \epsilon \right\}}{\ln \left(\frac{v^t}{m-1} \right)}.$$

When $m = v^t$, an ϵ-almost covering array exists with $N \leq v^t \ln \left(\frac{v^t}{\epsilon} \right)$ rows. Improvements result by focussing on covering arrays in which the symbols are acted on by a finite group. In this setting, one chooses orbit representatives of rows that collectively cover orbit representatives of t-way interactions under the group action; see [9], for example. Such group actions have been used in direct and computational methods for covering arrays [6, 25], and in randomized and derandomized methods [9, 27, 28].

We employ the sharply transitive action of the cyclic group of order v, adapting the earlier arguments using methods from [28]:

Theorem 4. *For integers t, k, v and real ϵ where $k \geq t \geq 2$, $v \geq 2$ and $0 \leq \epsilon \leq 1$ there exists an APCA$(N; t, k, v, v^t, \epsilon)$ with*

$$N \leq v^t \ln \left(\frac{v^{t-1}}{\epsilon} \right). \tag{10}$$

Proof. The action of the cyclic group of order v partitions $[v]^t$ into v^{t-1} orbits, each of length v. Let $n = \lfloor \frac{N}{v} \rfloor$ and let A be an $n \times k$ random array where each entry is chosen independently from the set $[v]$ with uniform probability. For $C \in \binom{[k]}{t}$, A_C *covers the orbit* X if at least one tuple $x \in X$ is present in A_C.

The probability that the orbit X is not covered in A is $\left(1 - \frac{v}{v^t}\right)^n = \left(1 - \frac{1}{v^{t-1}}\right)^n$. Let D_C denote the event that A_C does not cover at least one orbit. Applying the union bound, $\Pr[D_C] \le v^{t-1}\left(1 - \frac{1}{v^{t-1}}\right)^n$. By linearity of expectation, the expected number of column t-sets C for which D_C occurs is at most $\binom{k}{t}v^{t-1}\left(1 - \frac{1}{v^{t-1}}\right)^n$. As earlier, set this expected value to be at most $\epsilon\binom{k}{t}$ and solve for n. An array exists that covers all orbits in at least $(1 - \epsilon)\binom{k}{t}$ column t-sets. Develop this array over the cyclic group to obtain the desired array.   □

As in [28], further improvements result by considering a group, like the Frobenius group, that acts sharply 2-transitively on $[v]$. When v is a prime power, the *Frobenius group* is the group of permutations of $\mathbb{F}_v$ of the form $\{x \mapsto ax + b : a, b \in \mathbb{F}_v, a \ne 0\}$.

Theorem 5. *For integers t, k, v and real ϵ where $k \ge t \ge 2$, $v \ge 2$, v is a prime power and $0 \le \epsilon \le 1$ there exists an* $\mathsf{APCA}(N; t, k, v, v^t, \epsilon)$ *with*

$$N \le v^t \ln\left(\frac{2v^{t-2}}{\epsilon}\right) + v. \tag{11}$$

Proof. The action of the Frobenius group partitions $[v]^t$ into $\frac{v^{t-1}-1}{v-1}$ orbits of length $v(v - 1)$ (full orbits) each and 1 orbit of length v (a short orbit). The short orbit consists of tuples of the form $(x_1, \ldots, x_t) \in [v]^t$ where $x_1 = \ldots = x_t$. Let $n = \lfloor \frac{N-v}{v(v-1)} \rfloor$ and let A be an $n \times k$ random array where each entry is chosen independently from the set $[v]$ with uniform probability. Our strategy is to construct A so that it covers all full orbits for the required number of arrays $\{A_C : C \in \binom{[k]}{t}\}$. Develop A over the Frobenius group and add v rows of the form $(x_1, \ldots, x_k) \in [v]^t$ with $x_1 = \ldots = x_k$ to obtain an $\mathsf{APCA}(N; t, k, v, v^t, \epsilon)$ with the desired value of N. Following the lines of the proof of Theorem 4, A covers all full orbits in at least $(1 - \epsilon)\binom{k}{t}$ column t-sets C when

$$\binom{k}{t}\frac{v^{t-1} - 1}{v - 1}\left(1 - \frac{v - 1}{v^{t-1}}\right)^n \le \epsilon\binom{k}{t}.$$

Because $\frac{v^{t-1}-1}{v-1} \le 2v^{t-2}$ for $v \ge 2$, we obtain the desired bound.   □

Using group action when $m = v^t$ affords useful improvements. Does this improvement extend to cases when $m < v^t$? Unfortunately, the answer appears to be no. Consider the case for $\mathsf{PCA}(N; t, k, v, m)$ when $m \le v^t$ using the action of the cyclic group of order v on $[v]^t$. Let A be a random $n \times k$ array over $[v]$. When $v^t - vs + 1 \le m \le v^t - v(s - 1)$ for $1 \le s \le v^{t-1}$, this implies that for all $C \in \binom{[k]}{t}$, A_C misses at most $s - 1$ orbits of $[v]^t$. Then we obtain that $n \le \left(1 + \ln\left(t\binom{k}{t-1}\binom{v^{t-1}}{s}\right)\right) / \ln\left(\frac{v^{t-1}}{v^{t-1}-s}\right)$. Developing A over the cyclic group we obtain a $\mathsf{PCA}(N; t, k, v, m)$ with

$$N \le v\frac{1 + \ln\left\{\binom{k}{t-1}\binom{v^{t-1}}{s}\right\}}{\ln\left(\frac{v^{t-1}}{v^{t-1}-s}\right)} \tag{12}$$

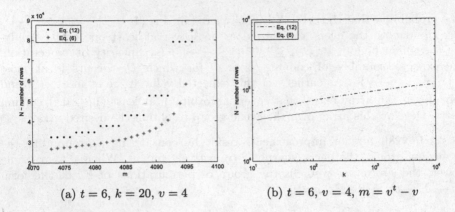

(a) $t = 6$, $k = 20$, $v = 4$ (b) $t = 6$, $v = 4$, $m = v^t - v$

Fig. 1. Comparison of (12) and (6). Figure (a) compares the sizes of the partial m-covering arrays when $v^t - 6v + 1 \leq m \leq v^t$. Except for $m = v^t = 4096$ the bound from (6) outperforms the bound obtained by assuming group action. Figure (b) shows that for $m = v^t - v = 4092$, (6) outperforms (12) for all values of k.

Figure 1 compares (12) and (6). In Fig. 1a we plot the size of the partial m-covering array as obtained by (12) and (6) for $v^t - 6v + 1 \leq m \leq v^t$ and $t = 6$, $k = 20$, $v = 4$. Except when $m = v^t = 4096$, the covering array case, (6) outperforms (12). Similarly, Fig. 1b shows that for $m = v^t - v = 4092$, (6) consistently outperforms (12) for all values of k when $t = 6$, $v = 4$. We observe similar behavior for different values of t and v.

Next we consider even stricter coverage restrictions, combining Theorems 2 and 4.

Theorem 6. *For integers t, k, v, m and real ϵ where $k \geq t \geq 2$, $v \geq 2$, $0 \leq \epsilon \leq 1$ and $m \leq v^t + 1 - \frac{\ln k}{\ln(v/\epsilon^{1/(t-1)})}$ there exists an $N \times k$ array A with entries from $[v]$ such that*

1. *for each $C \in \binom{[k]}{t}$, A_C covers at least m tuples $x \in [v]^t$,*
2. *for at least $(1 - \epsilon)\binom{k}{t}$ column t-sets C, A_C covers all tuples $x \in [v]^t$,*
3. *$N = O(v^t \ln\left(\frac{v^{t-1}}{\epsilon}\right))$.*

Proof. We vertically juxtapose a partial m-covering array and an ϵ-almost v^t-covering array. For $r = \frac{\ln k}{\ln(v/\epsilon^{1/(t-1)})}$ and $m = v^t - r + 1$, (8) guarantees the existence of a partial m-covering array with $v^t \ln\left(\frac{v^{t-1}}{\epsilon}\right)\{1 + o(1)\}$ rows. Theorem 4 guarantees the existence of an ϵ-almost v^t-covering array with at most $v^t \ln\left(\frac{v^{t-1}}{\epsilon}\right)$ rows.   □

Corollary 3. *There exists an $N \times k$ array A such that:*

1. *for any t-set of columns $C \in \binom{[k]}{t}$, A_C covers at least $m \leq v^t + 1 - v(t - 1)$ distinct t-tuples $x \in [v]^t$,*

2. for at least $\left(1 - \frac{v^{t-1}}{k^{1/v}}\right)\binom{k}{t}$ column t-sets C, A_C covers all the distinct t-tuples $x \in [v]^t$.

3. $N = O(v^{t-1}\ln k)$.

Proof. Apply Theorem 6 with $m = v^t + 1 - \frac{\ln k}{\ln(v/\epsilon^{1/(t-1)})}$. There are at most $\frac{\ln k}{\ln(v/\epsilon^{1/(t-1)})} - 1$ missing t-tuples $x \in [v]^t$ in the A_C for each of the at most $\epsilon\binom{k}{t}$ column t-sets C that do not satisfy the second condition of Theorem 6. To bound from above the number of missing tuples to a certain small function $f(t)$ of t, it is sufficient that $\epsilon \le v^{t-1}\left(\frac{1}{k}\right)^{\frac{t-1}{f(t)+1}}$. Then the number of missing t-tuples $x \in [v]^t$ in A_C is bounded from above by $f(t)$ whenever ϵ is not larger than

$$v^{t-1}\left(\frac{1}{k}\right)^{\frac{t-1}{f(t)+1}}\tag{13}$$

On the other hand, in order for the number $N = O\left(v^{t-1}\ln\left(\frac{v^{t-1}}{\epsilon}\right)\right)$ of rows of A to be asymptotically equal to the lower bound (4), it suffices that ϵ is not smaller than

$$\frac{v^{t-1}}{k^{\frac{1}{v}}}.\tag{14}$$

When $f(t) = v(t-1) - 1$, (13) and (14) agree asymptotically, completing the proof. $\quad\square$

Once again we obtain a size that is $O(v^{t-1}\log k)$, a goal that has not been reached for covering arrays. This is evidence that even a small relaxation of covering arrays provides arrays of the best sizes one can hope for.

Next we consider the efficient construction of the arrays whose existence is ensured by Theorem 6. Algorithm 2 is a randomized method to construct an APCA$(N; t, k, v, m, \epsilon)$ of a size N that is very close to the bound of Theorem 3. By Markov's inequality the condition in line 9 of Algorithm 2 is met with probability at most $1/2$. Therefore, the expected number of times the loop in line 2 repeats is at most 2.

To prove Theorem 3, t-wise independence among the variables is sufficient. Hence, Algorithm 2 can be derandomized using t-wise independent random variables. We can also derandomize the algorithm using the method of conditional expectation. In this method we construct A by considering the k columns one by one and fixing all N entries of a column. Given a set of already fixed columns, to fix the entries of the next column we consider all possible v^N choices, and choose one that provides the maximum conditional expectation of the number of column t-sets $C \in \binom{[k]}{t}$ such that A_C covers at least m tuples $x \in [v]^t$. Because $v^N = O(\mathsf{poly}(1/\epsilon))$, this derandomized algorithm constructs the desired array in polynomial time. Similar randomized and derandomized strategies can be applied to construct the array guaranteed by Theorem 4. Together with Algorithm 1 this implies that the array in Theorem 6 is also efficiently constructible.

Algorithm 2. Randomized algorithm for ϵ-almost partial m-covering arrays.

Input: Integers N, t, k, v and m where $v, t \geq 2$, $k \geq t$ and $1 \leq m \leq v^t$, and real $0 < \epsilon < 1$

Output: A : an APCA$(N; t, k, v, m, \epsilon)$

1 Let $N := \dfrac{\ln\left\{2\binom{v^t}{m-1}/\epsilon\right\}}{\ln\left(\frac{v^t}{m-1}\right)}$;

2 **repeat**

3 | Construct an $N \times k$ array A where each entry is chosen independently and uniformly at random from $[v]$;

4 | Set *isAPCA*:= true;

5 | Set *defectiveCount*:= 0;

6 | **for** *each column t-set* $C \in \binom{[k]}{t}$ **do**

7 | | **if** A_C *does not cover at least m distinct t-tuples* $x \in [v]^t$ **then**

8 | | | Set *defectiveCount*:= *defectiveCount* + 1;

9 | | | **if** defectiveCount $> \lfloor \epsilon \binom{k}{t} \rfloor$ **then**

10 | | | | Set *isAPCA*:= false;

11 | | | | **break** ;

12 | | | **end**

13 | | **end**

14 | **end**

15 **until** *isAPCA = true*;

16 Output A;

---

5   Final Remarks

We have shown that by relaxing the coverage requirement of a covering array somewhat, powerful upper bounds on the sizes of the arrays can be established. Indeed the upper bounds are substantially smaller than the best known bounds for a covering array; they are of the same order as the *lower* bound for CAN(t, k, v). As importantly, the techniques not only provide asymptotic bounds but also randomized polynomial time construction algorithms for such arrays.

Our approach seems flexible enough to handle variations of these problems. For instance, some applications require arrays that satisfy, for different subsets of columns, different coverage or separation requirements [8]. In [16] several interesting examples of combinatorial problems are presented that can be unified and expressed in the framework of S-constrained matrices. Given a set of vectors S each of length t, an $N \times k$ matrix M is S-*constrained* if for every t-set $C \in \binom{[k]}{t}$, M_C contains as a row each of the vectors in S. The parameter to optimize is, as usual, the number of rows of M. One potential direction is to ask for arrays that, in every t-tuple of columns, cover at least m of the vectors in S, or that all vectors in S are covered by all but a small number of t-tuples of columns. Exploiting the structure of the members of S appears to require an extension of the results developed here.

Acknowledgements. Research of KS and CJC was supported in part by the National Science Foundation under Grant No. 1421058.

References

1. Alon, N., Spencer, J.H.: The Probabilistic Method. Wiley-Interscience Series in Discrete Mathematics and Optimization, 3rd edn. John Wiley & Sons Inc, Hoboken, NJ (2008)
2. Becker, B., Simon, H.-U.: How robust is the n-cube? Inform. and Comput. **77**, 162–178 (1988)
3. Bryce, R.C., Chen, Y., Colbourn, C.J.: Biased covering arrays for progressive ranking and composition of web services. Int. J. Simul. Process Model. **3**(1/2), 80–87 (2007)
4. Cawse, J.N.: Experimental design for combinatorial and high throughput materials development. GE Global Res. Techn. Report **29**, 769–781 (2002)
5. Chandra, A.K., Kou, L.T., Markowsky, G., Zaks, S.: On sets of boolean n-vectors with all k-projections surjective. Acta Inf. **20**(1), 103–111 (1983)
6. Chateauneuf, M.A., Colbourn, C.J., Kreher, D.L.: Covering arrays of strength 3. Des. Codes Crypt. **16**, 235–242 (1999)
7. Chen, B., Zhang, J.: Tuple density: a new metric for combinatorial test suites. In: Proceedings of the 33rd International Conference on Software Engineering, ICSE, Waikiki, Honolulu, HI, USA, May 21–28, pp. 876–879 (2011)
8. Colbourn, C.J.: Combinatorial aspects of covering arrays. Le Mat. (Catania) **58**, 121–167 (2004)
9. Colbourn, C.J.: Conditional expectation algorithms for covering arrays. J. Comb. Math. Comb. Comput. **90**, 97–115 (2014)
10. Colbourn, C.J.: Covering arrays and hash families. In: Crnković, D., Tonchev, V. (eds.), Information Security, Coding Theory, and Related Combinatorics, NATO Science for Peace and Security Series, pp. 99–135. IOS Press (2011)
11. Damaschke, P.: Adaptive versus nonadaptive attribute-efficient learning. Mach. Learn. **41**(2), 197–215 (2000)
12. Francetić, N., Stevens, B.: Asymptotic size of covering arrays: an application of entropy compression. ArXiv e-prints (March 2015)
13. Gargano, L., Körner, J., Vaccaro, U.: Sperner capacities. Graphs Comb. **9**, 31–46 (1993)
14. Godbole, A.P., Skipper, D.E., Sunley, R.A.: t-covering arrays: upper bounds and Poisson approximations. Comb. Probab. Comput. **5**, 105–118 (1996)
15. Graham, N., Harary, F., Livingston, M., Stout, Q.F.: Subcube fault-tolerance in hypercubes. Inf. Comput. **102**(2), 280–314 (1993)
16. Gravier, S., Ycart, B.: S-constrained random matrices. DMTCS In: Proceedings (1) (2006)
17. Hartman, A.: Software and hardware testing using combinatorial covering suites. In: Golumbic, M.C., Hartman, I.B.-A. (eds.) Graph Theory, Combinatorics and Algorithms. OR/CSIS, pp. 237–266. Springer, Heidelberg (2005)
18. Hartman, A., Raskin, L.: Problems and algorithms for covering arrays. Discrete Math. **284**(13), 149–156 (2004)
19. Jukna, S.: Extremal Combinatorics: With Applications in Computer Science, 1st edn. Springer Publishing Company, Incorporated (2010)
20. Katona, G.O.H.: Two applications (for search theory and truth functions) of Sperner type theorems. Periodica Math. **3**, 19–26 (1973)

21. Kleitman, D., Spencer, J.: Families of k-independent sets. Discrete Math. **6**, 255–262 (1973)
22. Kuhn, D.R., Kacker, R., Lei, Y.: Introduction to Combinatorial Testing. CRC Press, Boca Raton (2013)
23. Kuhn, D.R., Mendoza, I.D., Kacker, R.N., Lei, Y.: Combinatorial coverage measurement concepts and applications. In: 2013 IEEE Sixth International Conference on Software Testing, Verification and Validation Workshops (ICSTW), pp. 352–361, March 2013
24. Maximoff, J.R., Trela, M.D., Kuhn, D.R., Kacker, R.: A method for analyzing system state-space coverage within a t-wise testing framework. In: 4th Annual IEEE Systems Conference, pp. 598–603 (2010)
25. Meagher, K., Stevens, B.: Group construction of covering arrays. J. Combin. Des. **13**, 70–77 (2005)
26. Moser, R.A., Tardos, G.: A constructive proof of the general Lovász local lemma. J. ACM **57**(2), 524–529 (2010). Art. 11, 15
27. Sarkar, K., Colbourn, C.J.: Two-stage algorithms for covering array construction. submitted for publication
28. Sarkar, K., Colbourn, C.J.: Upper bounds on the size of covering arrays. ArXiv e-prints (March 2016)
29. Shasha, D.E., Kouranov, A.Y., Lejay, L.V., Chou, M.F., Coruzzi, G.M.: Using combinatorial design to study regulation by multiple input signals: A tool for parsimony in the post-genomics era. Plant Physiol. **127**, 1590–2594 (2001)
30. Tong, A.J., Wu, Y.G., Li, L.D.: Room-temperature phosphorimetry studies of some addictive drugs following dansyl chloride labelling. Talanta **43**(9), 14291436 (1996)

Querying Probabilistic Neighborhoods in Spatial Data Sets Efficiently

Moritz von Looz[✉] and Henning Meyerhenke

Institute of Theoretical Informatics, Karlsruhe Institute of Technology (KIT),
Karlsruhe, Germany
{moritz.looz-corswarem,meyerhenke}@kit.edu

Abstract. The probability that two spatial objects establish some kind of mutual connection often depends on their proximity. To formalize this concept, we define the notion of a *probabilistic neighborhood*: Let P be a set of n points in $\mathbb{R}^d$, $q \in \mathbb{R}^d$ a query point, dist a distance metric, and $f : \mathbb{R}^+ \to [0,1]$ a monotonically decreasing function. Then, the probabilistic neighborhood $N(q, f)$ of q with respect to f is a random subset of P and each point $p \in P$ belongs to $N(q, f)$ with probability $f(\text{dist}(p, q))$. Possible applications include query sampling and the simulation of probabilistic spreading phenomena, as well as other scenarios where the probability of a connection between two entities decreases with their distance. We present a fast, sublinear-time query algorithm to sample probabilistic neighborhoods from planar point sets. For certain distributions of planar P, we prove that our algorithm answers a query in $O((|N(q, f)| + \sqrt{n}) \log n)$ time with high probability. In experiments this yields a speedup over pairwise distance probing of at least one order of magnitude, even for rather small data sets with $n = 10^5$ and also for other point distributions not covered by the theoretical results.

1 Introduction

In many scenarios, connections between spatial objects are not certain but probabilistic, with the probability depending on the distance between them: The probability that a customer shops at a certain physical store shrinks with increasing distance to it. In disease simulations, if the social interaction graph is unknown but locations are available, disease transmission can be modeled as a random process with infection risk decreasing with distance. Moreover, the wireless connections between units in an ad-hoc network are fragile and collapse more frequently with higher distance.

For these and similar scenarios, we define the notion of a *probabilistic neighborhood* in spatial data sets: Let a set P of n points in $\mathbb{R}^d$, a query point $q \in \mathbb{R}^d$, a distance metric dist, and a monotonically decreasing function $f : \mathbb{R}^+ \to [0,1]$ be given. Then, the probabilistic neighborhood $N(q, f)$ of q with respect to f is a random subset of P and each point $p \in P$ belongs to $N(q, f)$ with probability $f(\text{dist}(p, q))$. A straightforward query algorithm for sampling a probabilistic neighborhood would iterate over each point $p \in P$ and sample for each whether

© Springer International Publishing Switzerland 2016
V. Mäkinen et al. (Eds.): IWOCA 2016, LNCS 9843, pp. 449–460, 2016.
DOI: 10.1007/978-3-319-44543-4_35

it is included in $N(q, f)$. This has a running time of $\Theta(n \cdot d)$ per query point, which is prohibitive for repeated queries in large data sets. Thus we are interested in a faster algorithm for such a *probabilistic neighborhood query* (PNQ, spoken as "pink"). We restrict ourselves to the planar case in this work, but the algorithmic principle is generalizable to higher dimensions.

While the linear-time approach has appeared before in the literature for a particular application [2] (without formulating the problem as a PNQ explicitly), we are not aware of previous work performing more efficient PNQs with an index structure. For example, the probabilistic quadtree introduced by Kraetzschmar et al. [10] is designed to store probabilistic occupancy data and gives deterministic results. Other range queries related to (yet different from) our work as well as deterministic index structures are described in Sect. 2.2. Proofs, details, further experiments, pseudocode and visualizations omitted due to space constraints can be found in the full version of this paper [16].

Contributions. We develop, analyze, implement, and evaluate an index structure and a query algorithm that together provide fast probabilistic neighborhood queries in the Euclidean and hyperbolic plane. Our key data structure for these fast PNQs is a polar quadtree which we adapt from our previous work [17]. Preprocessing for quadtree construction requires $O(n \log n)$ time with high probability[1] (whp).

To answer PNQs, we first present a simple query algorithm (Sect. 3). We then improve its time complexity by treating whole subtrees as so-called virtual leaves, see Sect. 4. As shown by our detailed theoretical analysis, the improved algorithm yields a query time complexity of $O((|N(q, f)| + \sqrt{n}) \log n)$ whp to find a probabilistic neighborhood $N(q, f)$ among n points, for n sufficiently large. This is sublinear if the returned neighborhood $N(q, f)$ is of size $o(n/ \log n)$ – an assumption we consider reasonable for most applications. For our theoretical results to hold, the quadtree structure needs to be able to partition the distribution of the point positions in P, i.e. not all of the probability mass may be concentrated on a single point or line. In our case of polar quadtrees, this is achieved if the distribution is continuous, integrable, rotationally invariant with respect to the origin and non-zero only for a finite area.

Experimental results are shown in Sect. 5: We apply our query algorithm to generate random graphs in the hyperbolic plane [12] in subquadratic time. Graphs with millions of edges can now be generated within a few minutes sequentially. This yields an acceleration of at least one order of magnitude in practice compared to a reference implementation [2] that uses linear-time queries. Compared to our previous work on graph generation [17], our new algorithm is able to generate a more extensive model. Even if the distribution of a given point set P is unknown in practice, running times are fast: As an example of probabilistic spreading behavior, we simulate a simple disease spreading mechanism on real population density geodata. In this scenario, our fast PNQs are at least two orders of magnitude faster than linear-time queries.

---

[1] We say "with high probability" (whp) when referring to a probability $\geq 1 - 1/n$ for sufficiently large n.

2   Preliminaries

2.1   Notation

Let the input be given as set P of n points. The points in P are distributed
in a disk $\mathbb{D}_R$ of radius R in the hyperbolic or Euclidean plane, the distribution
is given by a probability density function $j(\phi, r)$ for an angle ϕ and a radius
r. Recall that, for our theoretical results to hold, we require j to be known,
continuous and integrable. Furthermore, j needs to be rotationally invariant –
meaning that $j(\phi_1, r) = j(\phi_2, r)$ for any radius r and any two angles ϕ_1 and
ϕ_2 – and positive within $\mathbb{D}_R$, so that $j(r) > 0 \Leftrightarrow r < R$. Due to the rotational
invariance, $j(\phi, r)$ is the same for every ϕ and we can write $j(r)$. Likewise, we
define $J(r)$ as the indefinite integral of $j(r)$ and normalize it so that $J(R) = 1$
(also implying $J(0) = 0$). The value $J(r)$ then gives the fraction of probability
mass inside radius r.

For the distance between two points p_1 and p_2, we use $\mathrm{dist}_\mathbb{H}(p_1, p_2)$ for the
hyperbolic and $\mathrm{dist}_\mathbb{E}(p_1, p_2)$ for the Euclidean case. We may omit the index
if a distinction is unnecessary. As mentioned, a point p is in the probabilistic
neighborhood of query point q with probability $f(\mathrm{dist}(p, q))$. Thus, a *query pair*
consists of a query point q and a function $f : \mathbb{R}^+ \rightarrow [0, 1]$ that maps distances
to probabilities. The function f needs to be monotonically decreasing but may
be discontinuous. Note that f can be defined differently for each query. The
query result, the probabilistic neighborhood of q w.r.t. f, is denoted by the set
$N(q, f) \subseteq P$.

For the algorithm analysis, we use two additional sets for each query (q, f):

– Candidates(q, f): neighbor candidates examined when executing such a query,
– Cells(q, f): quadtree cells examined during execution of the query.

Note that the sets $N(q, f)$, Candidates(q, f) and Cells(q, f) are probabilistic, thus
theoretical results about their size are usually only with high probability.

2.2   Related Work

Fast deterministic range queries. Numerous index structures for fast range
queries on spatial data exist. Many such index structures are based on trees or
variations thereof, see Samet's book [14] for a comprehensive overview. I/O efficient worst case analysis is usually performed using the EM model, see e.g. [3]. In
more applied settings, average-case performance is of higher importance, which
popularized R-trees or newer variants thereof, e.g. [9]. Concerning (balanced)
quadtrees for spatial dimension d, it is known that queries require $O(d \cdot n^{1-1/d})$
time (thus $O(\sqrt{n})$ in the planar case) [14, Ch. 1.4]. Regarding PNQs our algorithm matches this query complexity up to a logarithmic factor. Yet note that,
since for general f and dist in our scenario all points in the set P could be
neighbors, data structures for deterministic queries cannot solve a PNQ efficiently without adaptations.

Hu et al. [8] give a query sampling algorithm for one-dimensional data that, given a set P of n points in $\mathbb{R}$, an interval $q = [x, y]$ and an integer, $t \geq 1$, returns t elements uniformly sampled from $P \cap q$. They describe a structure of $O(n)$ space that answers a query in $O(\log n + t)$ time and supports updates in $O(\log n)$ time. While also offering query sampling, PNQs differ from the problem considered by Hu et al. in two aspects: We consider two dimensions instead of one and our sampling probabilities are not necessarily uniform, but can be set by the user by a distance-dependent function.

Range queries on uncertain data. During the previous decade probabilistic queries *different* from PNQs have become popular. The main scenarios can be put into two categories [13]: (i) Probabilistic databases contain entries that come with a specified confidence (e. g. sensor data whose accuracy is uncertain) and (ii) objects with an uncertain location, i. e. the location is specified by a probability distribution. Both scenarios differ under typical and reasonable assumptions from ours: Queries for uncertain data are usually formulated to return *all* points in the neighborhood whose confidence/probability exceeds a certain threshold [11], or computing points that are possibly nearest neighbors [1].

In our model, in turn, the choice of inclusion of a point p is a random choice for every different p. In particular, depending on the probability distribution, *all* nodes in the plane can have positive probability to be part of some other's neighborhood. In the related scenarios this would only be true with extremely small confidence values or extremely large query circles.

Applications in fast graph generation. One application for PNQs as introduced in Sect. 1 is the hyperbolic random graph model by Krioukov et al. [12]. The n graph nodes are represented by points thrown into the hyperbolic plane at random[2] and two nodes are connected by an edge with a probability that decreases with the distance between them. An implementation of this generative model is available [2], it performs $\Theta(n^2)$ neighborhood tests. Bringmann et al. provide an algorithm to generate hyperbolic random graphs in expected linear time [5]; to our knowledge no implementation of it exists yet.

In previous work we designed a generator [17] faster than [2] for a restricted model; it runs in $O((n^{3/2} + m) \log n)$ time whp for the whole graph with m edges. The range queries discussed there are facilitated by a quadtree which supports only deterministic queries. Consequently, the queries result in unit-disk graphs in the hyperbolic plane and can be considered as a special case of the current work (a step function f with values 0 and 1 results in a deterministic query).

Our major technical inspiration for enhancing the quadtree for probabilistic neighborhoods is the work of Batagelj and Brandes [4]. They were the first to present a random sampling method to generate Erdős-Rényi-graphs with n nodes and m edges in $O(n + m)$ time complexity. Faced with a similar problem of selecting each of n elements with a constant probability p, they designed an efficient algorithm. Instead of sampling each element separately, they use random

---

[2] The probability density in the polar model depends only on radii r and R as well as a growth parameter α and is given by $g(r) := \alpha \frac{\sinh(\alpha r)}{\cosh(\alpha R) - 1}$.

jumps of length $\delta(p)$, $\delta(p) = \ln(1 - rand)/\ln(1 - p)$, with $rand$ being a random number uniformly distributed in $[0, 1)$.

2.3 Quadtree Specifics

Our key data structure is a polar region quadtree in the Euclidean or hyperbolic plane. While they are less suited to higher dimensions as for example k-d-trees, the complexity is comparable in the plane. For the (circular) range queries we discuss, quadtrees have the significant advantage of a bounded aspect ratio: A cell in a k-d-tree might extend arbitrarily far in one direction, rendering theoretical guarantees about the area affected by the query circle difficult to impossible. In contrast, the region covered by a quadtree cell is determined by its position and level.

We mostly reuse our previous definition [17] of the quadtree: A node in the quadtree is defined as a tuple $(\min_\phi, \max_\phi, \min_r, \max_r)$ with $\min_\phi \leq \max_\phi$ and $\min_r \leq \max_r$. It is responsible for a point $p = (\phi_p, r_p)$ exactly if $(\min_\phi \leq \phi_p < \max_\phi)$ and $(\min_r \leq r_p < \max_r)$. We call the region represented by a particular quadtree node its quadtree *cell*. The quadtree is parametrized by its radius R, the $\max_r$ of the root cell. If the probability distribution j is known (which we assume for our theoretical results), we set the radius R to $\arg\min_r J(r) = 1$, i.e. to the minimum radius that contains the full probability mass. If only the points are known, the radius is set to include all of them. While in this latter case the complexity analysis of Sects. 3 and 4 does not hold, fast running times in practice can still be achieved (see Sect. 5).

3 Baseline Query Algorithm

We begin the main technical part by describing adaptations in the quadtree construction as well as a baseline query algorithm. This latter algorithm introduces the main idea, but is asymptotically not faster than the straightforward approach. In Sect. 4 it is then refined to support faster queries.

3.1 Quadtree Construction

At each quadtree node v, we store the size of the subtree rooted there. We then generalize the rule for node splitting to handle point distributions j as defined in Sect. 2.1: As is usual for quadtrees, a leaf cell c is split into four children when it exceeds its fixed capacity. Since our quadtree is polar, this split happens once in the angular and once in the radial direction. Due to the rotational symmetry of j, splitting in the angular direction is straightforward as the angle range is halved: $\text{mid}_\phi := \frac{\max_\phi + \min_\phi}{2}$. For the radial direction, we choose the splitting radius to result in an equal division of probability mass. The total probability mass in a ring delimited by $\min_r$ and $\max_r$ is $J(\max_r) - J(\min_r)$. Since $j(r)$ is positive for r between R and 0, the restricted function $J|_{[0,R]}$ defined above is a bijection. The inverse $(J|_{[0,R]})^{-1}$ thus exists and we set the splitting radius mid_r to $(J|_{[0,R]})^{-1}\left(\frac{J(\max_r)+J(\min_r)}{2}\right)$.

Figure 1 visualizes a point distribution on a hyperbolic disk with 200 points and Fig. 2 its corresponding quadtree.

Two results on quadtree properties help to establish the time complexity of quadtree operations. They are generalized versions of our previous work [17, Lemmas 1 and 2] and state that each quadtree cell contains the same expected number of points and that the quadtree height is $O(\log n)$ whp.

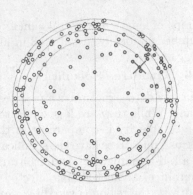

Lemma 1. *Let $\mathbb{D}_R$ be a hyperbolic or Euclidean disk of radius R, j a probability distribution on $\mathbb{D}_R$ which fulfills the properties defined in Sect. 2.1, p a point in $\mathbb{D}_R$ which is sampled from j, and T be a polar quadtree on $\mathbb{D}_R$. Let C be a quadtree cell at depth i. Then, the probability that p is in C is 4^{-i}.*

Fig. 1. Query over 200 points in a polar hyperbolic quadtree, with $f(d) := 1/(e^{(d-7.78)} + 1)$ and the query point q marked by a red cross. Points are colored according to the probability that they are included in the result. Blue represents a high probability, white a probability of zero. (Color figure online)

Proposition 1. *Let $\mathbb{D}_R$ and j be as in Lemma 1. Let T be a polar quadtree on $\mathbb{D}_R$ constructed to fit j. Then, for n sufficiently large, $\text{height}(T) \in O(\log n)$ whp.*

A direct consequence from the results above and our previous work [17] is the preprocessing time for the quadtree construction. The generalized splitting rule and storing the subtree sizes only change constant factors.

Corollary 1. *Since a point insertion takes $O(\log n)$ time whp, constructing a quadtree on n points distributed as in Sect. 2.1 takes $O(n \log n)$ time whp.*

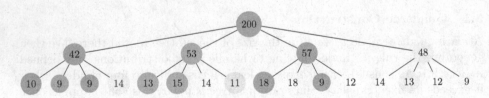

Fig. 2. Visualization of the data structure used in Fig. 1. Quadtree nodes are colored according to the upper probability bound for points contained in them. The color of a quadtree node c is the darkest possible shade (dark = high probability) of any point contained in the subtree rooted at c. Each node is marked with the number of points in its subtree.

Algorithm 1. QuadNode.getProbabilisticNeighborhood

Input: query point q, prob. function f, quadtree node c
Output: probabilistic neighborhood of q

1 N = {};
2 $\underline{b}$ = dist(q,c);
   /* Distance between point and cell   */
3 $\overline{b}$=$f(\underline{b})$;
   /* Since f is monotonically decreasing, a lower bound for the
      distance gives an upper bound $\overline{b}$ for the probability.   */
4 s = number of points in c;
5 **if** c *is not leaf* **then**
      /* internal node: descend, add recursive result to local set   */
6   **for** *child* $\in$ *children(c)* **do**
7   add getProbabilisticNeighborhood$(q, f, $child$)$ to N;
8 **else**
      /* leaf case: apply idea of Batagelj and Brandes [4]   */
9   **for** $i=0; i < s ; i++$ **do**
10       $\delta = \ln(1 - rand)/\ln(1 - \overline{b})$;
11   i += δ;
12   **if** $i \geq s$ **then**
13   break;
14       $prob = f(\text{dist}(q, c.\text{points}[i]))/\overline{b}$;
15   add c.points$[i]$ to N with probability $prob$
16 **return** N

3.2 Algorithm

The baseline version of our query (Algorithm 1) has unfortunately a time complexity of $\Theta(n)$, but serves as a foundation for the fast version (Sect. 4). It takes as input a query point q, a function f and a quadtree cell c. Initially, it is called with the root node of the quadtree and recursively descends the tree. The algorithm returns a point set $N(q, f) \subseteq P$ with

$$\mathbf{Pr}\,[\,p \in N(q,f)\,] = f(\text{dist}(q,p)). \tag{1}$$

Algorithm 1 descends the quadtree recursively until it reaches the leaves. Once a leaf l is reached, a lower bound $\underline{b}$ for the distance between the query point q and all the points in l is computed (Line 2). Such distance calculations are detailed in the full version [16]. Since f is monotonically decreasing, this lower bound for the distance gives an upper bound $\overline{b}$ for the probability that a given point in l is a member of the returned point set (Line 3). This bound is used to select *neighbor candidates* in a similar manner as Bategelj and Brandes [4]: In Line 10, a random number of vertices is skipped, so that every vertex in l is selected as a neighbor candidate with probability $\overline{b}$. The actual distance dist(q, a) between a candidate a and the query point q is at least $\underline{b}$ and the probability of $a \in N(q, f)$ thus at most $\overline{b}$. For each candidate, this actual distance dist(q, a)

is then calculated and a neighbor candidate is confirmed as a neighbor with probability $f(\text{dist}(q, a))/\overline{b}$ in Line 14.

Regarding correctness and time complexity of Algorithm 1, we can state:

Proposition 2. *Let T be a quadtree as defined above, q be a query point and $f : \mathbb{R}^+ \rightarrow [0,1]$ a monotonically decreasing function which maps distances to probabilities. The probability that a point p is returned by a PNQ (q, f) from Algorithm 1 is $f(\text{dist}(q,p))$, independently from whether other points are returned.*

Proposition 3. *Let T be a quadtree with n points. The running time of Algorithm 1 per query on T is $\Theta(n)$ in expectation.*

4   Queries in Sublinear Time by Subtree Aggregation

One reason for the linear time complexity of the baseline query is the fact that every quadtree node is visited. To reach a sublinear time complexity, we thus aggregate subtrees into *virtual leaf cells* whenever doing so reduces the number of examined cells and does not increase the number of candidates too much.

To this end, let S be a subtree starting at depth l of a quadtree T. During the execution of Algorithm 1, a lower bound $\underline{b}$ for the distance between S and the query point q is calculated, yielding also an upper bound $\overline{b}$ for the neighbor probability of each point in S. At this step, it is possible to treat S as a *virtual leaf cell*, sample jumping widths using $\overline{b}$ as upper bound and use these widths to select candidates within S. Aggregating a subtree to a virtual leaf cell allows skipping leaf cells which do not contain candidates, but uses a weaker bound $\overline{b}$ and thus a potentially larger candidate set. Thus, a fast algorithm requires an aggregation criterion which keeps both the number of candidates and the number of examined quadtree cells low.

As stated before, we record the number of points in each subtree during quadtree construction. This information is now used for the query algorithm: We aggregate a subtree S to a virtual leaf cell exactly if $|S|$, the number of points contained in S, is below $1/f(\text{dist}(S, q))$. This corresponds to less than one expected candidate within S. The changes required in Algorithm 1 to use the subtree aggregation are minor. Lines 5, 14 and 15 are changed to:

---

5 **if** *c is inner node and* $|c| \cdot \overline{b} \geq 1$ **then**

---

14 neighbor = maybeGetKthElement($q, f, i, \overline{b}, c$);
15 add neighbor to N if not null

The main change consists in the use of the function maybeGetKthElement. Given a subtree S, an index k, q, f, and $\overline{b}$, this function descends S to the leaf cell containing the kth element. This element p_k is then accepted with probability $f(\text{dist}(q, p_k))/\overline{b}$.

Since the upper bound calculated at the root of the aggregated subtree is not smaller than the individual upper bounds at the original leaf cells, Proposition 2 also holds for the virtual leaf cells. This establishes the correctness.

The time complexity is given by the following theorem, whose proof can be found in the full version [16].

Theorem 1. *Let T be a quadtree with n points and (q, f) a query pair. A query (q, f) using subtree aggregation has time complexity $O((|N(q, f)| + \sqrt{n}) \log n)$ whp.*

5 Application Case Studies

In order to test our algorithm for PNQs, we apply it in two application case studies, one for Euclidean, the other one for hyperbolic geometry. For the Euclidean case study we build a simple disease spread simulation as an example for a probabilistic spreading process. The probability distribution of points is in this case non-uniform and unknown. The hyperbolic application, in turn, is a generator for complex networks with a known point distribution.

5.1 Probabilistic Spreading

When both contact graph and travel patterns of a susceptible population are not known in detail, the resulting spreading behavior of an infectious disease seems probabilistic. Contagious diseases usually spread to people in the vicinity of infected persons, but an infectious person occasionally bridges larger distances by travel and spreads the disease this way. We model this effect with our probabilistic neighborhood function f, giving a higher probability for small distances and a lower but non-zero probability for larger distances. Note that this scenario is meant as an example of the probabilistic spreading simulations possible with our algorithm and not as highly realistic from an epidemiological point of view.

In the simulation, the population is given as a set P of points in the Euclidean plane. In the initial step, exactly one point (= person) from P is marked as infected. Then, in each round, a PNQ is performed for each infected person q. All points in $N(q, f)$ become infected in the next round. We use an SIR model [7], i.e. previously infected persons recover with a certain probability in each round and stay infectious otherwise. In our simulation, persons recover with a rate of 0.8 and are then immune.

5.2 Random Hyperbolic Graph Generation

Random hyperbolic graphs (RHGs, also see Sect. 2.2) are a generative graph model for complex networks. For graph generation one places n points (= vertices) randomly in a hyperbolic disk. The radius R of the disk can be used to control the average degree of the network. A pair of vertices is connected by an edge with a probability that depends on the vertices' hyperbolic distance.

This connection probability is given in [12, Eq. (41)] and parametrized by a temperature $T \geq 0$:

$$f(x) = \frac{1}{e^{(1/T)\cdot(x-R)/2} + 1} \tag{2}$$

This definition of random hyperbolic graphs is a generalized version of the one considered in our previous work, which was restricted to the special case of $T = 0$.

5.3   Experimental Settings and Results

Our implementation is included in the open-source toolkit NetworKit [15] and is written in C++ 11. Running time measurements were made with g++ 4.8 -O3 on a machine with 128 GB RAM and an Intel Xeon E5-1630 v3 CPU with four cores at 3.7 GHz base frequency. Our code is sequential, as is the reference implementation for random hyperbolic graph generation [2].

Disease Spread Simulation. We experimented on three data sets taken from NASA population density raster data [6] for Germany, France and the USA. They consist of rectangles with small square cells (geographic areas) where for each cell the population from the year 2000 is given. To obtain a set of points, we randomly distribute points in each cell to fit 1/20th of the population density. The data sets of France and USA have roughly 3 and 14 million points, respectively.

The number of required queries naturally depends heavily on the simulated disease. For our parameters, a number of 5000 queries is typically reached within the first dozen steps. To evaluate the algorithmic speedup, Table 1 compares running times for 5000 pairwise distance probing (PDP) queries against 5000 fast PNQs on the three country datasets. To obtain a similar total number of infections, we use a slightly different probabilistic neighborhood function for each country and divide by the population: $f(x) := (1/x) \cdot e^7/n$. This results in a slower initial progression for the US. Our algorithm achieves a speedup factor of at least two orders of magnitude, even including the quadtree construction time.

Table 1. Running time results for polar Euclidean quadtrees on population data. The query points were selected uniformly at random from P, the probabilistic neighborhood function is $f(x) := (1/x) \cdot e^7/n$.

Country	5000 PDP queries	Construction QT	5000 QT queries
France	1007 s	1.6 s	1.2 s
Germany	1395 s	2.8 s	1.3 s
USA	4804 s	8.7 s	0.7 s

Random Hyperbolic Graph Generation. We compare our generator using PNQs with the only (to our knowledge) previously existing generator for general random hyperbolic graphs [2], i.e. those not only following the threshold model.

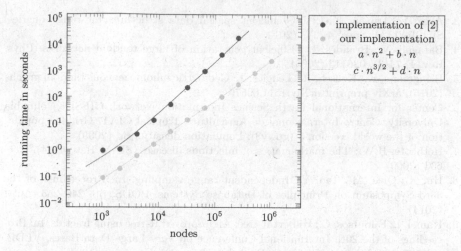

Fig. 3. Comparison of running times to generate networks with 2^{10}-2^{20} vertices, $\alpha = 1$, $T = 0.5$ and average degree $\bar{k} = 6$. The gap between the running times widens, which in the loglog-plot implies a different exponent in the time complexities. Running times are fitted with $a = 2.089 \cdot 10^{-7}$, $b = 3.311 \cdot 10^{-4}$, $c = 2.18 \cdot 10^{-6}$ and $d = 5.6 \cdot 10^{-6}$.

As seen in Fig. 3, our implementation is faster by at least one order of magnitude and the experimental running times support our theoretical time complexity of $O((n^{3/2} + m) \log n)$.

6   Conclusions

After formally defining the notion of probabilistic neighborhoods, we have presented a quadtree-based query algorithm for such neighborhoods in the Euclidean and hyperbolic plane. Our analysis shows a time complexity of $O(((|N(q, f)| + \sqrt{n}) \log n)$, our algorithm is to the best of our knowledge the first to solve the problem asymptotically faster than pairwise distance probing. With two example applications we have shown that our algorithm is also faster in practice by at least one order of magnitude.

Acknowledgements. This work is partially supported by German Research Foundation (DFG) grant ME 3619/3-1 within the Priority Programme 1736 *Algorithms for Big Data*. The authors thank Mark Ortmann for helpful discussions.

References

1. Agarwal, P.K., Aronov, B., Har-Peled, S., Phillips, J.M., Yi, K., Zhang, W.: Nearest neighbor searching under uncertainty II. In Proceedings of the 32nd Symposium on Principles of Database Systems, PODS, pp. 115–126. ACM (2013)
2. Aldecoa, R., Orsini, C., Krioukov, D.: Hyperbolic graph generator. Comput. Phys. Commun. **196**, 492–496 (2015). Elsevier, Amsterdam

3. Arge, L., Larsen, K.G.: I/O-efficient spatial data structures for range queries. SIGSPATIAL Spec. **4**, 2–7 (2012)
4. Batagelj, V., Brandes, U.: Efficient generation of large random networks. Phys. Rev. E **71**(3), 036113 (2005)
5. Bringmann, K., Keusch, R., Lengler, J.: Geometric inhomogeneous random graphs (2015). arXiv preprint arXiv:1511.00576
6. Center for International Earth Science Information Network CIESIN Columbia University; Centro Internacional de Agricultura Tropical CIAT. Gridded population of the world, version 3 (gpwv3): Population density grid (2005)
7. Hethcote, H.W.: The mathematics of infectious diseases. SIAM Rev. **42**(4), 599–653 (2000)
8. Hu, X., Qiao, M., Tao, Y.: Independent range sampling. In: Proceedings of the 33rd Symposium on Principles of Database Systems, PODS, pp. 246–255. ACM (2014)
9. Kamel, I., Faloutsos, C.: Hilbert R-tree: An improved R-tree using fractals. In: Proceedings of the 20th International Conference on Very Large Data Bases, VLDB, pp. 500–509. Morgan Kaufmann Publishers Inc., San Francisco (1994)
10. Kraetzschmar, G.K., Gassull, G.P., Uhl, K.: Probabilistic quadtrees for variable-resolution mapping of large environments. In: Proceedings of the 5th IFAC/EURON Symposium on Intelligent Autonomous Vehicles (2004)
11. Kriegel, H.-P., Kunath, P., Renz, M.: Probabilistic nearest-neighbor query on uncertain objects. In: Kotagiri, R., Radha Krishna, P., Mohania, M., Nantajeewarawat, E. (eds.) DASFAA 2007. LNCS, vol. 4443, pp. 337–348. Springer, Heidelberg (2007)
12. Krioukov, D., Papadopoulos, F., Kitsak, M., Vahdat, A., Boguñá, M.: Hyperbolic geometry of complex networks. Phys. Rev. E **82**(3), 036106 (2010)
13. Pei, J., Hua, M., Tao, Y., Lin, X.: Query answering techniques on uncertain, probabilistic data: tutorial summary. In: Proceedings of the ACM SIGMOD International Conference on Management of Data, pp. 1357–1364. ACM (2008)
14. Samet, H.: Foundations of Multidimensional and Metric Data Structures. Morgan Kaufmann Publishers Inc., San Francisco (2005)
15. Staudt, C.L., Sazonovs, A., Meyerhenke, H.: NetworKit: A tool suite for large-scale complex network analysis. In: Network Science. Cambridge University Press (2016, to appear)
16. von Looz, M., Meyerhenke, H.: Querying Probabilistic Neighborhoods in Spatial Data Sets Efficiently. ArXiv preprint arXiv:1509.01990
17. von Looz, M., Prutkin, R., Meyerhenke, H.: Generating random hyperbolic graphs in subquadratic time. In: Elbassioni, K., Makino, K. (eds.) ISAAC 2015. LNCS, vol. 9472, pp. 467–478. Springer, Heidelberg (2015)

Author Index

Printed in the United States
by Baker & Taylor

Printed in the United States
By Bookmasters